ADVANCES IN CHEMICAL PHYSICS

VOLUME XLI

Advances in CHEMICAL PHYSICS

EDITED BY

I. PRIGOGINE

University of Brussels
Brussels, Belgium
and
University of Texas
Austin, Texas

AND

STUART A. RICE

Department of Chemistry
and
The James Franck Institute
The University of Chicago
Chicago, Illinois

VOLUME XLI

AN INTERSCIENCE® PUBLICATION
JOHN WILEY AND SONS
NEW YORK · CHICHESTER · BRISBANE · TORONTO

An Interscience® Publication

Library of Congress Catalog Number: 58-9935

ISBN 0-471-05742-8

Printed in the United States of America

10 9 8 7 6 5 4 3 2 1

CONTRIBUTORS TO VOLUME XLI

JAMES B. ANDERSON, Department of Chemistry, Pennsylvania State University, University Park, Pennsylvania

CLAUDE ASLANGUL, Centré de Mécanique Ondulatoire Appliquée, Laboratoire du C.N.R.S., Paris, France

STEVEN L. BERNASEK, Department of Chemistry, Princeton University, Princeton, New Jersey

V. E. BONDYBEY, Bell Laboratories, Murray Hill, New Jersey

L. E. BRUS, Bell Laboratories, Murray Hill, New Jersey

PHILEMON KOTTIS, Laboratoire d'Optique Moléculaire, Université de Bordeaux I, Talence, France

A. J. KOX, Institute of Theoretical Physics, University of Amsterdam, Amsterdam, The Netherlands

A. K. RAJAGOPAL, Department of Physics and Astronomy, Louisiana State University, Baton Rouge, Louisiana

J. S. ROWLINSON, Physical Chemistry Laboratory, Oxford, England

F. W. WIEGEL, Department of Applied Physics, Twente University of Technology, Enschede, The Netherlands

INTRODUCTION

Few of us can any longer keep up with the flood of scientific literature, even in specialized subfields. Any attempt to do more, and be broadly educated with respect to a large domain of science, has the appearance of tilting at windmills. Yet the synthesis of ideas drawn from different subjects into new, powerful, general concepts is as valuable as ever, and the desire to remain educated persists in all scientists. This series, *Advances in Chemical Physics*, is devoted to helping the reader obtain general information about a wide variety of topics in chemical physics, which field we interpret very broadly. Our intent is to have experts present comprehensive analyses of subjects of interest and to encourage the expression of individual points of view. We hope that this approach to the presentation of an overview of a subject will both stimulate new research and serve as a personalized learning text for beginners in a field.

ILYA PRIGOGINE

STUART A. RICE

CONTENTS

ADVANCES IN CHEMICAL PHYSICS

VOLUME XLI

PENETRABLE SPHERE MODELS OF LIQUID–VAPOR EQUILIBRIUM

J. S. ROWLINSON

Physical Chemistry Laboratory
South Parks Road
Oxford, England

CONTENTS

I am most indebted to Professor B. Widom and Dr. D. B. Abraham for their comments on many parts of the manuscript.

I. INTRODUCTION

If a physical problem is too difficult for an exact theoretical solution then there are two options open to us: Either we seek an approximate solution to the problem, or we replace the real system which gives rise to the problem by a model system that can be handled more rigorously. Both methods of attack have been used in the theory of phase transitions, but the second has proved to be the more enlightening. It is exemplified by the Ising model in which the Hamiltonian of a real ferromagnet is represented by that of a lattice of two-state spin systems each of which interacts only with its nearest neighbors.[1] The interaction favors configurations with neighboring spins pointing in the same direction. At high temperatures this model forms one phase with, in zero magnetic field, an equal number of spins pointing up and spins pointing down, and so with zero magnetization. Below a critical temperature, or Curie point, this one homogeneous phase separates into two, each with a net magnetization arising from a preponderance of spins up, in one phase, and down, in the other. By changing the meaning of the symbols the Hamiltonian of this model can represent equally well a lattice gas, in which a spin up becomes equivalent to a cell containing a molecule, and a spin down to an empty cell. The lower energy of configurations with an abundance of parallel spins translates into a lower energy of configurations in which occupied cells are next to occupied cells and empty cells to empty cells.[2] Since the partition functions of the two models are isomorphous the lattice gas has also a gas–liquid critical point, and the line of zero net magnetization becomes the isochore through this point.

Two symmetry properties of the Ising or lattice-gas model have made it possible to analyze its partition function in greater depth than is the case for more realistic models of a gas–liquid transition, for example, an assembly of molecules with Lennard-Jones intermolecular pair potentials.[3] The first is the obvious transverse symmetry between spin-up and spin-down, which becomes a hole–particle symmetry in the lattice gas. Every point in the phase diagram of the latter (Fig. 1) has its conjugate point (e.g., 1 and 1′), orthobaric states are mutually conjugate (2 with 2′), states on the critical isochore $\rho = \rho_c$ are self-conjugate (a with itself, a' with itself, etc.). This hole–particle symmetry is of help in analyzing the partition function but does not lead to an exact determination of the critical temperature, nor of the singularities of the thermodynamic functions at that point. If, however, the molecules are confined to the cells of a two-dimensional square lattice then there exists also the so-called *dual lattice*,[4] which implies a longitudinal symmetry on the axis $\rho = \rho_c$. That is, point a is not only self-conjugate with respect to the

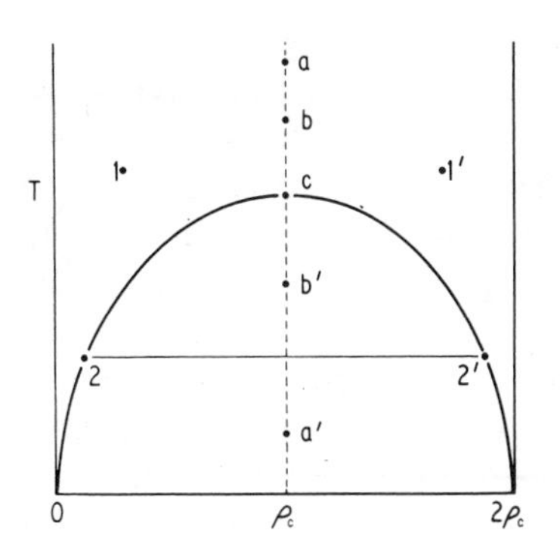

Fig. 1. Sketch of the phase diagram of the (Ising) lattice gas. The points 1 and 1′, 2 and 2′, etc. are conjugate pairs (see text), the pair 2, 2′,being orthobaric gas and liquid states. The point c is the critical point and the dashed line is the critical isochore, the line of symmetry, and, in the two-phase region, also the rectilinear diameter.

hole–particle symmetry, but is, in a different sense, conjugate also with point a', and similarly b with b'. The critical point (it is assumed that there is not more than one) is the only state which is self-conjugate in both senses, and in consequence can be shown to lie at a temperature determined by

$$\exp\left(\frac{w}{2kT_c}\right) = \sqrt{2} - 1 \tag{1.1}$$

where $-w$ is the energy of two molecules in neighboring cells. It was the presence of this longitudinal symmetry that permitted Kramers and Wannier to obtain (1.1); a few years later Onsager obtained the partition function for the Ising model on a square lattice in zero magnetic field, or, equivalently, for the square lattice gas on its critical isochore.[5] Other exactly soluble lattices with critical points have been discovered since Onsager's work; all are two-dimensional. In one dimension the critical point retreats to $T = 0$; in three diminsions the problem can be solved only by approximations or by numerical computations, but for a system of infinite dimensionality the problem can usually be solved without difficulty.

The vast literature on these models testifies to their value to the theory of phase transitions and critical points, but nevertheless they have also some serious disadvantages as representations of the gas–liquid transition. The first is the artificiality of the lattice itself which clearly has no place in a description of a fluid. This disadvantage is most obvious if we wish to discuss the structure of the fluid in the interface between the gas and liquid phases (Section VII). A second problem is that we have no reason to suppose that a real fluid exhibits the strict transverse symmetry of a lattice gas, although the phase diagram is empirically found to be asymptotically symmetrical near the critical point if it is plotted in appropriate thermodynamic variables.[6] Strict hole–particle symmetry imposes, however, a behavior on the thermodynamic functions that is not found in real fluids (Section IV.F).

Both these disadvantages are avoided in a continuum model proposed in 1970, the penetrable-sphere model,[7] and in its later generalizations.[8,9] The Hamiltonian is now a continuous function of the positions of the molecules, so there is no lattice to impose a spurious scale of length on the system and to make unrealistic any discussion of the gas–liquid surface. There is still a transverse symmetry, and indeed some of the few exact results known for this class of models depend on its retention, but it is now not as restrictive as the hole–particle symmetry of the lattice gas, in the sense that it admits of singularities in the thermodynamic functions at the critical point that were previously ruled out by the symmetry itself (Section IV.F).

The loss of the lattice implies the loss of longitudinal symmetry in two, or in any other number of dimensions (as far as we know), and Onsager's evaluation of the partition function for a square lattice has no analog here. The critical temperature is known only in one dimension ($T = 0$) and in infinite dimensions (Section V.A). Even numerical calculations are more difficult; this is partly because integrals over continuous space are more difficult to evaluate than lattice sums, but there is also a more subtle reason. It is often convenient to expand a partition function, or its derivative, as a series in the reciprocal temperature; the coefficients of these series are integrals or sums that can be represented by graphs. The pth term of the expansion for the penetrable-sphere model (Section V.B) is represented by graphs with p points connected by between $(p-1)$ and $\frac{1}{2}p(p-1)$ lines, whereas the pth term of the expansion for a lattice model is composed of graphs with only p connecting lines.[10] The second set is both simpler and much less numerous than the first when p is large. Such drawbacks are the price to be paid for moving from a discrete to a continuous system but they have not precluded the rich harvest of results which are the subject of this review.

II. MODELS

The models are specified by their classical Hamiltonian, which comprises a translational or kinetic part (about which no more will be said) and a configurational part. Each model comes in two forms which can be called the *primitive* and the *transcribed* forms. Let us start with the former in which the configurational energy is composed of pairwise additive intermolecular potentials in a system of n components ($n \geqslant 2$). The pair potentials $u_{\alpha\beta}$ are functions of the separation R of molecules of species α and β. For $1 \leqslant \alpha, \beta \leqslant n$, we have

$$u_{\alpha\alpha}(R) = 0; \qquad u_{\alpha\beta}(R) \geqslant 0 \qquad \alpha \neq \beta \tag{2.1}$$

The absence of any interaction between molecules of the same species implies that each pure component is a perfect gas at all densities and temperatures. The energy between unlike molecules is everywhere positive or zero, and is sufficiently "strong" to drive the system into n separate phases as all particle densities N_α/V become infinite. At low densities the n-component system forms one phase, and at zero density is a perfect gas mixture. The separation at high densities into n phases is a consequence of the mutual repulsion of molecules of different species; at infinite density each molecule must be surrounded only by its own kind. The sequence of phases and critical points between zero and infinite density is one of the features of interest of the model (Section VI). Subject to these requirements, the potentials $u_{\alpha\beta}(R)$ can have any thermodynamically admissible form.[11] The different versions of the primitive forms of the model arise from different choices of n, the number of components, and of the potentials $u_{\alpha\beta}$.

In the simplest case $u_{\alpha\beta}(R)$ is the potential between a pair of hard spheres of diameter a,

$$u_{\alpha\beta}(R) = \infty \quad (R < a_{\alpha\beta}); \qquad u_{\alpha\beta}(R) = 0 \quad (R \geqslant a_{\alpha\beta}) \tag{2.2}$$

In general we need $\frac{1}{2}n(n-1)$ diameters $a_{\alpha\beta}$ to determine the Helmholtz free energy for a given N_α, V, and T. In practice we make all diameters equal to have the greatest possible symmetry in the model; henceforth we drop the suffixes $\alpha\beta$. We can describe (2.2) in a different way by saying that there is an excluded or co-volume v_0 between molecules of different species, where

$$v_0 = \tfrac{4}{3}\pi a^3 \tag{2.3}$$

that is, the volume of a sphere of radius equal to the collision diameter of (2.2).

Other forms of positive potential are equally admissible[8,9] and we generalize the co-volume (2.3) as follows. Let $f(R)$ be the Mayer function of the cross-potential; that is

$$f(R) = \exp\left[\frac{-u(R)}{kT}\right] - 1 \tag{2.4}$$

then the generalization of (2.3) is

$$v_0 = -\int f(R)d\mathbf{R} \tag{2.5}$$

A particularly useful form of $u(R)$ was proposed for the primitive form of the model by Helfand and Stillinger[12,13] namely

$$u(R) = -kT \ln[1 - \exp(-l^{-2}R^2)] \tag{2.6}$$

$$f(R) = -\exp(-l^{-2}R^2) \tag{2.7}$$

$$v_0 = (\pi^{1/2}l)^d \tag{2.8}$$

where d is the dimensionality of the system. This form of the model can be interpreted either as one with a temperature-dependent potential (2.6), or, perhaps more satisfactorily, as a model that is to be studied only at a fixed temperature at which $f(R)$ has the particularly tractable Gaussian form (2.7).

The symmetry of these primitive forms of the model is the obvious n-fold symmetry of the permutation of the n components. In a binary system (Fig. 2) there is a transverse symmetry between the two components which is similar to that of the lattice gas. The ternary mixture has a symmetry similar to that of the three-states Potts model,[14–16] which is a generalization of the Ising model in which the spins can occupy three, rather than two, different states.

The symmetry of the primitive forms of the models is one feature of interest, and is exploited in Section IV.D. Another is the existence of a symmetry-preserving transcription of an n-component primitive model to an $(n-1)$-component transcribed model; to this we now turn.

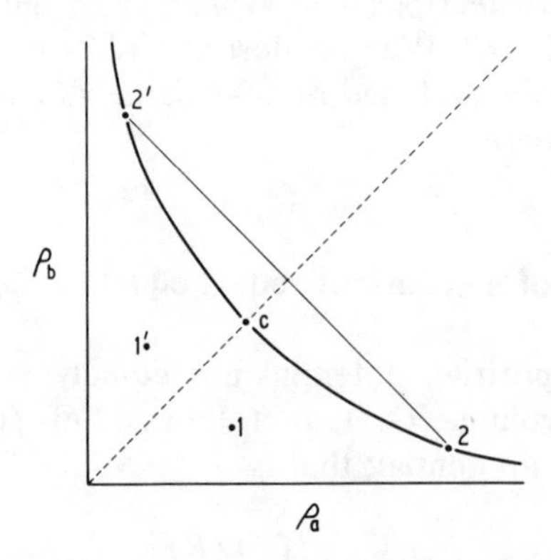

Fig. 2. Sketch of the phase diagram of the binary primitive system.[7] Two pairs of conjugate points, and the critical point are shown. The dashed line is the line of symmetry.

III. TRANSCRIPTION FROM n TO $(n-1)$ COMPONENTS

Let us start with the simplest primitive model, namely that described by (2.2), and consider a state in which only the first $(n-1)$ components are present in a configuration that satisfies (2.2); that is, no molecule is within a distance a of another of different species. We can now add the molecules of the nth component successively and independently to the existing system, while all the molecules of species 1 to $(n-1)$ remain fixed.

The configuration space open to the last molecule of the nth component is the same as that open to the first since there is no interaction between like molecules. We can express this property of the model by saying also that, having added all the molecules of the nth component, we can integrate, one by one, over their coordinates in the grand partition function, and so "remove" them from the integrals. The function that remains is found to have still the form of a grand partition function, but for a system of $(n-1)$ components and with quite a different intermolecular potential. This reduction of an n-component or primitive system to one of $(n-1)$ components preserves the original n-fold symmetry but hides it in such a way that it would not have been suspected were it not for this isomorphism of the primitive and transcribed forms.

The so-called *decorated lattice-gas* models have a similar property in that the occupancy of their secondary cells is mutually independent and determined only by the occupancy of the primary cells. The isomorphism of the decorated and underlying lattices has been exploited[17–19] to derive thermodynamic properties of multicomponent lattice systems in a way that is similar to that described for continuum models.

The primitive versions that have received most study have been those of two and three components ($n = 2, 3$). The one-component system obtained from the former is the system to which the name of penetrable-sphere model was first attached,[7] which is the prototype of this class of continuum models, and with which we start. Let the two components be a and b and introduce the reduced variables,

$$\rho_a = \frac{N_a v_0}{V}; \qquad \rho_b = \frac{N_b v_0}{V}; \qquad \pi = \frac{p v_0}{kT} \tag{3.1}$$

Let λ_a be the dimensionless activity of component a which is normalized to approach ρ_a as π tends to zero. The grand partition function is

$$\Xi_{(2)} = \exp\left[\frac{\pi_{(2)} V}{v_0}\right] = \sum_{N_a} \sum_{N_b} \frac{(\lambda_a/v_0)^{N_a}}{N_a!} \frac{(\lambda_b/v_0)^{N_b}}{N_b!} \int_V \int_V d\mathbf{N}_a d\mathbf{N}_b \exp\left[-\sum \frac{u_{ab}}{kT}\right] \tag{3.2}$$

The subscripts on Ξ and π denote that we have a two-component system, and the sum within the integral is taken over all unlike pairs. The symbol $\mathbf{N}_a$ denotes the set of positions defined by the vectors $\mathbf{R}_1 \ldots \mathbf{R}_{N_a}$. The exponential function is zero if any unlike separation is less than the diameter a, and unity otherwise. For a fixed configuration of the a molecules we can integrate over all allowed positions of the b molecules. They are each independently excluded from a volume $W(\mathbf{N}_a)$ which is the volume covered by N_a interpenetrating spheres of *radius a* centered on the a molecules. Thus

$$\int_V d\mathbf{N}_b \exp\left[-\sum \frac{u_{ab}}{kT}\right] = [V - W(\mathbf{N}_a)]^{N_b} \tag{3.3}$$

We can now sum over N_b in (3.2) to give

$$\begin{aligned}
\Xi_{(2)} &= \sum_{N_a} \frac{(\lambda_a/v_0)^{N_a}}{N_a!} \int_V d\mathbf{N}_a \exp\left[\frac{\lambda_b[V - W(\mathbf{N}_a)]}{v_0}\right] \\
&= \exp\left(\frac{V\lambda_b}{v_0}\right) \sum_{N_a} \frac{[\lambda_a \exp(-\lambda_b)]^{N_a}}{N_a!} \\
&\quad \times \int_V d\mathbf{N}_a \exp\left[\frac{[N_a v_0 - W(\mathbf{N}_a)]\lambda_b}{v_0}\right] &(3.4)\\
&= \exp\left(\frac{V\lambda_b}{v_0}\right)\Xi_{(1)} &(3.5)\\
&= \exp\left[\frac{(\pi_{(1)} + \lambda_b)V}{v_0}\right] &(3.6)
\end{aligned}$$

where $\Xi_{(1)}$ and $\pi_{(1)}$ are the grand partition function and pressure-to-temperature ratio of a one-component system whose number of molecules N, temperature T, activity λ, and configuration energy $U(\mathbf{N})$ are related to those of the original two-component system by the equations

$$N = N_a; \qquad \rho = \frac{N v_0}{V} \tag{3.7}$$

$$\frac{\epsilon}{kT} \equiv \theta = \lambda_b \tag{3.8}$$

$$\lambda = \lambda_a \exp(-\lambda_b) \tag{3.9}$$

$$U(\mathbf{N}) = \frac{[W(\mathbf{N}_a) - N v_0]\epsilon}{v_0} \leqslant 0 \tag{3.10}$$

These equations are the rules for translating the thermodynamic specification of the primitive model $(N_a, \lambda_a, \lambda_b)$ to that of the transcribed model (N, λ, θ). The configuration energy of the latter $U(\mathbf{N})$ can be described in words by saying that, if about the center of each molecule we draw a sphere of radius a, then $U(\mathbf{N})$ is proportional to the difference between the total volume covered by these freely penetrating spheres, $W(\mathbf{N})$, and the sum of the volumes of the spheres, Nv_0. Clearly U is never positive and constitutes an attractive potential between the molecules. It results in a separation of the system into two phases, gas and liquid, at sufficiently low temperatures, or large values of the reduced reciprocal temperature θ. The intermolecular potentials of the primitive version, u_{ab}, were pairwise additive, but U of the transcribed version is not.

The penetrable sphere model has all the a molecules of the original primitive system (3.7); its behavior could be described as that of the primitive system in which the b molecules have become ghosts. If we look at such a system through a glass through which the b molecules were invisible we should conclude that the observed tendency of the a molecules to associate was the result of the attractive potential (3.10), and not of their being driven together by the b molecules. The association would become more pronounced at higher activity of the b molecules, which we should therefore interpret as a low temperature, (3.8).

It follows from (3.2) and (3.6) that

$$\pi_{(1)}[\lambda, \theta] + \theta = \pi_{(2)}[\lambda_a, \lambda_b] \tag{3.11}$$

where λ and θ are related to λ_a and λ_b by (3.8) and (3.9). The symmetry of the primitive version can be expressed as

$$\pi_{(2)}[\lambda_a, \lambda_b] = \pi_{(2)}[\lambda_b, \lambda_a] \tag{3.12}$$

and, by virtue of the isomorphism (3.11), this symmetry is preserved in the transcribed version

$$\pi_{(1)}[\lambda, \theta] + \theta = \pi_{(1)}[\theta \exp(-\lambda e^{\theta}), \lambda e^{\theta}] + \lambda e^{\theta} \tag{3.13}$$

Thus to each point in the phase diagram of the field variables,[20] $\pi(\theta, \lambda)$, there is a conjugate point $\pi'(\theta', \lambda')$, where

$$\theta' = \lambda e^{\theta} \tag{3.14}$$

$$\lambda' = \theta \exp(-\lambda e^{\theta}) \tag{3.15}$$

and at which

$$\pi' = \pi + \theta - \lambda e^{\theta} \tag{3.16}$$

Equations 3.14 to 3.16 are similar to,[7] but not trivially isomorphous with, the transverse symmetry relations of a lattice gas.

These results can be extended at once to any intermolecular potential

$u(R)$ of the primitive version by introducing the Mayer function f and the associated co-volume volume v_0 of (2.4) and (2.5). The transcribed version is described again by (3.7) to (3.13) but with the covered volume $W(\mathbf{N}_a)$ now described by[8,9]

$$W(\mathbf{N}_a) = V - \int_V d\mathbf{R}_0 \prod_{i=1}^{N_a} [1 + f_{i0}(R)] \tag{3.17}$$

where $\mathbf{R}_0$ is the position of a test particle of species b which interacts with molecule i of species a via Mayer function $f_{i0}(R)$ of (2.4). The configuration energy is given again by (3.10).

If we start with a primitive version with more than two components then we can again integrate freely over the coordinates of molecules of one species, say the nth, and so obtain a system of $(n-1)$ components. The transcribed version now has a configuration energy that can be described[21] by saying that each pure component has the penetrable-sphere potential of (3.10), and in the mixture the energy is described by a similar function of the position of all $N_a, N_b, \ldots N_{n-1}$ molecules but with the proviso that U becomes positive infinite if any separation of two molecules of different species is less than the distance a; that is, for unlike interactions the spheres are penetrable only for half their diameters. Such a system can have up to $(n-1)$ different fluid phases, and both critical and multicritical points (Section VI).

IV. EXACT RESULTS

A. Thermodynamic Limit

Perhaps the first question to be asked of any new model is whether the potential satisfies the conditions necessary for the model to behave in a proper, or thermodynamic way, as the number of molecules N and the volume V become infinite at a constant density N/V. These conditions are those of stability

$$U(\mathbf{N}) \geqslant -Nu_0 \tag{4.1}$$

where u_0 is a positive constant, and of weak-tempering or short-range of the intermolecular potential.[11] If these conditions are satisfied by the potentials of a primitive version of the model (2.1) then it can be proved[7,9,22] that they are satisfied also for the transcribed version. Real molecules and realistic models (e.g., Lennard-Jones potentials) satisfy (4.1) by having a repulsive core which prevents any pair approaching too closely. The penetrable-sphere model has no such core but satisfies (4.1) by reason of the multibody nature of the potential (3.10), which can be resolved[8,23] into a sum of negative terms over sets of even numbers of molecules (pair, quadruples, etc.) and of positive terms for odd numbers (triplet, etc.). The smallest value of U is reached when all molecules are at one point when

$$U(\mathbf{N}) = -(N-1)\epsilon \tag{4.2}$$

which satisfies (4.1).

B. Existence of a Phase Transition

It seems intuitively obvious that the hard-core potential (2.2) must drive the primitive version of the model into two phases (if $n = 2$) as the densities ρ_a and ρ_b increase indefinitely. Intuition, however, is a dangerous guide to the behavior of partition functions in the thermodynamic limit and so rigorous proof of the existence or otherwise of phase transitions is desirable. As always, the behavior of the models depends on the dimensionality of the space.

We start with the one-dimensional system ($d = 1$); that is, in the primitive system, with a binary mixture of molecules moving on a line in which those of unlike species see each other as hard rods. This becomes, on transcription, a one-component system of penetrable rods moving on a line, and this potential is entirely equivalent[7] to a different pairwise additive negative potential in which each molecule interacts only with its two nearest neighbors. The partition function and thermodynamic properties of this system have been obtained exactly by two quite different methods[7,24] and, as expected, it has no phase transition at nonzero temperatures. If the dimensionality d can be treated as a continuous variable then it can also be shown[25] that there is a transition at $T > 0$ if $d = 1 + \delta$, where $\delta > 0$. Thus the one-dimensional case can be regarded as the limit in which the critical point has apparently retreated to $T = 0$. The corresponding result in the primitive version is a separation into two phases which occurs only in the limit of infinite activities, λ_a and λ_b.

In two dimensions it has not been possible to find the exact critical temperature since there is apparently no duality transformation of the kind that yields this result for the lattice gas. There are, however, arguments that can prove that, at sufficiently low temperature, there is a transition; these arise from consideration of the behavior of the system when a particular pattern of phase behavior is imposed on the boundary of the system. The arguments were put forward originally by Peierls[26] for the two-dimensional Ising model, and some defects in them remedied by Griffiths.[27] They were adapted first to a lattice form of the primitive version of this model,[28] and then by Ruelle[29] to the continuum form; that is, to (2.2). Ruelle's arguments apply equally to all dimensions of two and above. There appears to be no other continuum model with short-range forces for which there is a rigorous proof of the existence of a phase transition in a system of finite dimensionality.

The limit $d = \infty$ is also one in which the partition function can be obtained exactly. In the primitive version the second virial coefficient between molecules of unlike species,

$$B_{ab} = \tfrac{1}{2}v_0 \tag{4.3}$$

is the only one that is of thermodynamic significance. The ratios of all higher coefficients to the appropriate powers of v_0 tend to zero as d becomes infinite. This was shown by Lie[30] for a particular form of the hard-core potential in which the spheres are replaced by oriented squares, cubes, and so on, according to the dimensionality d. Such forms of the model can be described by saying that the potentials are all one-dimensional, which we denote by writing $s = 1$. The dimensionality of the potential, s, is to be distinguished from that of the space it inhabits, d. The value of s is determined by the form of the equation for the volume of overlap of two molecules as a function of the components of their separation $\mathbf{R}$. If this is a linear function, as for rods on a line, for oriented squares on a plane, or for oriented cubes in space, then we speak of a one-dimensional potential; these three cases are ($s = d = 1$), ($s = 1, d = 2$), and ($s = 1, d = 3$), respectively. The overlap of discs on a plane is $s = d = 2$, and of spheres in space $s = d = 3$. This last case is the original penetrable-sphere model or its primitive version (2.2). The infinite limit examined by Lie was $s = 1$ and $d = \infty$, but his arguments have been extended also[22] to the limit $s = d = \infty$, and to the $d = \infty$ limit of the Gaussian potential (2.6) to (2.8). The infinite-dimensional limit of neglecting all virial coefficients beyond the second becomes a mean-field approximation in the transcribed or penetrable-sphere version (Section V.A).

Ruelle's and Lie's arguments are for the two-component primitive versions ($n = 2$), but the former have been extended to an arbitrary number of components by Runnels and Lebowitz[31]; they were, however, unable to make any statement for continuum models about the behavior in the limit of n becoming infinite.

C. Geometric Probability

The construction of the configurational energy of the transcribed model from the volume common to a set of interpenetrating spheres leads to some interesting questions in geometric probability. If N spheres each of volume v_0 are distributed at random then it is easy to show that in the thermodynamic limit the mean covered volume is

$$\frac{\langle W(\mathbf{N})\rangle}{V} = 1 - e^{-\rho} \tag{4.4}$$

where ρ is the reduced density of (3.7). This is the equilibrium distribution at infinite temperature, or $\theta = 0$ (3.8), and so is a mean-field approximation at nonzero values of θ (Section V.A). At $\theta > 0$ the true equilibrium mean distribution is found by weighting each distribution $W(\mathbf{N})$ by its Boltzmann factor $\exp[-U(\mathbf{N})/kT]$, where the configurational energy U is related to W by (3.10). These canonical averages $\bar{W}$ and $\bar{U}$ cannot be calculated exactly but can be expressed as a cumulant expansion in powers of θ, in which $\langle W\rangle$ and $\langle U\rangle$ are the leading terms, that is, the terms independent of θ.

$$\frac{\langle U\rangle v_0}{\epsilon V} \equiv \langle\phi\rangle = 1 - \rho - e^{-\rho} \tag{4.5}$$

where $\phi(\mathbf{N})$ is the reduced energy density of $U(\mathbf{N})$.

The statistical mechanical problem can therefore be replaced by the problem in geometric probability of calculating the moments or cumulants of the volume covered by a random distribution of spheres in space. A one-dimensional version of this problem (in time, not space) is related to the calculation of the dead-time lost by a counter recording randomly distributed radioactive emissions.[32] A two-dimensional version arose during World War II in attempts to predict the area devasted by bombs if it were assumed that each bomb caused total destruction within a fixed radius of its point of impact.[33] The introduction of the penetrable-sphere model has stimulated further work both on the moments of the random distribution and on the equivalent problem of the Boltzmann distribution.

The key to the calculation of the moments is (3.13) from which it follows that the behavior of $\pi(\lambda, \theta)$ as θ tends to zero at fixed λ is related to its behavior as λ tends to zero at fixed. θ. The latter can be expressed as a virial series in powers of λ in which the coefficients are the Mayer cluster integrals $b_l(\theta)$,

$$\pi(\lambda, \theta) = \sum_{l=1}^{\infty} b_l(\theta)\lambda^l \tag{4.6}$$

The symmetry inherent in (3.13) to (3.16) allows (4.6) to be transformed to an expansion of π in powers of θ—a typical example of the use of the symmetry of the model to obtain a result that would be difficult to reach directly. The series for $\pi(\lambda)$ in powers of θ leads to a series for the energy density $\phi(\rho)$ in powers of θ by the usual thermodynamic equation

$$\left(\frac{\partial \pi}{\partial \theta}\right)_\lambda = -\phi \tag{4.7}$$

We obtain[7,8]

$$\phi + \rho = \sum_{l=0} (-1)^l k_{l+1}(\rho) \cdot \theta^l \tag{4.8}$$

where

$$k_{l+1}(\rho) = \frac{K_{l+1}(\rho)}{l!\,Vv_0^l} \tag{4.9}$$

and $K_l(\rho)$ is the lth cumulant of W for a random distribution,

$$K_1 = \langle W \rangle; \qquad K_2 = \langle W^2 \rangle - \langle W \rangle^2 \quad \text{etc.} \tag{4.10}$$

The reduced cumulants k_l can be expressed in terms of Mayer's cluster integrals $b_1(\rho), b_2(\rho) \ldots b_l(\rho)$ and their derivatives, where the arguments of b_l are the density, and not, as is usual, the reciprocal temperature. Thus

$$k_1(\rho) = 1 - e^{-\rho}; \qquad k_2(\rho) = e^{-2\rho}(b_2(\rho) - \rho) \quad \text{etc.} \tag{4.11}$$

where

$$v_0 b_2(\rho) = \frac{s\pi^{s/2}}{2(s/2)!}\int_0^{2a} d\mathbf{R}R^{s-1}\left\{\exp\left[\frac{-\rho[w(R) - 2v_0]}{v_0}\right] - 1\right\} \tag{4.12}$$

where $w(R)$ is the volume covered by two spheres at separation R, and $s = d$ is the dimension of the potential. The formal expressions for the first six terms of (4.8) are known.[8] Monte Carlo calculations[24,34] have confirmed the size of $k_2(\rho)$ but have told us little of the higher cumulants.

The calculation of the canonical average $\bar{W}$ at a nonzero value of θ is equivalent not only to the cumulant expansion above but also to the calculation of the expectation or random value of its exponential, that is, $\exp(-\theta W(\mathrm{N})/v_0)$. The latter has been studied both for a *multinomial* and for a *Poisson* distribution. In the former a fixed number of molecules N are placed at random in a fixed volume V to give a reduced density $\rho = Nv_0/V$, and the expectation value of the exponential calculated in the thermodynamic limit. In the latter the fixed volume V can contain a varying number of molecules and the probability that it has a reduced density ρ is $[(cV/v_0)^N/N!]\exp(-cV/v_0)$, where c is a new reduced density which determines this Poisson distribution. The relation between the multinomial and Poisson distributions is similar to that between the canonical and the grand canonical ensembles, and the penetrable-sphere model has been used to explore this similarity.[24] However, the densities

c and ρ are not the same, as first became apparent from some results of Moran.[35] The general relation between the two densities has now been found[24]; for the penetrable-sphere model c is proportional to the activity λ appropriate to the density ρ. We have

$$c = \lambda(\rho, \theta) \cdot e^{\theta} \tag{4.13}$$

D. Consequences of Symmetry

The symmetry of the binary primitive version is the obvious one between the two components a and b. This transcribes into a less obvious symmetry in the one-component or penetrable-sphere version, as was shown in (3.13) to (3.16). In this section we explore the thermodynamic consequences of these equations.[7]

In the primitive version we have the two thermodynamic identities

$$\rho_a = \lambda_a \left(\frac{\partial \pi_{(2)}}{\partial \lambda_a}\right)_{\lambda_b}; \qquad \rho_b = \lambda_b \left(\frac{\partial \pi_{(2)}}{\partial \lambda_b}\right)_{\lambda_a} \tag{4.14}$$

and in the transcribed version

$$\rho = \lambda \left(\frac{\partial \pi_{(1)}}{\partial \lambda}\right)_{\theta}; \qquad \phi = -\left(\frac{\partial \pi_{(1)}}{\partial \theta}\right)_{\lambda} \tag{4.15}$$

The application of these equations to the results of Section III leads to the two relations

$$\rho = \rho_a \tag{4.16}$$

$$1 - \rho - \phi = \frac{\rho_b}{\lambda_b} \tag{4.17}$$

where the properties of the primitive version are on the right and of the transcribed version on the left. The first, (4.16), has already been derived, (3.7), but the second is new. We can now supplement the three equations between the field variables at pairs of conjugate points, (3.14) to (3.16), with two others between the densities ρ and ϕ.

$$\rho' = \theta(1 - \rho - \phi) \tag{4.18}$$

$$\phi' = 1 - \theta(1 - \rho - \phi) - \left(\frac{\rho}{\lambda}\right) e^{-\theta} \tag{4.19}$$

The five equations are recipes for calculating the thermodynamic properties of a state (θ', λ', π', ρ', and ϕ') from the (unprimed) properties of the conjugate state.

Since orthobaric states are conjugate, as is obvious from the primitive version Fig. 2, and since ϕ is a linear function of ρ in the gas–liquid region, we have

$$\phi^{(2)}(\rho) = 1 - 2\theta^{-1}\bar{\rho} + \rho(\theta^{-1} - 1) \qquad (\rho_g \leq \rho \leq \rho_l) \tag{4.20}$$

where $\phi^{(2)}$ denotes the energy density of a two-phase system of overall density ρ, and where

$$\bar{\rho} = \tfrac{1}{2}(\rho_g + \rho_l) \tag{4.21}$$

On the diameter of the two-phase region $\rho = \bar{\rho}$ and

$$\phi = 1 - \bar{\rho}(1 + \theta^{-1}) \qquad \text{(diameter)} \tag{4.22}$$

The set of states that are self-conjugate ($\theta' = \theta$ etc.) defines the line of symmetry. The five equations above yield only the two independent equations.

$$\lambda = \theta e^{-\theta} \qquad \text{(line of symmetry)} \tag{4.23}$$

$$\phi = 1 - \rho(1 + \theta^{-1}) \qquad \text{(line of symmetry)} \tag{4.24}$$

It follows from (4.22) and (4.24) that the diameter is the line of symmetry in the two-phase region, and hence that the line passes through the critical point, exactly as in the primitive version. Moreover, the activity (4.23) is an analytic function of the temperature everywhere on the line of symmetry, including the critical point. In this the model resembles the lattice gas.

Clapeyron's equation and (4.20) yield the further result that the slope of the vapor–pressure curve is

$$\frac{d\pi^{(2)}}{d\theta} = 2\bar{\rho}\theta^{-1} - 1 \qquad (\theta \geq \theta_c) \tag{4.25}$$

Since this slope is everywhere negative and becomes zero at zero temperature we have

$$2\bar{\rho} = \theta \quad \text{as} \quad \theta \to \infty \tag{4.26}$$

and

$$2\rho_c \leq \theta_c \tag{4.27}$$

More subtle inequalities are discussed in the next section. Figure 3 shows the phase diagram in the field variables λ and θ, and Fig. 4 in ρ and θ.

There is a virial relation between the volume covered by N spheres, $W(\mathbf{N})$, and their exposed surface area $A(\mathbf{N})$ where

$$A(\mathbf{N}) = \frac{dW(\mathbf{N})}{da} \tag{4.28}$$

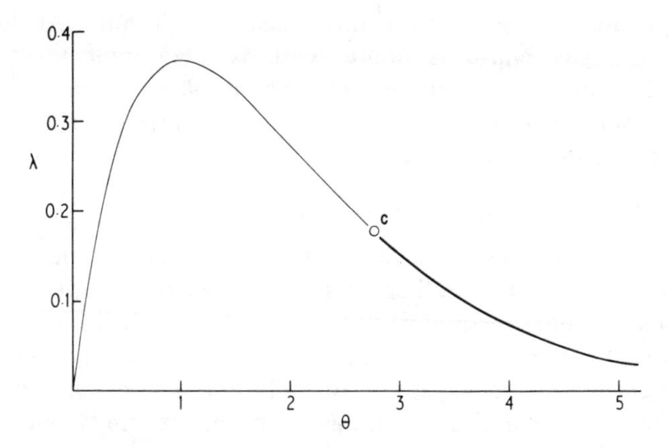

Fig. 3. Activity λ as a function of reduced reciprocal temperature θ on the line of symmetry in the transcribed or penetrable-sphere form of the model.[7] The curve is exact, but it is not known which part is the two-phase region, and which is the one-phase. The critical point shown is for the mean-field approximation and the two-phase region is then the heavy part of the line, $\theta > \theta_c$.

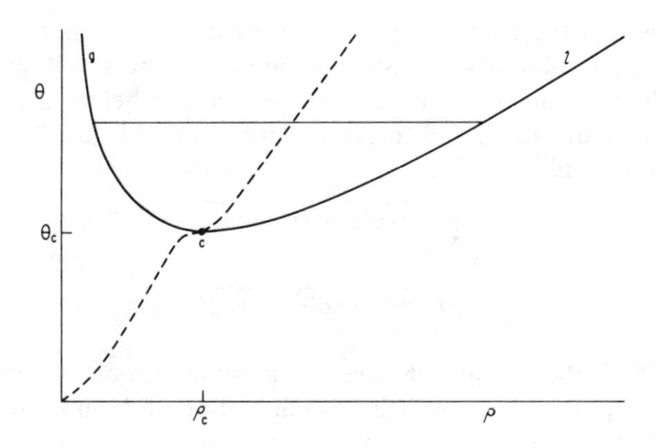

Fig. 4. Sketch of the phase diagram for the penetrable-sphere model.[7] The full curve is the orthobaric line and c the critical point. The dashed line is the line of symmetry which is not the critical isochore, but which is the diameter in the two-phase region. This line is singular on both sides of the critical point.

This virial relation and the fact that orthobaric states are conjugate leads[23,36] to the interesting result that

$$\overline{(A/V)_g} = \overline{(A/V)_l} \tag{4.29}$$

where the bars denote the equilibrium value. This result leads to a striking geometrical picture of the phase equilibrium. At low temperatures the gas phase is dilute, with isolated molecules sparsely distributed through the volume. The liquid phase is dense and has therefore a small number of polyhedral holes of different size and shape sparsely distributed through it.

E. Some Inequalities

The results that flow from symmetry do not determine the critical temperature but lead, together with some necessary thermodynamic restrictions, to some inequalities for π, ρ, and θ, and, in particular, for their values at the critical point. Since the thermodynamic restrictions do not distinguish between systems of different dimensionality, d, they must allow of θ_c becoming infinite as d approaches unity, and so many of the inequalities are only one-sided.

The first thermodynamic limitation is that the canonical average of the energy density cannot exceed that of a random configuration, and cannot be less than the value implied by (4.1). Hence

$$1-\rho-e^{-\rho} \geqslant \phi \geqslant -\rho \tag{4.30}$$

A second is that the configurational entropy density cannot exceed that of a perfect gas of the same number density. Others follow from the potential distribution theorem[37] and virial relations between the volume W, the area A of (4.28), and the next derivative (dA/da). The theorem implies two results[23]

$$\frac{\rho}{\lambda} = \overline{\exp[-\delta U(\mathrm{N})/kT]} \tag{4.31}$$

$$\frac{\phi}{\lambda} = \overline{\delta U^*(\mathrm{N}) \exp[-\delta U(\mathrm{N})/kT]} \tag{4.32}$$

where $\delta U(\mathrm{N})$ is the average change of energy in introducing an extra, or $(N+1)$th molecule into an existing configuration of N molecules, which maintain their fixed positions as the extra molecule moves over the whole volume. The ensemble average, denoted by the bar, is then taken over all Boltzmann-weighted distributions of $U(\mathrm{N})$. The energy $U^*(\mathrm{N})$ differs from $U(\mathrm{N})$ in that each m-body potential is weighted with a factor of m^{-1}, when $U(\mathrm{N})$ is written as a sum of multibody potentials (Section IV.A). It can be shown that[23]

$$0 \geqslant \tfrac{1}{2}\delta U(\mathrm{N}) \geqslant \delta U^*(\mathrm{N}) \geqslant \delta U(\mathrm{N}) \geqslant -\epsilon \tag{4.33}$$

The inequalities to which (4.27) and these results lead are[7,23]

$$3.146\ldots \geqslant \frac{\theta_c}{\rho_c} \geqslant 2 \tag{4.34}$$

$$\exp(\rho_c) \geqslant \frac{\theta_c}{\rho_c} \tag{4.35}$$

and that in all equilibrium states

$$\rho \geqslant \pi \geqslant \lambda; \qquad \rho \geqslant \chi^{-1} \tag{4.36}$$

where 3.146 . . . is the larger of the two roots of the equation $\ln x = x - 2$, and where χ is the reduced isothermal compressibility. It follows from (4.34) and (4.35) that

$$\rho_c \geqslant \ln 2; \qquad \theta_c \geqslant 2 \ln 2 \tag{4.37}$$

Thus as the dimensionality approaches unity, not only θ_c but also ρ_c tends to infinity.

These inequalities are a little weaker if (4.27) is not assumed. It seems, however, to be almost certainly the case that $(\partial \bar{W}/\partial V)_\theta > 0$, or, on the vapor-pressure curve, that $(\Delta \bar{W}/\Delta V) > 0$, where $\Delta \bar{W}$ $(= \Delta \bar{U})$ and ΔV are the changes on passing from gas to liquid. Either assumption leads to the stronger form of the inequalities in (4.34) to (4.37).

F. Critical Point

Perhaps the most important consequence of the symmetry relations of Section V.D are their implications for the behavior of the thermodynamic functions near the gas–liquid critical point. Let c_v denote the reduced heat capacity density, that is, $(C_v v_0/kV)$ where C_v is the heat capacity per molecule. We can differentiate (4.24) along the line of symmetry and express the result in terms of c_v by means of the identities

$$\left(\frac{\partial \phi}{\partial \theta^{-1}}\right)_\rho = c_v; \qquad \left(\frac{\partial \phi}{\partial \rho}\right)_\theta = \left(\frac{\partial \ln \lambda}{\partial \theta}\right)_\rho \tag{4.38}$$

We obtain

$$\frac{d\rho}{d\theta^{-1}} = \frac{c_v + \rho}{1 + \theta^{-1} + (\partial \ln \lambda/\partial \theta)_\rho} \qquad \text{(line of symmetry)} \tag{4.39}$$

Now $(\partial \ln \lambda/\partial \theta)_\rho$ is an analytic function of θ, (4.22), in the whole of the two-phase region $(\theta \geqslant \theta_c)$, and becomes $(\theta_c^{-1} - 1)$ at the critical point. Moreover the line of symmetry is also the orthobaric diameter, and so

$$\frac{d\bar{\rho}}{d\theta} \approx \tfrac{1}{2}\theta_c^{-1}(c_v + \rho)_c \tag{4.40}$$

near the critical point. Similarly, by differentiation of (4.25)

$$\frac{d^2\pi^{(2)}}{d\theta^2} \approx \left(\frac{c_v}{\theta^2}\right)_c \tag{4.41}$$

For this model, as for all others of finite dimensionality, and as for all real fluids, it may be presumed that c_v diverges weakly to infinity at the critical point; that is,

$$c_v \sim \left(\frac{a_-}{\alpha}\right)(\theta - \theta_c)^{-\alpha} \qquad (4.42)$$

where a_- is a positive constant and α is an index for which $1 \gg \alpha \geqslant 0$. (The limit $\alpha = 0$ implies a logarithmic divergence of c_v). If the presumption (4.42) is correct then it follows from (4.40) and (4.41) that the slope of the diameter and the curvature of the vapor-pressure curve are both infinite at the critical point.

Similar but less simple arguments lead to the same conclusions for the behavior of $(d\rho/d\theta)$ on the line of symmetry in the one-phase region and for $d^2\pi^{(1)}/d\theta^2$ on the critical isochore. The minor complication is that $(d\ln\lambda/d\theta)_\rho$ is itself weakly singular on the critical isochore as the critical point is approached from temperatures above the critical.

We have[7]

$$\lambda \sim \theta e^{-\theta} + b(\theta_c - \theta)^{1+\gamma-a} \qquad (\rho = \rho_c, \theta \leqslant \theta_c) \qquad (4.43)$$

where $b > 0$, and γ is the index which describes the divergence of the compressibility. In three dimensions $(1 + \gamma - \alpha) > 2$, and so λ and its first two derivatives are continuous at the critical point. This continuity suffices to ensure that $(d\rho/d\theta)$ on the line of symmetry and $(d^2\pi^{(1)}/d\theta^2)_\rho$ on $\rho = \rho_c$ have the same singularities above the critical temperature as they have below. The singularities on the line of symmetry are sketched in Fig. 4.

The heat capacity density can be written as a sum of two terms,

$$c_v = \theta^2\left(\frac{\partial^2\pi}{\partial\theta^2}\right)_\rho - \rho^2\left(\frac{\partial^2\ln\lambda}{\partial\theta^2}\right)_\rho \qquad (4.44)$$

We deduce that for this model the singularity in c_v arises solely from that in π and that there is no contribution from the term in λ. In this it resembles the lattice gas,[2] but not the classical theories (e.g., van der Waals[38]) where both terms in (4.44) contribute to the discontinuity in c_v. Conversely classical theories and the lattice gas have no singularity in their orthobaric diameter, whereas this model exhibits one that is proportional to that in c_v.

This singularity in the diameter was undoubtedly the most striking new result from this model when it was first published in 1970. (That in the vapor-pressure curve was already known.) Three other models with the same feature were discovered very shortly afterwards, two of which are decorated lattice gases,[19] and one of which is of quite a different type.[39]

As far as we know, real fluids and fluid mixtures have no exact thermodynamic symmetry at their critical points, although the former exhibit at least a rough asymptotic symmetry in a density–activity space,[6] and the latter in a temperature–volume fraction space.[40] It is therefore to be expected that the singularity in the diameter exhibited by these four models is also a feature of a real fluid. If so, it is proving difficult to establish this experimentally. Until 1970 all measurements had suggested that $\bar{\rho}$, although not independent of temperature (as in a lattice gas), was but a linear function of it near the critical point (the law of rectilinear diameters). However few measurements had been made close enough to the critical point to provide a searching test. At present the position is open; there are several sets of results that support the existence of a singularity, both for the pure fluids[41] and for liquid mixtures,[42] but the analysis and interpretation of these results is far from simple. Scott[43] has emphasized how difficult it is likely to be to get meaningful results for mixtures, since if an arbitrarily chosen density is used (e.g., mole fraction or volume fraction) then it is to be expected that the diameters will behave not as $(T_c - T)^{1-\alpha}$ but as $(T_c - T)^{2\beta}$, where $\beta \sim 0.3$ is the index that governs the change of orthobaric densities with temperature. Similarly in a one-component system any nonlinear function of ρ (e.g., ρ^{-1} or V) has an orthobaric diameter with a singularity stronger than α. If the α singularity in the diameter of ρ is confirmed then conventional ideas on the scaling of thermodynamic functions near critical points, which implicitly assume the law of rectilinear diameters, will need modification along the lines proposed by Widom and Stillinger.[9]

G. Low-temperature Limit

At low temperatures, θ large, the system comprises two phases, one an almost perfect gas, and the other a liquid phase so dense that the molecules cover almost the whole of the volume. The properties of these phases can be obtained exactly, in the limit of infinite θ, since we know the activity of both phases from (4.23). They are[7]

$$\begin{aligned} \lambda \sim \pi \sim \rho_g &\sim \theta e^{-\theta} \\ \rho_l &\sim \theta - \theta^2 e^{-\theta} \end{aligned} \tag{4.45}$$

These limits are valid only for $d > 1$. For d close to unity the limiting behavior is complicated[25] and depends on the relative magnitudes of θ^{-1} and $(d-1)$. At d of exactly unity

$$\rho \sim \tfrac{1}{2}\theta; \qquad \pi \sim \theta e^{-\theta/2} \qquad \text{(line of symmetry)} \tag{4.46}$$

V. APPROXIMATE RESULTS

A. Mean-field Approximation

The results that follow from symmetry tell us much about the general features of the phase diagram, and particularly of its behavior near the critical point, but do not themselves lead to an evaluation of the partition function and the thermodynamic properties. These can be obtained exactly only for $d = 1$ and $d = \infty$ (Section IV.C); for all other dimensions we must use approximations. We start first by supposing that the results that are exact for $d = \infty$ can be used as an approximation for all values of d greater than unity; that is, that the equation of state of the primitive version is that obtained by the neglect of all virial coefficients above the second,

$$\pi_{(2)} = \rho_a + \rho_b + \rho_a\rho_b \tag{5.1}$$

$$\lambda_a = \rho_a \exp(\rho_b); \qquad \lambda_b = \rho_b \exp(\rho_a) \tag{5.2}$$

The transcription of these equations to those of the penetrable sphere model, according to the rules of Section III, gives

$$\phi = 1 - \rho - e^{-\rho} \tag{5.3}$$

$$\pi_{(1)} = \rho + \theta(\rho e^{-\rho} + e^{-\rho} - 1) \tag{5.4}$$

$$\ln \lambda = \ln \rho - \theta(1 - e^{-\rho}) \tag{5.5}$$

As is usual we call this the *mean-field approximation* to ϕ, and so on, since it replaces the canonical average $\overline{U(\theta)}/V$ by its random average $\langle U \rangle/V$ or $\overline{U(\theta = 0)}/V$. From the cumulant expansion of Section IV.C we see that (5.3) is the zeroth-order term in the expansion of $\phi(\theta)$ in powers of θ, and hence that (5.4) and (5.5) include only the zeroth (or perfect-gas) and first-order terms in π and $\ln \lambda$. The mean-field approximation is therefore not only the $d = \infty$ limit for all nonzero values of θ, but also the $\theta = 0$ limit for all finite values of d. It is the first of these limits which ensures that the approximation satisfies exactly all the symmetry requirements of Section IV.D, even when d is finite. The second limit makes ϕ a function only of density—a feature this approximation shares with other mean-field approximations, such as the van der Waals equation for a fluid or the Weiss and Bragg–Williams approximations for lattices.[44]

From (3.14), (4.18), and (5.3) we have that conjugate states are related by

$$\theta' = \rho \exp(\theta e^{-\rho}) \tag{5.6}$$

$$\rho' = \theta e^{-\rho} \tag{5.7}$$

Since orthobaric states are conjugate, a particular case of (5.7) is

$$\theta = \rho_g \exp(\rho_l) = \rho_l \exp(\rho_g) \tag{5.8}$$

These equations can be solved parametrically[45] by introducing the difference Δ,

$$\Delta = \tfrac{1}{2}(\rho_l - \rho_g) \tag{5.9}$$

We have

$$\rho_g = \Delta \coth \Delta - \Delta; \qquad \rho_l = \Delta \coth \Delta + \Delta \tag{5.10}$$

$$\bar{\rho} = \Delta \coth \Delta; \qquad \theta = \Delta \operatorname{cosech} \Delta \cdot \exp(\Delta \coth \Delta) \tag{5.11}$$

The critical point is at $\Delta = 0$, or

$$\rho_c = 1; \qquad \theta_c = e; \qquad \pi_c = 3 - e; \qquad \phi_c = -e^{-1} \tag{5.12}$$

These results satisfy the inequalities of Section IV.E, and the value of the critical ratio $\pi_c = 0.2817$ is remarkably close to the value for the inert gases[3] (0.293).

The line of symmetry in the one-phase region is

$$\theta = \rho e^{\rho} \qquad (\rho \leq 1, \text{line of symmetry}) \tag{5.13}$$

In the two-phase region it is the equation $\rho = \bar{\rho}$, where $\bar{\rho}$ is related parametrically to θ by (5.11), which implies that the law of rectilinear diameters is obeyed near the critical point. Since (5.4) is an analytic function of ρ, it follows that all other thermodynamic properties conform qualitatively to the classical or van der Waals pattern at the critical point; Fig. 5 shows the phase diagram.

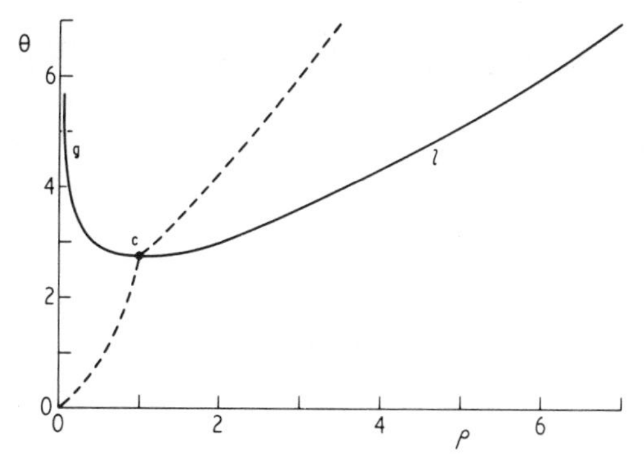

Fig. 5. The phase diagram for the penetrable-sphere model in the mean-field approximation.[7] The line of symmetry has a discontinuity in its slope at the critical point c, but is an analytic function of θ on each side of this point.

The heat capacity density c_v now has only a simple jump discontinuity on the critical isochore, $\rho = 1$. It rises from zero in the one-phase region to 3 in the two-phase region; the two terms of (4.44) contribute +2 and +1 to this discontinuity.

There are three principal routes from the molecular distribution functions to the thermodynamic properties,[46] the *energy equation*, the *virial* or *pressure equation*, and the *compressibility equation*. The first is an expression of the canonical average energy $\bar{U}$ in terms of the intermolecular energy and the most probable distribution of molecules, and the second a similar average of the intermolecular forces to give the thermodynamic virial function $(pV - NkT)$. The third expresses the compressibility (or its reciprocal) as an integral over the total (or direct) pair correlation function. In an exact treatment of any model the three routes lead to the same values of the thermodynamic functions, but if approximations are made then the three sets of values are generally different. Tests of the degree of consistency of the three sets of thermodynamic functions have long been one of the measures of the success or otherwise of approximations.[46,47] We apply this test to the mean-field approximation.

The equation of state (5.4) or (5.5) was derived from ϕ of (5.3) which is the average energy of an assumed distribution, in this case a random one. These equations are therefore the results obtained by the *energy equation*. Widom[23] has shown that the *virial equation* leads to the same result. The third route requires that we know either the total or the direct pair correlation function. The total function $h(R)$ is $g(R) - 1$, where $g(R)$ is the pair distribution function in the grand canonical ensemble. The total function is related to the direct function $c(R)$ by the Ornstein–Zernike equation, which can be written

$$h(R_{12}) = c(R_{12}) + \int d\mathbf{R}_3 \cdot n(\mathbf{R}_3) h(R_{13}) c(R_{23}) \tag{5.14}$$

where $n(\mathbf{R}_3)$ is the density at $\mathbf{R}_3$. In a homogeneous fluid this is independent of the position of molecule 3, and so can be taken outside the integral, but the more general form (5.14) is needed for the discussion of the gas–liquid surface in Section VII.

The total pair function $h(R)$ in the penetrable-sphere model is the same as that of component a in the primitive version,[22] $h_{aa}(R)$, and similarly for the higher correlation functions, although we do not need them here. The mean-field approximation to the direct correlation function of the primitive version,[8,22,48] $c_{ab}(R)$, is simply the Mayer function $f_{ab}(R)$. From these two results it follows that the direct correlation function for the penetrable-sphere model, in the mean-field approximation, is[22]

$$c(R_{12}) = \left(\frac{\theta}{v_0 \rho e^{\rho}}\right) \int_{v_{12}} d\mathbf{R}_3 \cdot \rho(\mathbf{R}_3) \tag{5.15}$$

where v_{12} is the volume common to two spheres of radius a placed at $\mathbf{R}_1$ and $\mathbf{R}_2$. In a homogeneous fluid,

$$c(R_{12}) = \left(\frac{\theta}{v_0 e^{\rho}}\right)[w(R_{12}) - 2v_0] \tag{5.16}$$

The range of $c(R)$ is that of the pair potential, namely the diameter of the penetrable sphere $2a$, since $w(R_{12})$, the volume covered by the two spheres, is $2v_0$ if $R_{12} > 2a$. The reciprocal of the compressibility is now obtained by integrating $c(R)$ over all R. In reduced units

$$\left(\frac{\partial \pi}{\partial \rho}\right)_{\theta} = 1 - \rho \int d\mathbf{r} \cdot c(r) \tag{5.17}$$

where $r = R/a$. Substitution of (5.16) into (5.17), and integration with respect to ρ leads again to (5.4).

Thus for the model the three routes to the equation of state are all consistent. This result could have been expected since the approximation becomes exact at $d = \infty$, and since the arguments leading to the energy, virial, and compressibility equations are essentially independent of d. An explicit demonstration is worth carrying out, however, since the three routes do not agree for similar approximations in other models of fluids, nor (Section V.B and C) for other approximations for this model. Moreover, there are still inconsistencies in the mean-field approximation in its handling of fluctuations,[23] which are closely related to the compressibility, and which have an explicit dependence on dimensionality.

The Fourier transform of (5.16) is

$$(2\pi)^{3/2}\tilde{c}(k) = \int d\mathbf{r} \cdot c(r) e^{i\mathbf{k}\cdot\mathbf{r}} \tag{5.18}$$

$$= 9\theta e^{-\rho} k^{-2} j_1^2(k) \qquad (d = 3) \tag{5.19}$$

where j_1 is a spherical Bessel function. The function $\tilde{h}(k)$ can be obtained from $\tilde{c}(k)$ and so, by means of an inverse transform,[22]

$$h(r) = \theta e^{-\rho} \frac{2}{3\pi} \int_0^{\infty} \frac{9k^2 j_0(rk) j_1^2(k)\, dk}{k^2 - 9\theta\rho e^{-\rho} j_1^2(k)} \tag{5.20}$$

Near the critical point this function behaves as

$$h(r) \sim \text{constant} \cdot r^{-1} e^{-\kappa r} \tag{5.21}$$

where the reduced inverse correlation length κ is given by[22]

$$\frac{\kappa^2}{5} = \rho - \ln(\rho\theta) \tag{5.22}$$

The reduced correlation length λ defined by means of the second moment of $h(r)$ is[22]

$$\frac{1}{5\lambda^2} = \exp\left(\frac{\kappa^2}{5}\right) - 1 \tag{5.23}$$

It follows that in the critical region there is but one correlation length, since

$$\lambda\kappa = 1 \tag{5.24}$$

Thus structurally as well as thermodynamically the mean-field approximation leads to classical behavior of the kind associated with the names of van der Waals, Ornstein, and Zernike.

At low temperatures the mean-field approximation becomes asymptotically correct, and leads to the limits in (4.45).

B. Virial and Cumulant Expansions

The mean-field approximation for the transcribed form of the model has been derived from the virial expansion of the primitive form by truncating $\pi(\rho)$ after the second virial coefficient. Inclusion of the higher terms in this virial expansion should lead to more accurate equations of state of both primitive and transcribed forms. Moreover, virial expansions in powers of activity λ or density ρ, in which the coefficients are connected or irreducible graphs, respectively, exhibit the symmetry of the primitive form in an obvious way.

In a two-component primitive system of hard spheres we denote the sum of all connected graphs with k distinguishable field points of species a and l of species b by $[(k+l)!v_0^{k+l-1}b_{kalb}]$. Transcription[7,8] to the one-component penetrable-sphere model gives the cluster integrals of (4.6) and so $\pi(\lambda, \theta)$.

$$b_l(\theta) = e^{l\theta} \sum_{k=1-\delta_{l1}} \binom{k+l}{l} \theta^k b_{kalb} \tag{5.25}$$

where $b_{ka1b} = (-)^k/(k+1)!$, and so $b_1(\theta) = 1$. We seek expansions of b_l in powers of θ, and so write

$$b_l(\theta) = \sum_{k=1-\delta_{l1}} \alpha_{lk}\theta^k; \qquad (\alpha_{1k} = \delta_{k0}) \tag{5.26}$$

Symmetry requires that $b_{kalb} = b_{lakb}$, and this result allows the coefficients a_{lk} to be divided into two classes, the *independent* coefficients, α_{10} and those for which $k \geqslant l$, and the *dependent* coefficients, for which $l > k \geqslant 1$. The latter can be calculated from the former by the equation,[8]

$$\alpha_{lk} = \sum_{m=1-\delta_{k1}}^{l} \frac{(-k)^{l-m}}{(l-m)!}\alpha_{km} - \sum_{m=1}^{k-1} \frac{(-l)^{k-m}}{(k-m)!}\alpha_{lm} \qquad (l > k \geqslant 1) \qquad (5.27)$$

The independent coefficients are found by performing the integration represented by the graphs in b_{kalb}. Any approximation to this set of coefficients (e.g., by truncating the series) yields an approximate set of dependent coefficients via (5.27), whose form ensures that the symmetry is maintained in the transcribed form. Thus the mean-field approximation arises from putting $\alpha_{10} = 1$ (its correct value) and all other independent coefficients equal to zero. This gives for the dependent coefficients[8]

$$\alpha_{lk} = (-)^{k+l-1}\left(\frac{l^{k-1}}{l!}\right)S_{l-1}^{(k)} \qquad (\text{mean-field}, l > k \geqslant 1) \qquad (5.28)$$

where $S_{l-1}^{(k)}$ is a Stirling number of the second kind.[49] It follows that

$$b_1(\theta) = 1; \qquad b_2(\theta) = \tfrac{1}{2}\theta; \qquad b_3(\theta) = -\tfrac{1}{6}\theta + \tfrac{1}{2}\theta^2 \qquad \text{etc.} \qquad (5.29)$$

The coefficients of the first power of θ are correct, namely $(-)^l/l!$, but the others are approximations.

Any symmetrical set of coefficients α generates a set of cluster integrals from which, via (4.11), we can obtain the cumulant expansion of $\phi(\theta)$. The energy equation then yields the equation of state. This route has been followed[8] but with disappointing results. The series are only slowly convergent and the only positive result is that in three dimensions ($s = d = 3$) the critical temperature and density are found to be somewhat above their values in the mean-field approximation.

Other attempts to go beyond the mean-field approximation have used not the expansion of π in powers of activity, whose coefficients are b_{kalb} in the primitive form, but the expansion in powers of the density, whose coefficients are the star graphs or irreducible cluster integrals. The configurational free energy density (per molecule) differs from that of the perfect gas by

$$kT \sum_{i=1}\sum_{j=1} \rho_a^i \rho_b^j C_{ij} \qquad (5.30)$$

where $[(i+j-1)v_0^{i+j-1}C_{ij}]$ is the sum of irreducible graphs with i molecules of species a and j of b. (The notation C_{ij} is used to conform with that of Helfand and Stillinger.[12]) The enumeration of the integrals in C_{ij} is a problem in bichromatic graph theory, that is, graphs in which the nodes have one of two colors, and only nodes of unlike color are joined. The calculation of the integrals represented by these graphs is prohibitively difficult beyond the seventh virial coefficient if the hard-sphere primitive model is used ($s = d = 3$). This is not a long enough series to yield accurate values of the critical constants, although the transcribed critical density,[8] $\rho_c = 1.30$, from the series of seven coefficients may not be greatly in error.

Straley et al.[48] were able to obtain more coefficients by using one-dimensional potentials ($s = 1$, $d = 2$, or 3), for which some of the cluster integrals had previously been calculated[50] by Zwanzig and by Runnels and Colvin. Again, no precise values of the critical constants were found.

The Gaussian potential, (2.6) to (2.8) is even more tractable than the one-dimensional hard-core potential since the problem of evaluating the integrals is now itself reducible to a problem in graph theory.[51] Helfand and Stillinger[12] were able to obtain the first eleven virial coefficients of the binary primitive form. The critical point and some of the critical indices can be found from this series by forming the appropriate Padé approximants. After transcribing their results from the primitive to the one-component system, and renormalizing the critical indices, their results are

$$\rho_c = 1.231; \qquad \alpha = 0.26 \pm 0.03; \qquad \gamma = 1.23 \pm 0.02 \tag{5.31}$$

where α and γ are the critical indices governing the singularities in c_v (as in Section IV.F) and in the compressibility. The value of α is considerably higher than that found for real fluids or for three-dimensional lattice models (~ 0.1), but the value of γ is similar.

The dimensionality d is an easily changed parameter for the Gaussian potential (indeed, it can readily be treated as a continuous variable[51]) and Helfand and Stillinger looked briefly also at dimensions other than 3. They give no quantitative results, but found, surprisingly, that some of their Padé approximants were erratic for d greater than 4. This is the region in which one would have expected classical behavior,[52] and so a steady convergence to the critical point with the classical indices $\alpha = 0$ and $\gamma = 1$. Such steadiness was found with the hyper-netted chain approximation to this model (Section V.C) and with the three-component primitive Gaussian model near its tricritical point (Section VI.B).

C. Percus–Yevick and Hyper-netted Chain Approximations

The truncation of virial expansions can be avoided by summing to all orders over certain classes of graphs in the expansions of the correlation functions. The result is still an approximation since other graphs must be neglected entirely, but may lead to a better result than the Padé approximation to a truncated series. The summation over certain graphs is often equivalent to the supplementing of the Ornstein–Zernike equation (5.14) with a closed approximation for the direct or total correlation function. Two such examples are the Percus–Yevick (PY) and hypernetted chain (HNC) approximations.

The former may be described[3] by saying that it restricts $c(r)$ to the range of the intermolecular potential. For the primitive model it shares this characteristic with the mean-field approximation which, more drastically, equates $c_{ab}(r)$ to $f_{ab}(r)$, the Mayer function of the cross-potential $u_{ab}(r)$. Both approximations require that $c_{aa}(r)$ and $c_{bb}(r)$ are zero at all separations. The PY graphs are readily evaluated for the two-component, hard-core primitive model in odd dimensions, when they are all rational fractions. The calculation of the virial coefficients up to the tenth[8] leads, via a Padé approximant, to a (transcribed) reduced critical density of 1.116 ± 0.002 and a classical critical index of $\gamma = 1$. This numerical result has, however, now been superseded by an exact solution by Ahn and Lebowitz[53] of the PY integral equation of the primitive model in one and three dimensions. They supplement the Ornstein–Zernike equation with the conditions

$$c_{aa}(r) = c_{bb}(r) = 0;$$

$$g_{ab}(r) = 0 \quad (r < 1); \qquad c_{ab}(r) = 0 \quad (r > 1) \tag{5.32}$$

In one dimension ($s = d = 1$)

$$(\rho_a \rho_b)^{1/2} c_{ab}(r) = -qJ_0[q(1 - r^2)^{1/2}] \tag{5.33}$$

where J_0 is a Bessel function and q is the root of the equation

$$q/\cos q = (\rho_a \rho_b)^{1/2} \tag{5.34}$$

The resulting equation of state is not exact; this is characteristic of PY equations in one dimension for mixtures of hard spheres in which the cross-diameters are not the arithmetic means of the like diameters.[54] In particular, the limiting pressure at infinite density is wrong, for here $(\pi/\rho)_{(2)}$ should approach unity, its value in a perfect gas, as the two species separate (without a transition) into two pure states. The PY values for this ratio are $(1 + \pi/4)$ from the pressure equation, and 2 from the compressibility equation. A scaled-particle treatment[55] is also inexact in one dimension, but it behaves better at high densities.

Ahn and Lebowitz[53] have obtained also an exact solution of the PY equation in three dimensions, but their results do not admit of easy summary. Here, however, they show that there is a phase transition (as had been foreshadowed from a Padé approximant) and they find the critical point to be classical. Their results for the critical density (after transcription) are 1.1157 via the compressibility equation and 1.1917 via the pressure equation. Other critical parameters are related to ρ_c by

$$\left(\frac{\theta}{\rho}\right)_c = e; \qquad \phi_c = 1 - \rho_c - e^{-1}; \qquad z_c = \theta_c \exp(-\theta_c) \qquad (5.35)$$

since the PY approximation shares these results with the mean-field approximation. (The equation for z_c is exact.)

The PY approximation does not, however, have the same high internal consistency as the mean-field approximation, as is apparent from the two different values of ρ_c quoted previously. The inconsistency grows worse if we try to develop the second-order PY approximation (PY2). There are two versions of this, one due to Verlet[56] and one to Wertheim,[57] which differ in how they generalize PY1 by an approximation for the triplet direct correlation function, rather than the pair function. Verlet's method is inherently unsymmetrical in the components of a mixture, and so cannot be used here. Wetheim's method does not have this fault but has some more subtle inconsistencies which mean that it can usefully be used for the primitive model only after choosing the most symmetrical version.[8] The resulting virial expansion is ill-behaved, as is the exact expansion, and the resemblance between them suggests that the PY2 approximation, unlike PY1, would lead to a nonclassical critical point.

The HNC approximation includes more graphs in the expansion of $h(r)$ than the PY, and Helfand and Wasserman[13] and Carley[58] have shown that it generates better approximations to the correlation functions for the Gaussian version of the binary primitive model. The calculation of the thermodynamic properties, and the transcription of these to the one-component version, have problems of inconsistency[22] similar to those that plague the PY2 approximation. By choosing the most symmetrical alternative and calculating the first eleven virial coefficients, it was found that the HNC critical point is not classical ($\gamma > 1$) for a three-dimentionsl system, in either the pressure or the compressibility equation. In four and five dimensions its behavior is either classical, or close to it ($\gamma \sim 1$).

The nonclassical behavior at the critical point of both the PY2 and the HNC equation is almost certainly connected with their more realistic treatment of $c(r > 1)$ than the drastic restriction of (5.32) which characterizes both mean-field and PY approximations. Neither PY2 nor HNC requires $c(r)$ to be zero beyond the range of the potential (indeed, the HNC approximation overestimates the correlation function in this range), and this tail on $c(r)$ seems to be a necessary feature of any approximation that aims to describe a nonclassical critical point.

This conclusion implies that PY2 and HNC should yield also nonclassical critical points for more realistic intermolecular potentials, such as the Lennard-Jones pair potential. No test has yet been made for the PY2 equation, but the critical point of the HNC equation of state for a Lennard-Jones fluid has been shown to be nonclassical,[59] but, alas, in an unrealistic way not found for the penetrable-sphere model.

VI. PRIMITIVE MIXTURES WITH THREE COMPONENTS

A. Mean-field Approximation

The primitive models described in Sections II and III have an arbitrary number of components, n, and the corresponding transcribed forms have one less, $(n-1)$. The examples discussed in Sections IV and V have, however, all been chosen from models with $n=2$. Mixtures with more components are worth study only if new features of interest appear. This is the case for $n=3$, for which tricritical points are found, and it is probable that multicritical points of higher order occur if n is greater than three, but no models of this kind have yet been studied.

We consider in this section the phase diagram of the three-component primitive model and its transcription to that of the two-component penetrable-sphere model. The equations are restricted initially to the mean-field approximation, but that restriction is lifted in the next section when we examine the nature of the tricritical points.

The pair potentials of the primitive form are (2.2) to (2.3) in which each unlike interaction is characterized by the same co-volume v_0. The configurational free energy density for a system of n components is[45]

$$\psi \equiv \frac{Av_0}{kTV} = \sum_{\alpha} \rho_\alpha \left[\frac{\mu_\alpha^\ominus}{kT} + \ln\left(\frac{\rho_\alpha}{\pi^\ominus}\right) - 1\right] + \sum_{\alpha<\beta}\sum \rho_\alpha \rho_\beta + 0(\rho^3) \tag{6.1}$$

where $\mu_\alpha^\ominus$ is the standard potential of pure species α at pressure $\pi^\ominus$, and when the terms of order ρ^3 are of higher order than the second virial coefficient and so are neglected in a mean-field approximation. The equation of state and activity of each component are given by

$$\pi = \sum_{\alpha} \rho_\alpha + \sum_{\alpha<\beta}\sum \rho_\alpha \rho_\beta + 0(\rho^2) \tag{6.2}$$

$$\ln \lambda_\alpha = \ln \rho_\alpha + \sum_{\beta \neq \alpha} \rho_\beta + 0(\rho^2) \tag{6.3}$$

In a ternary mixture the phase diagram can be found from (6.1) to (6.3) by using the usual thermodynamic criteria for equilibrium between the

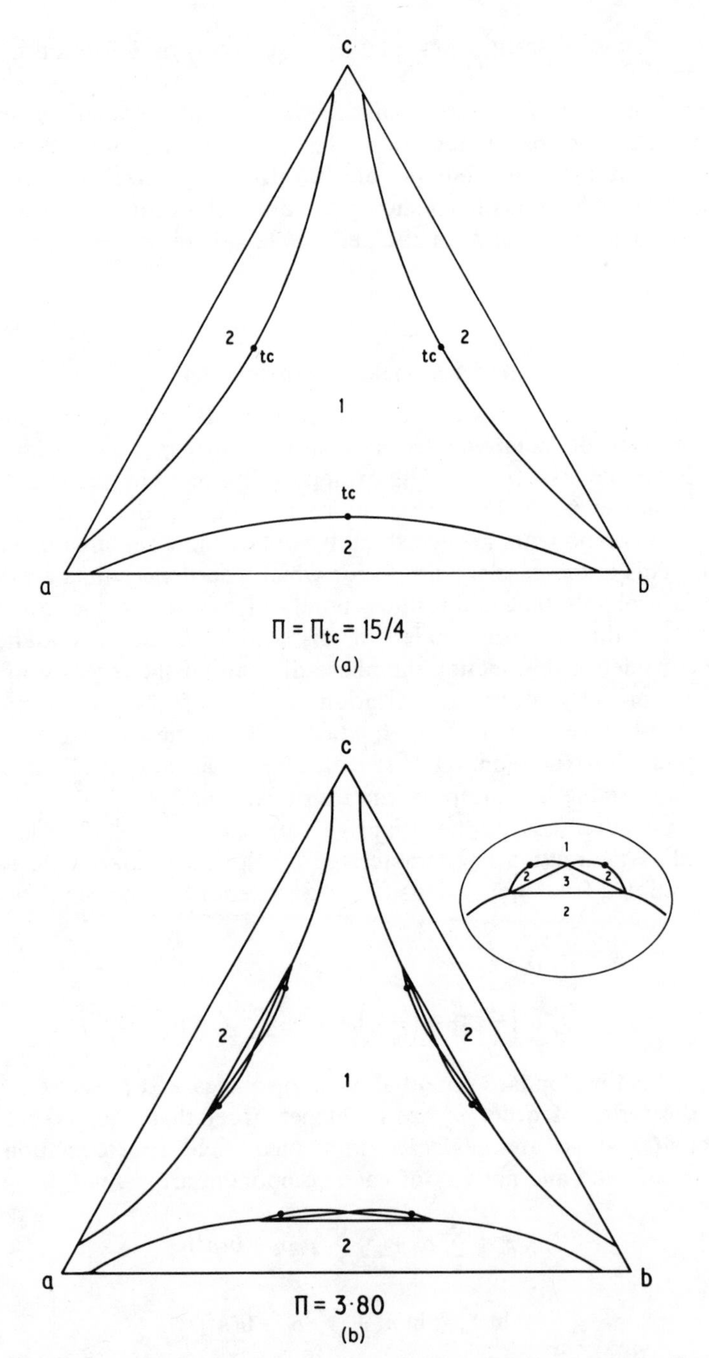

Fig. 6. The three-component primitive model in the mean-field approximation.[45] The four diagrams show the compositions of the one-, two-, and three-phase systems for four values of π. (*a*) The lowest, $\pi = \frac{15}{4}$, is that of the three tricritical points. (*b*) The second is at a slightly

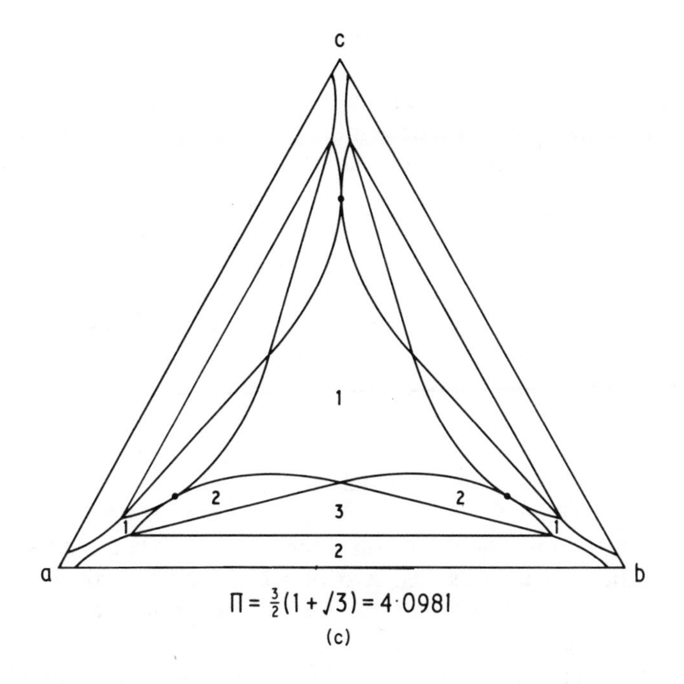

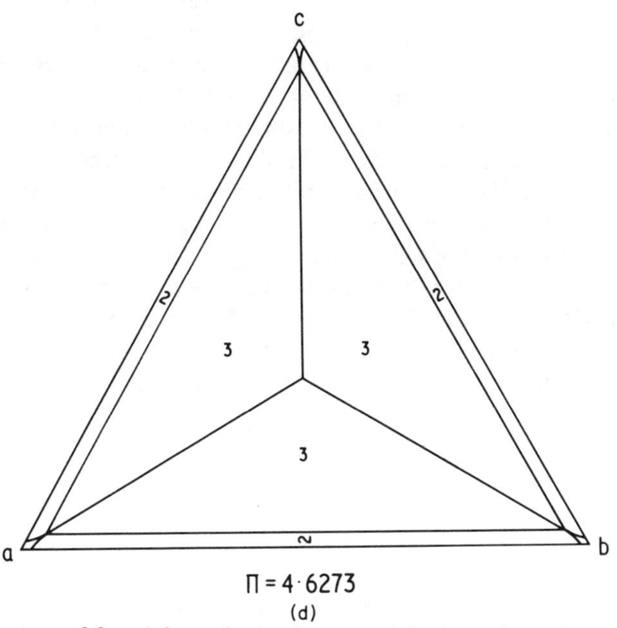

higher pressure $\pi = 3.8$, and shows the development of the three-phase triangles and the six ordinary critical points. (The inset sketch shows these more clearly). (*c*) The six critical points merge in pairs at $\pi = 4.0981$. (*d*) The three-phase triangles have expanded until they coalesce at the quadruple point, $\pi = 4.6273$.

phases. Critical points are points of contact of binodal and spinodal curves, where the former are determined by the equality of π and λ_α in each phase, and the latter from the zeros of the determinant whose elements are $(\partial^2\psi/\partial\rho_\alpha\partial\rho_\beta)$.

The results[45] are shown in Fig. 6 as a sequence of graphs in a symmetrical composition space at increasing values of π. If $\pi < 3$ then there is one phase at all compositions. At $\pi = 3$ each binary system has a critical point ($\rho_a = \rho_b = 1$, $\rho_c = 0$, and by permutation of the indices), and at higher pressures separates into two phases. The three critical lines starting from these points move in toward the center of the composition triangle ($\rho_a = \rho_b = 1$, $0 < \rho_c < \frac{1}{4}$) until a tricritical point is reached at $\pi_{tc} = \frac{15}{4} = 3.75$ ($\rho_a = \rho_b = 1, \rho_c = \frac{1}{4}$). This is a point of intersection of three critical lines, or a point at which there are three incipient phases of the same composition.[60] For $\pi > \frac{15}{4}$ there are three very flat three-phase triangles in the composition diagram, each of which has three two-phase regions abutting on to its sides, two of which end in ordinary critical points. At $\pi = \frac{3}{2}(1+\sqrt{3}) = 4.0981$ these six critical points merge in pairs and lead to a diagram in which there are four regions of one-phase, six of two-phase, and three of three-phase, but no critical point. The three-phase regions grow as the pressure rises until at $\pi = 4.6273$ they merge at a (noncritical) quadruple point. Here the mixture of equimolar composition is in equilibrium with three other phases, each of which is rich in one of the components. At all higher pressures the composition diagram is the common one of three almost wholly immiscible liquids. A sequence of phase diagrams of this kind was discovered some years ago by Meijering,[61] on the basis of a purely empirical form of the Gibbs free energy of mixing but there was no model (i.e., no Hamiltonian) behind his proposed form of the free energy.

The transcription of these results to those of the two-component penetrable-sphere model follows from the equations of Section III, where the intermolecular potential of this model is described. The equation of state in the mean-field approximation is

$$\pi = \rho_a + \rho_b + \rho_a\rho_b + \theta[(1 + \rho_a + \rho_b)e^{-\rho_a-\rho_b} - 1)] \tag{6.4}$$

At $\theta = 0$ this reduces to the equation of state of two-component hard-core model, truncated at the second virial coefficient with $B_{ab} = \frac{1}{2}v_0$. At $\rho_b = 0$ it reduces to the one-component penetrable-sphere model (5.4).

The phase diagram to which (6.4) gives rise is extremely complicated,[21] for it is a mapping of Fig. 6 into the transcribed variables with the preservation of all the topological features, including the three tricritical points and the quadruple point. Figure 7 shows the pressure–temperature projection. The critical line starting at $\pi_c = 3(\rho_a = \rho_b = 1)$ is

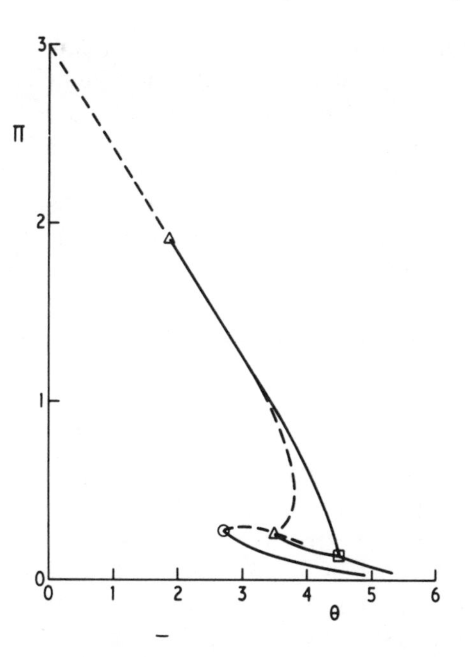

Fig. 7. The π, θ projection of the phase diagram of the two-component penetrable-sphere model in the mean-field approximation.[21] Full lines are projections of two or more phases in equilibrium; dashed lines are critical loci; the circle is the projection of critical points, the triangles of tricritical points, and the square of the quadruple point.

equivalent to one of the three critical lines that come in from the sides of the primitive composition triangle in Fig. 6. Its equation is

$$\pi_c = 3 - (1 - 3e^{-2})\theta = 3 - 0.594\theta \qquad \left(\theta \leq \frac{e^2}{4}\right) \tag{6.5}$$

and it ends at one of the tricritical points at $\theta_{tc} = e^2/4 = 1.847$, at which $\pi_{tc} = (15 - e^2)/4 = 1.903$. The other two lines of critical points start from those of the two-components ($\rho_a = 1$ or $\rho_b = 1$, $\theta_c = e$, $\pi_c = 3 - e$) and move along a pair of lines which have the same $\pi - \theta$ projection, namely

$$\pi_c = 3 \ln \theta - \theta; \qquad (e \leq \theta \leq e^{5/4}) \tag{6.6}$$

On these lines

$$\rho_a = 1; \qquad \rho_b = \ln \theta - 1 \tag{6.7}$$

or vice versa. They have a maximum of $\pi_c = 0.296$ at $\theta = 3$ and end at the other two tricritical points at $\theta_{tc} = e^{5/4} = 3.490$ and $\pi_{tc} = \frac{15}{4} - e^{5/4} = 0.260$, and densities of $\rho_a = 1$, $\rho_b = \frac{1}{4}$, or vice versa. The three tricritical points are linked to the quadruple point, $\theta_q = 4.4901$, $\pi_q = 0.1369$, where there are three liquid phases, one rich in component a, one rich in b, and one equimolar, all in equilibrium with an equimolar gas mixture. At lower temperatures, $\theta > 4.4901$, the phase diagram is the familiar one of two volatile but scarcely miscible liquids which form a hetero-azeotrope.

The presence of tricritical points in a binary system is a matter of comment since a critical point at which p phases become identical must, in general, have at least $(2p-3)$ components.[62] However, such restrictions are modified when there are elements of symmetry in the problem,[63] and it has been shown that this is the root of the paradox here.[21]

The reduced partial molar volume derived from the equation of state (6.4) is

$$v_a = \frac{1+\rho_a-\theta(\rho_a+\rho_b)e^{-\rho_a-\rho_b}}{\rho_a+\rho_b+2\rho_a\rho_b-\theta(\rho_a+\rho_b)^2e^{-\rho_a-\rho_b}} \tag{6.8}$$

If we put $\rho_b = 0$ then, naturally, $v_a = \rho_a^{-1}$, and at the critical point of pure a this gives the critical volume as $v_a = 1$. If, however, we first approach the critical line that starts from the critical point of pure a by substituting (6.7) into (6.8), then we find that $v_a = 0$ everywhere on that line except possibly at $\rho_b = 0$. This paradoxical result, that the value of partial molar volumes at a critical point can depend on the direction from which we approach that point, has been found also by Wheeler[64] for the decorated lattice gas models.

B. Tricritical Points

The existence of tricritical points was first suggested by van der Waals and Kohnstamm[65] and they have been demonstrated in at least eight fluid mixtures by two Russian groups in papers published since 1963. This experimental work is reviewed by Griffiths and Widom[60] in a paper in which they also point out the close analogy between such fluid tricritical points and some of the phase transitions in solid ammonium chloride, in metamagnets, and in $^3He + {}^4He$ mixtures. Lang and Widom[66] have studied in some detail the nature of the coalescence of the three phases and the critical indices that describe the collapse of the three-phase region. The analogy between the penetrable-sphere model and the Ising lattice model is matched by a similar analogy between the system described in the last section and the three-state Potts lattice model[14–16] which also exhibits a tricritical point.

The principal points of interest in tricritical points in fluid mixtures are, first, whether the generalized densities of the three coexisting states approach each other in a symmetrical (or "anomalous") or in an essentially unsymmetrical way, and whether the critical indices associated with a tricritical point have their classical values or not. On the theoretical side it would be interesting to be able to study these problems for systems of dimensionality other than three.

In principle, the model of the last section should furnish answers to these questions. In the mean-field approximation the tricritical points are essentially unsymmetrical in that the three-phase triangle collapses to a "line," and not symmetrically to a "point." The line is formed by the coordinates of the corners of the collapsing triangle and it singles out a particular direction in composition space[60]; in this case it is parallel to one of the sides of the composition triangle. It is easy to show also that the critical indices are classical.[45] This behavior is that of a system of infinite dimensionality, for which the mean-field approximation becomes exact.

The behavior of systems of finite dimensionality can be studied by adding to the free energy density (6.1) the missing terms arising from the virial coefficients above the second. The calculation of these is a tedious exercise in the enumeration of trichromatic irreducible graphs and only the first six coefficients are known in two, three, and four dimensions for the primitive Gaussian potential. The results[21] can be summarized, as follows

a. Four Dimensions. The tricritical point retains, as expected,[60,67] the unsymmetrical and classical form found at $d = \infty$. The indices β_1 and β_2 which describe, respectively, the rate at which the breadth and height of the three-phase triangle go to zero with $(\pi - \pi_{tc})$ and $\beta_1 = \frac{1}{2}$ and $\beta_2 = 1$. Since $\beta_2 > \beta_1$ the limiting shape of the triangle is a straight line.

b. Three Dimensions. This case is more difficult to decide on the basis of only six virial coefficients, and the usual method of "extending" the expansion by fitting it with a Padé approximant could not be used since the thermodynamic properties are functions of three variables, not one. (Fisher's proposal of partial differential approximants[68] or Chisholm's of multivariate approximants[69] might be of value here, but have not yet been tried.) Each approximation, up to the sixth, has an unsymmetrical tricritical point with the classical indices, $\beta_1 = \frac{1}{2}$ and $\beta_2 = 1$, but, with each successive virial coefficient, the tricritical pressure and densities approach more closely to those of the quadruple point. Should they reach this point with an infinite series then the tricritical point would be symmetrical or "anomalous," since it would be at the equimolar composition. The question remains open. The experimental evidence[66] and phenomenological arguments[60,67] suggest that in three dimensions tricritical points retain their classical character, but that $d = 3$ is the boundary between this and nonclassical behavior, rather as $d = 4$ is the boundary for ordinary critical points. The evidence from this model is that it is also the boundary between asymmetrical (or normal) and symmetrical (or anomalous) tricritical behavior.

c. Two Dimensions. Here there is strong evidence that the fluid tricritical point does not remain normal. It is almost certainly symmetrical, and the critical indices seem to be departing from their classical values. Again this result is in line with the phenomenological arguments.[60,67]

VII. THE GAS–LIQUID SURFACE

A. Symmetry and the Mean-field Approximation

The determination of the molecular structure of the surface between two phases is one of the unsolved problems of physical chemistry. Only in rare cases are there experimental methods that give any but the crudest information, and consequently much attention has been given recently to computer simulation, and to the devising of tractable molecular models. Of the latter, lattice models are, as usual, the most tractable and have received most attention.[70–73] Their use to represent the gas–liquid surface has, however, the obvious disadvantage of their inability to describe realistically a situation in which the density of the fluid is changing smoothly but rapidly over lengths comparable with the size of the molecules or the spacing of the lattice. The penetrable-sphere model, although less tractable, is free from this drawback, and one of its principal recent uses has been to test statistical theories of the surface tension and other properties of the gas–liquid surface.

As before, the key to progress is the symmetry of the partition function and hence of the phase diagram. We start[74] with the two-component hard-core primitive model, at an overall density above that of the critical point. There will be two phases of densities ρ'_a, ρ'_b and ρ''_a, ρ''_b; where, by symmetry $\rho''_a = \rho'_b$ and $\rho''_b = \rho'_a$. Moreover, this symmetry must persist at each reduced height z on a line normal to the surface (Fig. 8). That is, if we take the origin of the z-coordinate to lie in the x, y-plane on which $\rho_a = \rho_b$, then we have

$$\rho_a(z) = \rho_b(-z) \tag{7.1}$$

Transcription of this result to the gas–liquid surface of the one-component penetrable-sphere model gives, from (3.8), (4.16), and (4.17)

$$\phi(z) = 1 - \rho(z) - \theta^{-1}\rho(-z) \tag{7.2}$$

The mean value of the energy density in a two-phase system of overall density ρ is given by (4.20). Subtraction of this from (7.2) gives the

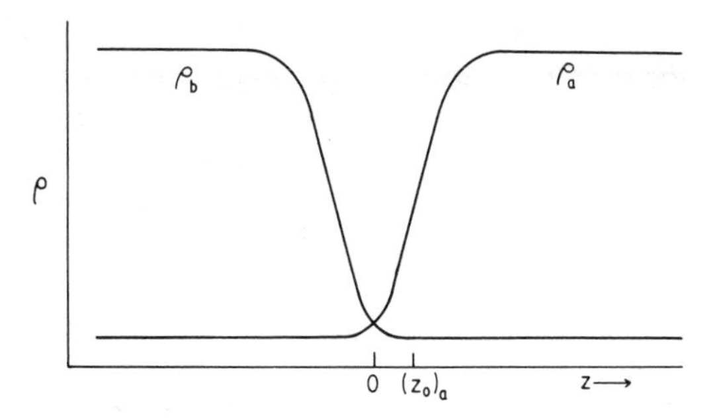

Fig. 8. Sketch of the densities of the two components of the primitive model along a line normal to the plane surface between the two phases.[74] The zero of the z-coordinate is taken at the point on this line at which $\rho_a = \rho_b$.

excess energy density

$$\phi_E^{(2)}(z) = \phi_E^{(2)}(-z) = \theta^{-1}[2\bar{\rho} - \rho(z) - \rho(-z)] \tag{7.3}$$

and integration over all z gives the total excess energy of the surface

$$\phi_E = \int_{-\infty}^{\infty} dz \phi_E^{(2)}(z) = \frac{4z_0\Delta}{\theta} \tag{7.4}$$

where $\Delta = \frac{1}{2}(\rho_l - \rho_g)$ and z_0 is the height of the Gibbs surface of zero excess number density. The reduced surface tension γ is equal to the excess free energy, ψ_E, and so, by integration of the Gibbs–Helmholtz equation

$$\psi_E = \gamma = \frac{1}{\theta}\int_{\theta_c}^{\theta} d\theta \phi_E = \frac{4}{\theta}\int_{\theta_c}^{\theta} d\theta \frac{z_0\Delta}{\theta}. \tag{7.5}$$

These equations are the consequences of symmetry, and so all are exact in the sense that they are independent of any of the approximations of the kind discussed in Section V. There is, however, uncertainty about the precise meaning of local thermodynamic functions such as $\phi(z)$. Thus at the surface of a Lennard-Jones fluid it is reasonable to calculate $\phi(z)$ by assigning half the intermolecular energy $u_{LJ}(r_{ij})$ to the position of molecule i and half to that molecule j (as has been proposed by Rao and Levesque[75]), but it would be equally reasonable to assign the whole energy to the midpoint of the vector $\mathbf{r}_i - \mathbf{r}_j$. A third choice would be to use the potential distribution theorem[37] and estimate $\phi(z)$ by the mean change of energy on inserting an additional molecule at a point of height z. In each case the same integral value ϕ_E is obtained in spite of the differences in the local function $\phi(z)$. Ono and Kondo have pointed out[76] that there is a similar ambiguity in the definition of the local value of the transverse component of the pressure, $p_T(z)$, but not of its integral, which gives γ.

In this model the total energy U is defined in terms of the volume covered by the penetrating spheres, and so a natural definition of $\phi(z)$ is that in terms of the mean area covered by the spheres at height z. We return to this in later text, but observe here only that this definition is also arbitrary; it is not, for example, the same function as is defined by the potential distribution theorem.[77]

A further feature of interest of these equations is the behavior of γ near the critical point. Let us introduce the usual critical indices β and μ to describe the zeros of Δ and γ, respectively, and an index χ to describe the behavior of z_0; that is,

$$\begin{aligned} &\gamma \sim (\theta - \theta_c)^{\mu}; \qquad \phi_E \sim (\theta - \theta_c)^{\mu - 1} \\ &\Delta \sim (\theta - \theta_c)^{\beta}; \qquad z_0 \sim (\theta - \theta_c)^{\chi} \end{aligned} \tag{7.6}$$

From (7.5) we have

$$\mu = 1 + \beta + \chi \tag{7.7}$$

For a real fluid[78] β and μ are about 0.32 and 1.28, respectively, so we should have $\chi \approx -0.04$, or a weak divergence of z_0 to infinity at the critical point. Such divergence may be connected with that in the curvature of the diameter of the orthobaric densities which this model exhibits. This is of order of $-\alpha$, the index which governs also the divergence of the heat capacity at constant volume, and which is believed to be between -0.00 and -0.08 for real fluids.

The statement that z_0 diverges weakly may be without meaning for a real fluid since there is, in general, no known plane of symmetry, $z = 0$, from which to measure the position of the Gibbs surface. What is meaningful and should lead to observable consequences is that at a two-component critical point the distance between the two Gibbs surfaces $(z_0)_a$ and $(z_0)_b$ should diverge weakly. We know of no experimental evidence on this point.

We can obtain some explicit results in the limit of zero temperature. Here the liquid phase is an increasingly dense array of penetrating spheres which cover the whole volume, and the gas phase is empty (4.45). It follows that the limiting shape of $\rho(z)$ is a step-function of height θ at $z = z_0 = \frac{1}{2}$, for all finite values of s. Hence, from (7.4) and (7.5) the limiting values of the reduced surface energy ϕ_E and tension γ are both equal to unity.

To make further progress we must invoke some approximation for $\phi(z)$ and the most useful, as before, is the mean-field approximation. If we have a density of molecules $\rho(z)$ on each x,y-plane parallel to the surface, and a random distribution over that plane, then we can calculate the mean covered area of the plane, which we now identify with $\phi(z)$. We can then deduce that[74]

$$\phi(z) = 1 - \rho(z) - e^{-\sigma(z)} \qquad \text{(cf. (5.3))} \tag{7.8}$$

where $\sigma(z)$ is an effective number density given, for a potential of dimensionality s by

$$\sigma(z) = \frac{s!}{2^s[((s-1)/2)!]^2} \int_{-1}^{1} du (1-u^2)^{(s-1)/2} \rho(z+u) \tag{7.9}$$

It is an average of $\rho(z)$ over a reduced distance equal to the diameter of a penetrable sphere; in a homogeneous fluid σ reduces to ρ. The energy density $\phi(z)$ can be eliminated between (7.2) and (7.8) to give, with (7.9), a nonlinear integral equation for $\rho(z)$. This equation has been solved explicitly only for $s = 1$, when[74]

$$z_0 = \tfrac{1}{2}; \qquad \rho(z) = \bar{\rho} + \Delta \tanh[\tfrac{1}{2}\Delta(z - \tfrac{1}{2})] \tag{7.10}$$

Potentials of other dimensions can be shown[79] to decay exponentially with distance at large values of $|z - z_0|$ but give a hyperbolic tangent only in the critical region, where, as $\Delta \to 0$,

$$(z_0)_c^2 = \frac{3}{4(s+2)}; \qquad \rho(z) = \bar{\rho} + \Delta \tanh\left[\frac{\Delta}{4z_0}(z - z_0)\right] \tag{7.11}$$

The critical value of z_0 is evidently finite in the mean-field approximation.

The effective density σ for the Gaussian potential (2.6) to (2.7) is given by

$$\sigma(z) = \frac{1}{\pi l^{1/2}} \int_{-\infty}^{\infty} du e^{-l^{-2}u^2} \rho(z+u) \tag{7.12}$$

where the parameter l, which measures the range of the potential, is now in reduced units. The integral equation for $\rho(z)$ has not been solved for this potential.

Not only $\phi(z)$ but also the activity $\lambda(z)$ can be obtained explicitly for this model, although now only in the mean-field approximation. By using again the potential distribution theorem, as in Section IV.E, but now for a particular plane at height z, it can be shown[74] that $\lambda(z)$ is determined by another effective number density $\tau(z)$, the average of $\sigma(z)$ over a distance equal to the diameter of a sphere, and so of $\rho(z)$ over twice this distance. (Woodbury[71] has assumed a similar result for a lattice model.)

$$e^{-\tau(z)} = \frac{s!}{2^s[((s-1)/2)!]^2} \int_{-1}^{1} du (1-u^2)^{(s-1)/2} e^{-\sigma(z+u)} \tag{7.13}$$

The three densities ρ, σ, and τ are related by a chain of equations

$$-\rho(z) = -\theta \exp[-\sigma(-z)] = -\theta \exp\{-\theta \exp[-\tau(z)]\} \qquad (7.14)$$

and are all equal in a homogeneous fluid. The activity at height z is given in terms of $\tau(z)$ by

$$\ln \lambda(z) = \ln \rho(z) - \theta(1 - e^{-\tau(z)}) \qquad \text{(cf. (5.5))} \qquad (7.15)$$

By substituting (7.13) or (7.14) into (7.15) we find explicitly that $\lambda(z)$ is a constant at each point in the surface layer, for potentials of all dimensionality s,

$$\ln \lambda(z) = \ln \theta - \theta \qquad (7.16)$$

From (4.23) we know that this result becomes exact as z approaches $\pm\infty$, and is not restricted to the mean-field approximation.

The constancy of activity throughout the surface layer was a tenet of Gibbs' treatment[80] of the thermodynamics of surfaces, but the preceding result seems to be the first case in which an explicit demonstration has been carried out via the potential distribution theorem. For this model the constancy of activity is, in a sense, equivalent to constancy of temperature, since the temperature in the penetrable-sphere model is obtained by the transcription of one of the activities, λ_b, of the underlaying primitive model. But of the constancy of temperature we are surely in no doubt since it may be defined in terms of a mean translational energy of the molecules, and since translational energy is rigorously separable from configurational energy in any classical partition function. So the preceding result may, alternatively, be regarded as the first demonstration of the validity of the potential distribution theorem in an inhomogeneous fluid.

The demonstration has relied upon the particular definition of $\phi(z)$ that led to (7.8) and (7.9), and the use of the condition of symmetry (7.2). We could, however, have made the constancy of $\lambda(z)$ a prior assumption, when we should have been led to the result that (7.8) and (7.9) was the only definition compatible with the constancy of λ and the symmetry of ϕ. We should then be forced to accept the validity of the potential distribution theorem for the calculation of λ in an inhomogeneous system while abandoning it for the definition of ϕ. This appears to be the most acceptable position in the case of a lattice gas.[73] The whole question of the definition of local thermodynamic functions in inhomogeneous systems needs further discussion. The status of the local Helmholtz free energy density is considered in Section VII.C.

B. Surface Tension

In Section V.A. we discussed the three routes from the molecular correlation functions to the pressure in a homogeneous fluid. We know that these are all equivalent[3,46] if no approximations are introduced into the calculations of the correlation functions, and it was shown there that the mean-field approximation for the penetrable-sphere model is self-consistent, for systems of all dimensions, in the sense that it gives the same value for the pressure via all three routes.

There are similarly three different routes to the surface tension of a fluid: (1) from the local energy density $\phi(z)$, (2) from the virial of the intermolecular forces, and (3) from the direct correlation function $c(\mathbf{r})$. There was until very recently no formal proof that these three routes are equivalent even for an exact treatment, and, indeed, doubts have been expressed[81] about the validity of the third route outside the critical region. The penetrable sphere model can be used to resolve these doubts since the surface tension can be calculated in all three ways in the mean-field approximation.

1. Energy Route

This is the route described in the last section, in which the surface tension is given in terms of the length z_0 by the integral in (7.5). For a one-dimensional potential ($s = 1$), and independent of the dimensionality of the space, d, we have $z_0 = \frac{1}{2}$ (7.10) and so, changing the variable of the integral from θ to Δ and using (5.11), we have

$$\gamma = \frac{2}{\theta} \int_0^{\Delta} d\Delta'(1 - \Delta'^2 \operatorname{cosech}^2 \Delta') \qquad (s = 1) \tag{7.17}$$

This integral can be expressed as a power series in Δ, for $\Delta < \pi$, or in $e^{-2\Delta}$, for large values of Δ. The result[74] is shown in Fig. 9.

For the original penetrable-sphere model, that is, for $s = 3$, we can calculate γ by solving numerically the integral equation for $\rho(z)$ and so determining z_0 as a function of θ. A further integration, (7.5), then gives the surface tension (Fig. 9). For all values of s the mean-field approximation gives the correct limiting values of unity to both ϕ_E and γ as the temperature tends to zero.

2. Virial Route

The surface tension is proportional to the isothermal change of free energy with the surface area $\mathcal{A}$.

$$\gamma = -kT\left(\frac{\partial \ln Q}{\partial \mathcal{A}}\right)_{N,V,T} \tag{7.18}$$

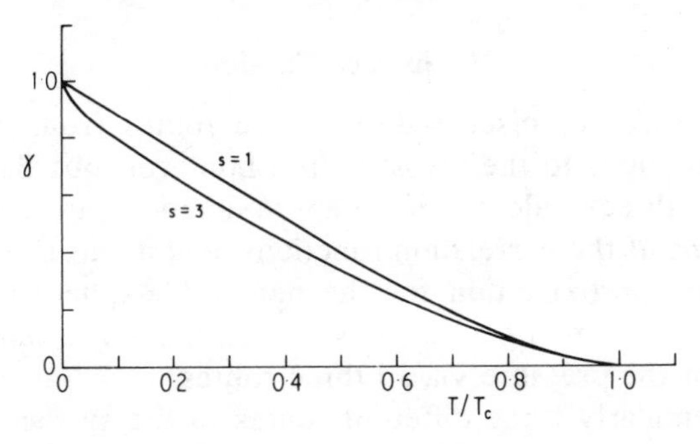

Fig. 9. The reduced surface tension as a function of T/T_c for the one- and three-dimensional potentials of the penetrable-sphere model in the mean-field approximation.[74]

where Q is the canonical partition function,

$$Q = \frac{1}{N!}\int d\mathbf{R}^N \cdot \exp\left(-\frac{U(\mathbf{R}^N)}{kT}\right) \tag{7.19}$$

Buff[82] and MacLellan[82] carried out this differentiation by scaling all the linear dimensions of the system so that the limits of the integral were fixed—a method used previously by Green[83] for calculating the pressure of a homogeneous fluid from the virial of the intermolecular potential. Green, Buff, and MacLellan all considered potentials that were the sum of pair interactions, but for this model the method can be adapted[74] to the present form of $U(\mathbf{R}^N)$ which is a sum of multibody potentials of all order.

The results depend on the dimensionality of the potential, and we consider only the two cases $s = 1$ and $s = 3$.

$$\gamma = \frac{1}{\theta}\int_{-\infty}^{\infty} dz[\tfrac{1}{2}\rho_g\rho_l + \tfrac{1}{2}\rho(-z)\rho(z+1) - \rho(-z)\sigma(z)] \qquad (s = 1) \tag{7.20}$$

and

$$\gamma = \frac{3}{4\theta}\int_{-\infty}^{\infty} dz\rho(z)\int_{-1}^{1} du(3u^2 - 1)\rho(-z+u) \qquad (s = 3) \tag{7.21}$$

where $\sigma(z)$ in (7.20) is the effective density of (7.9) with $s = 1$.

The integral (7.20) can be simplified by introducing the profile $\rho(z)$, (7.10), which satisfies the integral equation for ($s = 1$), and the corresponding expression for $\sigma(z)$. The resulting function can be expanded in powers of tanh Δ and re-summed, using the identity

$$2\sum_{i=0}^{\infty}\sum_{j=0}^{i}\frac{(\tanh\Delta)^{2i+1}}{(2i+1)(2j+1)} = \Delta\bar{\rho} + \int_0^{\Delta} d\Delta' \cdot \Delta'^2 \operatorname{cosech}^2\Delta' \tag{7.22}$$

This result establishes the identity of the energy and virial expressions (7.17) and (7.20) for $s = 1$. The case $s = 3$ cannot be treated so satisfactorily since we do not have an explicit solution of the integral equation for $\rho(z)$. However, numerical integration[74] establishes that (7.5) and (7.21) lead to the same value of γ at all temperatures.

3. Route via the Direct Correlation Function

More interesting, since the identity of the energy and virial routes was not in serious doubt, is the comparison with the values of γ obtained from the direct correlation function. This function is defined by the Ornstein–Zernike equation (5.14) in which the density $\rho(\mathbf{r}_3)$ is retained within the integral for an inhomogeneous fluid. In the underlying primitive model this density is that of the second or "ghost" component which vanishes on transcription. But we know from (7.1) that the density of the ghost component at $\mathbf{r}_3$ is that of the retained component at $-\mathbf{r}_3$, and so for the penetrable-sphere model in the mean-field approximation

$$c(\mathbf{r}_1, \mathbf{r}_2) = \frac{1}{v_0} \int_{v_{12}} d\mathbf{r}_3 \cdot \rho(-z_3) \qquad \text{(cf. (5.15))} \tag{7.23}$$

where v_{12} is again the volume common to two spheres of unit radius at $\mathbf{r}_1$ and $\mathbf{r}_2$.

Now Lovett, Mou, and Buff[84] and Wertheim[84] have recently obtained an integral equation that relates $c(\mathbf{r}_1, \mathbf{r}_2)$ to the gradients of the profile at these points. For a system of dimensionality $s = d$ their equation is

$$\frac{d \ln \rho(z_1)}{dz_1} = \frac{(\frac{1}{2}s)!}{\pi^{(1/2)s}} \int d\mathbf{r}_2 \cdot c(\mathbf{r}_1, \mathbf{r}_2) \rho'(z_2) \tag{7.24}$$

where the prime denotes differentiation. Substitution of (7.23) into (7.24) converts this to the integral equation for $\rho(z)$ obtained in the last section from symmetry arguments, thus establishing the consistency of the mean-field approximation.

The integral (7.23) can be obtained explicitly only in a one-dimensional system ($s = d = 1$), when

$$c(\bar{z}, r) = \bar{\rho}(1 - \tfrac{1}{2}r) - 2 \operatorname{arctanh}[\tanh(\tfrac{1}{2}\Delta(\bar{z} + \tfrac{1}{2})) \tanh(\tfrac{1}{2}\Delta(1 - \tfrac{1}{2}r))] \tag{7.25}$$

where

$$\bar{z} = \tfrac{1}{2}(z_1 + z_2) \quad \text{and} \quad 2 \geqslant r = z_2 - z_1 \geqslant 0 \tag{7.26}$$

The limits of $c(\bar{z}, r)$ in the gas and liquid phases are[45]

$$\begin{aligned} c_g(r) &= c(-\infty, r) = (\bar{\rho} + \Delta)(1 - \tfrac{1}{2}r) = \rho_l(1 - \tfrac{1}{2}r) \\ c_l(r) &= c(+\infty, r) = (\bar{\rho} - \Delta)(1 - \tfrac{1}{2}r) = \rho_g(1 - \tfrac{1}{2}r) \end{aligned} \tag{7.27}$$

These results show that, at least in a one-dimensional system, $c(\bar{z}, r)$ is not well represented by the simpler function $c([\rho(\bar{z})], r)$, the direct correlation function of the homogeneous fluid whose density is equal to that at the midpoint of the line joining the two molecules,[74]

$$c([\rho(\bar{z})], r) = [\bar{\rho} + \Delta \tanh(\tfrac{1}{2}\Delta(\bar{z} + \tfrac{1}{2}))](1 - \tfrac{1}{2}r) \tag{7.27}$$

(A one-dimensional system should, however, not have a "surface" between two phases (Section IV.B); expressions (7.25) to (7.28) are valid only in the mean-field approximation.)

The surface tension can be expressed in terms of the direct correlation function by using arguments based on fluctuation theory,[85] or based on the change of pressure in the liquid phase when the surface is curved.[86] The result is, in reduced units, $(s = d = 3)$

$$\gamma = \frac{1}{4\theta}\frac{3}{4\pi}\int_{-\infty}^{\infty} dz_1 \int d\mathbf{r}_2 \cdot \rho'(z_1)\rho'(z_2)(x_{12}^2 + y_{12}^2)c(r_{12}, z_1, z_2) \tag{7.29}$$

After much manipulation[74] this can be reduced to the double integral (7.21), thus showing the identity at all temperatures of the virial route and that via the direct correlation function. We cannot simplify either expression beyond this double integral without an explicit expression for $\rho(z)$ in three dimensions.

A proof of the identity of the three routes for a particular model is not a proof of their identity in general, such as we have for the parallel case of the pressure of a homogeneous fluid, but there seems to be no particular features of this model that would lead us to think that it was not typical of fluids in general. Since this work was completed, Jhon, Desai, and Dahler[87] have published an outline of a proof, for pair potentials, that the second and third routes are identical in all cases; full details are awaited.

C. Free Energy Density

The difficulties of defining local thermodynamic functions such as $\phi(z)$, $\lambda(z)$, and $p_T(z)$ have been touched upon in Section VII.A. A local Helmholtz free energy density was defined by van der Waals[88] in 1894 and used as the basis of a theory of surface tension. It was put forward again by Cahn and Hilliard in 1958 and is now very widely used. The form chosen for this local free energy density originally owed more to classical than to statistical mechanical arguments, but its statistical foundation is now established. A reduced free energy density for a homogeneous fluid can be defined for this model by

$$\psi = \frac{Av_0}{V\epsilon} \tag{7.30}$$

and the basis of the van der Waals theory is that the free energy density at height z, $\Psi(z)$, can be written as the sum of two terms,

$$\Psi(z) = \psi[\rho(z)] + \tfrac{1}{2}m[\rho'(z)]^2 \tag{7.31}$$

where $\psi[\rho(z)]$ is the free energy density of (7.30) for a hypothetical fluid of homogeneous density $\rho(z)$, where m is a constant coefficient, and where $\rho'(z)$ is the gradient of the profile. Lebowitz and Percus,[90] Yang, Fleming, and Gibbs,[91] and others[85,86] have shown that the free energy of an inhomogeneous system can indeed be expanded in powers of derivatives of the number density ρ, that the leading correction to $\psi[\rho(z)]$ is of the form of (7.31), and that the coefficient m is proportional to the square of Debye's persistence length, that is, to the second moment of the direct correlation function. In the reduced units of this model[74]

$$m = \frac{1}{\theta} \cdot \frac{1}{2d}[f(d)]^{-1} \int d\mathbf{r}r^2c(r) \tag{7.32}$$

where $d = s$ is the dimensionality of both space and potential, and where $f(d)$ is the volume of a d-dimensional sphere of unit radius.

A general discussion of the validity and value of (7.31) is beyond the scope of this review, but the equation is now so widely used[85,86,91–98] for the calculation of surface tension and other thermodynamic properties that it is worth examining the results that can be obtained for the penetrable-sphere model, which throw some light on the problems of using this equation. There is no attempt here to examine the status of metastable states for the penetrable-sphere model although Cassandro and Olivieri[99] have considered this for the primitive form.

The derivations of (7.31) show that the term in $[\rho'(z)]^2$ is but the first of an expansion in powers of derivatives of $\rho(z)$. When $\rho'(z)$ is small, that is, near the critical point, it seems reasonable to include only the term in $[\rho'(z)]^2$, to recognize that a mean-field approximation underlies all theories of the van der Waals kind, and so to assign to $m\theta$ its finite critical value in the mean-field approximation for this model, namely[74]

$$m^{-1} = m_c^{-1} = (d+2)\theta \tag{7.33}$$

Such assumptions underlie most of the work cited, although it is admitted that if the mean-field approximation is abandoned then the second moment of $c(r)$ is weakly divergent at the critical point.[92]

It is not, however, this aspect of the theory that the penetrable-sphere model is best able to test, but rather the increasing use of (7.31), without higher terms, at temperatures well below the critical, where $\rho'(z)$ is not small. We do this by comparing the surface tension calculated from (7.31) with that obtained from the mean-field approximation without further assumptions; that is, from (7.5) or more explicitly from (7.20) for $s = 1$ and from (7.21) for $s = 3$. The surface tension is the total excess free energy per unit area and so, from (7.31), is given by

$$\gamma = \int_{-\infty}^{\infty} dz[\Psi(z) - \psi^{(2)}(\rho(z))] \tag{7.34}$$

$$= \frac{1}{(d+2)\theta} \int_{-\infty}^{\infty} dz[\rho'(z)]^2 \tag{7.35}$$

$$= \frac{1}{(d+2)\theta} \int_{\rho_g}^{\rho_l} d\rho[z'(\rho)]^{-1} \tag{7.36}$$

The equivalence of (7.34) and (7.35) is a consequence of choosing a profile $\rho(z)$ which minimises γ, and such minimization gives at the same time a recipe for determining $\rho(z)$. We can use (7.34) to (7.36) in several ways.[77]

First, we substitute into (7.34) or (7.35) the "correct" profile found by solving the integral equation obtained from $\phi(z)$ (Section VII.A), and compare these values of γ with that obtained from $\phi(z)$ without further approximation, (7.5). Figure 10 shows that all agree well in the critical region, where $\rho'(z)$ is small, but not at low temperatures. The limiting value from (7.34) with this form of $\rho(z)$ is zero at zero temperature, whereas that from (7.35) is infinite. We know that the exact value, without any mean-field approximation, is unity. Nevertheless the error from (7.34) is not large down to a reduced temperature of about 0.5, and this is sufficiently low for most real fluids whose triple points are generally around $T/T_c \sim 0.5$.

A second, and more consistent way of using the van der Waals theory is to determine $\rho'(z)$ from the minimization of the free energy, and to substitute this result into either (7.34) or (7.35). The error is now less (Fig. 10) but γ is still infinite at zero temperature. The error at $T/T_c \sim 0.5$ is still reasonably small.

A third and more sophisticated route to γ is to follow the proposal of Yang, Fleming, and Gibbs[91] and assign to $m\theta$ not the constant value (7.33), but at each point to give m the value of $m[\rho(z)]$ for a homogeneous fluid of density $\rho(z)$, where $m[\rho(z)]$ is calculated from (7.32) by using the direct correlation function for the same homogeneous density $\rho(z)$. In the calculation of γ from (7.35) the function $m[\rho(z)]$ now appears within the integral. Such a calculation can be made explicitly for this model and leads to a finite nonzero value of γ at zero temperature (Fig. 10). The error at all temperatures is much less than for the more

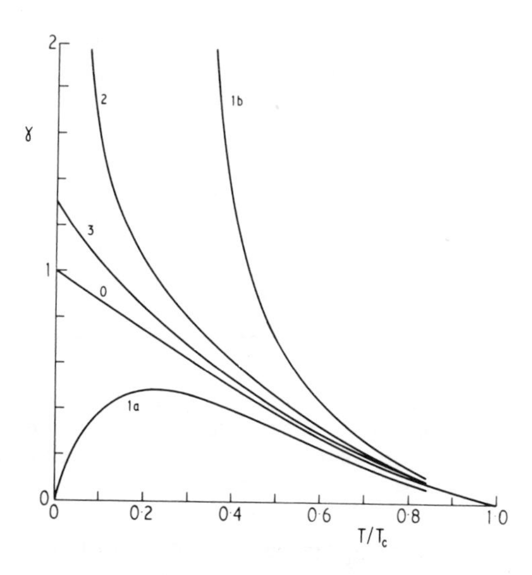

Fig. 10. The reduced surface tension as a function of T/T_c, for $s = 1$. The curve 0 is the mean-field approximation of Fig. 9; curve $1a$ is calculated from (7.34) and $1b$ from (7.35) with the same profile $\rho(z)$; curve 2 is calculated[77] from (7.34) or (7.35) with a profile chosen to minimize the van der Waals free energy; curve 3 is calculated[77] by replacing the constant coefficient by the function $m[\rho(z)]$ within the integral of (7.35), following the proposal of Yang, Fleming, and Gibbs.[91]

naive assumption, so the increased complexity of these calculations may well be justified for real fluids.

Finally, it should be noted that $\Psi(z)$ of (7.31) is not related to $\phi(z)$ of Section VII.A by the Gibbs–Helmholtz equation.

D. Mixtures

Little work has yet been done on the surface properties of mixtures of penetrable spheres, or indeed on any other continuum molecular model. It is clear, however, that there is here a fruitful way of studying such macroscopic laws as Young's equation, or the Gibbs adsorption isotherm, at a molecular level. Consider, for example, the primitive three-component system introduced in Section VI, and let the densities of components a and c be large and that of component b small. The system will form two phases, one rich in a, one rich in c, and each containing, by symmetry, the same low concentration of b. At the surface the profiles $\rho_a(z)$ and $\rho_c(z)$ have the form shown in Fig. 11 (cf.

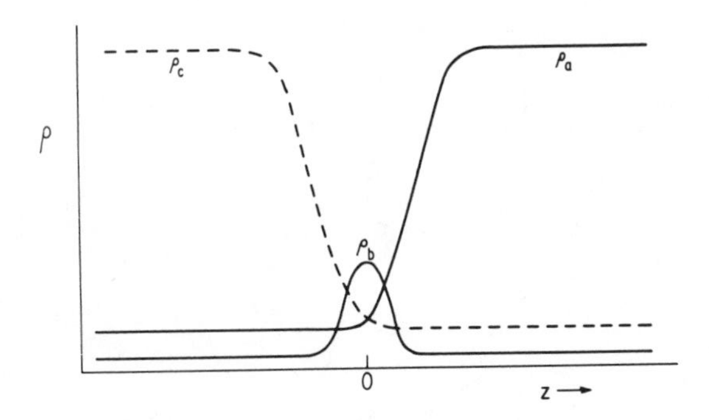

Fig. 11. Sketch of the densities of the three components of the primitive form of the model when ρ_a and ρ_c are large and ρ_b is small.[100] On integration over all coordinates of molecules of component c in the grand partition function a system is obtained in which component b is absorbed at the gas–liquid surface of component a.

Fig. 8). Molecules of component b see those of a and c indifferently; that is, they distribute themselves according to the sum of local densities $[\rho_a(z)+\rho_c(z)]$, and not according to $\rho_a(z)$ and $\rho_c(z)$ independently. It is clear from Fig. 11 and, follows from calculations similar to those already described, that the sum is less than $[\rho_a(\pm\infty)+\rho_c(\pm\infty)]$ when z is small; by symmetry, the difference is greatest when $z=0$. Since the $a-b$ and $c-b$ potentials are repulsive it follows that $\rho_b(z)$ is small at $z=\pm\infty$ and has its maximum at $z=0$ (Fig. 11).

Let us now integrate over the positions of all molecules of component c in the grand partition function, and so obtain this function for the transcribed two-component form discussed at the end of Section VI.A. The local densities of a and b are unaffected by the transcription and so we now have a model in which component a has a gas–liquid surface at which a small amount of component b is adsorbed. The density of b is equally low in the gas and liquid phases and has its maximum on the gas side of $(z_0)_a$, the Gibbs surface of component a.

If $\rho_b(z)$ is everywhere so small that we can, as a first approximation, assume that it does not perturb the existing distribution $\rho_a(z)$ then we can obtain, in a mean-field approximation,[100]

$$\rho_b(z)=\lambda_b\exp[-\sigma_a(z)-\sigma_a(-z)] \tag{7.37}$$

$$=\rho_b\frac{\cosh(\Delta_a z)+\cosh(3\Delta_a/2)}{\cosh(\Delta_a z)+\cosh(\Delta_a/2)} \tag{7.38}$$

where λ_b and $\rho_b\leqslant 1$ are, respectively, the constant activity of b throughout the system and its concentration in either bulk phase, and where Δ_a is $\frac{1}{2}(\rho_a^l-\rho_a^g)$. The function (7.38) resembles that shown in Fig.

11, with a maximum at $z = 0$. The surface excess is found by integration

$$\Gamma_b = \int_{-\infty}^{\infty} dz[\rho_b(z) - \rho_b] = \rho_b \cdot 2 \sinh \Delta_a \tag{7.39}$$

To obtain $\rho_b(z)$ at higher concentrations, or to calculate other thermodynamic properties, requires the solution of two coupled integral equations. The first-order corrections to (7.37) to (7.39) have been obtained analytically,[100] but the general solution will, no doubt, have to be found numerically. This is probably worthwhile since we know from the results of Section VI that the phase diagram of this a–b mixture is very rich, with critical, tricritical, and quadruple points. A molecular study of surface properties near these would be valuable.

VIII. CONCLUSIONS

Model systems are designed for the study of those physical and chemical phenomena that depend on certain broad features of the molecular behavior and not on the precise details. Preeminent among these is the study of phase equilibria, where the dimensionality of the system and the range of the intermolecular forces are the important broad features. The aim of this review has been to show that the penetrable-sphere model, where the forces are short-ranged, but the dimensionality and number of components can be varied at will, is a suitable model for the study of the gas–liquid transition. It lacks some of the exact results known for lattice models, and so much reliance is put on the mean-field approximation in the later sections of the review. Fortunately this approximation is found to have here an unusually high degree of internal consistency. The principal virtues of the model are that it allows truly continuous motion of the molecules, which have a continuously varying energy of interaction. It retains, however, an important symmetry that often leads to explicit results. These virtues are most obvious in the handling of critical phenomena and of the structure of the gas–liquid surface. The results in Section VII show that further progress can be expected at least on the second of these fronts. The model seems to be well-adapted for studying the vexed question of the definition of local thermodynamic functions, and for the study of adsorption at a molecular level.

PRINCIPAL SYMBOLS

A — Surface area of an array of penetrating spheres (Section IV.D); configurational Helmholtz free energy (Sections VI to VII)

$\mathscr{A}$	Area of the gas–liquid surface (Section VII)
a	Diameter of a hard sphere (primitive form), or radius of a penetrable sphere (transcribed form)
a, b, c	Components of a mixture
B	Second virial coefficient
b_l	Mayer's lth cluster integral
C_{ij}, C_{ijk}	Star cluster integrals (Sections V.B and VI)
$c(R)$	Direct correlation function at separation R
c_v	Reduced heat capacity density, $C_v v_0/kV$
d	Dimensionality of the model
$f(R)$	Mayer's f-function (2.4)
$h(R)$	Total correlation function
K, k	Cumulants of the volume distribution W (4.8) to (4.10)
m	van der Waals parameter (Section VII)
N	Number of molecules
n	Number of components (primitive form)
r	Reduced separation, R/a
s	Dimensionality of the potential
T	Temperature
$U, \bar{U}, \langle U \rangle$	Configurational energy, canonical and random averages of configurational energy
u	Pair potential (primitive form)
V	Volume
v_0	Co-volume (2.3) to (2.5) (primitive form), or volume of a penetrable sphere (transcribed form)
v_{12}	Volume common to two spheres of radius a with centers at $\mathbf{R}_1$ and $\mathbf{R}_2$
W	Volume covered by penetrable spheres, or its generalization (3.17)
$w(R)$	Volume covered by two spheres at separation R
X, Y, Z	Cartesian coordinates; the gas–liquid surface is in the X,Y plane (Section VII)
x, y, z	Reduced Cartesian coordinates, X/a, etc.
z_0	Gibbs surface of zero excess density (Section VII)
α	Critical index that governs the singularity in c_v
γ	Critical index that governs the singularity in the compressibility; surface tension in units of $\epsilon a/v_0$ (Section VII)
Δ	$=\frac{1}{2}(\rho_l - \rho_g)$

ϵ	A characteristic energy (3.8)
θ	The reduced reciprocal temperature, ϵ/kT
λ	Activity
Ξ	Grand partition function
π	Pressure-to-temperature ratio, pv_0/kT
ρ	$= Nv_0/V$; the reduced density
$\rho(z)$	Reduced density at height z (Section VII)
ρ_g, ρ_l	Reduced orthobaric gas and liquid densities
$\bar{\rho}$	$= \frac{1}{2}(\rho_l + \rho_g)$
$\bar{\bar{\rho}}$	$= (\rho_l \rho_g)^{1/2}$
$\sigma(z)$	The effective density at height z that determines the energy density $\phi(z)$ (Section VII)
$\tau(z)$	The effective density at height z that determines the activity $\lambda(z)$
ϕ	$= \bar{U}v_0/V\epsilon$; the reduced configurational energy density
ϕ_g, ϕ_l	Energy densities of orthobaric gas and liquid
$\phi(z)$	Energy density at height z
$\phi(\rho)$	Energy density of a homogeneous fluid of density ρ
$\phi[\rho(z)]$	Energy density of a hypothetical homogeneous fluid of density $\rho(z)$
$\phi^{(2)}(\rho)$	The mean value of ϕ in a two-phase system of separate densities ρ_g and ρ_l whose overall density is ρ
$\phi_E^{(1)}(z)$	The excess or difference between $\phi(z)$ and either ϕ_g (if $z < z_0$) or ϕ_l (if $z > z_0$)
$\phi_E^{(2)}(z)$	The excess or difference between $\phi(z)$ and $\phi^{(2)}(\rho)$, where $\rho = \rho(z)$
ϕ_E	The integral of $\phi_E^{(1)}$ or $\phi_E^{(2)}(z)$ over all z
ψ	$= Av_0/V\epsilon$; the reduced configurational free energy density. This symbol takes the same arguments, suffixes, etc., as ϕ.
Ψ	van der Waals free energy density
Subscripts (1), (2), (3)	number of components
Superscripts (1), (2)	number of phases

Notes added in proof

Section VI.B. The symmetrical tricritical point here refers to the so-called "anomalous" point, which has threefold symmetry,[15] and not to the symmetrical tricritical point of Griffiths,[67] which has twofold symmetry. The seventh virial coefficients of the primitive gaussian version are now known (A. P. Laughton, unpublished results).

Section VII.A. For a general proof of the constancy of λ, defined by the potential distribution theorem throughout an inhomogeneous system at equilibrium, see Widom.[73]

Section VII.B. The demonstration of Jhon et al. [ref. 74, and *J. Chem. Phys.* **70**, 5228 (1979)] of the equivalence of the virial (v) and direct correlation function (c) routes to the surface tension is strictly only an inequality, $\gamma_v \geqslant \gamma_c$. P. Schofield [*Chem. Phys. Lett.*, **62**, 413 (1979)] has given, in outline, a proof of the equality for a system with pair potentials.

J. Bricmont, J. L. Lebowitz, C. E. Pfister, and E. Olivieri[*Commun. Math. Phys.*, **66**, 1, 21 (1979) and preprint] have proved the existence of a sharp interface in three and more dimensions for a lattice version of the primitive model at high activities λ_a and λ_b, when the phases are stabilized by pinning at the boundaries. The proof does not imply the existence of a sharp interface for the continuum model in the thermodynamic limit.

T. Karkheck and G. Stell (preprint) have obtained the transport properties of the primitive model in the Enskog approximation.

References

1. For the history of the Ising model, see S. G. Brush, *Rev. Mod. Phys.*, **39**, 883 (1967). For recent work, see H. N. V. Temperley (Vol. 1, Chap. 6), I. Syozi (Vol. 1, Chap. 7), and C. Domb (Vol. 3, Chap. 6) of C. Domb and M. S. Green (eds.), *Phase Transitions and Critical Phenomena*, Academic, London, 1972–76; and A. Münster, *Statistical Thermodynamics*, Berlin, 1969, 1974, Vol. 2, Chap. 14.
2. T. D. Lee and C. N. Yang, *Phys. Rev.*, **87**, 410 (1952).
3. J. O. Hirschfelder, C. F. Curtiss, and R. B. Bird, *Molecular Theory of Gases and Liquids*, Wiley, New York, 1954; J. S. Rowlinson, *Liquids and Liquid Mixtures*, 2nd ed., Butterworth, London, 1969; J. P. Hansen and I. R. McDonald, *Theory of Simple Liquids*, Academic, London, 1976.
4. H. N. V. Temperley, loc. cit. Ref. 1; A. Münster, loc. cit. Ref. 1.
5. H. A. Kramers and G. H. Wannier, *Phys. Rev.*, **60**, 252 (1941); L. Onsager, *Phys. Rev.*, **65**, 117 (1944).
6. M. Vicentini-Missoni (Vol. 2, Chap. 2) in C. Domb and M. S. Green, loc. cit. Ref. 1; J. M. H. Levelt Sengers, *Physica*, **73**, 73 (1974) and Vol. 2, Chap. 14 of B. Le Neindre and B. Vodar (eds.) *Experimental Thermodynamics*, Butterworth, London, 1975.
7. B. Widom and J. S. Rowlinson, *J. Chem. Phys.*, **52**, 1670 (1970).
8. T. W. Melnyk, J. S. Rowlinson, and B. L. Sawford, *Mol. Phys.*, **24**, 809 (1972).
9. B. Widom and F. H. Stillinger, *J. Chem. Phys.*, **58**, 616 (1973).
10. C. Domb (Vol. 3, Chap. 1) in C. Domb and M. S. Green, loc. cit. Ref. 1.
11. A. Münster (Vol. 1, Chap. 14), loc. cit. Ref. 1.
12. E. Helfand and F. H. Stillinger, *J. Chem. Phys.*, **49**, 1232 (1968).
13. E. Helfand and Z. Wasserman, *J. Chem. Phys.*, **57**, 1480 (1972).
14. R. B. Potts, *Proc. Camb. Phil. Soc.*, **48**, 106 (1952).
15. J. P. Straley and M. E. Fisher, *J. Phys.*, **A6**, 1310 (1973).

16. R. J. Baxter, *J. Phys.*, **C6**, L445 (1973); G. R. Goldner, *Phys. Rev.*, **B8**, 3419 (1973); I. G. Enting, *J. Phys.*, **A7**, 1617 (1974); J. P. Straley, *J. Phys.*, **A7**, 2173 (1974); D. J. Amit and A. Shcherbakov, *J. Phys.*, **C7**, L96 (1974); S. Alexander and D. J. Amit, *J. Phys.*, **A8**, 1988 (1975); Y. Yamazaki and M. Suzuki, *Prog. Theor. Phys.*, **58**, 516, 1367 (1977).
17. B. Widom, *J. Chem. Phys.*, **46**, 3324 (1967); G. A. Neece, *J. Chem. Phys.*, **47**, 4112 (1967); R. K. Clark, *J. Chem. Phys.*, **48**, 741 (1968); R. K. Clark and G. A. Neece, *J. Chem. Phys.*, **48**, 2575 (1968); J. T. Bartis and C. K. Hall, *Physica*, **78**, 1 (1975).
18. J. C. Wheeler and B. Widom, *J. Amer. Chem. Soc.*, **90**, 3064 (1968); J. C. Wheeler, *J. Chem. Phys.*, **62**, 433 (1975); J. C. Wheeler, *Ann. Rev. Phys. Chem.*, **28**, 411 (1977).
19. N. D. Mermin, *Phys., Rev. Lett.*, **26**, 957 (1971); J. A. Zollweg and G. W. Mulholland, *J. Chem. Phys.*, **57**, 1021 (1972); J. J. Rehr and N. D. Mermin, *Phys. Rev.*, **A7**, 379 (1973); G. W. Mulholland and J. J. Rehr, *J. Chem. Phys.*, **60**, 1297 (1974).
20. R. B. Griffiths and J. C. Wheeler, *Phys. Rev.*, **A2**, 1047 (1970).
21. N. Desrosiers, M. I. Guerrero, J. S. Rowlinson, and D. Stubley, *J. Chem. Soc. Faraday Trans. II*, **73**, 1632 (1977).
22. M. I. Guerrero, J. S. Rowlinson, and B. L. Sawford, *Mol. Phys.*, **28**, 1603 (1974).
23. B. Widom, *J. Chem. Phys.*, **54**, 3950 (1971).
24. J. M. Hammersley, J. W. E. Lewis, and J. S. Rowlinson, *Sankhya* (*Ind. J. Stat. A*), **37**, 457 (1975).
25. B. Widom, *Mol. Phys.*, **25**, 657 (1973).
26. R. B. Peierls, *Proc. Camb. Phil. Soc.*, **32**, 477 (1936).
27. R. B. Griffiths, *Phys. Rev.*, **A136**, 437 (1964).
28. J. L. Lebowitz and G. Gallavotti, *J. Math. Phys.*, **12**, 1129 (1971); J. L. Lebowitz and E. G. Lieb, *Phys. Lett.*, **A39**, 98 (1972); L. K. Runnels and B. C. Freasier, *Phys. Rev.*, **A8**, 2126 (1973); See also L. K. Runnels, *J. Math. Phys.*, **15**, 984 (1974) for the lattice-gas transcription from two components to one.
29. D. Ruelle, *Phys. Rev. Lett.*, **27**, 1040 (1971).
30. T. J. Lie, *J. Chem. Phys.*, **56**, 332 (1972).
31. L. K. Runnels and J. L. Lebowitz, *J. Math. Phys.*, **15**, 1712 (1974).
32. L. I. Schiff, *Phys. Rev.*, **50**, 88 (1936); H. Lifschutz and O. S. Duffendak, *Phys. Rev.*, **54**, 714 (1938); D. F. Votaw, *Ann. Math. Stat.*, **17**, 240 (1946).
33. H. E. Robbins, *Ann. Math. Stat.*, **15**, 70 (1944); **16**, 342 (1945); L. Santalo, *Ann. Math. Stat.*, **18**, 37 (1947); F. Garwood, *Biometrika*, **34**, 1 (1947).
34. T. W. Melnyk and J. S. Rowlinson, *J. Comput. Phys.*, **7**, 385 (1971).
35. P. A. P. Moran, *J. Appl. Prob.*, **10**, 483, 837 (1973).
36. N. D. Mermin, *J. Chem. Phys.*, **54**, 3958 (1971).
37. B. Widom, *J. Chem. Phys.*, **39**, 2808 (1963); J. L. Jackson and L. S. Klein, *Phys. Fluids*, **7**, 228 (1964); N. S. Snider, *J. Chem. Phys.*, **55**, 1481 (1971).
38. R. E. Barrieau, *J. Chem. Phys.*, **49**, 2279 (1968).
39. P. C. Hemmer and G. Stell, *Phys. Rev. Lett.*, **24**, 1284 (1970).
40. J. H. Hildebrand and R. L. Scott, *Regular Solutions*, Chap. 10, Prentice Hall, Englewood Cliffs, N.J., 1962; J. B. Gilmour, J. O. Zwicker, J. Katz, and R. L. Scott, *J. Phys. Chem.*, **71**, 3259 (1967).
41. J. Weiner, K. H. Langley, and N. C. Ford, *Phys. Rev. Lett.*, **32**, 879 (1974); D. Y. Ivanov, L. A. Makerevitch, and O. N. Sokolova, *JETP Lett.*, **20**, 121 (1974); M. Ley-Koo and M. S. Green, *Phys. Rev.*, **A16**, 2483 (1977).
42. E. S. R. Gopal, R. Ramachandra, P. Chandra Sekhar, K. Govandrarachan, and S. V. Subramanyam, *Phys. Rev. Lett.*, **32**, 284 (1974); E. S. R. Gopal, P. Chandra Sekhar, G. Ananthakrishna, R. Ramachandra, and S. V. Subramanyam, *Proc. Roy. Soc.*, **A350**, 91 (1976).

43. R. L. Scott, *Chemical Thermodynamics*, Vol. 2 (Specialist Periodical Report, Chemical Society, London) 1978; See also M. J. Buckingham in C. Domb and M. S. Green, Vol. 2, p. 18, 1972, loc. cit. Ref. 1.
44. R. H. Fowler and E. A. Guggenheim, *Statistical Thermodynamics*, Chaps. 13 and 14, Cambridge University Press, 1939.
45. M. I. Guerrero, J. S. Rowlinson, and G. Morrison, *J. Chem. Soc. Faraday Trans. II*, **72**, 1970 (1976).
46. J. P. Hansen and I. R. McDonald, p. 130, loc. cit. Ref. 3; J. A. Barker and D. Henderson, *Rev. Mod. Phys.*, **48**, 587 (1976) (see eqs. 1.20, 1.22, and 1.24).
47. J. S. Rowlinson, Chap. 3 in H. N. V. Temperley, J. S. Rowlinson, and G. S. Rushbrooke (eds.), *Physics of Simple Liquids*, North-Holland, Amsterdam, 1968.
48. J. P. Straley, M. A. Cotter, T. J. Lie, and B. Widom, *J. Chem. Phys.*, **57**, 4484 (1972).
49. M. Abramowitz and I. A. Stegun, *Handbook of Mathematical Functions*, p. 824, National Bureau of Standards, Washington, 1964.
50. R. Zwanzig, *J. Chem. Phys.*, **39**, 1917 (1963); L. K. Runnels and C. Colvin, *J. Chem. Phys.*, **53**, 4219 (1970).
51. G. E. Uhlenbeck and G. W. Ford, in J. de Boer and G. E. Uhlenbeck (eds.) *Studies in Statistical Mechanics*, Vol. 1, Part 3, North-Holland, Amsterdam, 1962.
52. C. Domb and M. S. Green, Vol. 6, 1976, loc. cit. Ref. 1.
53. A. Ahn and J. L. Lebowitz, *Phys. Lett.*, **44A**, 424 (1973); *J. Chem. Phys.*, **60**, 523 (1974).
54. J. L. Lebowitz and D. Zomick, *J. Chem. Phys.*, **54**, 3355 (1971).
55. E. Bergmann, *Mol. Phys.*, **32**, 237 (1976); R. Tenne and E. Bergmann, *Phys. Rev.*, **A17**, 2036 (1978).
56. L. Verlet, *Physica*, **30**, 95 (1964); **31**, 959 (1965).
57. M. S. Wertheim, *J. Math. Phys.*, **8**, 927 (1967).
58. D. D. Carley, *J. Chem. Phys.*, **55**, 3426 (1971).
59. M. I. Guerrero, G. Saville, and J. S. Rowlinson, *Mol. Phys.*, **29**, 1941 (1975).
60. R. B. Griffiths and B. Widom, *Phys. Rev.*, **A8**, 2173 (1973); R. B. Griffiths, *J. Chem. Phys.*, **60**, 195 (1974).
61. J. L. Meijering, *Phillips Res. Rep.*, **5**, 333 (1950); **6**, 183 (1951); See also D. Furman, S. Dattagupta, and R. B. Griffiths, *Phys. Rev.*, **B15**, 441 (1977).
62. J. Zernike, *Rec. Trav. Chim.*, **68**, 585 (1949).
63. R. L. Scott, *J. Chem. Soc. Faraday Trans. II*, **73**, 356 (1977).
64. J. C. Wheeler, *Ber. Bunsen-Ges. Phys. Chem.*, **76**, 308 (1972).
65. J. D. van der Waals and P. Kohnstamm, *Lehrbuch der Thermodynamik*, Part 2, p. 39, Barth, Leipzig, 1912; P. Kohnstamm in H. Geiger and K. Scheele (eds.), *Handbuch der Physik*, Vol. 10, p. 223, Springer, Berlin, 1926.
66. J. C. Lang and B. Widom, *Physica*, **81A**, 190 (1975).
67. R. B. Griffiths, *Phys. Rev.*, **B7**, 545 (1973).
68. M. E. Fisher in U. Landman (ed.), *Statistical Mechanics and Statistical Methods in Theory and Application*, p. 3, Plenum, New York, 1977.
69. J. S. R. Chisholm, *Proc. Roy. Soc.*, **A358**, 351 (1977); **A362**, 43 (1978).
70. J. D. Weeks, G. H. Gilmer, and H. J. Leamy, *Phys. Rev. Lett.*, **31**, 549 (1973); R. H. Swendsen, *Phys. Rev. Lett.*, **38**, 615 (1977).
71. G. W. Woodbury, *J. Chem. Phys.*, **60**, 3674 (1974).
72. D. B. Abraham, G. Gallavotti, and A. Martin-Löf, *Physica*, **65**, 73 (1973); D. B. Abraham and P. Reed, *Phys. Rev. Lett.*, **33**, 377 (1974); *Commun. Math. Phys.*, **49**, 35 (1976); *J. Phys.*, **A10**, L121 (1977); D. B. Abraham and D. J. Heilmann, *Physica*, **85A**, 589 (1976).

73. B. Widom, *J. Stat. Phys.*, **19**, 563 (1978).
74. C. A. Leng, J. S. Rowlinson, and S. M. Thompson, *Proc. Roy. Soc.*, **A352**, 1 (1976); **A358**, 267 (1977).
75. M. Rao and D. Levesque, *J. Chem. Phys.*, **65**, 3233 (1976).
76. S. Ono and S. Kondo, in S. Flügge (ed.) *Handbuch der Physik*, Vol. 10, p. 134, Springer, Berlin, 1960; See also A. Harasima, *Adv. Chem. Phys.*, **1**, 203 (1958).
77. J. M. Harrington and J. S. Rowlinson, *Proc. Roy. Soc.*, **A367**, 15 (1979).
78. F. P. Buff and R. A. Lovett, in H. L. Frisch and Z. W. Salsburg (eds.) *Simple Dense Fluids*, Academic, New York, 1968; B. Widom, Vol. 2, Chap. 3, 1972 in C. Domb and M. S. Green, loc. cit. Ref. 1.
79. J. M. Hammersley, private communication.
80. J. W. Gibbs, *Trans. Connecticut Acad.*, **3**, 108, 343, 1875, reprinted in *Collected Works*, Vol. 1, p. 219, Longman, New York, 1928.
81. S. Toxvaerd, *Statistical Mechanics*, Vol. 2 (Specialist Periodical Report, Chemical Society, London) 1975; N. H. March and M. P. Tosi, *Atomic Dynamics in Liquids*, p. 260, Macmillan, London, 1976.
82. F. P. Buff, *Zeit. Elektrochem.*, **56**, 311 (1952); A. G. MacLellan, *Proc. Roy. Soc.*, **A213**, 274 (1952).
83. H. S. Green, *Proc. Roy. Soc.*, **A189**, 103 (1947).
84. R. A. Lovett, C. Y. Mou, and F. P. Buff, *J. Chem. Phys.*, **65**, 570 (1976); M. S. Wertheim, *J. Chem. Phys.*, **65**, 2377 (1976).
85. D. G. Triezenberg and R. Zwanzig, *Phys. Rev. Lett.*, **28**, 1183 (1972).
86. R. A. Lovett, P. W. DeHaven, J. J. Vieceli, and F. P. Buff, *J. Chem. Phys.*, **58**, 1880 (1973).
87. M. S. Jhon, R. C. Desai, and J. S. Dahler, *Chem. Phys. Lett.*, **56**, 151 (1978).
88. J. D. van der Waals, *Arch. Néerl.*, **28**, 121 (1894); *Zeit. Phys. Chem.*, **13**, 657 (1894); translation in *J. Stat. Phys.*, **20**, 197 (1979).
89. J. W. Cahn and J. E. Hilliard, *J. Chem. Phys.*, **28**, 258 (1958).
90. J. L. Lebowitz and J. K. Percus, *J. Math. Phys.*, **4**, 116 (1963).
91. A. J. M. Yang, P. D. Fleming, and J. H. Gibbs, *J. Chem. Phys.*, **64**, 3732, (1976); **65**, 7, (1976); **67**, 74 (1977).
92. S. Fisk and B. Widom, *J. Chem. Phys.*, **50**, 3219 (1969); B. Widom (Vol. 2, Chap. 3), in C. Domb and M. S. Green, loc. cit. Ref. 1.
93. B. Widom (p. 33), in U. Landmann, loc. cit. Ref. 68.
94. V. Bongiorno and H. T. Davis, *Phys. Rev.*, **A12**, 2213 (1975); V. Bongiorno, L. E. Scriven, and H. T. Davis, *J. Coll. Interface Sci.*, **57**, 462 (1976); H. T. Davis, *J. Chem. Phys.*, **67**, 3636 (1977).
95. F. F. Abraham, *J. Chem. Phys.*, **63**, 157, 1316 (1975).
96. J. W. Cahn, *J. Chem. Phys.*, **66**, 3667 (1977).
97. J. D. Weeks, *J. Chem. Phys.*, **67**, 3106 (1977).
98. L. Lekner and J. R. Henderson, *Mol. Phys.*, **34**, 333 (1977).
99. M. Cassandro and E. Olivieri, *J. Stat. Phys.*, **17**, 229 (1977).
00. C. A. Leng, J. S. Rowlinson and P. Turq, unpublished results.

THEORY OF INHOMOGENEOUS ELECTRON SYSTEMS: SPIN-DENSITY-FUNCTIONAL FORMALISM

A. K. RAJAGOPAL

Department of Physics and Astronomy
Louisiana State University
Baton Rouge, Louisiana 70803

CONTENTS

The author thanks IBM Thomas J. Watson Research Center, Yorktown Heights, New York for its warm hospitality during his sabbatical year when much of the ideas for this review took shape. In particular, he thanks Dr. J. F. Janak for many conversations on the subject matter discussed here. The author thanks Dr. Ulf von Barth for extensive discussions of many of the questions raised in this work. The author wishes to place on record fruitful discussions he had with Professor Walter Kohn, Dr. Art Williams, and Dr. Norton Lang on various aspects of density functional theory—both nonrelativistic and relativistic formulations. M. V. Ramana read the manuscript and made useful remarks. Martha Prather typed the entire chapter from the handwritten pages and many thanks are due to her for the excellent job.

I. INTRODUCTION

A. Statement of the Problem: The Hartree and the Thomas–Fermi Methods

Soon after the advent of quantum mechanics, one of the most important and interesting areas of research has been, even to this day, to provide a theory of atoms, molecules, and solids. It was at once realized that this many-body problem cannot be solved exactly and hence reliable approximation schemes had to be devised. The search for these methods continues even now and will continue for many years to come, as sophistication increases in computing facilities and techniques. One of the important requirements of these schemes is to fulfill as many exact statements as can be found for the actual problem while retaining a certain amount of simplicity in the scheme which will enable one to understand intuitively the system properties, such as trends across the periodic table. To be specific, let us define the problem and enumerate the known exact statements concerning its solution before giving a brief outline of the methods that are currently being employed.

The quantum mechanical problem that concerns us here is a system of N electrons of mass m and charge $-e$, moving about *fixed positive charges* of magnitudes $z_1e, z_2e, \ldots, z_ke$ at positions $\mathbf{R}_1, \ldots, \mathbf{R}_k$ in a three-dimensional space. Most often, we consider a neutral system so that the sum of the positive charges equals the total number of electrons, even though this can be relaxed when the situation demands. Also, for the most part, we will consider *fixed* positive charges; we have occasion to consider when they vibrate about their equilibrium positions. We are mainly concerned in this review with the electronic properties of this system. It is clear that the system described is general enough to cover atoms, molecules, and solids. The quantum mechanical Hamiltonian operator acting on the space of this system of particles is

$$H = -\frac{\hbar^2}{2m}\sum_{i=1}^{N}\nabla_i^2 - \sum_{i=1}^{N} V(\mathbf{r}_i) + \tfrac{1}{2}\sum_{i\neq j=1}^{N} W(\mathbf{r}_i - \mathbf{r}_j) + \tfrac{1}{2}\sum_{i\neq j=1}^{k} z_i z_j e^2 |\mathbf{R}_i - \mathbf{R}_j|^{-1} \tag{1.1}$$

Here $\hbar$ is Planck's constant divided by 2π. The first term is the kinetic energy operator for the electrons; the second represents the attractive electron–nuclear interaction

$$V(r) = e^2 \sum_{j=1}^{k} z_j |\mathbf{r} - \mathbf{R}_j|^{-1} \tag{1.2a}$$

The third term represents the mutual electron repulsion whereas the fourth term is the mutual nuclear repulsion:

$$W(\mathbf{r}) = e^2 |\mathbf{r}|^{-1} \tag{1.2b}$$

The first term is the kinetic energy $\mathcal{T}$ of the system and the last three terms are together the potential energy $\mathcal{V}$ of the system. Note that we do not include any explicit spin-dependent terms in our Hamiltonian; however, the electron is assumed to possess an intrinsic spin quantum number. Let $x = (\mathbf{r}, \sigma)$ stand for spatial coordinate $\mathbf{r}$ and spin quantum number σ. The Pauli exclusion principle plays a major role in our considerations, and it is expressed by requiring that the wave function, Ψ, for the many-particle system be antisymmetric under exchange of a pair of electron coordinates and spins.

We now have an eigenvalue problem, that of determining the eigenvalues and the associated eigenfunctions of the Hamiltonian, (1.1). Notice that this is a linear problem in a large number of variables of the system and as such very difficult to solve. Instead of seeking a full solution to the eigenvalue problem, one might ask if something can be learned about the ground state of the system, which is the lowest energy eigenstate of the Hamiltonian operator. Here we have several exact statements that can be made.

a. The ground-state wave function has a variational character to it in that it gives the minimum of the expectation value of the Hamiltonian. In other words, we have here a scheme where a trial ground-state wave function that is not the true eigenstate of H always gives a larger expectation value than the true ground state. This Rayleigh–Ritz principle is an extremely useful theorem, in developing approximation schemes.

b. If $|\Psi\rangle$ is the exact ground state of the Hamiltonian and is normalized to unity, then one has the *Hellmann–Feynman theorem*, which states that the force component $\partial\langle\Psi|H|\Psi\rangle/\partial\mathbf{R}_i$, acting on a nuclear coordinate $\mathbf{R}_i$ where $\langle\Psi|H|\Psi\rangle$ is the total energy of the system calculated quantum mechanically, is the same as the force that would be calculated by pure classical electrostatics acting on the nuclear coordinate, resulting from the electronic charge cloud and all other nuclei.

c. If we scale all the coordinates by the same parameter and used the minimum principle (*a*) and the Hellmann–Feynman theorem (*b*), we obtain the *virial theorem* which relates the average kinetic energy in the ground state and the average potential energy defined earlier:

$$\langle\Psi|\mathcal{T}|\Psi\rangle = -\tfrac{1}{2}\langle\Psi|\mathcal{V}|\Psi\rangle - \tfrac{1}{2}\sum_{i=1}^{k}\mathbf{R}_i\cdot\partial\langle\Psi|H|\Psi\rangle/\partial\mathbf{R}_i \tag{1.3}$$

These two theorems serve as very useful checks on the approximate theories as well as providing a simple picture of the system which involves essentially only the classical concepts!

d. Once one has obtained some idea of the nature of the ground state, one wants to know, at the least, some features of the low-lying excited states close to the ground state. To obtain such information, one often can use weak external time-dependent forces to perturb the ground state and examine its response. This can be done by means of an action principle, which is a stationary principle, in contrast to the minimum principle quoted in (*a*). If $|\Phi(t)\rangle$ is the state associated with the Hamiltonian $H + H_I(t)$, where $H_I(t)$ is the perturbing time-dependent potential, the action principle states that

$$A = \int_{t_1}^{t_2} dt \left\langle \Phi(t) \middle| \left\{ i\hbar\frac{\partial}{\partial t} - H - H_I(t) \right\} \middle| \Phi(t) \right\rangle \tag{1.4}$$

is stationary with respect to variations of $|\Phi\rangle$ such that $\delta|\Phi(t)\rangle$ at the end points $t = t_1$ and $t = t_2$ are zero. When $|\Phi(t)\rangle$ is assumed to be the ground state $|\Psi\rangle$ at the initial time $t = t_1$, one may then develop a scheme to examine the nature of the low-lying excited states. Alternatively, one can use a finite temperature extension of the theory involving imaginary time, which is sometimes useful.

These are the general principles that we can employ in developing approximation schemes. Around 1927–28, two important developments took place, both of which offered intuitive and heuristic models for attacking this many-body problem. One, due to Hartree,[1] is a mean field theory in which each electron is assumed to move in an average classical electrostatic field of all other electrons and the nuclei. In this way, the problem was reduced to solving a one-electron Schrödinger equation moving in a self-consistent potential, which depends on the electron density of the system. In this way, the original linear problem in a large number of variables was reduced to nonlinear equations in essentially smaller number of variables. The other, due to Thomas[2] and Fermi[3] who almost simultaneously published their theory, took an entirely different point of view by relating the local electron density, $n(\mathbf{r})$, to the total average potential, $\mathcal{V}(\mathbf{r})$, experienced by the electron. This relationship took into account the Pauli exclusion principle in balancing the kinetic energy of the electrons and the total average potential energy. One then employed the classical electrostatics to relate the potential to the density via the Poisson equation. In this scheme then, one has density as the basic variable in the problem. Thus, Thomas–Fermi (TF) theory is very attractive since it is much easier to handle and is physically appealing. Dirac,[4] from a semiclassical approximation to the Hartree scheme, derived the TF scheme. Almost at about the same time, Slater[5] and Fock[6] obtained the Hartree equations via the variational principle for the ground-state energy previously mentioned in item (*a*), with the trial variational ground-state wave function chosen to be just a product of the one-electron wave functions. This variational derivation of the Hartree equations was an important step because it showed the precise approximation made in the derivation. Since the Hartree equation did not take into account the exclusion principle, Slater and Fock suggested that instead of a product wave function, one ought to use a properly antisymmetrized (Slater determinant) form for the ground-state trial wave function in the variational scheme. This led them to a more complicated nonlocal integro-differential equation for the one-electron wave functions which are now known as the Hartree–Fock–Slater scheme (HFS). Dirac[4] recognized this aspect of lack of antisymmetry and by applying the semiclassical approximation to the HFS equations, he obtained a modified version of the TF equation, the now celebrated Thomas–Fermi–Dirac (TFD) equation. In simplifying the exchange contribution, Dirac used semiclassical plane wave scheme. Lenz[7] gave a variational formulation of the TF equation which showed clearly that the energy of the system was a functional of density in the TF scheme. Moreover, it becomes clear from this that it is only the average kinetic energy of the system which is different in appearance, being proportional to $\int n^{5/3}(\mathbf{r})d\mathbf{r}$ in the TF theory in contrast to the Hartree theory. This brief historical account then serves as a quick preview of the theme of this review article on the density-functional theory. We have tried to point out here that the variational formulation plays a very important role in exhibiting approximations involved and hence also shows a way for further improvement in the schemes. The HFS and TFD theories have been extensively used in the literature for examining many varieties of questions concerning the electronic structure of matter. We may also mention that the self-consistent time-dependent version of the HFS and TFD employing the stationary action principle (*d*) described previously have been used. The reader is referred to recent reviews by March[8] and by Kirzhnitz et al.[9] for a description of the results obtained by the methods and other areas of generalization and extension of the preceding schemes. It is very interesting to point out that Teller[10] gave compelling arguments to show that molecular binding is not possible in either TF or TFD schemes. This result is a great blow to the TFD scheme. Some doubts expressed concerning Teller's proof have been dispelled by a rigorous proof of the absence of molecular binding in the TF and TFD schemes given by Lieb and Simon.[11] This paper places the TF model of atoms, molecules, and solids on a firm mathematical footing. The reader is referred to this article for a complete mathematical account of many results concerning TF and TFD models hitherto known only heuristically. In particular, these authors establish that the TF scheme approaches the actual many-body solution asymptotically in a certain sense. One purpose here is not to review these theories but to give a brief outline of this phase of the development only. In the next section, we outline, again briefly, another successful scheme due to Slater,[12] called the $X\alpha$ method.

B. The $X\alpha$ Method

A grafting of the HF and TFD schemes was made by Slater[12] in his $X\alpha$ method, where the exchange potential is calculated as in TFD (proportional to $\frac{1}{3}$ power of the density) but with a numerical prefactor, α, which appears as a parameter in the theory but employs the kinetic energy part of the HF theory. This leads to a differential equation for the one-electron states which now has an approximate exchange potential in addition to the Hartree potential. This model was suggested by Slater[12] as a simplification of the HFS integro-differential equation for the one-electron states. This model has generated a large number of investigations and also a few novel ideas have emerged from it. We describe some of it here. Slater[13] and Connolly[14] have recently reviewed these works.

The major deficiency of the TFD model was in its treatment of the average kinetic energy of an electron, which in the statistical treatment is proportional to the $\frac{5}{3}$ power of the charge density. Slater's idea in 1951 was to keep the kinetic energy as in HFS model, but employ the statistical approximation to the exchange part of the potential instead. This corrects the deficiency of TFD but at the same time achieves a drastic simplification of the HFS model. In the Dirac derivation,[4] the parameter α would be $\frac{2}{3}$, because the statistical approximation is made in calculating the average exchange energy, whereas employing the statistical approximation to the exchange potential Slater obtained $\alpha = 1$. But Slater suggested the use of the parameter, α, in front of the exchange potential as an approximation scheme, so that the approximate aspect is corrected for parametrically. Slater[15] has compared the TFD and the $X\alpha$ methods in detail, and the reader is referred to it for an illuminating account of these theories. It turns out that $\alpha = 1$ is consistent with the variational principle as was discovered by Gaspar.[16]

Slater and co-workers[13,14] have developed the $X\alpha$ method to a great extent. In the Hartree–Fock–Slater scheme, the Koopmans theorem[17] showed that the eigenvalues appearing in the HFS equations really had an important physical significance: It is the difference of total HFS energy computed for the case when the particular electron state is occupied and when it is unoccupied. In the $X\alpha$ scheme this interpretation is lost but a different interpretation is necessary: It is the partial derivative of the total $X\alpha$ energy with respect to the occupation number of the state under consideration. In a strictly one-particle theory, the Fermi statistics would have required that the occupation numbers be either zero or one. Slater argued that the fractional occupation of a state can arise when one does not use a single determinant ground-state wave function, but rather employs a whole set, because one is really dealing with a real many-electron problem and the one-electron approximation is only a scheme. This then leads to multiplet problems with many determinantal functions since a single determinant scheme becomes inadequate. Slater invented ingeneous tricks to extract physically meaningful quantities from the $X\alpha$ theory by introducing the idea of "transition state" which is a state in which the occupation numbers are halfway between those of the initial and final states. In this way, an estimation of excitation energies in the $X\alpha$ scheme can be given. Slater and co-workers also devised ways of optimizing the value of α, which was found to lie between $\frac{2}{3}$ and 1, generally of the order of 0.7 for best results. It is important to realize, as Slater emphasized often, that $X\alpha$ method may give a good starting point for further computation. It should also be remarked that although the $X\alpha$ method is not variational in character, it obeys both the Hellmann–Feynman and the virial theorems, as does the HFS scheme, independent of the value of α. When many determinantal functions are used instead of a single one, which Slater called the Hyper Hartree–Fock method (HHFS), to deal with partially occupied states, he was able to make a comparison between the $X\alpha$ method and the HHFS and expressed the opinion that the $X\alpha$ method is superior being capable of handling larger systems than atoms, such as molecules and crystals. The method was also extended to study the magnetic elements such as Ni, Fe, and Co.

The primary objection to the $X\alpha$ method is that the total energy functional used does not correspond to a ground-state wave function of the actual many-electron problem, and because of this, this method is best used as a generator of a convenient basis set to be employed in an actual many-body approach. We emphasize this point along with the arbitrariness in the choice of α only to warn the reader that the method should be considered as a model scheme. In the work of Lieb and Simon,[11] it was shown mathematically that the TF model has some bearing to the actual quantum mechanical problem in some sense, whereas this is not even expected of the $X\alpha$ scheme. We should draw the attention of the reader to the article of Slater in Ref. 15 where he makes a careful comparison of the TFD and $X\alpha$ schemes. We should add that his comments on molecular binding in TFD should be modified in view of the Lieb–Simon work.[11] It should also be pointed out that Slater pioneered in the application and analysis of the $X\alpha$ method, providing much insight and guidance in the development of the theory of structure of matter. His efforts have become a cornerstone in the further development of the density-functional theory which we describe in detail in this review, a brief preview of which is given in the next subsection.

C. The Density-functional Theory

The brief description given in the previous two subsections serve to illustrate the need for a more formal and general framework for dealing with the actual many-body problem. In 1964, Hohenberg and Kohn[18] proved a remarkable theorem (HK) that the ground-state energy of a many-body system is a unique functional of density, $n(\mathbf{r})$, and is a minimum when evaluated for the true ground-state density. The proof was independent of whether the particles obeyed Bose or Fermi statistics so that the theorem has now become the touchstone of a vast number of investigations of a very large variety of many-body problems. The object of the present review is to give an account of this remarkable development.

In two important papers, Kohn and Sham,[19,20] developed various aspects of the HK theorem, providing a basis for further practical applications of the density-functional theory. In particular, they showed how a general one-particle equation can be set up, which would include in principle *all* the effects of correlations among the particles of the system. They proved in particular, that the theory gives correctly all the ground-state properties as well as suggested schemes for actual calculation of properties. The three papers[18–20] together form the basis of much of the development of electronic structure of matter today. In particular, very impressive and remarkable results have been obtained for atoms and elemental metals, including their magnetic properties, by the IBM group, which have now been collected together in the form of a book.[21] Kohn with Lang developed the theory of the electronic structure of metallic surfaces based on the density-functional theory; a masterly review of this phase of the development is found in an article by Lang.[22] Much of this work depends on various versions of the local-density scheme.[19,20] On completely general grounds as in HKS development, Janak has extended[23] Slater's proof of the Hellmann–Feynman and the virial theorems in the $X\alpha$ method as well as well as the Slater's demonstration that the eigenvalues appearing in the one-particle equations are the derivative of the total energy in the density-functional theory with respect to the occupation number of the corresponding state.[24] We thus note that the first three requirements of the quantum mechanics of the many-electron system are actually met. The finite temperature extension of the theory was given by Mermin,[25] which in some ways covers the last requirement stated in Section I.A. The stationary action principle has been suggested recently by Peuckert[26] and its imaginary-time counterpart by Rajagopal,[27] for studying the low-lying excited states of the system. But these developments have not become numerically feasible as yet and will, hopefully do so, when one can obtain some idea as to the nature of the low-lying excited states of the many-electron system.

The general philosophy of all the above methods is to approach the ground-state of the many-body problem via a one-particle picture. The HKS theory shows that the one-body effective potential can in principle contain all the effects of electron correlations, even though in practice, some schemes have to be proposed, for actual calculations. The theory is such that at least in principle, one may be able to employ more and more sophisticated extensions of the effective one-body potential so that one may begin with the $X\alpha$ method with $\alpha = 1$ (which is the variational answer) and proceed with increasingly complicated forms for it. Some of the "successful" suggestions employed in the literature will be indicated in later chapters. We must point out that numerical solution of the Schrödinger equation with a one-body potential has progressed to such a large extent that the incorporation of self-consistent one-body potentials that enter into the density functional scheme is relatively easily accomplished. In this manner, both the inhomogeneity in the system and the electron interactions are taken into account in the present scheme. In the earlier developments, one almost always took account of the inhomogeneities more accurately and treated the electron interactions as an afterthought. It should be mentioned that Roy Gordon[28] and his associates tried to incorporate the electron interactions on the same basis as the one-body potentials, but used an electron-gas approach as far the the electron interactions are concerned. We discuss this work later. In the next section we discuss in detail the Hohenberg–Kohn theorem and various generalizations of it along with a discussion of the purpose for each generalization. This section is a purely formal one, containing most of the exact theorems. In Section III, we develop practical schemes, suitable for computation. In Section IV, we develop a scheme for a time-dependent version of the theory. In Section V, we discuss applications of the method to nuclear theory, atoms, and molecules, intermolecular potentials, bulk properties of solids, surface properties of solids, chemisorption, motion of proton and muon in solids, and liquid-state properties. The last section contains some concluding remarks.

II. BASIC THEOREMS

In this section, we collect all the various theorems that have been proved in the spirit of the original work of Hohenberg and Kohn.[18] These theorems enable us to understand the extent of generality of the framework of the functional theory. In the next section, we present practical schemes that are currently being used in calculations of physical properties of systems. From the brief historical account given in Section I, it should be clear that, at least as far as the ground state of a many-body system is concerned, a density-functional theory in the spirit of Thomas–Fermi theory may be generally valid, not only because of the successes of such a scheme but also because of the Lieb–Simon demonstration[11] that the TF theory is approached asymptotically by the solution of the full quantum mechanical problem in a certain sense. The theorem of Hohenberg and Kohn, established quite generally that for a static, scalar, external, local potential, the ground-state energy of the many-body system with a nondegenerate ground state is a unique functional of density and that the correct ground-state density minimizes the energy. This is a very important development, because a large variety of questions can be posed in terms of a static scalar external potential, viz., atoms, molecules, solids, semiinfinite solid leading to studies of surface properties, and so on. It is in this very important aspect that the HK theorem goes well beyond the TFD theory, besides incorporating correctly the kinetic energy part of the ground-state energy. Although the theorem assures us of the existence of such a functional, it does not provide us with the actual form for the functional itself. Approximations and schemes are proposed and it is in this area, much ingenuity and novelty will be required in making the formalism successful. The formalism has therefore enabled us to treat the inhomogeneity and the correlations among the particles of the system on an equal footing, an extremely important step in the theoretical development of inhomogeneous systems. Various generalizations come to mind; that of static spin-dependent external potential, static velocity-dependent external potential, static nonlocal external potential fall into one class, which we take up in that order. Another class of generalization involves a generalization of the preceding schemes to finite temperatures,[25] which follows almost immediately, because, instead of the Rayleigh–Ritz minimum principle, one now employs the Gibbs-minimum principle of the equilibrium statistical mechanics. Much of the analysis concerns the ground state of the system. To examine the low-lying excited states, one may employ the general time-dependent extremum principle—the action principle or for that matter the finite-temperature theory. This is dealt with in Section 4. We ouline the theorems concerning static external potentials in this section. Much of the analysis, thus far, has been confined to nonrelativistic many-body theory. In practice, it has been found that relativistic corrections become important for elements with nuclear charge bigger than 36 (krypton). A relativistic generalization is therefore required and we present this also in this section. We may mention that the finite temperature theory, the time-dependent theory, and the relativistic generalization have not yet been fully exploited so far. Unless otherwise stated, all the external potentials refer to static situations.

We first introduce the second-quantized representation of the Hamiltonian (1.1) and establish some basic notations. Let $\psi(xt)$ and $\psi^{\dagger}(xt)$ be the destruction and creation operators for the particles; they obey the usual equal time commutation rules, which incorporate the fundamental statistics the particles obey. Thus

$$[\psi(xt), \psi(x't)]_{\pm} = 0 = [\psi^{\dagger}(xt), \psi^{\dagger}(x't)]_{\pm}$$

and

$$[\psi(xt), \psi^{\dagger}(x't)]_{\pm} = \delta_{\sigma\sigma'}\delta^{(3)}(\mathbf{r} - \mathbf{r}') \tag{2.1}$$

where $[A, B]_{\pm} = AB \pm BA$. The upper signs are for the Fermi–Dirac system (antisymmetric statistics) and the lower signs are for the Bose–Einstein system (symmetric statistics). The Hamiltonian operator H of (1.1) is then rewritten as

$$H = \mathcal{T} + V + W \tag{2.2}$$

where

$$\mathcal{T} \equiv \frac{\hbar^2}{2m} \sum_{\sigma} \int \nabla \psi^{\dagger}(xt) \cdot \nabla \psi(xt) d\mathbf{r} \tag{2.3a}$$

$$V \equiv \sum_{\sigma} \int v(r) \psi^{\dagger}(xt) \psi(xt) d\mathbf{r} \tag{2.3b}$$

$$W \equiv \tfrac{1}{2} \sum_{\sigma\sigma'} \frac{e^2}{|\mathbf{r} - \mathbf{r}'|} \psi^{\dagger}(xt) \psi^{\dagger}(x't) \psi(x't) \psi(xt)\, d\mathbf{r}\, d\mathbf{r}' + \tfrac{1}{2} \sum_{i \neq j = 1}^{k} z_i z_j e^2 |\mathbf{R}_i - \mathbf{R}_j|^{-1} \tag{2.3c}$$

$\mathcal{T}$, as before, stands for the kinetic energy operator; V *represents the external potential*, which we often think of arising from the nuclei as in (1.2). Often, we may leave it in more general form. W represents the repulsion terms, both among the electrons and the positive charges. We assume as before that the positive charges are stationary and are fixed in some preassigned equilibrium configuration. We assume this in most of our description. Very often, the repulsion energy among the positive charges is dropped from explicit consideration, because in this framework it acts as a constant. When questions concerning force on a given positive charge and the like are raised, its presence is needed explicitly, as in (1.4). The particle-density operator $\rho(\mathbf{r}t)$ is given by

$$\rho(\mathbf{r}t) = \sum_{\sigma} \psi^{\dagger}(xt) \psi(xt) \tag{2.4}$$

If $|\Psi\rangle$ is the ground state of the system, the particle density, $n(\mathbf{r})$, in this state is given by the expectation value,

$$\langle \Psi | \rho(\mathbf{r}t) | \Psi \rangle \equiv n(\mathbf{r}) \tag{2.5}$$

When there are no time-dependent forces in the system, all the ground-state averages are independent of time. The ground-state energy E is given by

$$\langle \Psi | H | \Psi \rangle \equiv E \tag{2.6}$$

The total number of particles is assumed to be fixed, N, given by

$$\int n(\mathbf{r}) d\mathbf{r} = \mathrm{N} \tag{2.7}$$

A. Scalar External Potential

Theorem 1 (Hohenberg–Kohn[18]). For a nondegenerate ground state $|\Psi\rangle$ of the system, $v(\mathbf{r})$ is (to within a constant) a unique functional of $n(\mathbf{r})$. For a given external potential $v(\mathbf{r})$, the correct $n(\mathbf{r})$ minimizes the ground-state energy E which is a unique functional of $n(\mathbf{r})$.

Proof. The proof is by *reductio ad absurdum*. Let there be another potential $v'(r)$, with ground state $|\Psi'\rangle$ but with the same particle density, $n(r)$. Since $|\Psi\rangle$ and $|\Psi'\rangle$ obey different Schrödinger equations, $|\Psi'\rangle$ is in general different from $|\Psi\rangle$. If H, H' are the Hamiltonians with potentials v, v' with associated ground states $|\Psi\rangle, |\Psi'\rangle$, respectively, with the corresponding ground-state energies E, E', then we have, from the minimum property of the ground state,

$$E' = \langle\Psi'|H'|\Psi'\rangle < \langle\Psi|H'|\Psi\rangle \tag{2.8}$$

when we employ $|\Psi\rangle$ as a trial state of the Hamiltonian H'. The strict inequality sign appears because of the assumption of nondegeneracy of the ground states. But

$$H' = \mathcal{T} + V' + W = H + V' - V \tag{2.9}$$

Using (2.5) and (2.3b), we obtain

$$E' < E + \int (v'(\mathbf{r}) - v(\mathbf{r}))n(\mathbf{r})d\mathbf{r} \tag{2.10}$$

The preceding argument is applicable equally, had we begun with E:

$$E = \langle\Psi|H|\Psi\rangle < \langle\Psi'|H|\Psi'\rangle$$

and hence

$$E < E' + \int (v(\mathbf{r}) - v'(\mathbf{r}))n(r)d\mathbf{r} \tag{2.11}$$

Adding (2.10) and (2.11) leads to the inconsistency

$$E + E' < E + E' \tag{2.12}$$

Thus, $v(r)$ to within an additive constant, is a unique functional of $n(r)$. Since $v(r)$ fixes H, we see that the full many-particle ground state is a unique functional of $n(r)$. This proves the first part of the theorem.

The second part of the theorem is variational in character and has practical significance. Since $|\Psi\rangle$ is a functional of $n(\mathbf{r})$, all many-particle ground-state expectation values of operators are functionals of $n(\mathbf{r})$. This is a very important consequence of the first part of the theorem. This enables us to state, in particular, that the ground-state expectation value of a two-particle correlation function, $\langle\Psi|\psi^\dagger(x_1)\psi^\dagger(x_2)\psi(x_2)\psi(x_1)|\Psi\rangle$, is a functional of density. This two-particle correlation function is related to the well-known pair correlation function $g(x_1, x_2)$, defined by

$$n(x_1)n(x_2)g(x_1, x_2) \equiv \langle\Psi|\psi^\dagger(x_1)\psi^\dagger(x_2)\psi(x_2)\psi(x_1)|\Psi\rangle \qquad (2.13)$$

$g(\mathbf{x}_1, \mathbf{x}_2)$ is the probability that if an electron with spin σ_1 is at $\mathbf{r}_1$ ($\equiv \mathbf{x}_1$), there will be another one at $\mathbf{r}_2$ with spin σ_2. The value $n(x_1)$ is the number of particles of spin σ_1 at $\mathbf{r}_1$. We discuss this important function later. We may thus define

$$F[n(\mathbf{r})] \equiv \langle\Psi|(\mathcal{T} + W)|\Psi\rangle \qquad (2.14)$$

where $F[n(\mathbf{r})]$ is a universal functional, valid for any number of particles and *any* local external scalar potential. Explicitly,

$$F[n(\mathbf{r})] = T[n(r)] + \frac{e^2}{2}\sum_{\sigma_1\sigma_2}\int\int d\mathbf{r}_1 d\mathbf{r}_2 \frac{n(x_1)n(x_2)}{|\mathbf{r}_1 - \mathbf{r}_2|} g(x_1, x_2) \qquad (2.15)$$

where

$$T[n] \equiv \langle\Psi|\mathcal{T}|\Psi\rangle$$

Now, for a given potential $v(r)$, the ground-state energy functional is

$$E_v[n(\mathbf{r})] \equiv \int v(\mathbf{r})n(\mathbf{r})d\mathbf{r} + F[n] \qquad (2.16)$$

We have the condition that the total number of particles is N:

$$N[n] \equiv \int n(\mathbf{r})d\mathbf{r} = N \qquad (2.17)$$

Subject to condition (2.17), we now show that $E_v[n]$ is a minimum for the correct $n(r)$. From the Rayleigh–Ritz principle, we know that the energy as a functional of the state $|\Psi'\rangle$,

$$\mathscr{E}_v[\Psi'] \equiv \langle\Psi'|V|\Psi'\rangle + \langle\Psi'|(\mathcal{T} + W)|\Psi'\rangle \qquad (2.18)$$

has a minimum for the correct ground state $|\Psi\rangle$, relative to arbitrary variations of $|\Psi'\rangle$ in which the total number of particles is held fixed. As a special case, if $|\Psi'\rangle$ be the ground state associated with a different potential $v'(\mathbf{r})$, then by (2.16) and (2.14),

$$\mathscr{E}_v[\Psi'] = \int v(\mathbf{r})n'(\mathbf{r})d\mathbf{r} + F[n'] > \mathscr{E}_v[\Psi] = \int v(\mathbf{r})n(\mathbf{r})d\mathbf{r} + F[n] \qquad (2.19)$$

that is, the minimal property of (2.16) is established relative to all the density functions $n'(r)$ associated with some other external potential $v'(\mathbf{r})$. This completes the proof of the theorem. Note that the fundamental statistics of the particles did not play any role in establishing the theorem.

We can make several remarks concerning this theorem. Münch et al.[29] have proved rigorously that the preceding theorem holds for many-particle Hamiltonians which are (*a*) essentially self-adjoint, (*b*) bounded from below, and (*c*) below a certain point the energy spectrum is discrete, and with ground states that are nondegenerate except for a permissible degeneracy due to time inversion. As stated earlier, all expectation values of operators in the ground state are functionals of density. In particular, the single-particle Green function, defined by[30]

$$G_1(xt; x't') = \left(\frac{-i}{\hbar}\right)\langle\Psi|T(\psi(xt)\psi^\dagger(x't'))|\Psi\rangle \tag{2.20}$$

or for that matter, the n-particle Green function

$$\begin{aligned} &G_n(x_1t_1\cdots x_nt_n; x_1't_1'\cdots x_n't_n') \\ &\quad = \left(\frac{-i}{\hbar}\right)^n \langle\Psi|(\psi(x_1t_1)\cdots\psi(x_nt_n)\psi^\dagger(x_n't_n')\cdots\psi^\dagger(x_1't_1'))|\Psi\rangle \end{aligned} \tag{2.21}$$

are functionals of $n(r)$. Here T is the usual time-ordering operator. This property is of importance in setting up the Green function formalism based on the preceding theorem. A similar statement can be made concerning n-particle density matrices which are special cases of the preceding statements concerning the zero-temperature Green functions. We now prove a finite temperature generalization of Theorem 1.

B. Finite Temperature Extension

Theorem 2 (Mermin[25]). In the grand canonical ensemble at a given temperature and chemical potential, the equilibrium density $n(r)$ is uniquely determined by the external potential $v(r)$. The correct equilibrium density, for a given $v(r)$, minimizes the Gibbs grand potential, Ω. (The theorem is valid for the canonical ensemble also.)

Proof. The Gibbs theorem of the minimum of grand potential for the equilibrium statistical mechanics is the analogue of the Rayleigh–Ritz minimum energy principle for the ground state of a many-body quantum mechanical system. The latter may be considered as the zero-temperature version of the Gibbs theorem. The Gibbs theorem is that the grand potential

$$\Omega[P] \equiv \mathrm{Tr}\left\{P\left(H - \mu N + \frac{1}{\beta}\ln P\right)\right\} \tag{2.22}$$

is a minimum as a functional of P, subject to the constraint

$$\mathrm{Tr}\, P = 1 \tag{2.23}$$

where P is the density matrix of the quantum mechanical system defined by the Hamiltonian H of the system, with chemical potential μ, and at a temperature $\tau(\beta = 1/k_B\tau$, k_B being the Boltzmann constant) and Tr denotes the trace over all the states of the Hamiltonian H. In fact, if P_0 is the equilibrium grand canonical density matrix,

$$P_0 = \frac{\exp\{-\beta(H-\mu N)\}}{\mathrm{Tr}\exp\{-\beta(H-\mu N)\}} \tag{2.24}$$

then

$$\Omega[P] > \Omega[P_0] \tag{2.25}$$

for $P \neq P_0$, and all positive definite P with unit trace. $\Omega[P_0]$ is given by

$$\Omega[P_0] = -\frac{1}{\beta}\ln \mathrm{Tr}\{\exp -\beta(H-\mu N)\} \tag{2.26}$$

Given this Gibbs theorem, the proof of Theorem 2 is immediate when the Hamiltonian H is given by (2.2), (2.3), and the particle density is defined by

$$n(\mathbf{r}) = \sum_{\sigma} \mathrm{Tr}\{P_0\psi^{\dagger}(xt)\psi(xt)\} \tag{2.27}$$

Clearly, by definition of P_0, $n(\mathbf{r})$ is a functional of $v(\mathbf{r})$, and in fact, it is a unique functional of it, as in Theorem 1. To prove this, we follow an argument similar to the one employed earlier, but, we use the minimum property (2.25) of the grand potential. Let the Hamiltonian, grand canonical density matrix, and the grand potential associated with $v'(\mathbf{r})$ be denoted by H', P_0', and Ω', respectively. Since $v' \neq v$, $P_0' \neq P_0$, we have from (2.25),

$$\Omega' = \mathrm{Tr}\left\{P_0'\left(H' - \mu N + \frac{1}{\beta}\ln P_0'\right)\right\} < \mathrm{Tr}\left\{P_0\left(H' - \mu N + \frac{1}{\beta}\ln P_0\right)\right\}$$

$$= \Omega + \mathrm{Tr}\, P_0(V' - V)$$

that is,

$$\Omega' < \Omega + \int d\mathbf{r}(v'(\mathbf{r}) - v(\mathbf{r}))n(\mathbf{r}) \tag{2.28}$$

Similarly, we obtain

$$\Omega < \Omega' + \int d\mathbf{r}(v(\mathbf{r}) - v'(\mathbf{r}))n(\mathbf{r}) \tag{2.29}$$

and by adding the preceding two results, we are faced with a contradiction, $\Omega + \Omega' < \Omega + \Omega'$.

Thus we find $n(r)$ uniquely determines $v(\mathbf{r})$ which in turn determines P_0. The equilibrium density matrix P_0 is a functional of $n(\mathbf{r})$ and hence as in the ground-state theory, all grand canonical ensemble averages, such as the finite temperature versions of the n-particle Green functions[30] are functionals of $n(r)$. In particular,

$$F[n(\mathbf{r})] = \mathrm{Tr}\left\{P_0\left(T + W + \frac{1}{\beta}\ln P_0\right)\right\} \tag{2.30}$$

is a functional of density alone with a universal form valid for all $v(\mathbf{r})$. For a given potential, $v(\mathbf{r})$, the functional defined by

$$\Omega_v[n] = \int v(\mathbf{r})n(\mathbf{r})d\mathbf{r} + F[n] - \mu \int n(\mathbf{r})d\mathbf{r} \tag{2.31}$$

equals the grand potential Ω when $n(\mathbf{r})$ is the correct equilibrium density associated with the potential, $v(\mathbf{r})$. If $n'(\mathbf{r})$ is the equilibrium potential associated with any other $v'(r)$, then

$$\Omega_v[n'] > \Omega_v[n] \tag{2.32}$$

because the right-hand side of (2.32) is the grand potential $\Omega[P_0]$, whereas the left-hand side is $\Omega[P_0']$. The correct density minimizes (2.31) over all density functions that can be associated with a potential, $v(\mathbf{r})$.

This is the generalization of Theorem 1 to finite temperatures. We can draw the following important consequence of this theorem. The finite temperature n-particle Green functions are defined by[30]

$$G_n(x_1t_1 \cdots x_nt_n; x_1't_1' \cdots x_n't_n') = \left(\frac{-i}{\hbar}\right)^n \mathrm{Tr}\{P_0T(\psi(x_1t_1) \cdots \psi(x_nt_n)\psi^\dagger(x_n't_n') \cdots \psi^\dagger(x_1't_1'))\} \tag{2.33}$$

where P_0 is defined by (2.24) or, equivalently, by the density matrix which minimizes the grand potential Ω, given by (2.22). One then employs a complex time-domain formulation[30] to obtain the equilibrium properties of the many-body system. This enables one to formally deal with some of the questions concerning the excited states of the system via this finite temperature extension of the theorem. For $n = 1$, we have the finite temperature version of (2.20), for the one-particle Green function. We return to this later.

C. Spinor External Potentials

The preceding theorems require a suitable generalization to cover the cases where the spin-dependent properties play an important role, for example, magnetic systems. In another sense, the problems considered thus far do depend on spin, only insofar as their wave functions or equilibrium density matrices are concerned and hence one could exploit the variational freedom a little more by making a suitable generalization where the spin-dependence is explicitly taken into account. A generalization of this kind was given by Kohn and Sham[19] whose interest was mainly a computation of the static spin susceptibility of the paramagnetic electron system. The full generalization to cover the magnetic states was given almost simultaneously by Pant and Rajagopal,[31] von Barth and Hedin,[32] and by Rajagopal and Callaway.[33] The generalization is straightforward but it is not obvious that there is a unique relationship between the spin-dependent potential and the spin density, a feature that was present in Theorems 1 and 2. In Ref. 32, it is shown that many potentials give the same spin density but we find the requirement of ground state seems to be quite severe and may restore the uniqueness somewhat. We raise this point again at the end of this subsection. Rajagopal and Callaway[33] approached the problem quite differently in that they formulated the problem for a relativistic situation and reduced it in the nonrelativistic limit for the case under consideration. This gives an elegant formulation for the spin-dependent as well as velocity-dependent cases. Pant and Rajagopal[31] formulated the problem in terms of the physical quantities; that is, particle density $n(\mathbf{r})$ and the spin-density vector $\boldsymbol{s}(\mathbf{r})$ and the associated external fields are the potential $v(\mathbf{r})$ associated with $n(\mathbf{r})$ and a static magnetic field $\mathbf{H}(r)$ associated with spin density, $\boldsymbol{s}(\mathbf{r})$. von Barth and Hedin,[32] in an entirely equivalent formulation, stated the theorem in terms of a spinor-density matrix, $P_{\alpha\beta}$, where α,β represent the two-spin states of the electron and an associated spinor external potential $v_{\alpha\beta}(\mathbf{r})$. The total number of external potentials in both cases is four and a correspondence between n, $\boldsymbol{s}$, and $P_{\alpha\beta}$ is immediately established:

$$n(\mathbf{r}) = \mathrm{tr}(\tau_0 P)$$

$$\boldsymbol{s}(\mathbf{r}) = \mathrm{tr}(\boldsymbol{\tau} P) \tag{2.34}$$

where tr denotes trace on spinor indices, τ_0 is a 2×2 unit matrix and $\boldsymbol{\tau}$ is the standard Pauli spin matrix, with components

$$\tau_x = \begin{pmatrix} 0 & 1 \\ 1 & 0 \end{pmatrix}; \qquad \tau_y = \begin{pmatrix} 0 & -i \\ +i & 0 \end{pmatrix}; \qquad \tau_z = \begin{pmatrix} 1 & 0 \\ 0 & -1 \end{pmatrix} \tag{2.34a}$$

And,

$$P_{\alpha\beta}(\mathbf{r}) = \langle \Psi | \psi^\dagger(\mathbf{r}\beta t)\psi(\mathbf{r}\alpha t)|\Psi\rangle \tag{2.35}$$

where $|\Psi\rangle$ as before stands for the ground state of the system; $P_{\alpha\beta}(\mathbf{r})$ are the elements of the 2×2 matrix P, in (2.34). We state the theorem as follows for finite temperatures for the sake of generality; the zero temperature version follows similarly.

Theorem 3. In the grand canonical ensemble at a given temperature and chemical potential, the equilibrium density $n(\mathbf{r})$ and the spin-density vector $\boldsymbol{s}(\mathbf{r})$ are determined by the scalar external potential $v(\mathbf{r})$ and the external magnetic field $\mathbf{H}(\mathbf{r})$. The correct equilibrium density and spin density vector for given $v(\mathbf{r})$, $\mathbf{H}(\mathbf{r})$ minimizes the Gibbs grand potential, Ω.

Equivalently,

Theorem 3′. In the grand canonical ensemble at a given temperature and chemical potential, the elements of the equilibrium spin-density matrix $P_{\alpha\beta}(\mathbf{r})$ are determined by the spinor external potential $v_{\alpha\beta}(\mathbf{r})$. The correct set of equilibrium $P_{\alpha\beta}(\mathbf{r})$ for given $v_{\alpha\beta}(\mathbf{r})$ minimizes the grand potential, Ω.

Proof. The proof proceeds exactly as before. The Hamiltonian in this case is given by

$$H = \mathcal{T} + V + W \tag{2.36}$$

where $\mathcal{T}$ and W are as given in (2.3a) and (2.3c), but

$$V = \sum_{\alpha\beta} \int v_{\alpha\beta}(\mathbf{r})\psi^\dagger(\mathbf{r}\beta t)\psi(\mathbf{r}\alpha t)d\mathbf{r} \tag{2.37}$$

Equivalently,

$$V = \sum_{\sigma} \int v(\mathbf{r})\psi^\dagger(\mathbf{r}\sigma t)\psi(\mathbf{r}\sigma t)d\mathbf{r} + \sum_{\sigma,\sigma'} \int \mathbf{H}(\mathbf{r}) \cdot (\psi^\dagger(\mathbf{r}\sigma t)\boldsymbol{\tau}_{\sigma\sigma'}\psi(\mathbf{r}\sigma' t))d\mathbf{r} \tag{2.38}$$

The grand potential $\Omega[P]$ is defined as in (2.23) but with the Hamiltonian defined by (2.36), which now includes therefore a proper spin sum in Tr. Instead of (2.33) and (2.35), we have, at finite temperatures,

$$P_{\alpha\beta}(\mathbf{r}) = \mathrm{Tr}\{P_0\psi^\dagger(\mathbf{r}\beta t)\psi(\mathbf{r}\alpha t)\}, \qquad \text{etc.} \tag{2.39}$$

Observe that the equilibrium density matrix P_0 is now a spinor, being defined by (2.24). The total number of particles is assumed to be fixed, as before, so that

$$N = \sum_{\alpha} \int P_{\alpha\alpha}(\mathbf{r})d\mathbf{r} = \sum_{\alpha} \int n(\mathbf{r}\alpha)d\mathbf{r} \tag{2.40}$$

In the proof, which proceeds exactly as before, we get, for the grand potential,

$$\Omega < \Omega' + \sum_{\alpha\beta} \int (v_{\alpha\beta}(\mathbf{r}) - v'_{\alpha\beta}(\mathbf{r}))P_{\beta\alpha}(\mathbf{r})d\mathbf{r} \tag{2.41}$$

instead of Eq. (2.29), where $v'_{\alpha\beta}$ is a set of different spinor potentials but such that they lead to the *same* $P_{\alpha\beta}(\mathbf{r})$.

von Barth and Hedin,[32] in the one-particle case show, at $T = 0°K$, that there are many potentials that give the same spin density. By explicit construction, they found a projection operator, which acting on the ground state of H gave zero, and so the ground state of H, is *an eigenstate* of $H' = H + V'$, where V' is an arbitrary function multiplying the projection operator. Clearly the ground state of H' is *not* the same as that of H. Hence, the requirement of the theorem that one must have a ground state of H' is not obeyed and the nonuniqueness proof is really not valid. It should be pointed out that von Barth and Hedin[32] realized that the ground state of H is not necessarily the ground state of H' but, argued heuristically that for small enough multiplicative function of the projection operator, it was indeed so. This is not at all convincing because the Hilbert space of H' seems to be quite different from that of H and it is not clear that there is a subspace of the ground state where the two operators can have simultaneously the same ground state.

We may end this subsection by stating that all equilibrium state properties like the grand potential, the one- and two-particle Green's functions, and so on, are functionals of the density matrix $P_{\alpha\beta}$ or equivalently, of $n(\mathbf{r})$ and $\mathcal{S}(\mathbf{r})$. By employing the imaginary-time formulation of the Green function formalism, we can develop a theory for the low-lying excited states of this system, as was attempted by Rajagopal.[27] We return to this question later.

D. Velocity-dependent External Potentials

For electrons in an electromagnetic field one generally has external scalar and vector potentials, viz., the scalar potential is $\phi(\mathbf{r}t)$ and the vector potential is $\mathbf{A}(\mathbf{r}, t)$. The Hamiltonian operator of the system, *including spin-dependent terms*, may be written as, assuming the electromagnetic potentials to be classical (this is enough for most purposes)

$$\begin{aligned} H = \mathcal{T} + W + \sum_{\sigma} \int d\mathbf{r}\left[-e\phi(\mathbf{r}t) + \frac{e^2}{2mc^2}\mathbf{A}^2(\mathbf{r}, t)\right] \cdot \psi^\dagger(\mathbf{r}\sigma t)\psi(\mathbf{r}\sigma t) \\ + \sum_{\sigma\sigma'} \int d\mathbf{r}\left(\frac{e\hbar}{mc}\right)\mathbf{A}(\mathbf{r}, t) \cdot \operatorname{curl}\{\psi^\dagger(\mathbf{r}\sigma t)\boldsymbol{\tau}_{\sigma\sigma'}\psi(\mathbf{r}\sigma' t)\} \\ + \sum_{\sigma} \int d\mathbf{r}\frac{1}{c}\mathbf{A}(\mathbf{r}t) \cdot \frac{e\hbar}{2mi}\{\psi^\dagger(\mathbf{r}\sigma t)(\nabla\psi(\mathbf{r}\sigma t)) - (\nabla\psi^\dagger(\mathbf{r}\sigma t))\psi(\mathbf{r}\sigma t)\} \\ + \sum_{\sigma\sigma'} \int d\mathbf{r}\, v_{\sigma\sigma'}(\mathbf{r})\psi^\dagger(\mathbf{r}\sigma' t)\psi(\mathbf{r}\sigma t) \end{aligned} \tag{2.42}$$

The last term containing $v_{\sigma\sigma'}(\mathbf{r})$ is added for completeness. If we assume that the *external potentials are static* (independent of time), one can prove a version of the theorems proved previously, which we quote as follows. The statement of the theorem is made for the finite temperature case to keep the discussion quite general.

Theorem 4. In the grand canonical ensemble at a given temperature and chemical potential, the equilibrium density, $n(\mathbf{r})$, the spin-density vector $\mathcal{S}(\mathbf{r})$ and the current density vector $(\mathbf{j}(\mathbf{r})+(e\hbar/mc)\,\mathrm{curl}\,\mathcal{S}(\mathbf{r}))$ are determined, respectively, by the total scalar potential $[-e\phi(\mathbf{r})+(e^2/2mc^2)\mathbf{A}^2(\mathbf{r})+\frac{1}{2}(v_{\uparrow\uparrow}(\mathbf{r})+v_{\downarrow\downarrow}(\mathbf{r}))]$, the part of the spinor potential, $v_{\sigma\sigma'}(\mathbf{r})$, with components $(v_{\uparrow\downarrow}(\mathbf{r}), v_{\downarrow\uparrow}(\mathbf{r}), \frac{1}{2}(v_{\uparrow\uparrow}(\mathbf{r})-v_{\downarrow\downarrow}(\mathbf{r}))$, and the vector potential $\mathbf{A}(\mathbf{r})$, respectively. The correct set of equilibrium density, spin-density vector, and the current density vector minimizes the grand potential Ω, for given $v_{\alpha\beta}(\mathbf{r})$, $\phi(\mathbf{r})$, and $\mathbf{A}(\mathbf{r})$. The particle current $\mathbf{j}(r)$ obeys the condition $\mathrm{div}\,\mathbf{j}(\mathbf{r})=0$.

Proof. The proof follows the same steps as in Theorem 3. The only thing to be noted is the definition of the current $\mathbf{j}(\mathbf{r})$, that is,

$$\mathbf{j}(\mathbf{r})=\frac{e\hbar}{2mi}\sum_{\sigma}\mathrm{Tr}\{P_0[\psi^{\dagger}(\mathbf{r}\sigma t)(\nabla\psi(\mathbf{r}\sigma t))-(\nabla\psi^{\dagger}(\mathbf{r}\sigma t))\cdot\psi(\mathbf{r}\sigma t)]\} \tag{2.43}$$

and the expression for Ω is

$$\begin{aligned}\Omega_{v,\phi;\mathbf{A}}[n(\mathbf{r}),\mathcal{S}(\mathbf{r}),\mathbf{j}(\mathbf{r})] &= F[n(\mathbf{r}),\mathcal{S}(\mathbf{r}),\mathbf{j}(\mathbf{r})]+\int d\mathbf{r}\left[-e\phi(\mathbf{r})+\frac{e^2}{2mc^2}\mathbf{A}^2(\mathbf{r})\right]n(\mathbf{r})\\ &\quad+\int d\mathbf{r}\left(\frac{e\hbar}{mc}\right)(\mathrm{curl}\,\mathcal{S}(r))\cdot\mathbf{A}(\mathbf{r})+\int d\mathbf{r}\frac{1}{c}\mathbf{A}(\mathbf{r})\cdot\mathbf{j}(\mathbf{r})\\ &\quad+\sum_{\sigma\sigma'}\int d\mathbf{r}\,v_{\sigma\sigma'}(\mathbf{r})P_{\sigma'\sigma}(\mathbf{r})\end{aligned} \tag{2.44}$$

where $P_{\sigma\sigma'}(r)$ is as defined in Theorem 3′, (2.39). Note that the universal functional F is a functional of the seven quantities.

Rajagopal and Callaway[33] approached this problem via the quantum electrodynamic formulation for the Dirac electrons where they showed that F is a functional of the four-current $j_{\mu}(\mathbf{r})$, which in the nonrelativistic limit goes over to the particle current and densities, and as the first nonvanishing relativistic correction, spin density makes its appearance, in addition to these. The nonlinearity in $\mathbf{A}$ in (2.44) does not invalidate the usual steps in the proof of the uniqueness. The question of gauge invariance is handled by noticing that $\mathrm{div}\,\mathbf{j}(\mathbf{r})=0$ in this theory always, because this is simply a statement of the charge conservation or, equivalently, the equation of continuity demands that the particle current $\mathbf{j}(\mathbf{r})$ be divergence free. That the spin is intrinsically relativistic in concept emerges in this relativistic formulation whereas in (2.42) we introduced it by definition. Had we dropped from consideration the spin terms, we would have obtained a simpler version of the preceding theorem, which we state here for completeness, without proof:

Theorem 4′. In the grand canonical ensemble at a given temperature and chemical potential, the equilibrium density $n(\mathbf{r})$ and the particle current density $\mathbf{j}(\mathbf{r})$ are determined by the potentials $(-e\phi(\mathbf{r}) + (e^2/2mc^2)\mathbf{A}(\mathbf{r}))$, and the vector potential $\mathbf{A}(\mathbf{r})$. The correct $n(r)$, $\mathbf{j}(r)$ minimizes the grand potential Ω, for given $\phi(\mathbf{r}), \mathbf{A}(r)$. The particle current obeys the divergenceless condition, $\operatorname{div} \mathbf{j}(\mathbf{r}) = 0$, to preserve the conservation of particles in the theory.

We end this subsection with a remark concerning the preceding result and the theory of Dashen and Sharp[34] wherein they propose to use particle density and current density operators as a complete set of operators to describe nonrelativistic quantum mechanical systems. These authors were able to express the nonrelativistic Hamiltonian operator in terms of the four-current operator expressions. Our theorem concerning the ground state in the presence of an external static four-vector potential seems to be another expression of the theory of Dashen and Sharp. These authors had, furthermore, observed that the expressions for the four-current operators, the algebra satisfied by these, and the expression for the Hamiltonian operator in terms of these were independent of statistics. To recover the underlying statistics obeyed by the particles, one has to rework the representations of the algebra of currents in symmetric and antisymmetric subspaces. In other words, they showed that the representation of the algebra of currents falls into two irreducible classes corresponding to the two known statistics of the particles. The only known use in the nonrelativistic many-body physics of the preceding theory is the work of Grest and Rajagopal[35] who applied this theory to a system of interacting Bose particles and derived the known results of Feynman and others.

The utility of Theorem 4 lies in calculating the static but space-dependent response functions of the system to static external electromagnetic fields within a functional formalism. One could also use this for studying the effects of strong external fields on the many-body systems which are currently of immense interest.

E. Nonlocal External Potential

The Hohenberg–Kohn theorem was extended to the case of nonlocal external potential by Gilbert.[36] The utility of this extension is not clear to us at present even though, as we point out, it seems to be related to another stationary principle in many-body theory. We state the theorem and the proof may be found in Gilbert's paper. We may remark that perhaps one may use this theorem in some way in dealing with the pseudopotential theory of simple metals. This is not clear to us because the nonlocal pseudopotential is a consequence of electron correlations and the purpose of the density-functional theory was to provide a framework for dealing with exactly that question.

Theorem 5. The nondegenerate ground state of a many-body system with nonlocal one body external potential, $v(\mathbf{r}, \mathbf{r}')$ is a functional of one-particle density matrix $P(\mathbf{r}, \mathbf{r}')$. The ground-state energy of the system for a given v is minimized by the correct $P(\mathbf{r}, \mathbf{r}')$.

The finite temperature version of this theorem states that the grand potential Ω is a functional of the temperature-dependent one-particle density matrix $P(\mathbf{r}, \mathbf{r}')$, and for a given $v(\mathbf{r}, \mathbf{r}')$, the correct P minimizes it. Baym,[37] in developing self-consistent approximations in the theory of many-particle systems, showed that the grand-partition function can be constructed from the one-particle Green function, $G_1(\mathbf{r}t; \mathbf{r}'t')$, introduced in (2.33), and thus it is a functional of $G_1(\mathbf{r}t; \mathbf{r}'t')$. Luttinger and Ward[38] had earlier shown that the grand partition function is stationary under variations of G_1, arising from the variations in self-energy, $\Sigma(\mathbf{r}t; \mathbf{r}'t')$ defined by $G_0^{-1} - G^{-1}$, where G_0^{-1} is the inverse of the one-particle Green function for the noninteracting system. This principle holds for the logarithm of the partition function for any nonlocal external disturbance. Baym[37] showed that approximation methods preserve the stationary property if the self-energy is Φ-derivable; that is, a functional Φ of G_1 exists such that Σ is $\delta\Phi/\delta G_1$. Since the nonlocal potential considered by Gilbert is a special case of the one used by Baym, we see that his proof of Theorem 5 is a variant of the Luttinger–Ward theorem and the procedure of Baym could perhaps be used to construct self-consistent approximation schemes in the functional theory.

F. Relativistic Systems

The purpose of relativistic generalization is twofold. As stated earlier, it provides a natural framework for dealing with problems in which spin densities are essential in describing the system. The second reason is that it is an important generalization for describing systems in which the relativistic effects cannot be considered as small perturbation, for example in atomic systems with nuclear charge larger than 36 and in solid state problems when one wants to consider the properties of heavy elements such as cesium and heavy rare-earths, or compounds such as lead–telluride. Such a generalization will also provide a natural framework of the relativistic Thomas–Fermi theory. The functional formalism was developed by Rajagopal and Callaway[33] with magnetic problems in mind. Here we make the scope quite general. We must emphasize that our interest *is not* in those aspects of the relativistic theory where the field-theoretic effects are important but rather in describing those relativistic effects that are much larger in magnitude than these. To this end, we may often consider *the electromagnetic fields in the theory to be described by classical means.* If one wants to consider questions of very high density matter, this description must be modified as well as other field effects such as pions included. This is outside the scope of our interest here. We may also point out that a formulation for the Klein–Gordon system can also be given but is not done so here. For electron systems of our interest, the Dirac theory is appropriate. We employ notations similar to the one used in Bjorken and Drell,[39] and for completeness we give here briefly an account of the basic formulas.

The spinor theory of Section II.C must be modified to accommodate Dirac spinors. The basic 4×4 matrices of Dirac defined by $\gamma^0 = \beta$, $\boldsymbol{\gamma} = \beta\boldsymbol{\alpha}$ replace the Pauli matrices of (2.34):

$$\gamma^0 = \beta = \begin{pmatrix} 1 & 0 \\ 0 & -1 \end{pmatrix}, \qquad \{\gamma^i\} = \boldsymbol{\gamma} = \begin{pmatrix} 0 & \boldsymbol{\tau} \\ -\boldsymbol{\tau} & 0 \end{pmatrix} \tag{2.45}$$

where 1 is a 2×2 unit matrix, $\boldsymbol{\tau}$ is the 2×2 Pauli spinor, defined by (2.34). Note that the 4×4 matrices are written in a partitioned form. The metric used is $(1, -1, -1, -1)$ so that the space–time coordinates $(t, \mathbf{r})$ are denoted by

$$x^\mu \equiv (x^0, x^1, x^2, x^3) \equiv (ct, x, y, z) \tag{2.46a}$$

and the covariant four-vector is

$$x_\mu \equiv (x_0, x_1, x_2, x_3) \equiv (ct, -x, -y, -z) \tag{2.46b}$$

The summation convention is used unless otherwise specified. The components of the momentum vector are

$$p^\mu \equiv (p^0, p^1, p^2, p^3) = \left(\frac{E}{c}, p_x, p_y, p_z\right) \tag{2.46c}$$

and

$$p_\mu p^\mu = \frac{E^2}{c^2} - \mathbf{p}^2 = m^2c^2 \tag{2.46d}$$

where E is the total energy and m is the rest mass of the particle. The momentum operator in coordinate representation is

$$p^\mu = -i\hbar\frac{\partial}{\partial x_\mu} \tag{2.46e}$$

We define a column vector Ψ as the relativistic wave function obeying the Dirac equation and its components are often indicated by a subscript: Ψ_σ with $\sigma = 1, \ldots, 4$ form the Dirac wave function. The second quantized one-particle destruction operator is ψ_σ and its Hermitian conjugate is the creation operator, $\psi_\sigma^\dagger$. They obey the equal time anticommutation rules

$$\begin{aligned} [\psi_\sigma(\mathbf{r}, t), \psi_{\sigma'}^\dagger(\mathbf{r}'t)]_+ &= \delta_{\sigma\sigma'}\delta^3(\mathbf{r}-\mathbf{r}') \\ [\psi_\sigma(\mathbf{r}, t), \psi_{\sigma'}(\mathbf{r}', t)]_+ &= 0 \\ [\psi_\sigma^\dagger(\mathbf{r}, t), \psi_{\sigma'}^\dagger(\mathbf{r}', t)]_+ &= 0 \end{aligned} \tag{2.47}$$

Here $\sigma, \sigma' = 1, \ldots, 4$. The particle density operator is

$$\rho(\mathbf{r}, t) = \psi^\dagger(\mathbf{r}, t)\psi(\mathbf{r}, t) \tag{2.48}$$

and the current-density operator is

$$j^k(\mathbf{r}, t) = c\psi^\dagger(\mathbf{r}, t)\alpha^k\psi(\mathbf{r}, t), \qquad (k = 1, 2, 3) \tag{2.49}$$

These two are together written as a four-vector

$$j^\mu(\mathbf{r}, t) = c\psi^\dagger(\mathbf{r}, t)\gamma^0\gamma^\mu\psi(\mathbf{r}, t) \tag{2.50}$$

with $j^0(\mathbf{r}, t) = c\rho(x)$ as the time component of a conserved four-vector j^μ and this obeys the constraint

$$\frac{\partial j^\mu(\mathbf{r}, t)}{\partial x^\mu} = 0 \tag{2.51}$$

Equation 2.51 represents just the equation of continuity, and note that it is an operator statement. One often introduces a new notation

$$\bar{\psi}(\mathbf{r}, t) = \psi^\dagger(\mathbf{r}, t)\gamma_0 \tag{2.52}$$

The electromagnetic fields are represented by the four-vector, $A^\mu(\mathbf{r}, t) = (\Phi, \mathbf{A})$, and the electromagnetic field strengths are given by

$$\mathbf{E} = -\frac{1}{c}\frac{\partial A}{\partial t} - \nabla\Phi, \qquad \mathbf{B} = \text{curl } \mathbf{A} \tag{2.53}$$

In the Coulomb gauge,

$$\nabla \cdot \mathbf{A} = 0 \tag{2.54}$$

which we employ in most of our discussions. After some analysis, the Hamiltonian operator for the system of electrons in the presence of an external four-potential is written in the form (operators are normal ordered as usual[39]):

$$H = \int d\mathbf{r}: \psi^\dagger(\mathbf{r}t)[-i\hbar c\boldsymbol{\alpha} \cdot \nabla + \beta mc^2]\psi(\mathbf{r}t):$$
$$+\frac{e^2}{2}\int\int d\mathbf{r}\, d\mathbf{r}': \psi^\dagger(\mathbf{r}t)\psi(\mathbf{r}t): \frac{1}{|\mathbf{r}-\mathbf{r}'|}: \psi^\dagger(\mathbf{r}'t)\psi(\mathbf{r}'t): + H_{\text{em}}$$
$$+\frac{e}{c}\int d\mathbf{r}: \mathbf{j}(\mathbf{r}t) \cdot \mathbf{A}(\mathbf{r}t): -\frac{e}{c}\int d\mathbf{r}: j_\mu(\mathbf{r}t): A^\mu_{\text{ext}}(\mathbf{r}t) \tag{2.55}$$

The first term in (2.55) is the energy of free Dirac electrons, the second term represents the instantaneous Coulomb interaction among the Dirac electrons arising from the elimination of the longitudinal part of the electromagnetic field, the third term, H_{em}, is the energy of the transverse part of the free electromagnetic radiation field, the fourth term is the interaction among the electrons with the transverse radiation field. The last term represents the interaction of the *unquantized external field*, $A^\mu_{\text{ext}}(r, t)$, with the electrons of the system. This representation is a useful one and for our purposes, we assume that $A^\mu_{\text{ext}}(\mathbf{r}, t)$ contains the Coulomb field produced by the nuclei of the system which are thought to be fixed as in our previous discussions. The quantized electromagnetic fields do not concern us here. Our purpose is to develop a scheme for handling the problem in the presence of $A^\mu_{\text{ext}}(\mathbf{r}, t)$, and as such it has no other role in the Hamiltonian, (2.55). For example, we do not write a term analogous to H_{em} or for that matter a term exhibiting the interaction of A^μ_{ext} and the transverse radiation field, because we do not think of A^μ_{ext} playing a similar role as the free radiation field. Our *model* is then represented by the Hamiltonian, (2.55). If the external field, A^μ_{ext} is *static*, then we may proceed as in the previous sections to prove the following theorem. Throughout our discussion, we *assume* that the *negative energy states are filled* so that the Hamiltonian H given previously is bounded from below and hence possesses a ground state. With this assumption then the theorem may be stated now and the proof proceeds exactly as before. We have made sure that our theory is gauge invariant, so that if $A^\mu_{\text{ext}} \to A^\mu_{\text{ext}} + \partial^\mu \Lambda$, the last term in (2.55) can be integrated by parts and assuming that the surface terms are zero, and noting that $\partial^\mu j_\mu(\mathbf{r}t) = 0$ is an identity, we observe that H is unaltered. The finite temperature version of the theorem proceeds exactly as before because one has a Gibbs theorem for this case also.

Theorem 6. In the grand canonical ensemble at a given temperature and chemical potential, the components of four-current density, $J_\mu(\mathbf{r}) = \text{Tr}\{P_0 j_\mu(\mathbf{r}t)\}$ are determined by the *static* external four-vector potential, $A^\mu_{\text{ext}}(\mathbf{r})$. The correct $J_\mu(\mathbf{r})$ minimizes the grand potential, Ω, for a given $A^\mu_{\text{ext}}(\mathbf{r})$. The current, $J_\mu(\mathbf{r})$, obeys the divergenceless condition.

Note that the grand potential Ω is defined by (2.22) in this case also. At zero temperature, we assume that there exists a nondegenerate ground state of the system, so that, as before the theorem holds. It is important to note that a minimum energy theorem holds here, too, once we fill the negative energy states. We write the grand potential in the form

$$\Omega[J_\mu] = F[J_\mu] - e\int J_\mu(\mathbf{r})A^\mu_{\text{ext}}(\mathbf{r})d\mathbf{r} \tag{2.56}$$

Here $F[J_\mu]$ is a universal functional of the four-current density which is subject to the constraint of being divergenceless:

$$F[J_\mu] = \text{Tr}\left\{P_0\left(H_0 + H_c + H_{\text{em}} + H_{\text{int}} - \mu N + \frac{1}{\beta}\ln P_0\right)\right\} \tag{2.57}$$

Here H_0 is the free-Dirac electron part, H_c is the instaneous Coulomb part, H_{em} is the free radiation part, and H_{int} is the electron-transverse radiation field part of the Hamiltonian of the system defined by (2.55). By following Rajagopal and Callaway,[33] further manipulation is needed to pass to the nonrelativistic regime, where we recover the result of Theorem 4, with the spin density and particle-current density appearing in a natural way.

It is worth mentioning that the many-electron wave function is antisymmetric under permutation of coordinates of the electrons as before. The Hamiltonian operator given by (2.55) without the electromagnetic terms and the external field term written as the Coulomb potential due to static nuclear charges is used in the literature to describe atoms, molecules, and solids. B. Swirles[40] set up the Hartree–Fock self-consistent scheme for this Hamiltonian, and many further works in atoms have appeared. One may refer to a comprehensive review of Lindgren and Rosen[41] on the subject of relativistic self-consistent-field calculations in atomic physics. The basic procedure is similar to the nonrelativistic variational scheme (Rayleigh–Ritz principle for the ground state, determinantal wave function, etc.). Kim[42] has also considered Breit corrections to the electron–electron Coulomb interaction. In our work, this can also be done. Furthermore, Kim derived a virial theorem in this case and found that the total energy is equal to the expectation value of the total rest-mass operator.

The development given until now only establishes some general principles. To obtain anything practical from these, one needs a general scheme, which is also based on some basic principles. In the next section we derive an important theorem which enables us to write an effective one-particle equation once the universal functional F of the theorems proved so far is presumed known. The next stage of development is the specification of F which is dealt with in Section III. This theorem was originally proved by Kohn and Sham.[19]

G. An Important Theorem—Derivation of One-Particle Equation

Theorem 7. The nondegenerate ground state of the inhomogeneous, interacting electron systems can be described by a system of one-particle self-consistent equations of the Schrodinger type with the effective self-consistent one-body potential given by

$$V_{\text{eff}}[\mathbf{r}; n] = v(\mathbf{r}) + e^2 \int \frac{n(\mathbf{r}')}{|\mathbf{r}-\mathbf{r}'|} d\mathbf{r}' + \frac{\delta E_{xc}[n]}{\delta n(\mathbf{r})} \tag{2.58}$$

where E_{xc} is defined by

$$E_{xc}[n] = F[n] - \frac{e^2}{2} \int\int \frac{n(\mathbf{r})n(\mathbf{r}')}{|\mathbf{r}-\mathbf{r}'|} d\mathbf{r}\, d\mathbf{r}' - T_s[n] \tag{2.59}$$

and is by this definition the exchange and correlation energy of the interacting system with density $n(\mathbf{r})$. $-T_s[n]$ is the kinetic energy of a system of noninteracting particles with the same density $n(\mathbf{r})$ as the system under consideration.

Proof. From Theorem 1, we know that the ground-state energy of a many-body system in a static external potential $v(\mathbf{r})$ is a functional of the density $n(\mathbf{r})$, given by (2.16). This functional is a minimum in $n(\mathbf{r})$ subject to the constraint that the total number $\int n(\mathbf{r})d\mathbf{r} = \text{N}[\text{n}]$ is fixed. This variational statement may be written down as

$$\int \delta n(\mathbf{r}) \frac{\delta E_v[n]}{\partial n(\mathbf{r})} d\mathbf{r} = 0 \tag{2.60a}$$

where the variations, $\delta n(\mathbf{r})$, in density are subject to the constraint

$$\int \delta n(\mathbf{r}) d\mathbf{r} = 0 \tag{2.60b}$$

Define the universal functional $F[n]$ of (2.16) following Kohn and Sham,[19] in the form

$$F[n] = T_s[n] + \frac{e^2}{2} \int\int \frac{n(\mathbf{r})n(\mathbf{r}')}{|\mathbf{r}-\mathbf{r}'|} d\mathbf{r}\, d\mathbf{r}' + E_{xc}[n] \tag{2.61}$$

The second term is the classical electrostatic self-energy of the charge densities, $n(\mathbf{r})$, of the system. The Euler equation is then written in the form

$$\int d\mathbf{r} \delta n(\mathbf{r}) \left\{ \frac{\delta T_s}{\delta n(\mathbf{r})} + v(\mathbf{r}) + e^2 \int \frac{n(\mathbf{r}')}{|\mathbf{r}-\mathbf{r}'|} d\mathbf{r}' + \frac{\delta E_{xc}}{\delta n(\mathbf{r})} \right\} = 0 \tag{2.62}$$

subject to the constraint, (2.60b). We thus obtain the equation

$$\frac{\delta E_v[n]}{\delta n(\mathbf{r})}=\frac{\delta T_s}{\delta n(\mathbf{r})}+v(\mathbf{r})+e^2\int\frac{n(\mathbf{r}')}{|\mathbf{r}-\mathbf{r}'|}d\mathbf{r}'+\frac{\delta E_{xc}}{\delta n(\mathbf{r})}=\mu \tag{2.63}$$

where μ is the Lagrange multiplier associated with the constraint. If there were no interactions among particles, one could have an equation of the form (2.61) but without the last two terms in it. Thus one has, in the interacting case, precisely the same equation as for noninteracting particles, moving in a given potential, $V_{\text{eff}}[\mathbf{r}; n]$ defined by (2.58). Therefore, we may define a system of one-particle self-consistent equations of the form

$$\left\{-\frac{\hbar}{2m}\nabla^2+V_{\text{eff}}[\mathbf{r}; n]\right\}\phi_i(\mathbf{r})=\epsilon_i\phi_i(\mathbf{r}) \tag{2.64}$$

with the density $n(\mathbf{r})$ given by

$$n(\mathbf{r})=\sum_{i(\text{occ})}|\phi_i(\mathbf{r})|^2 \tag{2.65}$$

the summation over the eigenstates go over the *occupied* states of the system such that we have the given total number N of particles:

$$\int n(\mathbf{r})d\mathbf{r}=N \tag{2.66}$$

This then establishes the theorem.

Several remarks are in order at this stage. The eigenvalues ϵ_i defined in (2.64) are introduced here purely as a mathematical trick and are not to be interpreted as the one-body energies of the interacting system. We shall show presently that those eigenvalues, ϵ_i, which are all equal to the Lagrange multiplier, μ, may be interpreted as the one-particle energies of the system; that is, for a Fermi system, μ would be the Fermi energy and the preceding statement is equivalent to saying that the one-electron states at the Fermi surface are correctly given by (2.64). To see that μ is the chemical potential, $\partial E_N/\partial N$, for large N, let us write the correct densities associated with total number of particles N and $N-1$ in the external potential $v(\mathbf{r})$ by $n_N(\mathbf{r})$ and $n_{N-1}(\mathbf{r})$. Then if E_N is the energy of the N-particle system, we have, following Lang,[22]

$$\left(\frac{\partial E_N}{\partial N}\right)_{v(\mathbf{r});\,T=0^\circ\text{K}}=E_v[n_N]-E_v[n_{N-1}]$$

$$=\int\left(\frac{\delta E_v[n]}{\delta n(\mathbf{r})}\right)_{n=n_N}[n_N(\mathbf{r})-n_{N-1}(\mathbf{r})]d\mathbf{r}=\mu \tag{2.67}$$

on using (2.63) and the definition preceding.

Equations 2.64 and 2.65 must be solved self-consistently. One starts with an assumed $n(r)$ and constructs $V_{\text{eff}}[\mathbf{r}, n]$ and finds a new $n(\mathbf{r})$ from (2.64) and (2.65). If the solutions $\phi_i(\mathbf{r})$ are normalized to unity, then the energy is given by, after some straightforward algebra,

$$E = \sum_{i(\text{occ})} \epsilon_i - \frac{e^2}{2} \int\int \frac{n(\mathbf{r})n(\mathbf{r}')}{|\mathbf{r}-\mathbf{r}'|} d\mathbf{r}\, d\mathbf{r}' + E_{xc}[n] - \int n(\mathbf{r}) \frac{\delta E_{xc}[n]}{\delta n(\mathbf{r})} d\mathbf{r} \quad (2.68)$$

We may note that if we introduce the quantity, following Janak,[24]

$$t_i = \int \phi_i^*(\mathbf{r}) \left(-\frac{\hbar^2 \nabla^2}{2m} \right) \phi_i(\mathbf{r}) d\mathbf{r}$$
$$= \epsilon_i - \int \phi_i^*(\mathbf{r}) V_{\text{eff}}[\mathbf{r}, n] \phi_i(\mathbf{r}) d\mathbf{r} \quad (2.69)$$

then, using (2.65), we define the kinetic energy functional $T_s[n]$ by

$$\sum_{i(\text{occ})} t_i = T_s[n]$$
$$= \sum_{i(\text{occ})} \epsilon_i - \int n(\mathbf{r}) V_{\text{eff}}[\mathbf{r}; n] d\mathbf{r} \quad (2.70)$$

Starting with (2.64), following Slater, Janak[23] proved the virial theorem for the general density-functional formalism. Operate on (2.64) with $(\mathbf{r} \cdot \nabla)$, multiply by $\phi_i^*(\mathbf{r})$, and integrate over all space. Using the conjugate of (2.64) to rewrite the term $\phi_i^*(\mathbf{r})(V_{\text{eff}} - \epsilon_i)(\mathbf{r} \cdot \nabla \phi_i(\mathbf{r}))$, one obtains

$$\frac{\hbar^2}{2m} \int d\mathbf{r} [\phi_i^*(\mathbf{r})(\mathbf{r} \cdot \nabla)(\nabla^2 \phi_i) - (\nabla^2 \phi_i^*)(\mathbf{r} \cdot \nabla \phi_i(\mathbf{r}))]$$
$$= \int \phi_i^*(r) \phi_i(\mathbf{r})(\mathbf{r} \cdot \nabla V_{\text{eff}}(\mathbf{r})) d^3r$$

Using Gauss's theorem, since all surface integrals are assumed to vanish at "large" distances, and summing over the occupied states, we obtain finally

$$\langle \Psi | \mathcal{T} | \Psi \rangle \equiv T_s[n] = \tfrac{1}{2} \int n(\mathbf{r})(\mathbf{r} \cdot \nabla V_{\text{eff}}[\mathbf{r}, n]) d\mathbf{r} \quad (2.71)$$

Using V_{eff} given by (2.58) and adding it to the contribution of repulsion between the positive charges as in (1.1), we obtain, on performing the algebra suggested by Slater and Janak, the virial theorem, (1.4) of Chapter 1.

Following Connolly,[14] using the complete orthonormal set of Lowdin's natural orbitals defined as the eigenstates of the one-particle density matrix, $P(\mathbf{r}, \mathbf{r}')$,

$$\int d\mathbf{r}' P(\mathbf{r}, \mathbf{r}') \xi_\lambda(\mathbf{r}') = n_\lambda \xi_\lambda(\mathbf{r}) \quad (2.72)$$

where

$$P(\mathbf{r}, \mathbf{r}') = N \int \Psi(\mathbf{r}, \mathbf{r}_2 \cdots \mathbf{r}_N)\Psi^*(\mathbf{r}', \mathbf{r}_2 \cdots \mathbf{r}_N) d\mathbf{r}_2 \ldots d\mathbf{r}_N \tag{2.73}$$

with Ψ being the ground state of the many-body problem, we may get an alternative set of equations for the one-particle states. The eigenvalues n_λ, for the Fermi system lie between 0 and 1. One then has, for the ground-state energy functional

$$E_v[n] = \sum_{\lambda(\mathrm{occ})} n_\lambda \int d\mathbf{r}\, \xi_\lambda^*(\mathbf{r})\left(-\frac{\hbar^2}{2m}\nabla^2\right)\xi_\lambda(\mathbf{r}) + \int v(\mathbf{r})n(\mathbf{r})d\mathbf{r} + E_{xc}[n] + \frac{e^2}{2}\int\int \frac{n(\mathbf{r})n(\mathbf{r}')d\mathbf{r}\, d\mathbf{r}'}{|\mathbf{r}-\mathbf{r}'|} \tag{2.74}$$

where

$$n(\mathbf{r}) = \sum_{\lambda(\mathrm{occ})} n_\lambda |\xi_\lambda(\mathbf{r})|^2 \tag{2.75}$$

and the constraint now is

$$\sum_{\lambda(\mathrm{occ})} n_\lambda = N, \quad \text{with} \quad 0 \leq n_\lambda \leq 1 \tag{2.76}$$

Note that the eigenfunctions are denoted by a different symbol than the one used in (2.64) for the purpose of clarity. Also, this exhibits the difference in the procedure clearly. One then obtains the eigenvalue equation

$$\left\{-\frac{\hbar^2}{2m}\nabla^2 + V_{\mathrm{eff}}[\mathbf{r}, n]\right\}\xi_\lambda(\mathbf{r}) = \epsilon_\lambda \xi_\lambda(\mathbf{r}) \tag{2.77}$$

as before but the construction of density, $n(\mathbf{r})$ is to be via (2.75). With this new construction, one has

$$T_s[n] = \sum_{\lambda(\mathrm{occ})} n_\lambda t_\lambda \tag{2.78}$$

where

$$t_\lambda = \int \xi_\lambda^*(\mathbf{r})\left(-\frac{\hbar^2}{2m}\nabla^2\right)\xi_\lambda(\mathbf{r})d\mathbf{r} \tag{2.79}$$

Following Janak's procedure[24] as before we obtain for the ground-state energy

$$E = \sum_{\lambda(\mathrm{occ})} n_\lambda \epsilon_\lambda - \frac{e^2}{2}\int\int \frac{n(\mathbf{r})n(\mathbf{r}')}{|\mathbf{r}-\mathbf{r}'|} d\mathbf{r}\, d\mathbf{r}' + E_{xc}[n] - \int n(\mathbf{r})\frac{\delta E_{\lambda c}[n]}{\delta n(\mathbf{r})} d\mathbf{r} \tag{2.80}$$

with $n(\mathbf{r})$ given by (2.75) and n_λ are subject to the conditions (2.76). This expression is slightly different from Janak's in that we have given here an interpretation of the generalized expression that Janak employed for E, where n_is were introduced with expression (2.68) for establishing a very important property concerning the eigenvalues ϵ_i. In this representation, we obtain

$$\frac{\partial E}{\partial n_\lambda} = \epsilon_\lambda \tag{2.81}$$

independent of the form of $E_{xc}[n]$. From the preceding discussion, it follows that the Fermi energy of the many-electron system and the chemical potential $(E_N - E_{N-1})$ are equal because ϵ_λ for the highest occupied state will change only very little on taking away an extra electron from a system that has already a large number N of electrons. The preceding demonstration is different from Janak's in that we introduced fractional occupation numbers via the Löwdin orbitals whereas Janak introduced them phenomenologically. In this way (2.80) represents the ground-state energy of an N-particle system from which we can find a continuous connection with that of an $(N+1)$-particle system, by means of a gradually increasing fraction of an electron into the lowest unoccupied level of the N-particle system. One may then follow the ideas of the Slater transition state, and so on, discussed earlier for the general density-functional theory. The preceding fractional occupancy n_λ takes on an interesting form when we extend our theory to the finite temperature, equilibrium situation as was done by Kohn and Sham.[19] We give this generalization now.

We first rewrite (2.31) for the grand potential in the following way.

$$\Omega = \int v(\mathbf{r})n(\mathbf{r})d\mathbf{r} + \tfrac{1}{2}e^2 \int\int \frac{n(\mathbf{r})n(\mathbf{r}')}{|\mathbf{r}-\mathbf{r}'|} d\mathbf{r}\, d\mathbf{r}' - \mu \int n(\mathbf{r})d\mathbf{r} + G[n] \tag{2.82}$$

where $G[n]$ is a unique functional of density at a given temperature τ. μ is the chemical potential. The variational principle now states that Ω is a minimum at the correct n. We now write

$$G[n] = G_s[n] + \Omega_{xc}[n] \tag{2.83}$$

where

$$G_s[n] \equiv T_s[n] - \tau S_s[n] \tag{2.84}$$

with $T_s[n]$ and $S_s[n]$ being, respectively, the kinetic energy and entropy of noninteracting electrons with density $n(\mathbf{r})$ at a temperature τ. The value $\Omega_{xc}[n]$ is the exchange and correlation contribution to the Gibbs free energy. The minimum principle now leads to the equation

$$0 = v(\mathbf{r}) + e^2 \int \frac{n(\mathbf{r}')}{|\mathbf{r}-\mathbf{r}'|} d\mathbf{r}' + \frac{\delta G_s[n]}{\delta n(\mathbf{r})} + \frac{\delta \Omega_{xc}[n]}{\delta n(\mathbf{r})} - \mu \tag{2.85}$$

This equation is identical to the corresponding equation for a system of noninteracting electrons moving in an effective potential, $V_{\text{eff}}[\mathbf{r}; n]$, defined by

$$V_{\text{eff}}[\mathbf{r}; n] = v(\mathbf{r}) + e^2 \int \frac{n(\mathbf{r}')d\mathbf{r}'}{|\mathbf{r}-\mathbf{r}'|} + \frac{\delta\Omega_{xc}[n]}{\delta n(\mathbf{r})} \tag{2.86}$$

Its solution, as before, is determined by the following set of equations,

$$\left\{-\frac{\hbar^2}{2m}\nabla^2 + V_{\text{eff}}[\mathbf{r}; n]\right\}\phi_i(\mathbf{r}) = \epsilon_i\phi_i(\mathbf{r}) \tag{2.87}$$

and

$$n(\mathbf{r}) = \sum_i \frac{|\phi_i(\mathbf{r})|^2}{\{\exp(\epsilon_i - \mu)/k_B\tau + 1\}} \tag{2.88}$$

μ is determined by the total number of particles from (2.88) as usual. This value also is our approximation for μ of the interacting system. A different scheme, using the finite temperature version of the Löwdin orbitals, can be set up as was done before, where now $\{\xi_\lambda(\mathbf{r})\}$ could be the eigen-solutions of the finite temperature version of (2.72) with the one-particle density matrix, $P(\mathbf{rr}')$, defined correspondingly at finite τ. A useful quantity of interest is the entropy functional $S[n]$, which can now be obtained from thermodynamics and (2.82). Let V be the volume of the system.

$$\begin{aligned} S[n] &\equiv -\frac{\partial}{\partial\tau}(\Omega + \mu N)_V \\ &= -\int \left(V_{\text{eff}}[\mathbf{r}; n] + \frac{\delta G_s[n]}{\delta n(\mathbf{r})}\right)\left(\frac{\partial n(\mathbf{r})}{\partial\tau}\right)_V d\mathbf{r} - \left(\frac{\partial G[n]}{\partial\tau}\right)_{n(\mathbf{r}),V} \end{aligned} \tag{2.89}$$

The stationary property of Ω expressed by (2.85), combined with the constraint that $N = \int n(\mathbf{r})d\mathbf{r}$, leads us to the result

$$S[n] = -\left(\frac{\partial G[n]}{\partial\tau}\right)_{n(\mathbf{r}),V} \tag{2.90}$$

The argument repeated for the case of noninteracting electrons of density $n(\mathbf{r})$ gives, for its entropy,

$$S_s[n] = -\left(\frac{\partial G_s[n]}{\partial n(\mathbf{r})}\right)_{n(\mathbf{r}),V} \tag{2.91}$$

Thus, finally we arrive at the formula

$$S[n] = S_s[n] - \left(\frac{\partial\Omega_{xc}[n]}{\partial\tau}\right)_{n(\mathbf{r}),V} \tag{2.92}$$

These results can be collected together in the form of a theorem.

Theorem 8. The equilibrium state of the inhomogeneous interacting electrons at a temperature τ may be described by a system of one-particle self-consistent equations of the Schrödinger type with the effective self-consistent, temperature-dependent, one-body potential $V_{\text{eff}}(\mathbf{r}; n]$ given by (2.86) where the density is determined by (2.88).

We have developed the general one-particle self-consistent equations in great detail for the simplest case. Instead of enumerating similar results for all the cases considered earlier, we merely state the results for two important cases, (*a*) the spin-density and (*b*) the relativistic situations. They are stated without proof for the zero temperature cases and the finite temperature results follow by similar procedures as previously.

Theorem 9. The nondegenerate ground state of an inhomogeneous interacting electron system can be described by a set of one-particle self-consistent equations of the Pauli type with the effective self-consistent one-body spinor potential given by

$$V_{\text{eff}}[\mathbf{r}; n, \boldsymbol{\sigma}] = \left(v(r) + e^2 \int \frac{n(\mathbf{r}')}{|\mathbf{r}-\mathbf{r}'|} d\mathbf{r}' + \frac{\delta E_{xc}[n, \boldsymbol{\sigma}]}{\delta n(\mathbf{r})}\right)\tau_0 + \left(\mathbf{H}(\mathbf{r}) + \frac{\delta E_{xc}[n, \boldsymbol{\sigma}]}{\delta \boldsymbol{\sigma}(\mathbf{r})}\right) \cdot \boldsymbol{\tau} \tag{2.93}$$

where

$$n(\mathbf{r}) = \sum_{i(\text{occ})} \text{tr}(\phi_i^*(r)\tau_0\phi_i(\mathbf{r})) \tag{2.94a}$$

$$\boldsymbol{\sigma}(\mathbf{r}) = \sum_{i(\text{occ})} \text{tr}(\phi_i^*(\mathbf{r})\boldsymbol{\tau}\phi_i(\mathbf{r})) \tag{2.94b}$$

The $\{\phi_i(\mathbf{r})\}$ obey the Pauli equation

$$\left\{-\frac{\hbar^2}{2m}\tau_0\nabla^2 + V_{\text{eff}}[\mathbf{r}; n, \boldsymbol{\sigma}]\right\}\phi_i(\mathbf{r}) = \epsilon_i\phi_i(\mathbf{r}) \tag{2.95}$$

Theorem 10. The nondegenerate ground state of an inhomogeneous interacting relativistic system of electrons can be described by a set of one-particle self-consistent equations of the Dirac type with the effective one-particle, self-consistent, one-body spinor potential given by

$$V_{\text{eff}}[\mathbf{r}; J_\mu(\mathbf{r})] = -e\left(v(\mathbf{r}) + \frac{\delta G[J_\mu(\mathbf{r})]}{\delta J_0(\mathbf{r})}\right) - \frac{e}{c}\left(\mathbf{A}(\mathbf{r}) + \frac{\delta G[J_\mu(\mathbf{r})]}{\delta \mathbf{J}(\mathbf{r})}\right) \cdot \alpha \tag{2.96}$$

where

$$J_\mu(\mathbf{r}) = c \sum_{i(\text{occ})} \text{tr}(\phi_i^*(r)\gamma_0\gamma_\mu\phi_i(\mathbf{r})) \tag{2.97}$$

and $\{\phi_i(\mathbf{r})\}$ obey the Dirac equation

$$\{(-i\hbar c\alpha \cdot \nabla + \beta mc^2) + V_{\text{eff}}[\mathbf{r}; J_\mu]\}\phi_i(\mathbf{r}) = \epsilon_i\phi_i(\mathbf{r}) \tag{2.98}$$

This construction maintains the continuity equation $\nabla_k \cdot \mathbf{J}_k(\mathbf{r}) = 0$ as required by Theorem 6. In (2.96) $G[\mathbf{r}; J_\mu]$ is a universal functional which stands for the expectation value of the interaction energy,

$$G[J_\mu] = \left\langle \Psi \middle| \left\{ H_{\text{em}} - \frac{e}{c} \int d\mathbf{r}\text{: } \mathbf{j}(\mathbf{r}t) \cdot \mathbf{A}(\mathbf{r}t)\text{:} \right.\right.$$
$$\left.\left. + \frac{e^2}{2} \int\int d\mathbf{r}\, d\mathbf{r}'\text{: } \psi^\dagger(\mathbf{r}t)\psi(\mathbf{r}t)\text{: } \frac{1}{|\mathbf{r}-\mathbf{r}'|}\text{: } \psi^\dagger(\mathbf{r}'t)\psi(\mathbf{r}'t)\text{: } \right\} \middle| \Psi \right\rangle \tag{2.99}$$

in the notation of (2.55).

It may be noted that the construction of the universal functional E_{xc} or Ω_{xc} is central to the formalism developed here. Its explicit construction is almost impossible and approximation schemes become essential at this stage. It would be useful to construct an expression for E_{xc} which has built in it at least some exact features of the quantum mechanics of the many-body system. In (2.15), we expressed the universal functional $F[n]$ in terms of the pair-correlation functions of the system. The pair-correlation function obeys some important properties in general and so it appears useful to exploit this further. The form given by (2.15) is inconvenient because of the appearance of $T[n]$ which is a very difficult quantity to calculate for an interacting system. For this reason, one requires an alternate expression for F. Such a scheme was developed by Ambladh[43] and independently by Gunnarsson and Lundquist.[44] We give here a generalization of this scheme to finite temperatures derived by Rajagopal and von Barth[45] for the spinor case. The derivation, however, will be given for the simple case. Similar results can be derived for other situations. The basis of the derivation is a complete use of the theorem, namely, that the correct density determines the external potential and vice versa. Let

$$H(\theta) = \mathcal{T} + V + \theta W \tag{2.100}$$

where θ denotes the strength of the electron interactions with $0 \leqslant \theta \leqslant 1$. Let $n(\mathbf{r})$ be the density for the full strength of the interaction, $\theta = 1$. According to Theorem 2, there exists a well-defined one-body external potential $V[n; \theta]$ (written here in operator form)

$$V[n; \theta] \equiv \int v[n, \theta; \mathbf{r}]\psi^\dagger(\mathbf{r})\psi(\mathbf{r})d\mathbf{r} \tag{2.101}$$

such that the Gibbs potential associated with

$$\tilde{H}(\theta) = H(\theta) + V[n; \theta] \tag{2.102}$$

is a minimum for the equilibrium statistical density matrix $P_0(\theta)$, associated with $H(\theta)$, and has the same density $n(\mathbf{r})$ for every θ. In other words, in order to maintain $n(r)$, we need to develop a one-body potential given by (2.101). Equivalently, $P_0(\theta)$ is such that, for all θ,

$$\mathrm{Tr}\, P_0(\theta) = 1$$

and

$$\mathrm{Tr}\{P_0(\theta)\psi^\dagger(\mathbf{r})\psi(\mathbf{r})\} = n(\mathbf{r}) \tag{2.103}$$

Clearly

$$V[n; \theta = 1] = 0 \quad \text{or} \quad v[n, \theta = 1; \mathbf{r}] \equiv 0 \tag{2.104}$$

by this construction. Then, the grand potential $\Omega(\theta)$ is

$$\begin{aligned}\Omega(\theta) &= \mathrm{Tr}\, P_0(\theta)\left\{\tilde{H}(\theta) - \mu N + \frac{1}{\beta}\ln P_0(\theta)\right\} \\ &= \frac{1}{\beta}\ln \mathrm{Tr}\{\exp - \beta(\tilde{H}(\theta) - \mu N)\}\end{aligned} \tag{2.105}$$

Now, from (2.105), we have

$$\begin{aligned}\frac{d}{d\theta}\Omega(\theta) &= \mathrm{Tr}\left\{\left(\frac{d}{d\theta}\tilde{H}(\theta)\right)P_0(\theta)\right\} \\ &= \mathrm{Tr}\left\{\left(W + \frac{dV[n;\theta]}{d\theta}\right)P_0(\theta)\right\} \\ &= \mathrm{Tr}(WP_0(\theta)) + \int \left(\frac{d}{d\theta}v[n, \theta; \mathbf{r}]\right)n(\mathbf{r})d\mathbf{r}\end{aligned}$$

after using (2.101) and the definition of $n(\mathbf{r})$. Thus,

$$\frac{d}{d\theta}\Omega(\theta) = \mathrm{Tr}(P_0(\theta)W) + \frac{d}{d\theta}\int v[n, \theta; \mathbf{r}]n(\mathbf{r})d\mathbf{r} \tag{2.106}$$

Integrate both sides of this equation w.r.t. θ from 0 to 1 and use definition (2.3c) for W to obtain

$$\begin{aligned}\Omega = \Omega(0) &+ \frac{e^2}{2}\int_0^1 d\theta' \iint \frac{d\mathbf{r}\, d\mathbf{r}'}{|\mathbf{r} - \mathbf{r}'|}\mathrm{Tr}(P_0(\theta')\psi^\dagger(\mathbf{r})\psi^\dagger(\mathbf{r}')\psi(\mathbf{r}')\psi(\mathbf{r})) \\ &- \int v[n, \theta = 0; \mathbf{r}]n(\mathbf{r})d\mathbf{r}\end{aligned} \tag{2.107}$$

because of (2.104). Now,

$$\Omega(0) = \mathrm{Tr}\, P_0(\theta = 0)\left(\mathcal{T} + V + V[n; \theta = 0] - \mu N + \frac{1}{\beta}\ln P_0(\theta = 0)\right) \tag{2.108}$$

By construction, $P_0(\theta = 0)$ is such that it gives the same density, $n(\mathbf{r})$, so that

$$\mathrm{Tr}\, P_0(\theta = 0)V[n; \theta = 0] = \int v[n, \theta = 0; \mathbf{r}]n(\mathbf{r})d\mathbf{r}$$

and thus we finally obtain,

$$\Omega[n] = \left\{T_0[n] + \frac{1}{\beta}\operatorname{Tr}(P_0(\theta=0)\ln P_0(\theta=0))\right\} + \int v(\mathbf{r})n(\mathbf{r})d\mathbf{r} - \mu\int n(\mathbf{r})d\mathbf{r}$$
$$+\frac{e^2}{2}\int_0^1 d\theta' \int\int \frac{d\mathbf{r}\,d\mathbf{r}'}{|\mathbf{r}-\mathbf{r}'|} n(\mathbf{r})n(\mathbf{r}')g_{\theta'}(\mathbf{r},\mathbf{r}') \tag{2.109}$$

The first two terms are the grand potential for the *noninteracting* system with density $n(\mathbf{r})$, and the pair-correlation function $g_{\theta'}(\mathbf{r},\mathbf{r}')$ is defined as before

$$n(\mathbf{r})n(\mathbf{r}')g_{\theta'}(\mathbf{r},\mathbf{r}') \equiv \operatorname{Tr}\{P_0(\theta')\psi^\dagger(\mathbf{r})\psi^\dagger(\mathbf{r}')\psi(\mathbf{r}')\psi(\mathbf{r})\} \tag{2.110}$$

The first two terms in (2.109) may be identified with $G_s[n]$ of (2.83) and $\Omega_{xc}[n]$ is given by

$$\Omega_{xc}[n] = \frac{e^2}{2}\int_0^1 d\theta' \int\int \frac{d\mathbf{r}\,d\mathbf{r}'}{|\mathbf{r}-\mathbf{r}'|} n(\mathbf{r})n(\mathbf{r}')(g_{\theta'}(\mathbf{r},\mathbf{r}')-1) \tag{2.111}$$

This result was given by Ambladh[43] and Gunnarsson and Lundqvist.[44] In the spin-density situation, Rajagopal and von Barth[45] obtained

$$\Omega_{xc}[n,\boldsymbol{s}] = \tfrac{1}{2}e^2\int_0^1 d\theta' \int\int \frac{d\mathbf{r}\,d\mathbf{r}'}{|\mathbf{r}-\mathbf{r}'|}\sum_{\sigma,\sigma'} n(\mathbf{r}\sigma)n(\mathbf{r}'\sigma')(g_{\theta'}(\mathbf{r}\sigma,\mathbf{r}'\sigma')-1) \tag{2.112}$$

There are a few general properties that the pair-correlation functions $g_\theta(\mathbf{r}\sigma,\mathbf{r}'\sigma')$ obey[45,46] and we list them here for convenience.

(1) $g_\theta(\mathbf{r}\sigma,\mathbf{r}'\sigma')$ is nonnegative. More explicitly

$$g_\theta(\mathbf{r}\sigma,\mathbf{r}\sigma)=0, \qquad \text{and}$$
$$g_\theta(\mathbf{r}\sigma,\mathbf{r}\sigma')\geqslant 0, \qquad \text{for } \sigma\neq\sigma' \tag{2.113}$$

(2) They obey simple sum rules at zero temperature (and a little more complicated ones at $\tau \neq 0$°K):

$$\int d\mathbf{r}'n(\mathbf{r}'\sigma')(g_\theta(\mathbf{r}\sigma,\mathbf{r}'\sigma')-1) = -\delta_{\sigma\sigma'}$$
$$\int d\mathbf{r}n(\mathbf{r}\sigma)(g_\theta(\mathbf{r}\sigma,\mathbf{r}'\sigma')-1) = -\delta_{\sigma\sigma'} \tag{2.114}$$

These relations are the bases for the concept of "Coulomb" and "exchange" holes which are discussed in the next section.

(3) They obey the symmetry property:

$$g_\theta(\mathbf{r}\sigma,\mathbf{r}'\sigma') = g_\theta(\mathbf{r}'\sigma',\mathbf{r}\sigma) \qquad \text{for all } \sigma,\sigma' \tag{2.115}$$

Essentially similar relations hold for the Boson case. Besides these, there are some "cusp" conditions that these functions obey.[46] Define $\mathbf{r}_1 - \mathbf{r}_2 = \mathbf{r}$, $\mathbf{r}_1 + \mathbf{r}_2 = 2\mathbf{R}$, and write

$$g_\theta(\mathbf{r}_1\sigma_1, \mathbf{r}_2\sigma_2) \equiv g_\theta(\mathbf{r}, \mathbf{R}; \sigma_1\sigma_2) \tag{2.116}$$

Then, we have

(4)

$$\frac{\partial}{\partial|\mathbf{r}|} g_\theta(\mathbf{r}, \mathbf{R}; \sigma_1\sigma_2)|_{|\mathbf{r}|\to 0} = 0 \tag{2.117a}$$

$$\frac{\partial^2}{\partial|\mathbf{r}|^2} g_\theta(\mathbf{r}, \mathbf{R}; \sigma_1\sigma_1)|_{|\mathbf{r}|\to 0} = \frac{2}{3}\frac{a_0}{\theta}\frac{\partial^3}{\partial|\mathbf{r}|^3} g_\theta(\mathbf{r}, \mathbf{R}; \sigma_1\sigma_1)|_{|\mathbf{r}|\to 0} \tag{2.117v}$$

where a_0 is the Bohr radius $= \hbar^2/me^2$.

(5) For $\sigma_1 \neq \sigma_2$,

$$g_\theta(\mathbf{r}, \mathbf{R}; \sigma_1\sigma_2)|_{|\mathbf{r}|\to 0} = \frac{a_0}{\theta}\frac{\partial}{\partial|\mathbf{r}|} g_\theta(\mathbf{r}, \mathbf{R}; \sigma_1\sigma_2)|_{|\mathbf{r}|\to 0} \tag{2.118}$$

These conditions are a consequence of the Coulomb repulsion between electrons.

In this section, we have collected together all the general results known to the author concerning the formalism, so that, in developing the approximation schemes and applying them for specific problems, the reader may become aware of the simplifications involved. In the next subsection, we make a few general comments about the formalism in relation to questions of many-body theory.

H. Relation to Many-body Theory

The formalism and the method developed in the last few subsections are exact for a many-body system in thermodynamic equilibrium at a finite temperature or for a nondegenerate ground state at zero temperature, as long as no approximation is made concerning the universal exchange-correlation energy of the system. The question then arises as to the nature of the connection of this theory with the other exactly known aspects of the many-body problem. Most importantly, in the density-functional formalism, the equilibrium density (or the spin-density or the four-current density) of the system plays the part of the total many-body wave function of the system. The many-body wave function obeys a linear Schrödinger equation in many variables whereas one has a nonlinear functional equation determining the density. In the density-functional approach, the necessary quantum mechanical operations have all been carried out and the result is expressed in terms of the density variable. The relationship of the approximate wave function to the corresponding approximation made in the density-functional theory becomes impossible to discern. There are other methods of obtaining approximate solutions which are commonly used in physics and chemistry which share this basic objection. Ohrn[47] has reviewed some of these. Two of the most important methods that share the popularity of the density-functional method are the density-matrix and the Green function method.[47] In the density-matrix method, the total energy of the Hamiltonian concerned here can be expressed in terms of a second-order reduced density matrix, whereas in the Green function method, it can be expressed in terms of a one-particle Green function. These functions themselves obey a hierarchical set of equations in general and one usually employs some kind of truncation of the hierarchy at some convenient stage. This can lead to basic troubles with positive definiteness of the underlying probabilities in the theory as has recently been pointed out by Rajagopal and Sudarshan.[48] For the energy expression to be the correct quantum mechanical expectation value, the second-order reduced density matrix must be derivable from a many-body wave function which in turn implies that there must be constraints on the functions employed for the purpose. Similar comments hold for the Green function method as was noted by Baym.[37] In the density-matrix method, this problem has become known as the N-representability problem, and so far no practical approach has been proposed for the resolution of the question even though much formal progress has been achieved.[47] In the density functional scheme, one employs a much smaller subset of functions compared to the more general functions such as the one-particle Green function or density matrix. If one does not want to employ variational methods and hence the boundedness property of the approximation scheme is given up, one may use schemes where the constraint conditions and the consistency criteria are obeyed by the suggested approximations. In the density-functional scheme too, one uses approximations where the total energy functional itself is approximated in some "reasonable" ways. One then gives up the original variational principle since the approximate energy functional no longer corresponds to the exact one. It is in the nature of the problem that if progress has to be made, one must resort to approximations which physically and mathematically make "reasonable" sense. We discuss some of these schemes in the next section.

It has been known for some time that the ground state of an interacting many-electron system has a well-defined Fermi surface and that one can define one-electron states with infinite lifetime at the Fermi energy. Away from the Fermi energy, the one-electron states have a finite lifetime which goes to infinity inversely as the square of the energy difference between the one-electron energy and the Fermi energy. Moreover, the occupation number for the one-electron states does not exactly obey the Fermi-distribution law as for the noninteracting electrons but exhibits a long tail for energies outside the Fermi surface and is less than unity just below it. In other words, the occupation numbers n_i, with $0 \leq n_i \leq 1$, introduced by Slater to represent the one-electron distribution in his $X\alpha$ scheme and by Janak in the density-functional theory takes partially into account this many-electron result. It is important to realize for this reason and others already mentioned, that the one-electron picture in the Kohn–Sham formalism is strictly a mathematical trick as was emphasized by them, and the one-electron eigenvalues ϵ_i defined in this theory have well-defined interpretation only at the Fermi energy. This question is considered in the next section where an approximation scheme will be outlined for handling this aspect of the one-electron states, and again in Section IV when we try to outline a time-dependent extension of the density-functional formalism.

Before closing this section, we would like to mention an interesting development in the density-functional formalism. Srebrenik and coworkers[49] have developed a quantum mechanical description of parts of a molecular system in a subspace of the whole system through the use of a quantum action principle. This idea is attractive because these authors noted that fragments of molecular systems have characteristic sets of properties which can be defined within the framework of quantum mechanical principles. In particular, these authors show that zero-flux surfaces define such subspaces where the virial theorem can be applied as well as the variational principles, and the Hohenberg–Kohn theorem is valid. It remains to be seen if one can develop schemes to exploit these interesting ideas into practical methods for calculating properties of systems.

III. PRACTICAL SCHEMES

In this section, we develop in some detail a few of the practical schemes that have been devised and employed in actual calculations of the properties of physical systems. In essence, one requires an approximation for the exchange-correlation contribution to the total energy of the system whose first variational derivative enters in determining the one-particle orbitals of the system. It is, according to the theorem proved, a universal functional. One of the most widely used approximation is the *local density scheme* first suggested by Kohn and Sham[19] which is strictly valid for a slowly varying charge density. Gradient corrections were soon incorporated because one felt that they were important; this correction is most exhaustively studied in the surface problems. We discuss this in Section V. The present consensus is that the series of gradient terms is too slowly convergent and unwieldy; other alternatives are being suggested, which are discussed in this section.

We do not review the various techniques employed by different groups in solving the self-consistent equations. For the sake of completeness we mention that Slater and his collaborators have developed the multiple-scattering method for the $X\alpha$ scheme and have used it for a large variety of problems, a review of which may be found in Connolly's article[14] or in Wood.[50] The IBM group (Janak, Williams, and Moruzzi) employs a Korringa–Kohn–Rostoker method and the details may be found in Ref. 21. Callaway and his co-workers employ a self-consistent linear combination of Gaussian orbitals method and some of their recent efforts are described in Refs. 51, 52. Andersen[53] has reviewed some of the advances made by his group.

A. Local Density Schemes

Kohn and Sham[19] suggested a particularly intuitive approximation for $E_{xc}[n]$

$$E_{xc}[n] \simeq \int d\mathbf{r}\, n(\mathbf{r})\epsilon_{xc}^{(h)}(n(r)) \tag{3.1}$$

where $\epsilon_{xc}^{(h)}(n)$ is the contribution of exchange and correlation to the total energy per electron in a homogeneous, interacting, electron gas of density n. The function $\epsilon_{xc}^{(h)}(n)$ is itself known only approximately. Almost all the calculations to date make this approximation with different forms for $\epsilon_{xc}^{(h)}(n)$. The approximation implied in (3.1) is that the energy of a given electron is lowered by the tendency of the other electrons to move out of its path. To determine this quantitatively involves many subtle processes, and the most important aspect of (3.1) is that it exploits the considerable knowledge of these effects in the homogeneous electron gas, some of which is recently reviewed by Rajagopal.[54] The local density approximation, (3.1), assumes that the exchange-correlation contribution to the total energy can be cumulatively added from each portion of a nonuniform gas as if it were locally uniform. It cannot be formally justified for inhomogeneous systems such as atoms. The only justification for its use in atoms is empirical, in that the numerical results obtained score impressive successes as is discussed in Section V. From (2.108), in the zero temperature limit, with which we are concerned mostly in the discussions to follow, we have, in general,

$$E_{xc}[n] = \frac{e^2}{2}\int_0^1 d\theta' \int\int \frac{d\mathbf{r}\, d\mathbf{r}'}{|\mathbf{r}-\mathbf{r}'|} n(\mathbf{r})n(\mathbf{r}')\cdot(g_{\theta'}(\mathbf{r},\mathbf{r}')-1)$$

The local density approximation consists of rewriting this in the form[44]

$$E_{xc}^{LD}[n] \simeq \frac{e^2}{2} \int d\mathbf{r} n^2(\mathbf{r}) \int \frac{d\mathbf{r}'}{|\mathbf{r}-\mathbf{r}'|} \int_0^1 d\theta' (g_{\theta'}^{(h)}(|\mathbf{r}-\mathbf{r}'|; n) - 1) \tag{3.2}$$

where $g_{\theta'}^{(h)}(|\mathbf{r}-\mathbf{r}'|; n)$ is the pair correlation function of the homogeneous electron gas of density n. In the homogeneous electron gas, the spatial dependence of $g_{\theta'}^{(h)}(|\mathbf{r}|; n)$ is such that $\mathbf{r}k_F$ is a dimensionless variable where k_F is the Fermi momentum of the electron gas such that the electron density is given by the usual formula $n = k_F^3/3\pi^2$. Then, a simple scaling of the integral involving $g_{\theta'}^{(h)}$ leads us to the following structure for $E_{xc}^{LD}[n]$:

$$E_{xc}^{LD}[n] \simeq \frac{e^2}{2} \int d\mathbf{r} n^{4/3}(\mathbf{r}) \alpha(n(\mathbf{r})) \tag{3.3}$$

where

$$\alpha(n) = \frac{4\pi}{(3\pi^2)^{2/3}} \int_0^\infty x\, dx \int_0^1 (g_\theta^{(h)}(x; n) - 1) d\theta \tag{3.4}$$

Specializing (2.114) for the paramagnetic case, and for the homogeneous gas, where the only pair-correlation function of interest introduced previously is the spin-averaged functions, given by $\frac{1}{4}\sum_{\sigma,\sigma'} (g_0(\mathbf{r}\sigma; \mathbf{r}'\sigma') - 1)$, and $n(\mathbf{r}, \sigma)$ is $\frac{1}{2}n(\mathbf{r})$, we have the condition

$$n \int d\mathbf{r}' (g_\theta^{(h)}(|\mathbf{r}-\mathbf{r}'|; n) - 1) = -1 \tag{3.5a}$$

In the notation used previously, this may be rewritten as

$$\int_0^\infty x^2\, dx (g_\theta^h(x; n) - 1) = -\tfrac{3}{4}\pi \qquad \text{for all } \theta \tag{3.5b}$$

We discuss the existing schemes in terms of these general expressions. They bring out clearly the basic assumptions of the local density approximation. The nonlocal character of the exchange-correlation contribution due to the probability distribution of a pair of electrons in the electron system is reduced to a *local* form. McWeeny[55] has recently discussed the correlation problem in atoms, molecules, and solids. Gunnarsson and Lundqvist[44] discuss the same question in relation to the density-functional methods. Some of these are discussed in Section V. For now, we make some general remarks concerning the theoretical basis for the local density approximation. In the Hartree–Fock–Slater approximation, $g_\theta^{(h)}(x; n) - 1$, can be calculated explicitly and we know that this approximation takes into account only parallel-spin correlations. The better approximations therefore include many other processes, most important of which is the antiparallel-spin contributions. The HFS expression is

$$(g_\theta^{(h)}(x; n) - 1)_{HFS} = -\frac{9\pi}{4x^3} J^2_{3/2}(x) \tag{3.6}$$

where $J_{3/2}(x)$ is the Bessel function of order 3/2. This function obeys the sum rule, (3.5*b*), and $\alpha(n)$ in this case is a constant independent of n, given by

$$\alpha_{HFS}(n) = -3(3\pi^2)^{1/3} \int_0^\infty \frac{dx}{x^2} J^2_{3/2}(x) = -\frac{3}{2\pi}(3\pi^2)^{1/3} \tag{3.7}$$

This leads to the celebrated $n^{1/3}$ approximation of Dirac,[4] Gaspar,[16] Kohn and Sham.[19] This differs from Slater's $n^{1/3}$ scheme in a numerical prefactor as discussed earlier. Slater's $X\alpha$ method is to take $\alpha(n(\mathbf{r}))$ in (3.3) to be an adjustable constant independent of n, to obtain the best fit. A very important physical picture emerges from the HFS result for the pair correlation function. The physical interpretation of $g(\mathbf{x}; n)$ is that it is the probability that if an electron is at the origin, there will be another one at $\mathbf{x}$ from it, irrespective of the spin of the two electrons. If the electrons are uncorrelated then $g(\mathbf{x}; n)$ will be unity for all $\mathbf{x}$. In HFS, one observes from (3.6) that $g(\mathbf{x}; n)$ is less than unity and the quantity $(g(\mathbf{x}; n) - 1)$ describes a deficiency of electrons at any relative position $\mathbf{x}$ of the pair of electrons. For $\mathbf{x} \to 0$, the value of $(g(\mathbf{x}; n) - 1)$ in the HFS scheme is half and so one has an "exchange hole" arising entirely out of antisymmetry of the wave functions. It is important to note that the sum rule (3.5*a*) is obeyed. The dynamical effect of Coulomb interaction is completely ignored at the HFS level. Slater has discussed this extensively[12] in setting up the ground work for his $X\alpha$ scheme. The "Coulomb hole" is not discussed as much because the long- and short-range correlation effects are difficult to incorporate in actual calculations. At the short-range level, one has the "cusp" conditions that the function $g(\mathbf{x}; n)$ obeys.[46] In systems such as molecules and certain types of solids, the long-range correlations are significant and are extremely difficult to handle. For practical calculations, one may refer to McWeeny's review.[55] From the structure of (3.6), notice that $(g(x; n) - 1)_{HFS}$ is essentially peaked at $x = 0$ and dies away beyond $x \simeq 1$. This feature perhaps continues in more sophisticated calculations of the pair-correlation function. The local density scheme thus involves making two basis assumptions: (*1*) the exchange-correlation hole is centered on the electron, and, (*2*) the electron density supporting the hole is electron-gas-like and approximately uniform on the scale of the size of the hole itself. In making the second comment, it is implied that a continuous spectrum of the energy states of electron is assumed to exist as in the uniform gas (plane waves) which go to screen the electron charge. Clearly in an atomic core this is not true,[56] even though atomic cores play a relatively minor role as atoms condense to form metals. The size of the exchange-correlation hole is about the size of the atom itself and so the preceding second assumption is not justifiable in systems made up of atoms. Gunnarsson and Lundqvist[44] made two important observations. Firstly, the quasiparticle of the local density scheme is perfectly neutral. This means that the energies associated with small charge imbalances can be quite large. Secondly, the total energy depends only on the spherical average of the exchange-correlation hole and thus is insensitive to many of the deformations of the hole caused by inhomogeneities of the system. But the total energy is quite sensitive to inhomogeneities that make the hole to be off-center, a point to note concerning the first assumption. An extreme example of this is an electron going through the surface of a metal. The hole that was centered on the electron as it approached the metal surface from within the metal separates from the electron completely, becoming the image of the electron, as the electron gets out of the metal. A similar phenomenon occurs when an electron moves to the outer parts of an atom. The hole remains in the interior of the atom where it attracts the electron with the usual inverse distance dependence of the Coulomb law. This is not obtained in local density schemes where the potential falls off exponentially instead of following the Coulomb law. This is discussed in a quantitative way in Section 5. The preceding analysis of the local density schemes was quite general and is given at the outset itself in order to bring out the underlying limitations of the scheme. We now discuss some of the actual forms used in computing $\alpha(n)$ in (3.3). Even here, one requires to make further approximations because the pair correlation functions of the electron gas are not known. This will contribute to the accuracy of the theory ultimately.

The exchange-correlation energy of an interacting electron gas is not known accurately for all densities, nor is the pair-correlation function. This problem is an old one and many people have tried to make progress in this area of many-particle physics. We do not review this here. We only quote the most significant of these results which have been adopted to the density-functional schemes. Since the spin-density-functional scheme offers a better starting point for most work in the sense that it has an additional degree of freedom for variation, we present the results for this. The paramagnetic case where only density appears as a variable is a special case of the spin-density scheme when we set the magnetization equal to zero identically. We should remark that the exchange-correlation energy of an interacting magnetic electron gas is much less well known than its paramagnetic counterpart. To make the discussion simple, we quote the local spin-density scheme, following the same steps as with the local density scheme given in detail at the beginning of this chapter. In the homogeneous electron gas, the exchange-correlation energy depends only on the magnitude of the spin-density vector, a feature that follows from time reversal and the translational symmetry of the system. Thus $E_{xc}^{LD}[n, \boldsymbol{s}] = E_{xc}^{LD}[n, |\boldsymbol{s}|]$, and we write it in the following form

$$E_{xc}^{LD}[n, |\boldsymbol{s}|] = \frac{e^2}{2} \sum_{\sigma\sigma'} \int d\mathbf{r} n_\sigma(\mathbf{r}) n_{\sigma'}(\mathbf{r}) \int \frac{d\mathbf{r}'}{|\mathbf{r}'|} H_{\sigma\sigma'}^{(h)}(|\mathbf{r}'|; n, |\boldsymbol{s}|) \tag{3.8}$$

where

$$H_{\sigma\sigma'}^{(h)}(|\mathbf{r}|; n, |\boldsymbol{s}|) \equiv \int_0^1 d\theta (g_\theta^{(h)}(|\mathbf{r}'|; \sigma\sigma'; n, |\boldsymbol{s}|) - 1) \tag{3.9}$$

The sum rules (2.114) now read

$$\int d\mathbf{r}' H_{\sigma\sigma'}^{(h)}(|\mathbf{r}'|; n, |\boldsymbol{s}|) = -\frac{1}{n_{\sigma'}} \delta_{\sigma\sigma'} \tag{3.10}$$

where n_σ is the number of electrons with their spins along the σ-direction. For $\sigma = \sigma'$, we have the result that an electron of spin σ digs a hole with its spin parallel to that of the electron and may be called the exchange hole of spin σ which is a consequence of the exclusion principle. For $\sigma \neq \sigma'$, one has a "Coulomb hole" where an electron of spin σ digs a hole of opposite spin in such a way that on the average there is an equal number of opposite spin holes. In HFS approximation, there is *no* hole of opposite spin because, in that approximation, $(H_{\sigma\sigma'}^{(h)})_{HFS}$ is zero identically for $\sigma \neq \sigma'$. This is a very important point. In HFS approximation, one then obtains merely a sum of $n_\uparrow^{4/3}$ and $n_\downarrow^{4/3}$ of the Dirac–Gaspar–Kohn–Sham form as before.

At this point, we very briefly summarize the known exact results in the electron gas theory. In the paramagnetic case, there exists a large body of work, and in the magnetic case, there is relatively speaking much less work. The work of Gell-Mann and Brueckner[57] established rigorous results in the high-density limit. Another exact result was derived by Wigner[58] in the low-density limit. Attempts to extend these to densities in the metallic range have not met with much success. One of the most important difficulties one faces in the development of the theory is that the pair correlation function violates the positivity condition in many of the proposed schemes in the metallic density region. Singwi and his collaborators[59] have recently succeeded in developing a self-consistent scheme that does not violate many of the exact conditions of the theory and also maintains positive definiteness of the pair correlation functions in the metallic density region in a reasonable manner. All this work is for the paramagnetic electron gas. The extension of this to ferromagnetic case has been proposed by Rajagopal[60] but no numerical work has so far been completed. Hedin and Lundqvist[61] used the early work of Singwi et al.[59] to parametrize the exchange-correlation energy of the electron gas in a manner suitable for its application to the density-functional scheme. von Barth and Hedin[32] extended this parametrization to the spin-density-functional case. This involved a further approximation because of the nonexistence of such calculations for magnetic electron gas. von Barth and Hedin[32] *assumed* the scaling relationship for the correlation part of the energy for the fully ferromagnetic and paramagnetic cases which exists in the Gell–Mann–Brueckner scheme to hold even in the Singwi scheme. The attractive feature of the von Barth–Hedin parametrization is that it captures the correct form of the correlation energy for the extreme high and low densities of the electron gas. In the absence of any better calculation, this parametrization has now become a standard one and several different fitting of the parameters which depend on different ways of estimating the correlation energy have been used in the literature. Introduce the dimensionless parameter r_s such that the density of the electron gas is given by

$$n = \frac{k_F^3}{3\pi^2} = \frac{1}{\frac{4}{3}\pi r_s^3 a_0^3} \tag{3.11}$$

where a_0 is the Bohr radius (= 0.5 Å). The values of r_s between 2 and 6 are appropriate for simple metals. Let ζ be the magnetization of the electron gas defined by $n_\uparrow = \frac{1}{2}n(1+\zeta)$, $n_\downarrow = \frac{1}{2}n(1-\zeta)$, with $0 \leqslant \zeta \leqslant 1$, where $n_\uparrow, n_\downarrow$ are the densities of electrons of up and down spins in the system. $\zeta = 0$ corresponds to the paramagnetic state and $\zeta = 1$ to that of the fully saturated ferromagnetic state. The parameterization is then made[32] in terms of four constants which are chosen to give a numerical fit to any approximate calculation of the exchange-correlation energy as follows:

$$E_{xc}[r_s, \zeta] = E_{xc}[r_s, \zeta = 0] + [E_{xc}[r_s, \zeta = 1) - E_{xc}[r_s, \zeta = 0]]g(\zeta) \tag{3.12}$$

Here

$$g(\zeta) = [(1+\zeta)^{4/3} + (1-\zeta)^{4/3} - 2]/(2^{4/3} - 2) \tag{3.13a}$$

$$E_{xc}(r_s, \zeta = 0) = E_x(r_s, \zeta = 0) - c_0 f\left(\frac{r_s}{r_0}\right) \tag{3.13b}$$

$$E_{xc}(r_s, \zeta = 1) = E_x(r_s, \zeta = 1) - c_1 f\left(\frac{r_s}{r_1}\right) \tag{3.13c}$$

$$f(z) = (1 + z^3) \ln\left(1 + \frac{1}{z}\right) + \tfrac{1}{2}z - z^2 - \tfrac{1}{3} \tag{3.13d}$$

$E_x(r_s, \zeta)$ is the HFS exchange energy of a magnetic electron gas given by

$$E_x(r_s, \zeta) = -\frac{3}{4\pi\alpha r_s}[(1 + \zeta)^{4/3} + (1 - \zeta)^{4/3}] \tag{3.13e}$$

with $\alpha = (4/9\pi)^{1/3}$. The Gell-Mann and Brueckner calculation involves an infinite sum of only a certain subclass of terms in the many-body perturbation series, known as the ring diagrams. The pair-correlation function can be calculated in this way, from which the correlation energy can be deduced. We have[62] numerically calculated this energy as a function of magnetization ζ and density r_s, *without* making the high-density approximation of Gell-Mann and Brueckner. Such a calculation is of interest in the present context. We have then fitted[62] this to the form given by (3.13) and the result is given in Table I. In the same table, we have given the same parameters used by other authors who employed other procedures of determining E_{xc} for the electron gas. The scaling relations mentioned earlier were used by von Barth–Hedin and Janak et al. and in the present context they are

$$\tfrac{1}{2}c_0 = c_1 \quad \text{and} \quad 2^{4/3}r_o = r_1 \tag{3.14}$$

The paramagnetic energy was then numerically fitted to the form given in (3.12) to determine the constants c_0 and c_1. von Barth–Hedin used the early work of Singwi et al. whereas Janak et al. used Vashishta–Singwi answers. Gunnarsson and Lundqvist used a plasmon model for computing the correlation energy of the magnetic electron gas, since the plasmon mode has a larger coefficient for the quadratic dispersion when the gas becomes magnetic compared to its paramagnetic state, so that the density response of the system becomes weaker. We[63] used the full

TABLE I
Parameters Obtained by Fitting Numerical Values of $E_{xc}(r_s, \zeta)^a$

	Ours (Ref. 63)	VH (Ref. 32)	GL (Ref. 44)	JMW (Ref. 64)
r_0	39.7	30	11.4	21
r_1	70.6	75	15.9	52.9
c_0	0.0461	0.0504	0.0666	0.045
c_1	0.0263	0.0254	0.0406	0.0225

[a]Present results are compared with those of von Barth and Hedin,[32] Gunnarsson and Lundqvist,[44] and Janak, Moruzzi, and Williams.[64]

ring diagram terms and unlike von Barth–Hedin, we calculated the correlation energy more carefully and accurately. We may point out that our parametrization has not been used so far in any numerical work. From Table I we observe that the coefficients differ but the actual computations of physical quantities do not seem to exhibit these differences.

We should point out that the pair-correlation functions used in the preceding calculations do not obey all the exact conditions that we exhibited earlier. As pointed out earlier, the ring diagram calculation violates the positive definiteness criterion for r_s of order 2.6 and beyond[62] whereas the Vashishta–Singwi calculation exhibits this defect for r_s of order 4 and above. The cusp conditions are not obeyed in the ring-diagram calculation[46] and we suspect this to be the case with Vashishta–Singwi calculation as well.

Recently McWeeny[55] proposed an interesting way of calculating the correlation energy of the electron gas in the paramagnetic state. This seems to hold some promise and we[46] have extended this scheme to the magnetic state, so that the scheme may be employed in the spin-density-functional formalism in the future. We present here a slightly different version of the McWeeny scheme. The main idea is to incorporate the cusp conditions as well as the sum rules that the pair-correlation functions obey. Also, we know that when the interaction strength θ is zero, we can calculate these pair-correlation functions explicitly using the determinant of plane waves which form the exact ground state of the many-electron system. This is the same as the HFS results for the pair-correlation functions, and they are:

$$\begin{aligned} g^{(h)}_{HFS}(|\mathbf{r}|;\sigma\sigma') &= 1 - 9\delta_{\sigma\sigma'}\left(\frac{\sin k_{F\sigma}|\mathbf{r}| - k_{F\sigma}r\cos k_{F\sigma}r}{k^3_{F\sigma}r^3}\right)^2 \\ &= 1 - \frac{9\pi}{2k^3_{F\sigma}r^3}J^2_{3/2}(k_{F\sigma}r)\delta_{\sigma\sigma'} \end{aligned} \tag{3.15}$$

We then suggest the following parametric forms for the pair-correlation functions with several parameters to be determined using the sum rules and the cusp conditions given earlier:

$$g^{(h)}(|\mathbf{r}|;\sigma\sigma;\theta) = g^{(h)}_{HFS}(|\mathbf{r}|;\sigma\sigma)\left[1 + e^{-\beta^2_{\sigma\sigma}r^2}\left(-1 + \alpha_{\sigma\sigma}(\theta)\left(1 + \frac{\theta r}{2a_0}\right)\right)\right] \tag{3.16}$$

$$g^{(h)}(|\mathbf{r}|;\uparrow\downarrow;\theta) = g^{(h)}_{HFS}(|\mathbf{r}|;\uparrow\downarrow)\left[1 + e^{-\beta^2_{\uparrow\downarrow}r^2}\left(-1 + \alpha_{\uparrow\downarrow}(\theta)\left(1 + \frac{\theta r}{a_0}\right)\right)\right] \tag{3.17}$$

$\beta_{\sigma\sigma}$ may be interpreted as the size of the Fermi hole of spin σ and $\beta_{\uparrow\downarrow}$ that of the Coulomb hole. We further assume $\beta_{\sigma\sigma'}(\theta)$ are independent of the strength of the interaction, θ. Equations 3.16 and 3.17 were chosen to obey the cusp conditions given in Section II. The sum rules given earlier, give us a connection between $\beta_{\sigma\sigma'}$ and $\alpha_{\sigma\sigma'}$ for a given σ, σ'. We thus have three parameters as yet undetermined. Note that

$$\alpha_{\sigma\sigma}(\theta) = \frac{I_0(\beta_{\sigma\sigma})}{I_0(\beta_{\sigma\sigma}) + \dfrac{\theta}{2a_0} I_1(\beta_{\sigma\sigma})} \tag{3.18}$$

$$\alpha_{\uparrow\downarrow}(\theta) = \frac{\frac{1}{2}\sqrt{\pi}\beta_{\uparrow\downarrow}}{\frac{1}{2}\sqrt{\pi}\beta_{\uparrow\downarrow} + \theta/a_0} \tag{3.19}$$

where

$$I_n(\beta_{\sigma\sigma}) = \int_0^\infty d\mathbf{r}\, e^{-\beta_{\sigma\sigma}^2|\mathbf{r}|^2} |\mathbf{r}|^n g_{HF}^{(h)}(\mathbf{r}; \sigma\sigma), \qquad (n = 0, 1) \tag{3.20}$$

These integrals can all be done analytically exactly. The exchange-correlation energy for the electron gas is then given by appropriate integrals over these functions:

$$E_{xc}^{(h)}(r_s; \zeta) = \frac{e^2}{2} \sum_{\sigma_1\sigma_2} n_{\sigma_1} n_{\sigma_2} \int \frac{d\mathbf{r}}{|\mathbf{r}|} \int_0^1 d\theta (g^{(h)}(\mathbf{r}; \sigma_1\sigma_2; \theta) - 1) \tag{3.21}$$

The integrals appearing in (3.21) can also be performed analytically and thus $E_{xc}^{(h)}$ is expressed in terms of three parameters. McWeeny[55] used a similar idea but did not use the θ-integration trick. He assumed that only the antiparallel spin contributions are important in determining the ground state of the paramagnetic gas and so has only one parameter to determine. He fitted the exactly known correlation energy of the helium atom to the formula derived in terms of the one parameter and thus determined the unknown quantity, $\beta_{\uparrow\downarrow}$. With this choice, he calculated the correlation energy of the electron gas for a wide range of densities which interpolated very impressively with the Gell-Mann–Brueckner high-density result and the low-density result of Wigner. In the fully saturated ferromagnetic case, one again has only one parameter and we suggest that it can be determined by fitting the correlation energy of lithium atom.[46] We have not carried out these numerical calculations yet. As noted by McWeeny,[55] the cusp behavior is important, owing to the singularity in the repulsion energy, and it is precisely this feature of some of the wave function calculations containing the proper behavior which leads to such impressive successes. McWeeny stressed that, to get really good energies, the cusp must be correct but longer-range behavior is unimportant. His model calculation of the correlation energy of the electron gas is another illustration of this. It is for this reason, we believe that our model for the correlation energy of the magnetic electron gas may be of value.

We may conclude this subsection by making the remark that most of the local density schemes do not use the θ-integration trick and therefore there have been some suggestions to improve the theory by incorporating corrections arising out of gradients of density. With the θ-integration trick, the kinetic energy contributions are taken into account exactly. The gradient corrections were shown by Hohenberg and Kohn[18] to be related to the density-response functions of all orders of the electron gas. Only the linear response function of the electron gas has been studied well and estimates of the corrections do exist in the literature. Since the consensus of opinion on the inclusion of gradient terms is that they are of dubious value because the gradient series does not seem to be convergent, one should rather put the efforts in improving the theory in the direction of a proper nonlocal density-functional scheme. We do not therefore review this aspect of the theory here. The only case where the gradient corrections were tested in some detail is in the theory of work functions, which we discuss in Section V. For a review of the various forms for the potential in the local density scheme including gradient corrections one may refer to the paper of Rennert.[65] We must add that, although the θ-integration trick helps to avoid the questions concerning the treatment of the kinetic energy contributions, in actual practice, the θ dependence of the pair-correlation function becomes the new task. This is not necessarily simple or easy.

B. Nonlocal Schemes

Any significant improvement of the local scheme perhaps lies in a nonlocal reformulation of the density-functional theory. Kohn and Sham[19] in trying to incorporate a certain class of density gradient terms suggested one such form. The observations of Gunnarsson and Lundqvist[44] concerning the necessity of treating the hole properly as discussed in the last section seems to be physically the most appealing. This involves the full use of the theorem and brings into consideration the pair-correlation function of the homogeneous system for varying interaction strength. Other exact statements concerning the pair-correlation functions listed in Section II can be incorporated, thus making the formulation theoretically satisfactory. The actual implementation of these ideas in carrying out practical calculations can be expected to appear in the near future. A different formulation of Chihara[66] in this connection may be mentioned where the author develops a formal integral equation for the density in the presence of an external potential involving the density-correlation function as an input. This work is similar in spirit to some of the approximation schemes in the theory of liquids (hypernetted approximation, etc.) and we return to a discussion of it in Section V. We here present a formulation which in our opinion contains all other similar attempts as special cases and as such enables us to discuss these other attempts in terms of this presentation. The paramagnetic case has been studied in detail by von Barth[66] and the spin-density version of it was given by Rajagopal and von Barth.[45] It should be mentioned that an early attempt in this direction with some numerical success was made by Gunnarsson et al.[67,68] and also by Alanso and Girifalco.[69,70]

Our starting point is the exact expression for the exchange-correlation energy[45] given by a slightly rewritten version of (2.112):

$$E_{xc}[n, \boldsymbol{s}] = \tfrac{1}{2}e^2 \int\int \frac{d\mathbf{r}\, d\mathbf{r}'}{|\mathbf{r}-\mathbf{r}'|}\{n(\mathbf{r})n(\mathbf{r}')H_0[\mathbf{r}, \mathbf{r}'; n, \boldsymbol{s}] + n(\mathbf{r})s_z(\mathbf{r}')H_2[\mathbf{r}, \mathbf{r}'; n, \boldsymbol{s}]$$
$$+ n(\mathbf{r}')s_z(\mathbf{r})H_1[\mathbf{r}, \mathbf{r}'; n, \boldsymbol{s}] + s_z(\mathbf{r})s_z(\mathbf{r}')H_3[\mathbf{r}, \mathbf{r}'; n, \boldsymbol{s}]\} \tag{3.22}$$

where

$$H_0 = \tfrac{1}{4}(H_{\uparrow\uparrow} + H_{\downarrow\downarrow} + H_{\uparrow\downarrow} + H_{\downarrow\uparrow}) \tag{3.23a}$$

$$H_1 = \tfrac{1}{4}(H_{\uparrow\uparrow} - H_{\downarrow\downarrow} + H_{\uparrow\downarrow} - H_{\downarrow\uparrow}) \tag{3.23b}$$

$$H_2 = \tfrac{1}{4}(H_{\uparrow\uparrow} - H_{\downarrow\downarrow} - H_{\uparrow\downarrow} + H_{\downarrow\uparrow}) \tag{3.23c}$$

$$H_3 = \tfrac{1}{4}(H_{\uparrow\uparrow} + H_{\downarrow\downarrow} - H_{\uparrow\downarrow} - H_{\downarrow\uparrow}) \tag{3.23d}$$

with $H_{\sigma\sigma'}$ related to the pair-correlation functions through the formula

$$H_{\sigma\sigma'}[\mathbf{r}\mathbf{r}'; n, \boldsymbol{s}] \equiv \int_0^1 d\theta (g_\theta[\mathbf{r}\sigma, \mathbf{r}'\sigma'; n, \boldsymbol{s}] - 1) \tag{3.24}$$

In this form, the first term represents the spin-singlet contribution to the energy whereas H_1, H_2, H_3 are the triplet contributions. In the paramagnetic case only H_0 and H_3 are nonzero whereas H_1 and H_2 are zero identically. From E_{xc}, according to the theorem, the one-particle effective potentials are given by

$$V_{xc}[\mathbf{r}; n, \boldsymbol{s}] = \frac{\delta E_{xc}}{\delta n(\mathbf{r})} \tag{3.25}$$

and

$$\mathbf{W}_{xc}[\mathbf{r}; n, \boldsymbol{s}] = \frac{\delta E_{xc}}{\delta \boldsymbol{s}(\mathbf{r})} \tag{3.26}$$

W_{xc} may be thought of as an effective space-dependent internal magnetic field due to spin correlations in the system. The various sum rules, symmetries, and other conditions on $g_{\sigma\sigma}$ can be rewritten in terms of H_i

defined previously. The functional derivatives appearing in (3.25), (3.26) involve the corresponding functional derivatives of H_i. Using the sum rules, these derivatives can be seen to obey functional integro-differential equations which when solved would give the one-particle potentials. These equations cannot be solved exactly, of course, and need a scheme for making further progress.

We illustrate our procedure by considering in some detail the calculation of V_{xc} in this scheme. This also serves as a reference scheme with which we can compare in a unified manner other similar schemes. The calculation of $\mathbf{W}_{xc}$ proceeds in a similar fashion. Thus,

$$\begin{aligned} V_{xc}[\mathbf{r}] = {} & e^2 \int \frac{1}{|\mathbf{r}-\mathbf{r}_{1'}|} n(\mathbf{r}_{1'}) H_0[\mathbf{r}, \mathbf{r}_1'] d\mathbf{r}_{1'} \\ & + e^2 \int \frac{1}{|\mathbf{r}-\mathbf{r}_{1'}|} \sigma_z(\mathbf{r}_{1'}) H_2[\mathbf{r}, \mathbf{r}_{1'}] d\mathbf{r}_{1'} \\ & + \frac{e^2}{2} \int\int \frac{d\mathbf{r}_1\, d\mathbf{r}_{1'}}{|\mathbf{r}_1-\mathbf{r}_{1'}|} \Big\{ n(\mathbf{r}_1) n(\mathbf{r}_{1'}) \frac{\delta H_0[\mathbf{r}_1, \mathbf{r}_{1'}]}{\delta n(\mathbf{r})} \\ & + n(\mathbf{r}_1) \sigma_z(\mathbf{r}_{1'}) \frac{\delta H_2[\mathbf{r}_1, \mathbf{r}_{1'}]}{\delta n(r)} + \sigma_z(\mathbf{r}_1) n(\mathbf{r}_{1'}) \frac{\delta H_1[\mathbf{r}_1, \mathbf{r}_{1'}]}{\delta n(\mathbf{r})} \\ & + \sigma(\mathbf{r}_1) \sigma_z(\mathbf{r}_{1'}) \frac{\delta H_3[\mathbf{r}_1, \mathbf{r}_{1'}]}{\delta n(\mathbf{r})} \Big\} \end{aligned} \tag{3.27}$$

The symmetry relations on the pair-correlation functions now become symmetry relations for H_is:

$$H_0[\mathbf{r}_1, \mathbf{r}_{1'}] = H_0[\mathbf{r}_{1'}, \mathbf{r}_1], \qquad H_3[\mathbf{r}_1, \mathbf{r}_{1'}] = H_3[\mathbf{r}_{1'}, \mathbf{r}_1],$$

and

$$H_1[\mathbf{r}_1, \mathbf{r}_{1'}] = H_2[\mathbf{r}_{1'}, \mathbf{r}_1] \tag{3.28}$$

and these have been used in deriving (3.27). We now use the sum rules, (2.114), after integrating both sides of these equations over θ, to obtain equations for $\delta H_i/\delta n(\mathbf{r})$. These equations cannot be solved exactly without precise knowledge of the pair-correlation functions and the densities for the inhomogeneous appearing there. To determine them, therefore, we begin with

$$\frac{\delta H_i[\mathbf{r}_1, \mathbf{r}_{1'}]}{\delta n(\mathbf{r})} = \frac{1}{2} \frac{\partial H_i^0[\mathbf{r}_1 - \mathbf{r}_{1'}]}{\partial n^0} [A_i^0[\mathbf{r}, \mathbf{r}_1] + A_i^0[\mathbf{r}, \mathbf{r}_{1'}]] \tag{3.29}$$

where $\partial H_i^0/\partial n^0$ is the partial derivative of the pair-correlation function of the homogeneous system (which is a function of the difference in coordinates) with respect to the density of the homogeneous system, n^0 (which is independent of the coordinates). The new quantities A_i^0 are now determined in terms of the known homogeneous quantities from the integral equation. The sum rules applied to the homogeneous system determine $\partial H_i^0/\partial n^0$ quite simply because we can use the simplifying features of the homogeneous system mentioned previously. In (3.29) we have introduced a symmetric combination of $\mathbf{r}_1$ and $\mathbf{r}_{1'}$. A similar scheme is made for $\partial H_i/\partial s$. Thus, we can now express V_{xc}, W_{xc} in terms of the pair-correlation functions and its partial derivatives of the homogeneous gas.

We can now make contact with other attempts to relate V_{xc} to the pair-correlation functions in approximate schemes. Sham[71] developed a scheme similar to the preceding but only for the paramagnetic case. Since he did not use the θ-integration trick, he introduced a kinetic energy contribution to the pair-correlation function for the interacting system, which is a major unknown entity. He did not use the sum rules on the pair-correlation functions but instead suggested an approximation:

$$\frac{\delta H_i[\mathbf{r}_1, \mathbf{r}_{1'}]}{\delta n(\mathbf{r})} \cong \frac{H_i^0[\mathbf{r}_1 - \mathbf{r}_{1'}]}{\partial n^0} \delta(\mathbf{r}_{1'} - \mathbf{r}) \qquad \text{etc.} \tag{3.30}$$

The preceding scheme was extended to the magnetic case by Stoddart and Hanks[72] and Kawazoe et al.[73] Stoddart and Hanks employ the Sham scheme in a straightforward way whereas Kawazoe et al. use the θ-integration trick, but these authors employ a scheme as in (3.30). It is clear from our discussion preceding that such approximations violate the sum rules required to conserve charge- and spin-densities. Niklasson et al.[74] showed that when the Sham scheme outlined previously is applied to the homogeneous system it is equivalent to the Vashishta–Singwi theory[59] of the homogeneous paramagnetic electron gas in the static long wavelength limit. Our scheme, when applied to the electron gas also goes to over the Vashishta–Singwi scheme[66] and our magnetic generalization then provides an extension of the Vashishta–Singwi scheme to the magnetic electron gas. Gunnarsson et al.[67,68] and Alanso and Girifalco[69,70] are closer in spirit to our scheme in that they employ the sum rules as an input in their theory but do not use it to obtain an equation for the functional derivative. Their development is for the paramagnetic (spinless) case only. They proceed by taking the Hartree–Fock result, (3.15), as their basic input, where they replace the Fermi momentum, k_f, by its local value, via the expression, $k_F^3/3\pi^2 = n(\mathbf{r})$. The sum rule is then used as a way of obtaining $n(\mathbf{r})$. This attractive scheme is made possible because one may introduce an auxiliary functional $\tilde{n}(\mathbf{r})$ related to $n(\mathbf{r})$ in a known way, so that by varying $\tilde{n}(\mathbf{r})$ as a separate function, one can satisfy the sum rule. The use of homogeneous pair-correlation function makes possible a simple scheme by which the sum rule can be exploited. This scheme violates the symmetry of the correlation function, (3.28); by using an unsymmetric form, these authors were able to solve for $\tilde{n}(\mathbf{r})$. However, using a symmetric form and our preceding scheme, von Barth[66] has succeeded in obtaining a better method of incorporating the nonlocal aspects of the correlations. All these authors have used their schemes for examining atomic problems and they are discussed in Section V.

The preceding schemes are all pertaining to the one-particle properties of the ground state of a general inhomogeneous system. The general theory developed in Section II is in principle capable of providing two-particle properties of the ground state also. Most importantly, it can give us the static, long-wavelength limit of the density correlations as well as the spin-density correlations which are respectively the linear dielectric response and the spin-susceptibility response functions of the inhomogeneous systems. A theory of this was outlined in the original work of Hohenberg–Kohn–Sham[18,19] and in the papers of von Barth–Hedin[32] and Rajagopal–Callaway.[33] These works have been generalized for calculating solid-state systems notably by Vosko and Perdew[75] and Janak.[76] Vosko and Perdew[75] developed a variational principle for calculating the paramagnetic susceptibility which is identical to the variational principle put forward by Rajagopal[77] for dealing with the solution of the vertex functions of the many-electron theory appropriate to the calculation of various linear response functions. We present the theory of static linear-response functions in the next section from a general viewpoint which incorporates some of the known Green function methods of the many-electron theory.[27] We also relate the parameters appearing in this theory to similar ones that appeared in this subsection, by expressing them in terms of the second-order functional derivatives of E_{xc}.

C. Static Linear Response Functions

The static linear response functions are calculated by applying static external field and computing the response of the system to first order in the field. In terms of the functional scheme, this implies that one may compute the density of the inhomogeneous system to first order in purturbation theory from an equation such as (2.64), for instance, when we add a contribution to V_{eff} from the applied external field. It is clear from this description that, once the self-consistent equation such as (2.64) is solved, the static linear response function can be expressed in terms of the solutions as well as a new additional quantity involving $(\delta^2 E_{xc}/\delta n(\mathbf{r})\delta n(\mathbf{r}'))$, calculated from the known E_{xc} in the absence of the applied external field. We now present this calculation in a compact way using the Green function language. This enables us to present the theory in a unified and succint fashion.

We begin with a brief resumé of the theory of Green functions.[30] This language has increasingly become popular with chemists as well[47] and so the presentation given here seems appropriate to the readers of this series of volumes. Since we are here interested in *equilibrium* properties only, we use the complex-time formalism. This gives a general framework for the next section as well. Let 1 stand for $(\mathbf{r}_1, \sigma_1, t_1)$ where t_1 is the imaginary time lying between 0 and $-i\beta$, $\beta = 1/k_{B}T$, as usual. We present the spin-dependent version of the theory as it is most pertinent for our discussions later on also. We follow Ref. 78, and introduce the 2×2 matrix Green function

$$G(1; 1') = (-i)\langle T(\psi(1)\psi^{+}(1'))\rangle \tag{3.31}$$

in the notation of Section II. For the general many-electron Hamiltonian, (2.2), the equation of motion that G obeys is

$$\left[\left\{i\frac{\partial}{\partial t_1} + \frac{\nabla_1^2}{2m} - v(r_1) - U_0(r_1, t_1)\right\}\tau_0 - \mathbf{U}(\mathbf{r}_1, t_1)\cdot\boldsymbol{\tau}\right]\mathbf{G}(11') - \int \Sigma(1\bar{1})\cdot\mathbf{G}(\bar{1}1')d\bar{1} = \tau_0\delta(11') \tag{3.32}$$

where $\tau_0, \boldsymbol{\tau}$ are the 2×2 unit and Pauli matrices, (2.34). $U_0(\mathbf{r}_1 t_1)$ is the external spin-independent perturbing potential whereas $\mathbf{U}(\mathbf{r}, t_1)$ is the external spin-dependent perturbation. The time-dependence here is purely fictitious and we make them time-independent at the end of the calculation. $\boldsymbol{\Sigma}(1\bar{1})$ is the so-called mass operator which is also a 2×2 matrix and it contains all the effects of electron interactions. The inverse of the Green function, $\mathbf{G}^{-1}$, is defined by

$$\int \mathbf{G}^{-1}(1\bar{1})\mathbf{G}(\bar{1}1')d\bar{1} = \tau_0\delta_0\delta(11') = \int \mathbf{G}(1\bar{1})\mathbf{G}^{-1}(\bar{1}1')d\bar{1} \tag{3.33}$$

The expressions for the density and spin density given by (2.33) can be expressed in terms of $\mathbf{G}$:

$$n(\mathbf{r}_1 t_1) = -i\ \mathrm{tr}(\tau_0\mathbf{G}(11^{+})) \tag{3.34}$$

$$\boldsymbol{\sigma}(\mathbf{r}_1 t_1) = -i\ \mathrm{tr}(\boldsymbol{\tau}\mathbf{G}(11^{+})) \tag{3.35}$$

From (3.34) and (3.35), we can define the linear response function because as stated earlier we need to compute the changes in $n, \boldsymbol{\sigma}$ to first order in the externally applied fields $U_0, \mathbf{U}$, thus:

$$n(\mathbf{r}_1t_1) = n_0(\mathbf{r}_1t_1) + \int \left(\frac{\delta n(\mathbf{r}_1t_1)}{\delta U_0(\mathbf{r}_2t_2)}\right)_0 U_0(\mathbf{r}_2t_2)d\mathbf{r}_2\,dt_2$$

$$+ \int \left(\frac{\delta n(\mathbf{r}_1t_1)}{\delta \mathbf{U}(\mathbf{r}_2t_2)}\right)_0 \cdot \mathbf{U}(\mathbf{r}_2t_2)d\mathbf{r}_2\,dt_2 \tag{3.36}$$

and a similar equation for $\mathscr{s}$. Here the subscript 0 represents the evaluation of the quantity under consideration for the zero value of the external fields. From (3.33) and (3.34), we note that the changes in linear order can be expressed in terms of the "vertex functions"

$$\mathbf{\Gamma}_j(12; r_3t_3) = \left(\frac{\delta \mathbf{G}^{-1}(12)}{\delta U_j(r_3t_3)}\right)_0 \tag{3.37}$$

Thus, introducing the response functions $\chi_{ij}(\mathbf{r}_1t_1; \mathbf{r}_2t_2)$, via the preceding relations, we obtain ($n_0 = n$; $n_1 = \mathscr{s}_x$, etc.)

$$\chi_{ij}(\mathbf{r}_1t_1; \mathbf{r}_2t_2) = \left(\frac{\delta n_i(\mathbf{r}_1t_1)}{\delta U_j(\mathbf{r}_2t_2)}\right)_0 = -i\,\mathrm{tr}\left\{\tau_i \frac{\delta \mathbf{G}(r_1\sigma_1t_1; r_1\sigma_1t_1^+)}{\delta U_j(\mathbf{r}_2t_2)}\right\}_0$$

$$= i\int \mathrm{tr}\{\tau_i \mathbf{G}_0(1\bar{1})\mathbf{\Gamma}_j(\bar{1}\bar{2}; \mathbf{r}_2t_2)\mathbf{G}_0(\bar{2}1^+)\}d\bar{1}\,d\bar{2} \tag{3.38}$$

Here $\mathbf{G}_0(12)$ is the Green function obtained by solving (3.32) when $U_0, \mathbf{U}$ set equal to zero. The equation obeyed by $\mathbf{\Gamma}$ is then obtained by using its definition, (3.33) and (3.32):

$$\mathbf{G}^{-1}(12) = \left\{i\frac{\partial}{\partial t_1} + \frac{\nabla_1^2}{2m} - v(\mathbf{r}_1) - U_0(\mathbf{r}_1, t_1)\tau_0 - \boldsymbol{\tau}\cdot\mathbf{U}(\mathbf{r}_1, t_1)\right\}$$

$$\times\, \delta(\mathbf{r}_1 - \mathbf{r}_2)\delta(t_1 - t_2) - \mathbf{\Sigma}(12) \tag{3.39}$$

and

$$\mathbf{\Gamma}_j(12; r_3t_3) = -\tau_j\delta(\mathbf{r}_1 - \mathbf{r}_2)\delta(t_1 - t_2)\delta(\mathbf{r}_1 - \mathbf{r}_3)\delta(t_1 - t_3) - \left(\frac{\delta \mathbf{\Sigma}(12)}{\delta U_j(\mathbf{r}_3t_3)}\right)_0 \tag{3.40}$$

The preceding formulation is quite general. Once the self-energy is known in some scheme, the indicated differentiation is explicitly carried out and integral equations for $\mathbf{\Gamma}_j$ are obtained. These equations are usually difficult to solve and Rajagopal[77] employed a variational scheme to solve them. We now apply the preceding formalism to the spin-density-functional theory, which enables us to deduce a form for the self-energy, $\mathbf{\Sigma}(12)$, in some appropriate scheme such as local density approximation. For the *equilibrium state*, as shown in Section II, $\mathbf{\Sigma}(12)$ is a functional of $n, \mathscr{s}$ and in the *static case*, we may write as an approximation,

$$\Sigma(12) \simeq \{V_{\text{eff}}[\mathbf{r}_1; n, \boldsymbol{\sigma}]\}\delta(\mathbf{r}_1 - \mathbf{r}_2)\delta(t_1 - t_2)$$

where V_{eff} is given by (2.93). For the sake of future convenience, we introduce (3.25, 3.26) to rewrite the preceding in the form

$$\Sigma(12) \simeq \{\tau_0(V_H[\mathbf{r}_1, n] + V_{xc}[\mathbf{r}_1, n, \boldsymbol{\sigma}]) + \boldsymbol{\tau} \cdot \mathbf{W}_{xc}[\mathbf{r}_1, n, \boldsymbol{\sigma}]\}\delta(\mathbf{r}_1 - \mathbf{r}_2)\delta(t_1 - t_2) \tag{3.41}$$

where V_H is the usual Hartree potential

$$V_H[\mathbf{r}_1, n] = e^2 \int \frac{1}{|\mathbf{r}_1 - \mathbf{r}_{1'}|} n(\mathbf{r}_{1'}) d\mathbf{r}_{1'} \tag{3.42}$$

The Green function in the absence of the external fields and for the equilibrium state in the HKS formalism is written in terms of the complete set of orthonormal spinor wave functions $\{\phi_{i\sigma}^{\lambda}\}$ given by (2.95), and the various densities are given by (2.94*a,b*). The Green function is

$$G_0(\mathbf{r}_1\sigma_1 t_1; \mathbf{r}_2\sigma_2 t_2) = \frac{1}{(-i\beta)} \sum_{n,i} \sum_{\lambda=1,2} e^{iE_n(t_1-t_2)} g_i^{\lambda}(E_n) \cdot \phi_{i\sigma_1}^{\lambda}(\mathbf{r}_1)\phi_{i\sigma_2}^{\lambda *}(\mathbf{r}_2) \tag{3.43}$$

where $E_n = (2n+1)\pi/(-i\beta) + \mu$, μ being the chemical potential, $n = 0, \pm 1, \ldots,$ and

$$g_i^{\lambda}(E_n) = \frac{1}{(E_n - \epsilon_i^{\lambda})} \tag{3.44}$$

The quantum numbers i introduced in (2.95) is written out in a more elaborate way here; i labels the appropriate quantum numbers and λ denotes the eigenspinor. Once the equations (2.95) are solved self-consistently, the Green function G_0 can be computed and hence the equation obeyed by $\boldsymbol{\Gamma}_j$ consistent with the approximation, (3.41), can in principle be solved and thus χ_{ij}s are determined. To be specific, the $\boldsymbol{\Gamma}_j$ equation is, using its definition and Σ of (3.41),

$$\begin{aligned}\boldsymbol{\Gamma}_k(12; \mathbf{r}_3 t_3) = &-\tau_k\delta(\mathbf{r}_1 - \mathbf{r}_2)\delta(t_1 - t_2)\delta(\mathbf{r}_1 - \mathbf{r}_3)\delta(t_1 - t_3) - \tau_0\delta(\mathbf{r}_1 - \mathbf{r}_2)\delta(t_1 - t_2) \\ &\times \Bigg\{ \int \left[\frac{e^2}{|\mathbf{r}_1 - \mathbf{r}_{\bar{1}}|} + K_{xc}[\mathbf{r}_1, \mathbf{r}_{\bar{1}}; n, \boldsymbol{\sigma}] \right] \chi_{0k}(r_{\bar{1}} - t_1; r_3 t_3) d\mathbf{r}_{\bar{1}} \\ &+ \sum_{\alpha=1}^{3} \int G_{xc\alpha}^{(1)}[\mathbf{r}_1, \mathbf{r}_{\bar{1}}; n, \boldsymbol{\sigma}] \chi_{\alpha k}(r_{\bar{1}} - t_1; r_3 t_3) d\mathbf{r}_{\bar{1}} \Bigg\} \\ &- \delta(\mathbf{r}_1 - \mathbf{r}_2)\delta(t_1 - t_2) \Bigg\{ \sum_{\alpha=1}^{3} \int \tau_\alpha G_{xc\alpha}^{(1)}[\mathbf{r}_1, \mathbf{r}_{\bar{1}}; n, \boldsymbol{\sigma}] \chi_{0k}(r_{\bar{1}} - t_1; r_3 t_3) d\mathbf{r}_{\bar{1}} \\ &+ \sum_{\alpha,\beta=1}^{3} \int \tau_\alpha G_{xc\alpha\beta}^{(2)}[\mathbf{r}_1, \mathbf{r}_{\bar{1}}; n, \boldsymbol{\sigma}] \chi_{\beta k}(r_{\bar{1}} t_1; r_3 t_3) d\mathbf{r}_{\bar{1}} \Bigg\}\end{aligned} \tag{3.45}$$

where we have introduced the notations:

$$K_{xc}[\mathbf{r}_1, \mathbf{r}_2; n, \phi] = \left(\frac{\delta^2 E_{xc}}{\delta n(\mathbf{r}_1)\delta n(\mathbf{r}_2)}\right)_0 \tag{3.46a}$$

$$G^{(1)}_{xc\alpha}[\mathbf{r}_1, \mathbf{r}_2; n, \phi] = \left(\frac{\delta^2 E_{xc}}{\delta n(\mathbf{r}_1)\delta \phi_\alpha(\mathbf{r}_2)}\right)_0 = \left(\frac{\delta^2 E_{xc}}{\delta \phi_\alpha(\mathbf{r}_2)\delta n(\mathbf{r}_1)}\right)_0 \tag{3.46b}$$

$$G^{(2)}_{xc\alpha\beta}[\mathbf{r}_1, \mathbf{r}_2; n, \phi] = \left(\frac{\delta^2 E_{xc}}{\delta \phi_\alpha(\mathbf{r}_1)\delta \phi_\beta(\mathbf{r}_2)}\right)_0 \tag{3.46c}$$

Thus the various suceptibilities are determined from the set of linear equations obtained by using (3.38) in conjunction with (3.45):

$$\begin{aligned}\chi_{ij}(\mathbf{r}_1 t_1; \mathbf{r}_2 t_2) = {} & \chi^{(0)}_{ij}(\mathbf{r}_1 t_1; \mathbf{r}_2 t_2) + \int \chi^{(0)}_{i0}(\mathbf{r}_1 t_1; \mathbf{r}_{\bar 1} t_{\bar 1}) \\ & \times \left[\frac{e^2}{|\mathbf{r}_{\bar 1} - \mathbf{r}_{\bar 2}|} + K_{xc}[\mathbf{r}_{\bar 1}, \mathbf{r}_{\bar 2}]\right]\chi_{0j}(\mathbf{r}_{\bar 2} t_{\bar 1}; \mathbf{r}_2 t_2)\, d\mathbf{r}_{\bar 1}\, dt_{\bar 1}\, d\mathbf{r}_{\bar 2} \\ & + \int \chi^{(0)}_{i0}(\mathbf{r}_1 t_1; \mathbf{r}_{\bar 1} t_{\bar 1}) \sum_{\alpha=1}^{3} G^{(1)}_{xc}[\mathbf{r}_{\bar 1}, \mathbf{r}_{\bar 2}]\chi_{\alpha j}(\mathbf{r}_{\bar 2} t_{\bar 1}; \mathbf{r}_2 t_2)\, d\mathbf{r}_{\bar 1}\, dt_{\bar 1}\, d\mathbf{r}_{\bar 2} \\ & + \int \sum_{\alpha=1}^{3} \chi^{(0)}_{i\alpha}(\mathbf{r}_1 t_1; \mathbf{r}_{\bar 1} t_{\bar 1}) G^{(1)}_{xc\alpha}[\mathbf{r}_{\bar 1}, \mathbf{r}_{\bar 2}]\chi_{0j}(\mathbf{r}_{\bar 2} t_{\bar 1}; \mathbf{r}_2 t_2)\, d\mathbf{r}_{\bar 1}\, dt_{\bar 1}\, d\mathbf{r}_{\bar 2} \\ & + \int \sum_{\alpha,\beta=1}^{3} \chi^{(0)}_{i\alpha}(\mathbf{r}_1 t_1; \mathbf{r}_{\bar 1} t_{\bar 1}) G^{(2)}_{xc\alpha\beta}[\mathbf{r}_{\bar 1}, \mathbf{r}_{\bar 2}]\chi_{\beta j}(\mathbf{r}_{\bar 2} t_{\bar 1}; \mathbf{r}_2 t_2)\, d\mathbf{r}_{\bar 1}\, dt_{\bar 1}\, d\mathbf{r}_{\bar 2}\end{aligned} \tag{3.47}$$

Here $\chi^{(0)}_{ij}(\mathbf{r}_1 t_1; \mathbf{r}_2 t_2)$ are the "noninteracting" correlation functions

$$\chi^{(0)}_{ij}(\mathbf{r}_1 t_1; \mathbf{r}_2 t_2) = -i \,\mathrm{tr}[\tau_i G_0(\mathbf{r}_1 t_1; \mathbf{r}_2 t_2)\tau_j G_0(\mathbf{r}_2 t_2; \mathbf{r}_1 t_1^+)] \tag{3.48}$$

These equations are the generalized version of the Vosko–Perdew equation for the magnetic vase. Vosko–Perdew then solve their equation analogous to (3.47) by a variational method and this is equivalent to our variational solution for the vertex, Γ_{ij}, for the electron gas.[77]

It should now be observed that various second derivatives of E_{xc} appear in the computation of the linear-response functions. The expression (3.22) for E_{xc} involving the various pair-correlation functions can now be used to obtain the second derivatives in terms of the derivatives of H_is. One then employs the sum rules once again and writes an ansatz similar to (3.29):

$$\begin{aligned}\frac{\delta^2 H_i[\mathbf{r}_1, \mathbf{r}_{1'}]}{\delta n_k(\mathbf{r})\delta_{nj}(\mathbf{r}')} = {} & \tfrac{1}{2}\frac{\partial H^0_i[\mathbf{r}_1 - \mathbf{r}_{1'}]}{\partial n^0_k}[P^{kj}_i(\mathbf{r}_1, \mathbf{r}, \mathbf{r}') + P^{kj}_i(\mathbf{r}_{1'}, \mathbf{r}, \mathbf{r}')] \\ & + \tfrac{1}{4}\frac{\partial^2 H^0_i[\mathbf{r}_1 - \mathbf{r}_{1'}]}{\partial n^0_k \partial n^0_j}(A^k_i(\mathbf{r}_1, \mathbf{r}) + A^k_i(\mathbf{r}_{1'}, \mathbf{r}))(A^j_i(\mathbf{r}_1, \mathbf{r}') + A^j_i(\mathbf{r}_{1'}, \mathbf{r}'))\end{aligned} \tag{3.49}$$

A_i^ks were already determined in the context of V_{xc}. P_i^{ij}s are now determined similarly. An explicit solution of these can be obtained for the homogeneous system. Sham's[71] approximation for this case is (for a paramagnetic system)

$$\frac{\delta^2 H_i[\mathbf{r}_1, \mathbf{r}_{1'}]}{\delta n(\mathbf{r})\delta n(\mathbf{r}')} \simeq \frac{\partial^2 H_i^0[\mathbf{r}_1 - \mathbf{r}_{1'}]}{\partial n^2}\delta(\mathbf{r}_1 - \mathbf{r})\delta(\mathbf{r}_{1'} - \mathbf{r}') \tag{3.50}$$

A similar scheme was used by Stoddart and Hanks[72] and Kawazoe et al.[73] We may point out that Gunnarsson et al.[67,68] and Alonso and Girifalco[69,79] did not consider the second-order terms as they did not examine the linear response functions.

In the paramagnetic case, using a *local density scheme*, one may further reduce the parameters encountered in the preceding general scheme considerably, and obtain the only two static correlation functions of interest in the system, that is, the density-response function and the spin susceptibility; this was done by Hedin and Lundqvist[61] and von Barth and Hedin,[32] respectively, for the case of the homogeneous gas. The theory of the spin susceptibility was given by Vosko and Perdew[75] and Janak[76] who used the orbitals generated by the one-particle equations, which is equivalent to the use of an expression of the form (3.43) for $G_0(\mathbf{r}_1\sigma_1 t_1; \mathbf{r}_2\sigma_2 t_2)$. Janak[76] computed the static long-wavelength limit of the spin-susceptibility for the entire $3d$-series of elements to determine which of them exhibits the ferromagnetic instability. He was successful in showing that the theory correctly predicts the magnetism of Fe, Co, Ni whereas Pd is almost ferromagnetic, in agreement with the known facts about these metals. We return to the calculation of the corresponding frequency dependent functions in the low-frequency, long-wavelength regions in the next section where we discuss the time-dependent theory. The scheme outlined previously can be taken over in that case if one makes an adiabatic approximation. The formalism presented by us in this section has the merit of being able to accommodate a formulation involving the pair-correlation functions which by the theorem of fluctuation dissipation are related themselves back to the frequency-dependent parts of the linear response functions themselves, integrated over all frequencies. A complete time-dependent theory would then determine K_{xc}, and so on, fully self-consistently. This goal is an ultimate aim, not yet accessible, since a rigorous time-dependent theory is lacking as of now.

We have thus shown how one could calculate various ground-state properties of the system in terms of the first and second functional derivatives of E_{xc}. In the next section, we discuss some recent attempts to generalize the preceding schemes to time-dependent properties. This enables us also to obtain information about the excited states of the system, not just the ground-state properties presently described by the formalism given thus far.

IV. TOWARD A TIME-DEPENDENT THEORY

In this section, we outline essentially three possible approaches to the time-dependent theory of the functional formalism. The first is based on an old theory of Bloch[79] and Jensen[80] who tried to develop a hydrodynamical extension of the Thomas–Fermi theory. The second is based on the time-dependent Hartree–Fock theory first developed by Dirac,[4] and recently exploited by nuclear physicists. The third approach is the use of the well-known stationary action principle stated in Section I. This approach is equivalent to the complex-time formulation of the equilibrium statistical mechanics.[30] We consider these three separately and critically evaluate their validity in light of the functional theory.

A. Time-dependent Thomas–Fermi Theory

In the first section, we noted that Lenz[7] reformulated the TF theory as a variational principle for the total energy of the system, which he was able to express in terms of the density of the system only. In analogy with classical mechanics, if one can postulate a variable canonically conjugate to the density, one can then employ the standard action principle to set up a time-dependent theory, by appropriately constructing the Lagrangian and hence the action. In classical hydrodynamics, such canonically conjugate variables as density and velocity fields are commonly postulated for developing the Hamiltonian theory of fluid motions. When passing on to the quantum theory following the usual procedure of converting all the canonically conjugate variables with quantum operators now obeying commutation rules, one encounters a formal difficulty. The density variable is, by definition, a positive quantity and in the quantum theory, it should retain this positivity property. The velocity field conjugate to the density variable satisfies an equal time commutation rule. The spectrum of such operators (e.g., coordinate and its conjugate linear momentum) then go from positive infinity to negative infinity. If one restricts one of the variables to positive values only, then one faces a serious dilemma, with the preceding observation. For a discussion of these questions, one may consult Kobe and Coomer.[82] For a more mathematical description of these questions, one may refer to the work of Primas et al.[83] Bloch[79] and Jensen[80] developed a hydrodynamic theory of an electron gas as an extension of the TF theory to time-dependent phenomena. In analogy with hydrodynamics, *assuming irrotational flow*, the velocity field is described by a scalar potential field, $u(\mathbf{r}, t)$, related to the velocity $\mathbf{v}(\mathbf{r}, t)$ via the formula,

$$\mathbf{v}(\mathbf{r}, t) = -\nabla u(\mathbf{r}, t) \tag{4.1}$$

$n(\mathbf{r}, t)$ and $\mathbf{v}(\mathbf{r}, t)$ are then our canonical variables in terms of which we now write a Hamiltonian and a Lagrangian, using the Hohenberg–Kohn expression for the ground-state energy as that part of the Hamiltonian determined by the density $n(\mathbf{r}, t)$ alone, in addition to which one adds the classical energy contribution from the velocity field:

$$H = \tfrac{1}{2} \int n(\mathbf{r}, t) |\nabla u(\mathbf{r}, t)|^2 \, d\mathbf{r} + F[n\mathbf{r}, t)] \tag{4.2}$$

where F is the universal functional of density of the HKS theory, (2.14). The Lagrangian then is

$$L = \int n(\mathbf{r}, t) \frac{\partial u(\mathbf{r}, t)}{\partial t} d\mathbf{r} - H \tag{4.3}$$

We must stress that these expressions are *introduced by analogy only* and not derived in any formal way. The stationary action principle

$$\delta \int_{t_1}^{t_2} L \, dt = 0 \tag{4.4}$$

then leads to two equations, one arising from varying $n(\mathbf{r}, t)$ and the other arising from variations in $u(\mathbf{r}, t)$. The variation with respect to $u(\mathbf{r}, t)$ leads to the usual continuity equation, whereas the variation with respect to $n(\mathbf{r}, t)$ leads to the Navier–Stokes type of equation involving $\delta E_{xc}/\delta n(\mathbf{r}t)$ for the equation of motion of $u(\mathbf{r}, t)$. Such equations were derived by Ying.[84] These equations are supposed to be valid in the hydrodynamic regime where the quasiparticles undergo many collisions during one period of oscillation of the perturbing field; that is, the theory given previously is expected to be valid for slow enough motions where the frequency is small compared to the inverse collision times, $\delta\tau \ll 1$. In this regime, one has local equilibrium everywhere in the electron system so that a description such as the one preceding in terms of density and velocity may be appropriate. In the opposite regime, $\omega\tau \gg 1$, this theory is incorrect. Lundqvist[85] proposed a linearized quantum description for density oscillations where he tries to retain the preceding theory but in addition tries to incorporate some aspects of quantum dynamics. This theory, too, is somewhat ad hoc. This theory is supposed to be valid for the entire region of frequencies from $\omega\tau \ll 1$ to $\omega\tau \gg 1$. The theory is exact within the linear response approximation and has been applied to surface problems and to atomic systems. In the Section IV.C we make contact with such schemes from a more formal standpoint.

B. Adiabatic Time-dependent Theory

This theory is principally developed by nuclear theorists, notably Kerman and Koonin[86] and Villars,[87] and is based on the early work of Dirac.[4] Kerman and Koonin[86] develop a canonical Hamiltonian formulation for the general time-dependent variational principle associated with the Schrödinger equation, stated in (1.4). Villars[87] uses a similar principle to develop, in a systematic way, the adiabatic limit of time-dependent HFS theory. The stationary principle is used to find a best possible approximation to a solution of the time-dependent HFS equations in terms of some time-dependent c-number parameters. These schemes have not been used in the theory of collective excitations in solid state physics even though for discussing certain types of collective motions such as spin waves in an itinerant ferromagnet, such a scheme may be set up. We follow the general description of Kerman and Koonin[86] in describing this theory.

In general, the complete description of our system is contained in the many-electron (N here) time-dependent wave function $\Psi(x_1 \cdots x_N; t)$ where the generalized coordinate x_i stands as before for the spatial coordinate $\mathbf{r}_i$ and spin quantum number, σ_i. Ψ is antisymmetric in any two coordinates x_i, x_j. The time-dependent Schrödinger equation,

$$i\hbar \frac{\partial \Psi}{\partial t} = H\Psi$$

where H is the Hamiltonian given by (1.1) for example, may be derived from a variational principle. Define a time-dependent Lagrangian which is a functional of Ψ and Ψ^*:

$$\begin{aligned}\mathscr{L}[\Psi, \Psi^*] &= \left\langle \Psi \middle| \left(i\hbar \frac{\partial}{\partial t} - H \right) \middle| \Psi \right\rangle \\ &= \int \cdots \int dx_1 \cdots dx_N \Psi^*(x_1 \cdots x_N; t)\left(i\hbar \frac{\partial}{\partial t} - H \right)\Psi x_1 \cdots x_N; t)\end{aligned} \tag{4.5}$$

where $\int dx_1$ stands for $\Sigma_{\sigma_1} \int d\mathbf{r}_1$, and the multiple integration in (4.5) runs over all the particles of the system. In analogy to classical mechanics, we may write (4.5) in the form

$$\mathscr{L}[\Psi, \Psi^*] = \int \cdots \int dx_1 \cdots dx_N i\hbar \Psi^* \left(\frac{\partial}{\partial t} \Psi \right) - \mathscr{H}[\Psi, \Psi^*] \tag{4.6}$$

where we have defined the real Hamiltonian $\mathscr{H}$ as a functional of Ψ and Ψ^*:

$$\mathscr{H}[\Psi, \Psi^*] = \int \cdots \int dx_1 \cdots dx_N \Psi^*(x_1 \cdots x_N; t) H \Psi(x_1 \cdots x_N; t) \tag{4.7}$$

The action A corresponding to the Lagrangian (4.6) for evolution between times t_1 and t_2 is

$$A = \int_{t_1}^{t_2} dt \mathscr{L}[\Psi, \Psi^*] \tag{4.8}$$

The equations of motion are then determined by the stationary action principle which states that A is stationary with respect to variations of Ψ and Ψ^* between the fixed end points t_1 and t_2. With an integration by parts, one obtains

$$i\hbar \frac{\partial \Psi}{\partial t} = \frac{\partial \mathscr{H}}{\partial \Psi^*} \tag{4.9a}$$

and

$$i\hbar \frac{\partial \Psi^*}{\partial t} = \frac{\partial \mathscr{H}}{\partial \Psi} \tag{4.9b}$$

where $\partial/\partial\Psi$ is the usual functional derivative. These are just the familiar Schrödinger equation and its adjoint. These equations clearly admit of a classical interpretation in terms of field coordinates and momenta, by defining real multicoordinate fields Φ and Π by

$$\begin{aligned} \Phi &= \sqrt{2}\,\mathrm{Re}\,\Psi \\ \Pi &= \sqrt{2}\,\mathrm{Im}\,\Psi \end{aligned} \tag{4.10}$$

so that (4.9) become the familiar looking Hamilton equations for the fields Φ and its conjugate momentum Π, with the Hamiltonian considered as a functional of Π and Φ.

Given the preceding general discussion, it appears obvious that one may seek *approximate solutions* by parametrizing the wave function in terms of *several time-dependent fields* or *parameters* and employing the variational principle to determine their time evolution. In the next section, we describe yet another approach, advocated by Peuckert.[26] The method suggested here is physically motivated and requires one to introduce in an intuitive fashion the time-dependent variables. In the ferromagnet, this could be the Euler angles that the spin-density points to, for example. Since Ψ is complex, the parametrized wave function must also be complex. It is very important to stress that, as with any variational procedure, the appropriateness of any suggested parametrization is difficult to assess. The true motion of Ψ will be approached as the wave function is chosen with the greatest of freedom of variation.

To illustrate the method, let us consider the time-dependent Hartree–Fock (TDHF) scheme. This consists in parametrizing Ψ as a determinental wave function containing single-particle (complex) wave functions $\phi_n(x, t)$. The determinantal form maintains the required antisymmetry of Ψ. Thus,

$$\Psi(x_1 \cdots x_N; t) = \det\{\phi_n(x_k, t)\} \tag{4.11}$$

With this form for Ψ, we can calculate the Lagrangian:

$$\mathscr{L}[\{\phi_n\}, \{\phi_n^*\}] = \sum_{n=1}^{N} \int dx \phi_n^*(x, t) i\hbar \frac{\partial \phi_n(x_1 t)}{\partial t} - \mathscr{H}[\{\phi_n\}, \{\phi_n^*\}] \tag{4.12}$$

where the Hamiltonian functional $\mathscr{H}$ is a quartic functional in ϕ_ns; that is, it contains terms like $(\phi_n^*(\nabla^2/2m)\phi_n)$, $\phi_n^* V_{nm}\phi_m$, and $\phi_n^*\phi_m^* v_{nm;mn}\phi_n\phi_m$ only. Considering the action A as a functional of $\{\phi_n\}$ and $\{\phi_n^*\}$, we arrive at the equations of motion

$$i\hbar \frac{\partial \phi_n}{\partial t} = \frac{\partial \mathscr{H}}{\partial \phi_n^*} \tag{4.13a}$$

and

$$i\hbar \frac{\partial \phi_n^*}{\partial t} = -\frac{\partial \mathscr{H}}{\partial \phi_n} \tag{4.13b}$$

One may then define real variables similar to the classical ones, (q_n, p_n),

$$\phi_n = \frac{q_n + ip_n}{\sqrt{2}} \tag{4.14a}$$

$$\phi_n^* = \frac{q_n - ip_n}{\sqrt{2}} \tag{4.14b}$$

and obtain equations like those of Hamiltonians for q_n and p_n. Note now that the Hamiltonian $\mathscr{H}$ will be quartic in $\{q_n, p_n\}$ and so the solution for p_n in terms of q_n and $\partial/\partial t(q_n)$ will not be an easy task, if not an impossible one. There is thus no simple way of obtaining the Lagrangian in its canonical form. As a consequence of the parametrization, (4.11), two conservation laws are immediately seen to be fulfilled. The Hamiltonian formulation enables us to verify the energy conservation. The second is the time-independence of the matric of the Hilbert space spanned by $\{\phi_n\}$. Using the equations of motion, we have

$$\begin{aligned} i\hbar \frac{d}{dt} g_{nm} &= i\hbar \frac{d}{dt} \int dx \phi_n^*(x, t)\phi_m(x, t) \\ &= \int dx \left(\frac{\partial \mathscr{H}}{\partial \phi_n} \phi_m - \phi_n^* \frac{\partial \mathscr{H}}{\partial \phi_m^*} \right) = 0 \end{aligned} \tag{4.15}$$

in view of the actual structure of $\mathscr{H}[\{\phi_n\}, \{\phi_n^*\}]$. In other words, $\partial\mathscr{H}/\partial\phi_n$ may be thought of as a result of operating on ϕ_n with the Hermitian

Hartree–Fock Hamiltonian operator, $\mathscr{H}_{HF}$:

$$\frac{\partial \mathscr{H}}{\partial \phi_n^*} = \mathscr{H}_{HF}\phi_n \tag{4.16}$$

with

$$\mathscr{H}_{HF}^+ = \mathscr{H}_{HF} \tag{4.17}$$

This conservation law is associated with the gauge invariance of $\mathscr{H}[\{\phi_n\}, \{\phi_n^*\}]$ under any unitary transformation among $\{\phi_n\}$.

Several comments are in order at this stage. The solutions ϕ_n which are normalized are the usual HFS solutions. But those that are not normalized to unity remain quantum mechanically unclear as they have lost their meaning as a single determinantal wave function. A single particle wave function of a static solution will evolve in time according to (4.13) as

$$\phi_n(x, t) = e^{-i\epsilon_n t/\hbar}\phi_n^{HF}(x) \tag{4.18}$$

where $\phi_n^{HF}(x)$ is the real, static wave function and ϵ_n is the single-particle HF energy given by the eigenvalue equation

$$\frac{\partial \mathscr{H}[\{\phi_n^{HF}\}]}{\partial \phi_n^{HF}} - \epsilon_n\phi_n^{HF} \equiv (\mathscr{H}_{HF} - \epsilon_n)\phi_n^{HF} = 0 \tag{4.19}$$

If the solutions (4.18) are used in the construction of the solution, Ψ, we see that even in the static solution each ϕ_n oscillates with frequency ϵ_n, and these merely produce a time-dependent overall phase of Ψ. In fact, any unitary transformation of the ϕ_ns leaves the physics of the system unchanged. Hence, it is important to retain only those changes that bring about a change in the physics. The transformation

$$\phi_n \rightarrow \sum_{m=1}^{N} \left(T \exp i \int_0^t \tilde{h}(t')dt'\right)_{nm} \phi_m \tag{4.20}$$

where $\tilde{h}$ has the matrix elements

$$\tilde{h}_{nm} = \int\int dx_1\, dx_2 \phi_n^*(x_1)\tilde{h}(x_1 x_2)\phi_m(x_2) \tag{4.21}$$

and if T denotes the time-ordered product of the exponential, ϕ_n satisfy the new equations

$$i\hbar\phi_n = \sum_{m=1}^{N} (\mathscr{H}_{HF}\delta_{mn} - \tilde{h}_{mm})\phi_m \tag{4.22}$$

From this it is readily proved that

$$\int dx \phi_n^*(xt)\dot{\phi}_m(xt) = 0$$

for all m, n. Thus, (4.20) removes physically irrelevant changes in the coordinates and momenta. Under this transformation, the stationary solutions are

$$q_n(x, t) = q_n^{HF}(x) \tag{4.23a}$$

and

$$p_n(x, t) = 0 \tag{4.23b}$$

as we usually require of coordinates and momenta.

The variational description given previously shows that the time-dependent HF scheme is designed to represent an approximation to the evolution of the many-electron wave function. Hence, a Fourier analysis of the approximate TDHF solution may give information about the *excitation energies*. The usual random-phase approximation (RPA) is obtained by considering wave-packets corresponding to infinitesimal motions about a stationary HF solution and finding eigenfrequencies of the one-particle density matrix. The case of finite amplitude motion, which is more interesting, raises the question of significant high-frequency components which will tend to give a finite lifetime to the time evolution of any one-particle operator.

Now, as a function of the field variables the Hamiltonian $\mathcal{H}$ possesses an absolute minimum at the ordinary HF ground state, and many other local minima corresponding to the excited HF configurations. As is well known, small amplitude motion of the system about any of these minima or for that matter, extrema, is describable by a linearized version of the equations of motion. The normal modes and eigenfrequencies are the RPA solutions, as is usually known. Much of the preceding discussion can be taken over when one makes other types of time-dependent approximation schemes such as the one suggested by Peuckert, discussed in the next subsection. The preceding detailed discussion is thought to be a useful introduction to their more complicated formulation.

To illustrate the RPA solutions which describe the near equilibrium dynamics, we begin by assuming that the static HF problem is solved. In the Green function language of Section III.C this amounts to knowning the self-energy Σ, and hence also the one-particle Green function. The near equilibrium motion is then investigated by writing

$$\phi_n(x, t) = \phi_n^{HF}(x) + \delta\phi_n(x, t) \tag{4.24}$$

where $\delta\phi_n$ is supposed to be small. This substituted into (4.13) leads to the RPA equations for $\delta\phi_n$ and $\delta\phi_n^*$. In the Green function language of Section III.C this procedure leads to the equation governed by the vertex function, calculated in RPA. The solutions of these equations for the case of harmonic time dependence of $\delta\phi_n(x, t)$ yields the eigenfrequencies and are usually identified with the excitation energies of the system in an RPA scheme. This scheme may be expected to be reliable only if the system is nearly harmonic. Any large-amplitude collective motion, even if slow, requires one to go beyond RPA. The RPA description is useful in describing the low-lying excitations of the many-electron system.[30,78] The preceding description makes contact with the Green function approach. There is another framework in which we are interested only in a restricted set of parameters, as in nuclear physics, for example, quadrupole moment or rms radius. The TDHF scheme is too elaborate for discussing those types of questions. This may be of interest in our context, too. It is therefore of importance to investigate the reduction of variables by parametrizing the many-body wave function not in terms of N continuous function, ϕ_n, but in terms of a restricted class. We follow again Kerman and Koonin[86] in outlining such a framework of restricted dynamical parameters. In the solid-state context, this may become the basis for discussing spin excitations in magnetic systems, even though such an approach has not yet been examined. In view of our discussion in the next subsection, this framework may prove to be of some interest. *It should be remarked that the scheme given in Ref. 27 is in the style of TDHF but generalized for the HKS scheme.*

Suppose then that the many-particle wave function may be parametrized by a set of continuous variables $\{u_i, i = 1, \ldots, m\}$. For example, such wave functions may be just a constrained HF solution although the parametrized wave function need not be a determinant. Collective motion is then imagined to be the motion in the space of many-body wave functions parametrized by $\{u_i\}$. These are then considered as time dependent which determine the time dependence of Ψ. The Langrangian, (4.6), is then a functional of $\{u_i\}$:

$$\mathscr{L}[\{u_i(t)\}] = i\hbar \int \cdots \int dx_1 \cdots dx_N \Psi^*(\{u_i\}) \sum_{j=1}^{M} \frac{\partial u_j}{\partial t} \frac{\partial \Psi(\{u_i\})}{\partial u_j} - \mathscr{H}[\{u_i\}] \tag{4.25}$$

where $\mathscr{H}[\{u_i\}]$ is defined as

$$\mathscr{H}[\{u_i\}] = \int \cdots \int dx_1 \cdots dx_N \Psi^*(\{u_i\}) \mathscr{H} \Psi(\{u_i\}) \tag{4.26}$$

A variation of the action A with respect to $\{u_i\}$ gives us the equations of motion:

$$\sum_j \{u_i, u_j\} \dot{u}_j = \frac{\partial \mathscr{H}}{\partial u_i} \tag{4.27}$$

where the real Lagrangian bracket is defined, as usual, by[81]

$$\{u_i, u_j\} = i\hbar \int \cdots \int dx_1 \cdots dx_N \left(\frac{\partial \Psi^*}{\partial u_i} \frac{\partial \Psi}{\partial u_j} - \frac{\partial \Psi^*}{\partial u_j} \frac{\partial \Psi}{\partial u_i} \right) \tag{4.28}$$

and it is *antisymmetric* in its indices, (i, j). A general discussion of (4.27) is given in Ref. 81 and we give here a brief account of it. Kerman and Koonin[86] discuss several special cases of interest to nuclear physics and they do not concern us here. Suppose the antisymmetric matrix, $\{u_i, u_j\}$, is nonsingular, and has an inverse $\Gamma^{ij}(\{u_k\})$. Then (4.27) may be rewritten as

$$\dot{u}_j = \sum_{r=1}^{N} \Gamma^{rj}(\{u_k\}) \frac{\partial \mathcal{H}}{\partial u_r} \tag{4.29}$$

and these equations can be expressed in terms of nonsingular generalized Poisson brackets in the variables $\{u_j\}$ and the usual classical mechanical theory holds. If the matrix $\{u_i, u_j\}$ is singular, we encounter a problem with constraints and when this is handled appropriately, one again obtains a generalized canonical description of the system in terms of u_is. Now, the antisymmetry of $\{u_i, u_j\}$ implies that it is singular for odd numbers of u_is. Suppose we had only one parameter u_1 in our theory. Then we obtain, at once, from (4.27) that

$$\frac{\partial \mathcal{H}}{\partial u_1} = 0 \tag{4.30}$$

This condition is the same as the extremum condition or, equivalently, the static condition that the energy of the system is an extremeum. All dynamics is then impossible because there is no evolution of u_1 in time! Hence, we conclude that *a parametrization of Ψ in order to deal with collective motion requires at least two* u*s*. In the case of an odd number of parameters, we must choose the largest even number of variables, and so on. For different types of parameterization in the case of two variables, see Ref. 86.

One important application of the preceding theory to a situation as in our problem is mentioned in Ref. 86 which throws some light on the discussion in Section IV.A about the time-dependent TF theory. Take

$$\Psi = \exp\left(-\frac{i}{\hbar} \int \phi(\mathbf{r}, t) \hat{\rho}(\mathbf{r}) d\mathbf{r}\right) \Psi_0 \tag{4.31}$$

as a trial function, with the operators

$$\hat{\rho}(\mathbf{r}) = \sum_{i=1}^{N} \delta(\mathbf{r} - \mathbf{r}_i) \tag{4.32}$$

and the real wavefunction Ψ_0 still to be determined. The density of the system is, of course,

$$n(\mathbf{r}, t) = \int \cdots \int dx_1 \cdots dx_N \Psi^* \hat{\rho} \Psi \tag{4.33}$$

The Lagrangian, (4.6), then is

$$\mathscr{L} = +\int d\mathbf{r}_1 n(r_1, t) \frac{\partial \phi(r_1, t)}{\partial t} - \int \cdots \int dx_1 \cdots dx_N \Psi^* H \Psi \tag{4.34}$$

In the case of the Hamiltonian considered by us, we have the exact form of the Hamiltonian, $\mathscr{H}[n, \phi]$, written formally for the second term in $\mathscr{L}$:

$$\mathscr{H}[n, \phi] = \frac{1}{2} \int d\mathbf{r} n(\mathbf{r})(\nabla \phi(\mathbf{r}))^2 + E[n(r)] \tag{4.35}$$

where $E[n(\mathbf{r})]$ is the minimized energy as a functional of $n(\mathbf{r})$, that is, *all other degrees of freedom in* Ψ_0 *are chosen to minimize* E *under the constraint of fixed* n($\boldsymbol{r}$). It is then clear that $\phi(\mathbf{r}, t)$ is the field momentum conjugate to $n(\mathbf{r}, t)$. Thus, the Hamiltonian's equations of motion for $n(\mathbf{r}, t)$ and $\phi(\mathbf{r}, t)$ are obtained as before and the preceding derivation coincides with Ying's[84] time-dependent HKS theory. Note that $E[n(\mathbf{r})]$ appeared here in a different guise with the assumptions on ψ_0 spelled out previously, in contrast to the HKS theory. The preceding theory then leads to the theory of *irrotational velocity field fluid flow* from the quantum principles. It is important to stress again that although no assumption of adiabaticity is invoked here, the many other degrees of freedom in Ψ_0 have been frozen to minimize $E[n]$. This, therefore, did not give rise to dissipative terms, and we obtained a conservative Hamiltonian theory. Somehow, a more realistic theory should include these other degrees of freedom in developing the theory of collective motions. The Hohenberg–Kohn–Sham theory is just a ground-state theory and as such a description of the low-lying excited states of the system require some more input than has been incorporated in Ying's approach. Using the Wigner distribution function approach, one could rewrite the TDHF equations in fluid dynamical form. Specific approximations are needed to close the hierarchy of moment equations to describe the dynamics of the fluid.[86] This takes us then outside the scope of the HKS formalism. In the next section, this formalism is given another twist so that we can deal with the problem of electron correlations in condensed matter in a manner analogous to HKS formalism. It should be clear to the reader, from the preceding description, that such an attempt is still at an infant stage but an account of the formalism may provide a more fruitful framework for future investigations.

C. Time-dependent Functional Theory Based on the Stationary Action Principle

The action principle for the time-dependent quantum theory developed in the last section has been used by Peuckert[26] to provide an appropriate generalization of Hohenberg–Kohn theory. It should be stressed that, unlike the HKS theory, this generalization is weaker because it is based on a stationary principle. In complete analogy with the ground-state theory, Peuckert suggested a scheme in which the time-dependent particle density, $n(\mathbf{r}, t)$, of a system of interacting electrons can be calculated from the solutions of a time-dependent Schrödinger equation for noninteracting fermions in an effective one-particle potential. The response function can then be found by functional differentiation with respect to an infinitesimal external potential, $v_{\text{ext}}(\mathbf{r}, t)$, as in the Green function theory outlined in Section III.C. This formalism, in contrast to the HKS theory, is applicable to the determination of excited states of the system. Peuckert also outlined a generalization of the scheme to a spin-density system and appropriate modification required for dealing with nonstationary, time-dependent problems. This development is described in detail here. We should remark that this theory has an entirely different flavor compared with the one described in the last section, in that it is more in the style of HKS theory than in the time-dependent Hartree–Fock–Slater formalism. An application of this to the excited states of a helium atom was made by Peuckert, which is discussed in the next section. At the end of this section, we show how the finite-temperature Green function theory can be used for the same purpose,[27] a point realized also by Peuckert. We state these results in the form of propositions.

Proposition 1. The action function A of a many-electron system in the presence of an external time-dependent potential $V(r, t)$ is a functional of the time-dependent density $n(r, t)$, and is stationary with respect to variations in $n(r, t)$.

Proof. The Hamiltonian is assumed to be of the form (2.2), except that we now write V to be time dependent in the form

$$v(\mathbf{r}, t) = v(\mathbf{r}) + v_{\text{ext}}(\mathbf{r}, t) \tag{4.36}$$

where we have separated the time-independent part from the time-dependent part. The action function is then written in the form

$$A[\Psi] = -\int_{t_1}^{t_2} dt \int d\mathbf{r}\, n(\mathbf{r}, t) v(\mathbf{r}, t) + A_0[\Psi] \tag{4.37}$$

where $n(\mathbf{r}, t)$ is the particle density given by

$$n(\mathbf{r}, t) = \langle \Psi(t)|\rho(\mathbf{r})|\Psi(t)\rangle \tag{4.38}$$

in which the density operator $\rho(r)$ is $\Sigma_\sigma\, \psi^\dagger(x)\psi(\mathbf{x})$ is in the Schrödinger representation. The time evolution is contained entirely in the state vector $|\Psi(t)\rangle$. $A_0[\Psi]$ in (4.37) is given by

$$A_0[\Psi] = \int_{t_1}^{t_2} dt \Big\{ \int d\mathbf{r} \sum_\sigma \Big\langle \Psi(t) \Big| \psi^\dagger(x)\Big(i\hbar\frac{\partial}{\partial t} + \frac{\hbar^2}{2m}\nabla^2\Big)\psi(x)\Big|\Psi(t)\Big\rangle - \langle \Psi(t)|W|\Psi(t)\rangle \Big\} \tag{4.39}$$

where W is given by (2.3c), which may be rewritten in terms of density operators in the form

$$W = \tfrac{1}{2} \int\int \frac{e^2}{|\mathbf{r}-\mathbf{r}'|}(\rho(\mathbf{r})\rho(\mathbf{r}') - \delta(\mathbf{r}-\mathbf{r}')\rho(\mathbf{r})) \tag{4.40}$$

with $\rho(\mathbf{r})$, the density operator given by $\Sigma_\sigma\, \psi^\dagger(x)\psi(x)$, the Schrödinger representation of (2.4). Now, let $|\Psi(t)\rangle$ be the solution of the time-dependent Schrödinger equation

$$H(t)|\Psi(t)\rangle = i\hbar\frac{\partial}{\partial t}|\Psi(t)\rangle \tag{4.41}$$

with $H(t) = \mathcal{T} + V(t) + W$, $\mathcal{T}$, $V(t)$, W are as in (2.3) along with (4.36), and $|\Psi(t)\rangle$ is such that the initial condition is

$$|\Psi(t_1)\rangle = |\Psi_0\rangle \tag{4.42}$$

Equation 4.41 is obtained by making the action A stationary subject to $|\delta\Psi(t_{1,2})\rangle = 0$ and the initial condition (4.42). If only state vectors of this form are admitted in the construction of the action function $A_0[\Psi]$, in (4.39), then A_0 can be understood as a functional of the external potential, $v_{\text{ext}}(\mathbf{r}, t)$. Note that A_0 depends on the initial state $|\Psi_0\rangle$, but, in the foregoing development only variations of A_0 with respect to v_{ext} will be allowed. In analogy with the HKS formalism, we would like A_0 to be a functional of the density, $n(\mathbf{r}, t)$. This is possible only if the mapping

$$v_{\text{exp}}(\mathbf{r}, t) \rightarrow n(\mathbf{r}, t) = \langle \Psi(t)|\rho(\mathbf{r})|\Psi(t)\rangle \tag{4.43}$$

is invertible. In the static case, one has the Rayleigh–Ritz minimum principle for the energy functional, which enabled Hohenberg and Kohn[18] to establish the inversion of the mapping as we discussed earlier in Section II.A. *The weaker stationary property of the action function, (4.37), does not suffice to establish an analogous theorem for time-dependent fields. It is therefore asserted that the mapping is invertible in a small neighborhood of* $v_{ext}(r, t) = 0$. This condition is met if the inverse of the linear response operator, L_0, exists. Because, for small enough v_{ext}, one has in first-order perturbation theory the change in density given by

$$\delta n(\mathbf{r}, t) = \int\int \left(\frac{\delta n(\mathbf{r}, t)}{\delta v_e(\mathbf{r}', t')}\right)_{v_{ext}=0} v_{ext}(\mathbf{r}', t')d\mathbf{r}'\, dt' \tag{4.44}$$

The linear response operator, L_0, is $(\delta n/\delta v_e)_{v_e=0}$. If we exclude from the set of admissible potentials v_{ext}, those which satisfy $L_0 v_{ext} = 0$, then (4.44) can be inverted to give v_{ext} in terms of $\delta n(\mathbf{r}, t)$. But this procedure is not a priori sufficient to allow the construction of the functional $A_0[n]$. The spirit of this argument is similar to the time-dependent HF theory discussed in the last section. It is now clear the reason for naming the preceding result as a proposition and not a theorem in contrast to the static case.

Now $A_0[n]$ can be regarded as a functional of n in a neighborhood of the density $n_0(\mathbf{r})$ corresponding to $v_{ext} = 0$. In addition, A_0 also depends on the coupling parameter e^2, so that $A_0 = A_0[e^2, n]$. Peuckert[26] proves the following statement, similar to the one obtained in Section II.G.

$$A_0[e^2, n] = A_0[0, n] - \int_0^1 \frac{d\theta'}{\theta'} \int_{t_1}^{t_2} dt' \langle W\rangle[\theta', n] \tag{4.45}$$

where $\langle W\rangle[\theta', n]$ is the expectation value of the operator representing the Coulomb interaction among the electrons, considered as a function of the coupling constant θ and as a functional of n. It is important to observe that θ and n are independent variables. To prove (4.45), we put θW for W, and calculate $dA_0/d\theta$ directly by using (4.39):

$$\frac{d}{d\theta} A_0 = \int_{t_1}^{t_2} dt \int d\mathbf{r} \sum_{\sigma} \left\{\left\langle \Psi(t;\theta)\middle| \psi^\dagger(x)\left(i\hbar\frac{\partial}{\partial t} + \frac{\hbar^2}{2m}\nabla^2\right)\psi(x) - W\middle|\frac{\partial\Psi}{\partial\theta}\right\rangle\right.$$
$$\left. + \left\langle\frac{\partial\Psi}{\partial\theta}\middle|\psi^\dagger(x)\left(i\hbar\frac{\partial}{\partial t} + \frac{\hbar^2}{2m}\nabla^2\right)\psi(x) - W\middle|\Psi\right\rangle - \frac{1}{\theta}\langle\Psi|W|\Psi\rangle\right\} \tag{4.46}$$

where we used the scaling of W as indicated previously. From the time-dependent Schrödinger equation obeyed by $|\Psi\rangle$, we arrive at

$$\left\langle\frac{\partial\Psi}{\partial\theta}\middle|\psi^\dagger(x)\left(i\hbar\frac{\partial}{\partial t} + \frac{\hbar^2}{2m}\nabla^2\right)\psi(x) - W\middle|\Psi\right\rangle = \left\langle\frac{\partial\Psi}{\partial\theta}\middle|V(\mathbf{r}, t)\middle|\Psi\right\rangle \tag{4.47}$$

Integrating the first term by parts in the time-domain and then going through a procedure similar to the one used for obtaining (4.47) we obtain

$$\left\langle\Psi\middle|\psi^\dagger\left(-i\hbar\frac{\partial}{\partial t} + \frac{\hbar^2}{2m}\nabla^2\right)\psi - W\middle|\frac{\partial\Psi}{\partial\theta}\right\rangle = \left\langle\Psi\middle|V(\mathbf{r}, t)\middle|\frac{\partial\Psi}{\partial\theta}\right\rangle \tag{4.48}$$

Thus we finally get from (4.46–48), the result

$$\frac{dA_0}{d\theta} = \int_{t_1}^{t_2} dt \int d\mathbf{r} v(\mathbf{r}, t) \frac{d}{d\theta} n(\mathbf{r}, t; \theta) - \int_{t_1}^{t_2} \frac{1}{\theta} \langle \Psi | W | \Psi \rangle \tag{4.49}$$

Here $n(\mathbf{r}, t; \theta)$ is the density associated with the Coulomb system with coupling strength θe^2. From the stationary principle for the total action A, given by (4.37), after making appropriate changes, we have

$$\frac{dA[\Psi]}{d\theta} = 0 = \frac{\partial A_0}{\partial \theta} - \int_{t_1}^{t_2} dt \int d\mathbf{r} v(\mathbf{r}, t) \frac{d}{d\theta} n(\mathbf{r}, t; \theta) \tag{4.50}$$

and hence, we obtain

$$\frac{dA_0}{d\theta} = \frac{\partial A_0}{\partial \theta} - \int_{t_1}^{t_2} dt \frac{1}{\theta} \langle \Psi | W | \Psi \rangle$$

which on integrating w.r.t. θ from 0 to 1 leads to the result quaoted in (4.45).

$A_0[0, n]$ is the action function of a system of *independent particles* in (4.45). Unlike in the static case, discussed in Section II, we can only *assume* that there is a system of N independent fermions in a suitable local potential, characterized by the state vector $|\Psi_0(t)\rangle$ such that

$$n(\mathbf{r}, t) = \langle \Psi_0(t) | \rho(\mathbf{r}) | \Psi_0(t) \rangle \tag{4.51}$$

and

$$A_0[0, n] = \int_{t_1}^{t_2} dt \int d\mathbf{r} \sum_{\sigma} \left\langle \Psi_0(t) \middle| \psi^{\dagger}(x) \left(i\hbar \frac{\partial}{\partial t} + \frac{\hbar^2}{2m} \nabla^2 \right) \psi(x) \middle| \Psi_0(t) \right\rangle \tag{4.52}$$

with $n(r, t)$ the density of the interacting system. In the simplest case where $|\Psi_0(t)\rangle$ is a determinant of spin orbitals, $\phi_{n\sigma}(\mathbf{r}, t)$, (4.51) and (4.52) become simply

$$n(\mathbf{r}, t) = \sum_{n=1}^{N} \sum_{\sigma} |\phi_{n\sigma}(\mathbf{r}, t)|^2 \tag{4.53}$$

and

$$A_0[0, n] = \int_{t_1}^{t_2} dt \int d\mathbf{r} \sum_{n=1}^{N} \sum_{\sigma} \phi_{n\sigma}^*(\mathbf{r}, t) \left(i\hbar \frac{\partial}{\partial t} + \frac{\hbar^2}{2m} \nabla^2 \right) \phi_{n\sigma}(\mathbf{r}, t) \tag{4.54}$$

Differentiating the total action function, (4.37), with respect to $\phi_{n\sigma}^*(\mathbf{r}, t)$ and using (4.45), (4.53), (4.54), and the stationary action principle, we find that the one-particle wave functions obey a time-dependent Schrödinger equation, which we state here as another proposition.

Proposition. 2. $\{\phi_n(\mathbf{r}, t)\}$ obey the equations

$$i\hbar \frac{\partial}{\partial t} \phi_n(\mathbf{r}, t) = \left(-\frac{\hbar^2}{2m} \nabla^2 + V_{\text{eff}}(\mathbf{r}, t) \right) \phi_n(\mathbf{r}, t) \tag{4.55}$$

with the effective potential

$$V_{\text{eff}}(r, t) = V_0(r) + V_{\text{ext}}(r, t) + V_c(r, t) \tag{4.56}$$

where V_c is due to the Coulomb interaction:

$$V_c(\mathbf{r}, t) = \frac{\delta}{\delta n(\mathbf{r}, t)} \int_{t_1}^{t_2} dt' \int_0^1 \frac{d\theta'}{\theta'} \langle W \rangle[\theta', n] \tag{4.57}$$

It is clear from this scheme that for a given approximation for $\langle W \rangle[\theta', n]$, the density can be computed from (4.53) and (4.55), and an improved approximation for $\langle W \rangle$ can be determined by functional differentiation of $n(\mathbf{r}, t)$ with respect to $v_{\text{ext}}(\mathbf{r}', t')$. This iteration process should be continued until a satisfactory convergence is achieved but, in practice, this may prove quite difficult. Peuckert used the Keldysh[88] real-time formalism to set up a theory of the type outlined previously. One could equally well use the imaginary-time formalism outlined in Section III.C, where one requires analytic continuation to real times.[30] As noted in (3.38), we introduce as was done by Peuckert, but in our notation, the density-response function

$$iL(\mathbf{r}_1 t_1; \mathbf{r}_2 t_2) = i\frac{\delta n(\mathbf{r}_1 t_1)}{\delta v_{\text{ext}}(\mathbf{r}_2 t_2)} = \langle T(\rho(\mathbf{r}_1 t_1)\rho(\mathbf{r}_2 t_2))\rangle - \langle \rho(\mathbf{r}_1 t_1)\rangle\langle \rho(r_2 t_2)\rangle \tag{4.58}$$

so that

$$\langle W \rangle[\theta', n] = \frac{e^2}{2}\theta' \int\int \frac{d\mathbf{r}\, d\mathbf{r}'}{|\mathbf{r} - \mathbf{r}'|}[iL(\mathbf{r}, t; \mathbf{r}'t) - \delta(\mathbf{r}' - \mathbf{r})n(\mathbf{r}, t) + n(\mathbf{r}, t)n(\mathbf{r}', t)] \tag{4.59}$$

where the response function L is a functional of v_{ext}. For the explicit determination of L, we must employ the system of independent particles, (4.55), (4.56), (4.57), to represent the particle density $n(\mathbf{r}, t)$ using again its definition via the Green function, (3.34) and set up a series for $n(\mathbf{r}, t) - n_0(\mathbf{r}, t)$. All this leads to complicated equations to determine L in terms of n. In its simplest form, where one uses a HF scheme for E_{xc}, the equation for L becomes that of RPA, in the first round of calculation and subsequently becomes very complicated.

Peuckert has given generalization of the preceding scheme to time-dependent and spin-dependent problems using the Keldysh formulation. His spin-dependent formulation does not include the general spin-dependent external perturbation such as that used in Theorem 4 of Section II, where we outlined a time-independent theory. Instead of presenting this, we here give a relativistic generalization of Peuckert's time-dependent theory from which the spin-dependent theory can be extracted as a nonrelativistic approximation. In the relativistic theory, the Dirac equation can also be deduced from an action principle for the Dirac particles:

$$A[\Psi] = \int_{t_1}^{t_2} dt \int d\mathbf{r} \left\{ \left\langle \Psi(t) \middle| \psi^\dagger(r, t) i\hbar \frac{\partial}{\partial t} \psi(\mathbf{r}, t) \middle| \Psi(t) \right\rangle - \langle \Psi(t) | H | \Psi(t) \rangle \right\} \tag{4.60}$$

where$H$ is the Hamiltonian operator given by (2.55). Separating out the external fields, we may write this in the form

$$A[\Psi] = -\frac{e}{c} \int_{t_1}^{t_2} dt \int d\mathbf{r} J_\mu(\mathbf{r}, t) A^\mu_{\text{ext}}(\mathbf{r}, t) + A_0[\Psi] \tag{4.61}$$

Here $J_\mu(\mathbf{r}, t)$ is the expectation value of the current operator $j_\mu(\mathbf{r}, t)$ in the state $|\Psi(t)\rangle$. $A_0[\Psi]$ is given by

$$A_0[\Psi] = \int_{t_1}^{t_2} dt \int d\mathbf{r} \left\{ \left\langle \Psi(t) \middle| \psi^\dagger(\mathbf{r}, t) i\hbar \frac{\partial}{\partial t} \psi(\mathbf{r}, t) \middle| \Psi(t) \right\rangle - \langle \Psi(t) | H_0 | \Psi(t) \rangle \right\} \tag{4.62}$$

where H_0 is the Hamiltonian containing no external fields but includes all other interactions of the Dirac particles with the radiation fields, and so on, as before. The external fields A^μ_{ext} are assumed to be unquantized as before. One then obtains the following propositions by proceeding exactly as in the nonrelativistic case:

Proposition 3. The action function A of a many-electron relativistic system in the presence of an external time-dependent four potentials, $A^\mu_{\text{ext}}(\mathbf{r}, t)$ is a functional of the time-dependent four-current density $J_\mu(\mathbf{r}, t)$ and is stationary with respect to variations in $J_\mu(\mathbf{r}, t)$.

Proposition 4. The set of 4-spinors $\{\phi_i(\mathbf{r}, t)\}$ obeys Dirac-like equations

$$i\hbar \frac{\partial}{\partial t} \phi_i(\mathbf{r}, t) = \{(-i\hbar c \boldsymbol{\alpha} \cdot \boldsymbol{\nabla} + \beta mc^2) + V_{\text{eff}}[\mathbf{r}, t; J_\mu]\} \phi_i(\mathbf{r}, t) \tag{4.63}$$

where

$$V_{\text{eff}}[\mathbf{r}, t; J_\mu] = -e\left(v_{\text{ext}}(\mathbf{r}, t) + \frac{\delta G[J_\mu(\mathbf{r}, t)]}{\delta J_0(\mathbf{r}, t)} \right) 1 - \frac{e}{c}\left(A_{\text{ext}}(\mathbf{r}, t) + \frac{\delta G[J_\mu(\mathbf{r}, t)]}{\delta \mathbf{J}(\mathbf{r}, t)} \right) \cdot \boldsymbol{\alpha} \tag{4.64}$$

where

$$J_\mu(\mathbf{r}, t) = c \sum_{i(\text{occ})} \text{tr}(\phi_i^\dagger(\mathbf{r}, t) \gamma_0 \gamma_\mu \phi_i(\mathbf{r}, t)) \tag{4.65}$$

G in (4.64) is given by the universal form,

$$G[J_\mu] = \int_{t_1}^{t_2} dt \Big\langle \Psi \Big| (t) \Big\{ H_{em} - \frac{e}{c} \int d\mathbf{r} \colon j_{tr}^i(\mathbf{r}, t) \colon \colon A_{tr\,i}(\mathbf{r}, t) \colon$$
$$+ \frac{e^2}{2} \int\int d\mathbf{r}\, d\mathbf{r}' \colon \psi^\dagger(\mathbf{r}, t)\psi(\mathbf{r}, t) \colon \frac{1}{|\mathbf{r}-\mathbf{r}'|} \colon \psi^\dagger(\mathbf{r}', t)\psi(\mathbf{r}', t) \colon \Big\} \Big| \Psi \Big\rangle \quad (4.66)$$

in the notation of (2.55).

The preceding analysis can be made only if the interactions represented by $G[J_\mu]$ is a functional of J_μ. This can be done if the mapping $A^\mu_{\text{ext}} \to J_\mu$ is invertible in a small neighborhood of $A^{\mu 0}_{\text{ext}}(\mathbf{r}, t)$ that we may be interested in. This problem is more difficult than in the scalar case of the nonrelativistic problem discussed earlier in this section because we now have to solve a coupled set of integral equations. The linear response function, $\partial J_\mu / \delta A^v_{\text{ext}} \equiv \Pi_{\mu\nu}$, can be calculated by the methods outlined before. In the static external field case, Rajagopal and Callaway[33] showed how the relativistic calculation can be carried through, in the sense that the mapping is invertible. In the time-dependent case, the assumption that such a scheme can be made plausible, has to be asserted. One observes again that the equation of continuity holds for J_μ, that is, $\partial^\mu J_\mu(\mathbf{r}, t) = 0$, from (4.63).

We conclude this section with several comments. It should be stated at the outset that a time-dependent theory is desirable, for this would make the formulation fully self-consistent and more widely applicable for all kinds of problems of interest, since the scheme is essentially a one-particle, self-consistent one. In particular, we can determine the excited states of the system also. However, from our discussion here, it is clear that the theory is not entirely based on rigorous foundation as was the case with the ground-state properties, discussed in earlier sections. It is for this reason, we put together the results in the form of propositions rather than theorems. In the context of self-energy of the system $\Sigma(\mathbf{r}, t; \mathbf{r}', t')$ introduced in the full many-body theory, we note that the time-dependent theory gives for it a "local" expression in both space and time variables from (3.32) and (4.55), we note that

$$\Sigma(\mathbf{r}, t; \mathbf{r}', t') = V_{\text{eff}}[\mathbf{r}, t; n(\mathbf{r}, t)]\delta(\mathbf{r}-\mathbf{r}')\delta(t-t') \quad (4.67)$$

where $V_{\text{eff}}[\mathbf{r}, t; n(\mathbf{r}, t)]$ is given by (4.56). In the many-body theory, one always obtains a nonlocal expression for Σ. An interesting feature of (4.67) is that it is *nonlinear* potential in that $n(\mathbf{r}, t)$ is defined back in terms of the Green function, via (3.34) and so the Green function equation is closed. The linear response functions can then be determined using the procedure outlined in Section III and thus, the time-dependent scheme gives in principle, a scheme for fully self-consistent theory of electronic structure of condensed matter. The actual application of this theory is yet to come. Peuckert has made an application of his scheme to the calculation of excited states of helium atom, with some success, and this is discussed in the next section. As observed in the section, one could also use the imaginary-time theory instead of the real-time formalism.

V. APPLICATIONS

The formalism developed in the previous sections is now discussed in light of the applications made by several groups of people to a large variety of problems. These applications are dealt with under five broad classifications—nuclear theory, atomic and molecular physics (with applications to chemistry), solid-state physics with bulk and surface properties discussed separately and, finally, liquid-state properties. The activity in this field is fierce and we have attempted here to review as much as possible. The purpose of our review is to exhibit the kinds of results people have been able to achieve in this formalism and give the reader thereby an overview of the scope of the theory. The questions that arise in such applications are many and we raise them where necessary. We do not review here any work of similar nature based on either the Thomas–Fermi theory for which the reader is referred to March's book[8] or the $X\alpha$ method for which one may see Connolly's review[14] and also Slater's remarkable book on the self-consistent fields,[89] and many papers appearing in the Quantum Theory Symposium series held every year at Sanibel, Florida, some of which have been mentioned in this article.

A. Nuclear Theory

Since nuclear theory is outside the scope of the present review, we content ourselves with a very brief account. Brueckner and his collaborators[90,91] applied the HKS theory to finite nuclei. The potential energy functional of density is derived from a nuclear-matter calculation. The ground-state density distribution is found through minimization with respect to the density as required by the HKS formalism. The main part of the inhomogeneity corrections comes from the finite range of the nuclear forces. This connection is the only phenomenological aspect of the theory, which is adjusted to reproduce the experimental binding energy of Ca^{40}. As a first step, in Ref. 90, neutron and proton densities are assumed to be proportional and thus the calculations were limited to light and medium nuclei. In Ref. 91 this assumption was relaxed and the calculations were thus extended to medium and heavy nuclei. Binding energy, radius, and surface thickness are found to be in good agreement with experiment for light nuclei[90] whereas surface thicknesses are too large and rms radii too small for heavy nuclei.[91] The HKS formalism goes beyond TF theories and as such these results are expected. A shell calculation seems to be needed to reproduce the detailed structure of the distribution of the nuclei[90] which the simple HKS theory does not incorporate. These authors conclude that the energy-density formalism is a very good method for studying the existence of superheavy nuclei. Improvement in the energy functional will enable one to obtain better answers. The authors are able to reproduce the binding energies of nuclei from a mass formula that is fundamentally based in theory. The formalism is seen to be a good substitute for the Hartree–Fock approximation when one is looking for a gross structure rather than a microscopic description. More work incorporating nonlocality of interaction and use of meson-exchange theory in developing the density-functional may prove useful. The time-dependent theory of Section IV may also be similarly useful as substitute for the time-dependent Thomas–Fermi theory used by Koonin and Rand[92] for studying one-body nuclear dynamics as well as the extended time-dependent HF scheme of Wong and Tang.[93]

B. Atomic and Molecular Physics—Chemical Applications

In this subsection, we review various aspects of atomic and molecular physics in the light of the HKS formalism. The types of questions raised fall into three classes: (*1*) physics of atoms where one is concerned with the ground state of the atomic many-electron system (multiplet structure) with one central attractive nuclear charge appearing as the external potential in the HKS theory; (*2*) physics of simple molecules where two-center potentials due to the two nuclear charges appear as the external potential in the HKS theory. Here one asks for not only the ground-state properties but also possible low-lying excitation energies. There are other questions of interest such as collisional polarizabilities of closed shell systems which can also be studied in this formalism. Lastly, recently a density-functional viewpoint of electronegativity has been put forward, which we discuss also. We take these up in the sequence stated.

1. Physics of Simple Atoms

Since the HKS formalism is so general, it ought to be applicable to a wide variety of inhomogeneous systems. Atoms represent the simplest such many-body inhomogeneous system with the Coulomb interaction of the point nuclear charge with the electrons appearing as an external potential. The use of the HKS formalism with the exchange-correlation energy E_{xc} of the homogeneous electron gas as input may seem to be a poor approximation on intuitive grounds. Gordon and Kim[28] were one of the first to investigate a model of this kind. The density-functional methods supercede these model calculations because they are based on certain general principles and not ad hoc as with Gordon–Kim models. The first atomic calculations were done by Tong and Sham,[94] based on the local density approximation. Such atomic calculations are of intrinsic importance and, moreover, are essential in the calculation of cohesive energies. The cohesive energy is the difference between the energy per atom of a metal and the energy of the isolated atom. The calculation of Tong and Sham showed that the correlation energies did not give good answers (being a factor of 2 too large) while the calculation with exchange contribution only were quite good (within 10% of known answers for quantities such as atomic ground-state energies and densities). Tong[56] studied a model finite system to investigate why such answers were obtained and showed that the major source of error comes from the fact that the low-lying levels of a finite system are discrete and have finite spacings. The replacement of summation by integration in a diagrammatic analysis showed an overestimate of correlation energy, but only a small error in the exchange energy of a finite system such as an atom. Although this observation seemed like an important point against the HKS scheme (where one assumes correlation contributions can be taken into account by the use of a local density scheme that is itself calculated from the known electron gas results where one has only a continuous set of energies of the system), there was still a hope in that one thought that by incorporating inhomogeneity corrections such as gradient terms, this could be overcome. This proved to be not the case. It is of interest to point out that Gordon and Kim[28] also examined how best finite systems with discrete energy levels can be represented by an electron gas with continuous energy levels and had come to similar conclusions. In a direct approach to the correlation energies of atoms, Kelly[95] found that in the calculation of correlation energy of atomic systems using many-body perturbation theory, there is a class of terms violating the exclusion principle that are among the major contributors and these terms do not contribute in the homogeneous interacting infinite electron gas systems. It was therefore thought that the HKS method is perhaps best suited for systems where electron gas is a good representation and so the ground-state properties of simple metals such as Na could perhaps be dealt with in this way. Tong[96] made such an application to sodium metal. In a full self-consistent computation, in which the only input parameter is the number of electrons per atom ($Z = 11$) Tong found the equilibrium lattice constant to be 1.3% of the experimental result and the compressibility about 11% of the experimental value whereas the cohesive energy was 23% larger than the experimental result. The first two quantities concern the solid state whereas the cohesive energy calculation requires the energies of the solid and the atomic states. The latter is a source of error as is now clear from Tong's earlier work.

An important observation was made at this crucial stage of the development by Gunnarsson, Lundqvist, and Wilkins.[97] They argued that for the hydrogen and sodium atoms and for other applications with unpaired valence electrons, a spin-density formulation should be used and not the density formalism as was the case with Tong's work. They showed that a local-spin-density (LSD) approximation gives a value of the energy of the hydrogen atom only 1.6% smaller in magnitude than the exact Rydberg energy as compared to 10% derivation in the local density (LD) approximation. In sodium, they obtained the cohesive energy to within 4% of the experimental value. These authors thus turned the tide in favor of the HKS-scheme with LSD approximation as the appropriate method. They argued that the spin-density formulation has in it the relevant additional physical input compared to the then used local density formalism since it takes proper account of the exchange-correlation effects of systems with unpaired electrons. The usefulness of the LSD formalism in other systems of physical interest was eloquently put forward by Gunnarsson and Lundqvist.[44] We describe the results obtained by Gunnarsson and co-workers in the areas of atomic and molecular physics in this section.

The main advantage of the spin-density functional (SDF) over the density functional (DF) formalism is that the spin dependence allows one greater flexibility in the variational calculation as well as allowing one to incorporate more of the actual physics into the approximate functionals. The SDF scheme allows one to have freedom in the specification of orbital and spin parts and hence can be employed in a calculation of a large class of excited states, too. Thus a whole class of functionals of more general order parameters could be constructed. Usually the exchange and correlation effects are nonlocal in character. In the SDF theory, they can be exactly described in terms of a local potential, which however will depend on the spin densities in a nonlocal way, in principle. One resorts to approximate functionals since the exact solution of the many-electron problem is not known. The LSD approximation seems to be a natural starting point. This approximation is exact in the limit of slow and weak spatial variation of the spin density.

It is assumed that the spin density dependence of E_{xc} is via $|\boldsymbol{s}(\mathbf{r})|$ only because this is so for the homogeneous gas. This is an important assumption in all subsequent discussion.

In practice by calculating the HKS ground states of a neutral atom or molecule and the ion, the ionization energies can be computed. To apply the HKS scheme to excited states, Gunnarsson and Lundqvist[44] proved a theorem which allows one to apply the HKS theory to a *large class of excited states*, namely the energetically lowest state of each symmetry. This has been used by von Barth[98] to calculate the multiplet structure of several light atoms in the HKS formalism. This is discussed presently. Let us first state the theorem.

Theorem A. Suppose that a set of observables {0} are constants of motion restricting the external spinor potentials $v_{\alpha\beta}$, so that the Hamiltonian of the inhomogeneous system commutes with 0. For such external potentials, the lowest state which has the specified eigenvalues $\{\lambda\}$ of {0} is a functional of the spin density. Consequently, the total energy is also a functional of the spin density

$$E_{v_{\alpha\beta,\lambda}}[n, \boldsymbol{s}] \equiv \sum_{\alpha\beta} \int v_{\alpha\beta}(\mathbf{r})p_{\alpha\beta}(\mathbf{r})d\mathbf{r} + \frac{e^2}{2}\int \frac{n(\mathbf{r})n(\mathbf{r}')}{|\mathbf{r}-\mathbf{r}'|}d\mathbf{r}\, d\mathbf{r}' + G_\lambda[n, \boldsymbol{s}] \tag{5.1}$$

Observe that the functional $E_{v_{\alpha\beta,\lambda}}$ depends on the quantum numbers $\{\lambda\}$ because of the symmetry requirement. The functional $G_\lambda[n, \boldsymbol{s}]$ is defined only for densities that can be realized with potentials $v_{\alpha\beta}$ in the class {0}.

In this scheme, it is clear that the quantum numbers $\{\lambda\}$ are well defined for both the noninteracting and interacting systems. The variational principle can now be used on the functional, (5.1), and we find the Schrödinger-like equation for noninteracting electrons moving in an effective one-body potential. To ensure that the total system has quantum numbers $\{\lambda\}$, we sometimes have to go beyond the simple product state of the HKS scheme. The procedure for doing this is well-known in atomic physics and we use this in our discussion of multiplet structures. The effective potential is a big unknown entity in this theory. As a practical matter, one employs the electron-gas results as before but uses a spin density constructed from orbitals of appropriate symmetry. Such a scheme provides a numerical way of approaching this problem.

For the open-shell atoms, the simplest way to construct the HKS state with the specified quantum numbers is provided by the well-known rules due to Slater.[99] The SDF scheme is applied to the lowest state of each symmetry; for light atoms, the quantum numbers LSM_LM_S characterize the states. The inclusion of spin orbit coupling is straightforward where now $LSJM$ are the quantum numbers to be used. The charge density is kept spherically symmetric in the case of light atoms by merely averaging the charge density of all M_l's. The electron configuration giving the lowest total energy is not always the same as the experimental ground-state configuration, and in two cases (Fe and Co), Moruzzi et al.[21] found it essential to introduce nonintegral configurations (i.e., $(3d\uparrow)^5(3d\downarrow)^{1.404}(4s\uparrow)^1(4s\downarrow)^{0.596}$ for Fe and $(3d\uparrow)^5(3d\downarrow)^{2.897}(4s\uparrow)^1(4s\downarrow)^{0.103}$ for Co). The use of nonintegral occupation numbers has been discussed by Slater.[39] The incorrect and nonintegral configurations are occurring because the $3d$ levels are dropping down through $4s$ levels too early in the transition series as can be found by the table of Moruzzi et al. which is reproduced here for convenience as Table II. This table is of interest in many ways since it gives a comparison of the LSD result with both HF and experimental results. It is not clear whether this is due to the approximations made in the LSD scheme or whether it is due to the spherical symmetry forced on the atom by the choice of the density as stated previously. From Table II, the differences between the Hartree–Fock calculation and LSD theory become evident. We should point out that, since the HF energies represent a minimization of the exact many-electron Hamiltonian within a set of approximate wave functions (determinants) these energies are upper bounds to the true energies. Since the LSD energies are systematically higher than their HF counterparts, they are more in error.

Gunnarsson et al.[44,97] have calculated results for the first-row atoms and some noble-gas and alkali atoms. They found that the deviation from the experimental results for the ionization energies is on the average smaller and has a smoother variation from atom to atom in their calculation than in other schemes. In particular, the LSD answers are substantially better for H, He, and O than those in the LD scheme of Tong and Sham. The results for Ne and Ar in LSD and LD schemes are similar. This is because in Ne and Ar the spin polarization is unimportant whereas in H, He, and O, it is of great importance. The two calculations of Moruzzi et al. and Gunnarsson et al. are for different set of atoms generally except for a few. The two sets of results are in general agreement and the two groups employ slightly different parametrixation of E_{xc} as discussed in Section III.

It is of interest to discuss two of the special atoms in some detail and they are H atom and H^- ion. In the case of H atom, the exchange-correlation contribution should essentially cancel out to give the correct result whereas the H^- ion, as is well known, is a consequence of the all-important correlation effects. The LSD scheme[44,97] for H atom gives -13.39 eV whereas the LD scheme[94] gives -12 eV, compared to the exact answer of -13.61 eV. For the H^-, the LSD answer is quite close to

TABLE II
Atomic Total Energies and Ground-State Configurations Obtained in the Spin-polarized Local-Density (LSD) Approximation, Compared to Configuration-Averaged Hartree–Fock Total Energies, and Experimental Ground-State Configurations[a]

Atom	LSD total energy (Ry)	HF total energy (Ry)	LSD ground config.	Exp. config.
H	−0.976	−1.000	$1s$	$1s$
Li	−14.709	−14.865	$1s^2 2s$	$1s^2 2s$
Be	−28.909	−29.146	$1s^2 2s^2$	$1s^2 2s^2$
Na	−322.902	−323.718	$3s$	$3s$
Mg	−398.274	−399.229	$3s^2$	$3s^2$
Al	−482.637	−483.754	$3s^2 3p$	$3s^2 3p$
K	−1196.382	−1198.330	$4s$	$4s$
Ca	−1351.442	−1353.517	$4s^2$	$4s^2$
Sc	−1517.322	−1519.472	$3d4s^2$	$3d4s^2$
Ti	−1694.591	−1696.740	$3d^3 4s$	$3d^2 4s^2$
V	−1883.529	−1885.607	$3d^4 4s$	$3d^3 4s^2$
Cr	−2084.386	−2086.284	$3d^5 4s$	$3d^5 4s$
Mn	−2297.211	−2299.252	$3d^5 4s^2$	$3d^5 4s^2$
Fe	−2522.369	−2524.582	$3d^{6.4} 4s^{1.6}$	$3d^6 4s^2$
Co	−2760.288	−2762.617	$3d^{7.9} 4s^{1.1}$	$3d^7 4s^2$
Ni	−3011.233	−3013.632	$3d^9 4s$	$3d^8 4s^2$
Cu	−3275.464	−3277.928	$3d^{10} 4s$	$3d^{10} 4s$
Zn	−3553.007	−3555.697	$3d^{10} 4s^2$	$3d^{10} 4s^2$
Ga	−3843.556	−3846.522	$3d^{10} 4s^2 4p$	$3d^{10} 4s^2 4p$
Rb	−5872.477	−5876.716	$5s$	$5s$
Sr	−6258.683	−6263.092	$5s^2$	$5s^2$
Y	−6658.819	−6663.369	$4d5s^2$	$4d5s^2$
Zr	−7073.319	−7077.938	$4d^3 5s$	$4d^2 5s^2$
Nb	−7502.361	−7506.984	$4d^4 5s$	$4d^4 5s$
Mo	−7946.087	−7950.739	$4d^5 5s$	$4d^5 5s$
Tc	−8404.434	−8409.215	$4d^6 5s$	$4d^5 5s^2$
Ru	−8877.807	−8882.913	$4d^7 5s$	$4d^7 5s$
Rh	−9366.418	−9371.675	$4d^9$	$4d^8 5s$
Pd	−9870.419	−9875.844	$4d^{10}$	$4d^{10}$
Ag	−10389.75	−10395.40	$4d^{10} 5s$	$4d^{10} 5s$
Cd	−10924.44	−10930.27	$4d^{10} 5s^2$	$4d^{10} 5s^2$
In	−11474.28	−11480.34	$4d^{10} 5s^2 5p$	$4d^{10} 5s^2 5p$

[a] Reproduced from Moruzzi et al.[21]

the exact answer, −14.4 eV, even though there is uncertainty due to slow convergence towards self-consistency. A detailed discussion of the H^- problem is contained in a paper of Shore et al.[100] The main point is that LD scheme completely fails in the case of H^- in that one cannot get the bound state in a straightforward way, whereas LSD gives essentially the correct answer. More importantly, the case of H^- should serve as an important test case for all schemes, such as nonlocal exchange-correlation schemes.

Recently Schwarz[101] has obtained the first ionization potentials of light atoms up to krypton by means of Slater's transition-state method with LD scheme. He has also computed the multiplet structure. In view of the comments made in the last few paragraphs, an LSD scheme is preferred to the LD scheme for such calculations. von Barth's calculation[98] provides such a scheme which we briefly discuss. It is found that a straightforward application of the theorem of Gunnarsson–Lundqvist stated in this section (Theorem A) with symmetry-independent E_{xc} does not work. He shows how the local density theory can be modified suitably to give energies of states of mixed symmetry and the multiplet splittings are obtained from these estimates. von Barth first notes that the symmetry dependence of the energy of a state, $E_{v_{\alpha\beta},\{\lambda\}}$ of a given symmetry $\{\lambda\}$ enters via the charge and spin densities of that state because of Theorem A. Consider, for example, the p^2 configuration of carbon (C) atom. The p^2 configuration gives rise to three terms, ^3P, ^1D, and ^1S. The singlet states ^1D and ^1S have no spin density. Also, the ^1S state has a spherical charge density and it is possible to form a linear combination of the five states of the ^1D to have spherical charge density. Since the states belonging to ^1D are degenerate, in the absence of spin-orbit coupling, the previous procedure would give the same energy for the ^1S and ^1D states. This is clearly wrong. The previous procedure applied to the excited $1s2s$ configuration of helium atom is also shown to be incorrect. von Barth made an important observation that any single determinental many-electron state will have the $g_{\uparrow\downarrow}(\mathbf{r}, \mathbf{r}') = g_{\downarrow\uparrow}(\mathbf{r}, \mathbf{r}') = 1$, in the notation of Section. III. Therefore, any scheme where a single determinant is used will have $E_{xc}^{\uparrow\downarrow} = E_{xc}^{\downarrow\uparrow} = 0$ and consequently this part of the exchange energy is exactly given by the local density theory. This is the reason for the agreement in experimental ^3P–^1D splitting for C and the simple LD energies outlined previously. He points out the importance of the sum rules on the pair correlation functions and the dependence of the E_{xc} on these correlation functions. Having shown that the symmetry independent approximation to E_{xc} suggested by Gunnarsson and Lundqvist[44] for the case of states of pure symmetry works only when the states in the noninteracting case reduces to a single Slater determinant, he goes on to set up a scheme that retains the symmetry-independent E_{xc} but constructs an HKS theory with states of mixed symmetry. As he showed, in the case of the p^2 configuration of C atom, one can get the energies of ^3P and ^1D states but not that of the ^1S state (being equal to that of ^1D). The obvious way out is to construct a symmetry dependent E_{xc} which of necessity would be outside the scope of the electron gas theory. The approach suggested retains the electron gas E_{xc} but chooses mixtures of states such that their noninteracting counterparts have single determinants because Theorem A applies only in this situation. Since a mixed symmetry state is not an eigenstate of the Hamiltonian, the corresponding charge density will oscillate in time, of necessity, by virture of the principles of quantum theory. von Barth then argues that the state with the smallest energy expectation value is a functional of the density at the initial time of preparation of the state. Essentially, he shows that the theory for mixed symmetries ought to be time-dependent by the very nature of the problem. In view of Section IV and Peuckert's time-dependent HKS theory using the stationary action principle, the von Barth demonstration is given here in some detail. Peuckert[26] has used the stationary action principle directly to examine the excited states of the helium atom, which we discuss after we complete our description of von Barth's work.

Suppose the lowest state of each symmetry S is $|S_i, 0\rangle$. For simplicity consider only two such states. Let

$$|\psi\rangle = \alpha_1|S_1, 0\rangle + \alpha_2|S_2, 0\rangle \tag{5.2}$$

be a mixed symmetry state with the real constants (for simplicity) α_1, α_2 such that $\alpha_1^2 + \alpha_2^2 = 1$. By this choice, it is hoped that $|\psi\rangle$ is not degenerate. The Hamiltonian H of the system is assumed to commute with S and hence in general

$$H|S_i, 0\rangle = E_0(S_i)|S_i, 0\rangle \qquad (i = 1, 2) \tag{5.3}$$

From this it is clear that $H|\psi\rangle \neq E_0|\psi\rangle$. However, the expectation value of H, given by

$$\langle\psi|H|\psi\rangle = \alpha_1^2 E_0(S_1) + \alpha_2^2 E_0(S_2) \tag{5.4}$$

with no cross terms because S_is are good quantum numbers for H, too. Suppose there exists a different state

$$|\psi\rangle' = \alpha_1|S_1, 0\rangle' + \alpha_2|S_2, 0\rangle' \tag{5.5}$$

which has the same density as $|\psi\rangle$. The states $|S, n\rangle'$ are obtained from another Hamiltonian $H' = H_0 + W'$ with a different external potential, W', which commutes with S as before. It is assumed that α_1, α_2 are the same as before, and they are determined uniquely by the requirement that $|\psi\rangle$ is a single determinant state. Then von Barth proves that one gets into an absurd situation as with the HKS theorem and hence $|\psi\rangle$ is a functional of $n(\mathbf{r})$ and so is the energy, (5.4). Moreover, $E[n]$ is a minimum for the correct density and will depend on the symmetries S_1

and S_2 and on the coefficients α_1 and α_2, and these *coefficiients must be chosen so that $|\psi\rangle$ is a single determinant in the noninteracting case.*

If a state $|\Psi\rangle$ is prepared at time $t = 0$ to be given by (5.2), then at some later instant, t, this state becomes

$$|\Psi_t\rangle = \alpha_1 e^{-iE_0(S_1)t}|S_1, 0\rangle + a_2 e^{-iE_0(S_2)t}|S_2, 0\rangle \tag{5.6}$$

The corresponding density is

$$\begin{aligned} n_t(\mathbf{r}) = \langle\Psi_t|\rho(\mathbf{r})|\Psi_t\rangle &= \alpha_1^2\langle S_1, 0|\rho(\mathbf{r})|S_1, 0\rangle + \alpha_2^2\langle S_2, 0|\rho(\mathbf{r})|S_2, 0\rangle \\ &\quad + 2\alpha_1\alpha_2\langle S_1, 0|\rho(\mathbf{r})|S_2, 0\rangle \cos(E_0(S_1) - E_0(S_2))T \\ &\neq n_{t=0}(\mathbf{r}) \end{aligned} \tag{5.7}$$

because

$$\langle S_1, 0|\rho(\mathbf{r})|S_2, 0\rangle \neq 0$$

Treating time t as a parameter and using the fact that the external potential v_{ext} commutes with S, and hence

$$0 = \langle S_1, 0|v_{\text{ext}}|S_2, 0\rangle = \int v_{\text{ext}}(\mathbf{r})\langle S_1, 0|\rho(\mathbf{r})|S_2, 0\rangle d\mathbf{r} \tag{5.8}$$

the new energy functional is shown to be a functional of n_t, and is minimized by the correct, n_t. Using the HKS equations, we would obtain $n(\mathbf{r}) = n_t(\mathbf{r})$ which is clearly wrong. The question then is which of the $E[n]$, $E_t[n_t]$ is appropriate. From Theorem A and the observation that α_is are so chosen to give a single determinant for $|\Psi\rangle_{t=0}$ in the noninteracting case, we observe that $E_t[n_t]$ is to be used for $t = 0$ only. This shows that the initial preparation of the state $|\Psi\rangle_{t=0}$ as a single determinant fixes the procedure to be followed. It should be clear from this description, that the theory of mixed symmetry requires a time-dependent formalism by its very nature and the prescription given by von Barth is to be thought of as a calculational scheme that retains the essence of HKS formalism. The time-dependent theory of Peuckert is on a different footing entirely and there are some points of similarity that will be brought out later. von Barth has applied his scheme to examine the p^2 configurations of the carbon and silicon atoms, the p^3 configuration of the nitrogen atom, and to the excited $1s2s$ configuration of the helium atom. Only the last case has been dealt with by Peuckert in his time-dependent theory. It is found that the local density theory gives results in good agreement with experiment and is better than the HF results. In the case of $1s2s$ excited states of helium atom, the HF results are very bad giving a splitting of 0.12 eV between the singlet and the triplet states whereas the local density approach gives it to be 0.96 eV, which compares very well with the experimental value of 0.80 eV. But the multiplet average energy $[=\frac{1}{4}E(^1S)+\frac{3}{4}E(^3S)]$ is not as good, the local density approximation giving 19.32 eV whereas the experimental value is 20.02 eV and the spherical paramagnetic calculation giving 21.90 eV.

In conclusion, it appears that the von Barth scheme offers a simple and useful technique which is at least as accurate as HF, or even better, for calculating multiplet structures of simple atoms and molecules. It is shown that the straightforward application of Theorem A for some states of pure symmetry fails because the exchange-correlation hole is misrepresented by the local density theory. Appealing to the pair-correlation functions, von Barth demonstrated that the use of a single-determinant form for the mixed symmetry state would give a proper description within a local-density scheme. In general there are more determinants than pure state energies and one obtains an overdetermined set of equations which are solved by a mini-max procedure. von Barth also showed that in the cases he considered (and mentioned earlier) except that of helium, the local density theory with exchange only reproduced the HF results to within 0.1 eV! He showed that the charge density of the electrons responsible for multiplet splittings is slowly varying in such a way that the local density approximation seems to be quite applicable. The scheme developed may be entirely wrong for the cases where the electron correlations become strong and so one may have to employ a symmetry-dependent exchange-correlation functional as Theorem A would demand.

We now discuss another approximation scheme based on the time-dependent theory of Peuckert.[26] This scheme is more complicated and so we content ourselves with a description of the method in a qualitative way. It is an iterative scheme beginning with the assumption that $V_c(r, t)$ in (4.56) is independent of $n(\mathbf{r}, t)$ and linear response in $\delta n = n(\mathbf{r}, t) - n_0(\mathbf{r})$, where $n_0(\mathbf{r})$ is the density of the state to be considered is treated in linear order in a time-dependent perturbation method. Unlike the von Barth scheme, this method does not use the electron gas data and as such, one has to construct the effective potentials in each case. The theory was described in Section IV.C and we therefore outline its application to the study of the ^1S and ^3S states of the He atom.

We make use of a complete set of *real* eigenfunctions $\{\zeta_n(r)\}$ of the time-independent Schrödinger equation

$$h_0\zeta_n(r) \equiv \left\{-\frac{\hbar^2}{2m}\nabla^2 + \left(-\frac{2e^2}{r} + V_c(r)\right)\right\}\zeta_n(r) = \epsilon_n\zeta_n(r) \qquad (5.9)$$

The annihilation operator, $\psi_\sigma(r)$, is then written in terms of it:

$$\psi_\sigma(r) = \sum_{n=1}^{\infty} C_{n\sigma}\zeta_n(r) \tag{5.10}$$

For the ground state assumed to be $|\Psi_0\rangle = C_{1\uparrow}^\dagger C_{1\downarrow}^\dagger|0\rangle$, we can compute the response function to be approximately

$$\begin{aligned} iL^{(0)}(r,t;r',t) &\equiv \langle\Psi_0|\rho(\mathbf{r}t)\rho(\mathbf{r}'t)|\Psi_0\rangle - \langle\Psi_0|\rho(\mathbf{r}t)|\Psi_0\rangle\langle\Psi_0|\rho(\mathbf{r}t)|\Psi_0\rangle \\ &= \delta(\mathbf{r}-\mathbf{r}')n^{(0)}(\mathbf{r},t) - \tfrac{1}{2}n^{(0)}(\mathbf{r},t)n^{(0)}(\mathbf{r},t) \end{aligned} \tag{5.11}$$

where

$$n^{(0)}(\mathbf{r},t) = \langle\Psi_0|\rho(\mathbf{r},t)|\Psi_0\rangle = 2\zeta_1^2(rt) \tag{5.12}$$

In (5.11) and (5.12) we use the form (5.10) for $|\Psi_0\rangle$ but the wave function $\zeta_1(\mathbf{r}t)$ is left in a more general time-dependent form which will be determined presently. The ground-state self-consistent potential in the lowest approximation is then

$$V_c(r) = \int d\mathbf{r}' \frac{e^2}{|\mathbf{r}-\mathbf{r}'|}\zeta_1^2(\mathbf{r}') \tag{5.13}$$

using (4.57) and the preceding expressions. This is the same as Hartree–Fock approximation.

To construct the first excited $1s2s$ states of helium, we begin with

$$|\Psi\rangle_\pm = \frac{1}{\sqrt{2}}(C_{1\uparrow}^\dagger C_{2\downarrow}^\dagger \pm C_{2\uparrow}^\dagger C_{1\downarrow}^\dagger)|0\rangle \tag{5.14}$$

where $+,-$ signs correspond to the states with spin $S = 0, 1$, respectively, with their z-component of total spin, zero. As before, we have

$$n^{(\pm)}(\mathbf{r},t) = \zeta_1^2(r,t) + \zeta_2^2(r,t) \tag{5.15}$$

$$\begin{aligned} iL^{(\pm)}(\mathbf{r}t;\mathbf{r}t) = \delta(\mathbf{r}-\mathbf{r}')n^{(\pm)}(\mathbf{r},t) &- (\zeta_1^2(rt)\zeta_2^2(r't) + \zeta_1^2(r't)\zeta_2^2(rt) \\ &\pm 2\zeta_1(rt)\zeta_2(rt)\zeta_1(r't)\zeta_2(r't)) \end{aligned} \tag{5.16}$$

The time-independent part is calculated with $t = 0$ in (5.16) and then using it in (4.57) to construct equations for the corresponding $\zeta_1(r)$, $\zeta_2(r)$. This amounts to using an HF expression for the energy

$$\begin{aligned} E_H[\zeta_1,\zeta_2] = &\sum_{n=1}^{2}\int d\mathbf{r}\zeta_n(r)\left(-\frac{\hbar^2}{2m}\nabla^2 - \frac{2e^2}{r}\right)\zeta_n(r) \\ &+ \iint d\mathbf{r}\,d\mathbf{r}'\frac{e^2}{|\mathbf{r}-\mathbf{r}'|}(\zeta_1^2(r)\zeta_2^2(r') \pm \zeta_1(r)\zeta_2(r)\zeta_1(r')\zeta_2(r')) \end{aligned} \tag{5.17}$$

The next step is to construct $\delta n(\mathbf{r},t) = n(r,t) - n(r)$ using a first-order time-dependent perturbation theory in the presence of a time-dependent external potential, $v_{ext}(\mathbf{r}, t)$. This is done by solving for $\zeta_1(rt)$, $\zeta_2(rt)$ from a time-dependent version of (5.9) where $v_{ext}(r, t)$ appears and $v_c(rt)$ is computed from (5.15), (5.16) as follows:

$$\left.\begin{aligned} \zeta_1(rt) &\cong \zeta_1(r) + \int G_0(\mathbf{r}t; \mathbf{r}_1 t_1) v_{ext}(\mathbf{r}_1 t_1) \zeta_1(r_1) d\mathbf{r}_1\, dt_1 \\ &\text{and} \\ \zeta_2(rt) &\cong \zeta_2(r) + \int G_0(\mathbf{r}t; \mathbf{r}_1 t_1) v_{ext}(\mathbf{r}_1 t_1) \zeta_2(r_1) d\mathbf{r}_1\, dt_1 \end{aligned}\right\} \tag{5.18}$$

where G_0 is the one-particle Green function in the absence of $v_{ext}(r, t)$, from which, we obtain, after some simple algebra

$$\int dt \delta n(\mathbf{r}, t) = \int (n^{(\pm)}(\mathbf{r}, t) - n^{(0)}(\mathbf{r})) \cong \int dt(\zeta_1(r) f_1(\mathbf{r}, t) + \zeta_2(r) f_2(\mathbf{r}, t)) \tag{5.19}$$

where

$$\int dt f_i(\mathbf{r}, t) = \sum_{k=3}^{\infty} \langle k | v_{ext} | i \rangle \zeta_k(r) \frac{-2}{\epsilon_k - \epsilon_i} \tag{5.20}$$

with

$$\langle k | v_{ext} | i \rangle = \int d\mathbf{r} v_{ext}(\mathbf{r}, t) \zeta_k(r) \zeta_i(r)$$

Hence

$$\int dt \int_0^1 \frac{d\theta}{\theta} (\langle W_c \rangle^{(\pm)} - \langle W \rangle_0^{(\pm)}) = \int dt\, d\mathbf{r}[\rho_1(r) f_1(rt) + \rho_2(r) f_2(rt)] \tag{5.21}$$

where we used (5.18) in (5.15), (5.16) to obtain the lowest order change in the energy due to the time-dependent perturbation. Here

$$\rho_i^{(\pm)}(r) = \zeta_i(r) v_{jj}(r) \pm \zeta_j(r) v_{ij}(r), \qquad (i, j) = (1, 2), (2, 1) \tag{5.22}$$

with

$$v_{ij}(\mathbf{r}) = e^2 \int d\mathbf{r}' \frac{1}{|\mathbf{r} - \mathbf{r}'|} \zeta_i(r') \zeta_j(r') \tag{5.23}$$

The t-integrations run from $-\infty$ to $+\infty$. Notice that the scheme resembles the RPA in that we are computing changes in the HF states due to a time-dependent one-particle perturbing potential. The only difference is in the choice of v_{ext}. In Section IV, we pointed out that we should be able to invert the mapping $\int dt v_{ext}(\mathbf{r}, t) \to \int dt \delta n(\mathbf{r}, t)$. In the present problem, we avoid this explicit construction of the inverse mapping but choose v_{ext} such that

$$\langle i|v_{\text{ext}}|j\rangle = 0 \qquad \text{for} \qquad i, j = 1, 2 \tag{5.24}$$

This choice is made because of the observation following from (5.20), (5.9) and the completeness of $\{\zeta_n\}$:

$$\int dt(h_0 - \epsilon_i)f_i(\mathbf{r}, t) = -2\Big(\int dt v_{\text{ext}}(\mathbf{r}, t)\zeta_i(r) - \langle 1|v_{\text{ext}}|i\rangle\zeta_1(r) - \langle 2|v_{\text{ext}}|i\rangle\zeta_2(r)\Big) \tag{5.25}$$

and with (5.24) we are led to a constraint

$$\int dt\zeta_2(r)(h_0 - \epsilon_1)f_1(\mathbf{r}, t) = \int dt\zeta_1(r)(h_0 - \epsilon_2)f_2(\mathbf{r}, t) \tag{5.26}$$

Because of (5.19), we may replace f_1 by $(\delta n - \zeta_2 f_2)/\zeta_1$, so that the interaction energy expression (5.21) takes the form

$$\int dt \int_0^1 \frac{d\theta}{\theta}(\langle W\rangle^{(\pm)} - \langle W\rangle_0^{(\pm)}) = \int dt\, d\mathbf{r}\Big[\frac{\rho_1^{(\pm)}}{\zeta_1}\delta n + \Big(\rho_2^{(\pm)} - \frac{\zeta_2}{\zeta_1}\rho_1^{(\pm)}\Big)f_2\Big]$$

and it is now treated as a functional of δn and f_2 with the constraint condition (5.26). These lead to a new v_c:

$$v_c(r) = \frac{\rho_1^{(\pm)}(r) + (h_0 - \epsilon_1)(\zeta_2\lambda^{\pm})}{\zeta_1} \tag{5.27}$$

where the Lagrange multiplier $\lambda(r)$ satisfies the equation

$$\zeta_2(h_0 - \epsilon_1)(\zeta_2\lambda^{(\pm)}) + \zeta_1(h_0 - \epsilon_1)(\zeta_1\lambda^{(\pm)}) = \zeta_2\rho_1^{(\pm)} - \zeta_1\rho_2^{(\pm)} \tag{5.28}$$

Since ζ_1, ζ_2 are s-states, $\lambda(r)$ in (5.28) also may be taken to depend only on $|\mathbf{r}| = r$ and the Equations (5.9) with v_c given by (5.27) and (5.28) can be solved numerically.

The result obtained this way for ^1s was very good whereas ^3s was poor compared with the exact results of Pekeris.[102] For ^3s state the HF result is much better than the present approach because the ζ_is satisfying the HF equations give the energy its minimum value and the approximations made here admit of only local potentials. In the ^1s case, however, the present approach gives a very good result, better than the HF value.

It appears that the Peuckert method holds some promise, even though the method does not seem to be as good as, if not better than, the HF calculation. The basic philosophy of the method is attractive in that one has only a simple problem of dealing with a noninteracting system of fermions moving in an effective time-dependent potential that can be dealt with by an iteration method. In contrast to the method suggested by von Barth, who found that the HKS theory for the mixed symmetry case requires a time-dependent theory which he connects to a time-independent scheme and uses an approximate estimate of correlation effects from the homogeneous magnetic electron gas, Peuckert's method is similar in approach to the time-dependent HF theory except that there is a stationary principle involved in the solution of the time-dependent problem. He does not use a spin-dependent theory nor does he use electron gas results as input. It should be clear from the preceding analysis, that this problem is of great interest and as yet has not been solved satisfactorily, either mathematically or in an intuitive way.

2. Physics of Simple Molecules in HKS Theory

This subsection falls into three topics neatly—the early work of Gordon and Kim[28] who studied an electron gas approach to intermolecular potentials; the more recent works of Gunnarsson and his co-workers[44,103–107] on the LSD theory of simple molecules; and the work of Harris and co-workers[108–112] on the calculation of polarizabilities of interacting atoms based on HKS theory. These topics are taken up in this order after a brief account of the shortcomings of previous attempts in these areas based on HF and other approaches.

A theory of intermolecular forces between closed shell atoms is of fundamental importance in understanding a wide variety of physical and chemical properties of molecules in their various states of aggregation. There are theories for various regions of the range of intermolecular forces, but they are of limited scope in their range of applicability as well as in their practical applicability to large systems. The self-consistent field and configuration interaction calculations handle only short-range repulsion; the long-range attraction is dealt with by means of perturbation theory; at intermediate distances where the intermolecular potential exhibits a minimum, the preceding methods are entirely unsatisfactory. To describe dissociation and chemical reaction correctly, an important requirement on the theory is that it gives the proper separation products. The HF approximation has well-known defects of not being able to describe the separated products correctly.[112] The main reason for the failure is the neglect of correlation which implies that much emphasis was placed on ionic configuration. The other traditional approach of Heitler–London[112] (HL) emphasizes the resonance between degenerate states involving localized electrons but does not incorporate electron hopping between the nuclear sites as well as charge transfer. The internuclear distance dependence of the pair polarizability is an important quantity with which one may describe a wide variety of collision-induced optical properties of gases and liquids. This is another aspect of intermolecular forces. Such calculations are usually done in an HF scheme and are tedious to perform for heavy atoms. The prediction of sign and magnitude of such quantities are of immense interest. Many of the recent advances in answering such questions successfully have come from the work of Gordon and Kim[28] which has now been superceded by the HKS method.

The work of Gordon and Kim develops a simple description of intermolecular forces which correctly predicts the region of the potential minimum as well as the repulsion at shorter distances. The theory incorporates both the overlap of the separate atomic densities and electron correlations. It was used successfully to predict the potential parameters for the rare-gas systems, over the whole range of separations stated previously. The theory was based on three main approximations. (*a*) No distortion of the separate atomic densities is supposed to take place when atoms are brought together; (this is based on HF calculations for He–He). This assumption of additive atomic electron densities is perhaps reasonable for systems that do not form strong chemical bonds and for distances of separation greater than about half the equilibrium internuclear separation, R_m. (*b*) An electron gas approach is employed in calculating the interaction from the additive atomic densities. In fact, Gordon and Kim use a local density scheme, which is reasonable in the outer portions of the atoms, where the electron density is slowly varying. This is where the HKS scheme fits in naturally. (*c*) The electron densities of separate atoms are chosen from the HF solutions. Here again the HKS would be better because a theory of atoms based on the HKS scheme would be the natural starting point of such calculations. Gordon and Kim showed how simple such a scheme of calculation is by numerically calculating the potential parameters for homonuclear (e.g., Ar–Ar), heteronuclear (e.g., Ne–Ar), and ionic pair (Na^+–Cl^- diatomic as well as crystalline) systems with signal success. Other theoretical methods such as self-consistent field and configuration interaction are much more involved and complex and yield worse results.

The work of Gunnarsson and co-workers concerns calculation of binding energies, equilibrium separations, vibration frequencies, and dipole moments of a series of diatomic molecules based on SDF. The calculation of ground-state properties of molecules is this scheme is computationally straightforward and involves much less labor in comparison to other methods. The SDF formalism provides a natural starting point for molecular calculations because it is a one-electron theory of the molecular orbital (MO) type with *correlation included.* It provides a natural bridge in the gap between the traditional HF–MO and HL methods. With the help of Theorem A, the SDF scheme applies not only to the absolute ground state but also to the lowest energy state of each completely specified symmetry of the molecule. Within a Born–Oppenheimer approximation, the scheme also can be employed to calculate the vibrations and rotational frequencies of the molecule. In the literature several calculational schemes exist that are adopted for the SDF formalism and we do not describe them here. The $X\alpha$ method has been applied to molecules and several methods developed there are useful here too.[89] Gunnarsson and Johansson[103] presented the first results for the total energies for the ground states of H_2^+, H_2, He_2^{2+}, and He_2, as well as for the excited states $^2\Sigma_u^+$ of H_2^+ and $^3\Sigma_u^+$ of H_2, and also the vibrational frequencies of H_2 in the SDF scheme. These molecules have known exact results[113,114] and so the comparison with these become important and illuminating. For E_{xc} they used the von Barth–Hedin parametrized form quoted in Section III. They used a difference approximation for the kinetic energy at discrete points (~250), treating the values of the wave function at the discrete points as unknown. The diagonalization of the resulting 250×250 matrix is avoided by a technique which projects the lowest eigenvalue. Typically a deviation of $\frac{1}{2}$eV from accurate calculations were found. For the ground state of H_2, for example, the error in the energy difference, $E(H_2)$–$E(2H)$, is less than 0.25 eV in the whole range of internuclear separations $0.9 \leqslant R \leqslant 3.0$ a.u. considered. On the other hand, for large separations in the case of H_2^+ and He_2^{2+} the errors were significant, indicating the limitation of the LSD scheme. It is important to stress that there is only a local E_{xc} in the HKS theory. This is important in the formulation of Gunnarsson and Johansson. All the calculations, for atoms as well as molecules, are made in the SDF scheme. In atoms, we found that the LSD approximation is less satisfactory for core electrons whereas the valence electrons are fairly well described. The molecules H_2 and H_2^+ treated here are examples of loosely bound electrons and Gunnarsson and Johansson found rather good results. The helium molecular ion is an example where the tightly bound electrons do participate in the bond formation. In this case, an error of 1.3 eV in the dissociation energy was found. It should be pointed out that the multiple-scattering $X\alpha$ method[89] is equivalent to the SDF formalism; as we showed in Section II, many of the concepts and techniques developed for the $X\alpha$ method are applicable to the SDF method. In contrast to these, the HF method is more complex, being nonlocal. For example, the HF approximation gives good results for the triplet state of H_2 but fails for the singlet state. The LSD results suggest that the difficulties of HF are overcome since both exchange and correlation effects are approximately incorporated. Gunnarsson, Harris, and Jones[104,105,106] have developed a method to solve the LSD equations which is hopefully good for systems with sizes between H_2 where discretization techniques are accurate and bulk close-packed metals where muffin-tin potential methods are adequate and have applied it for several simple molecular systems. See Ref. 105 for detailed description of the method. When applied to H_2, this new method gave results as good as the work of Gunnarsson and Johansson. The binding-energy curve parameters of the first-row diatomic molecules B_2, C_2, N_2, O_2, F_2, CO, and BF were obtained in Ref. 105. The trends across the series are given correctly and the differences between theory and experiment are systematic. The calculated equilibrium separations are too large by $(0.1–0.2)a_0$, vibration frequencies and binding energies are too small by 100 to 200 cm^{-1} and 1 to 2 eV, respectively. The dipole moments of CO and BF were also computed and these compared favorably with the experimental results. In Ref. 106, these authors reported work on four low-lying excited states of C_2, which were also in good agreement with experiment. They also compared their LSD results with a calculation based on LDF (local-density-functional) scheme. This showed that the two functionals give remarkably similar results for the ground-state propertrties of N_2, F_2, and CO. These molecules have no spin polarization in the ground state. If spin-polarization effects are *excluded*, binding energies for $X\alpha$ and LDF are also similar and since the molecular and separated atomic energies are widely different, this exhibits cancellation of errors from a consistent use of a single functional be it $X\alpha$, LDF, or SDF. In Ref. 107, Harris and Jones have reported results of calculation of the binding energy curves for the ${}^1\Sigma_g^+$ ground state of the alkali dimers Li_2–Fr_2 by the methods of Refs. 105 and 106, except the core densities being large and extended, require careful treatment. For this purpose, they introduced a simple "renormalized frozen core" approximation. They also computed properties of Li_2, Na_2, K_2, Rb_2, Cs_2, and Fr_2 systems. The results are very satisfactory and thus provide further evidence of the usefulness of the SDF scheme in energy and electron density calculations.

We conclude this discussion by stressing the different relative importance of the exchange and correlation for the molecule than for the constituent atoms. The latter have often a net spin and the exclusion principle keeps the valence electrons apart. As the atoms approach each other, electrons with different spins can pair up with strong correlation effects. In simple molecules, correlation amounts to 25 to 50% of the binding energy which shows the importance of proper treatment of correlation in molecular calculations. It appears the LSD scheme scores impressively in this regard and seems to be very well suited for dealing with molecular problems in an efficient and simple way.

The last part of the discussion on the theory of molecular interactions goes one step beyond the calculations described so far, in that, the HKS theory will be now shown to apply for a priori calculations of the pair polarizability tensor, of simple systems such as $(He)_2$, $(Ne)_2$, and $(Ar)_2$ as a function of internuclear distance. Such a calculation based on LDF theory was made by Harris et al.[108] who used in addition the Gordon and Kim approximation of additive electron densities. This is equivalent to computing the static linear response of the system to external electric fields. A general discussion of static linear response functions was given in Section III. The new feature here is a calculation of the static linear response function as a function of internuclear distance. There are a few assumptions that must be made in this context, which were made earlier by Gordon and Kim.[28] The electronic density $n(\mathbf{r})$ of the molecule is assumed to be the sum of the densities of the isolated atoms, $\Sigma_i\, n_i(\mathbf{r})$, plus the density $\Delta n(\mathbf{r})$, representing the density in the regions away from the atoms in their molecular configuration. For closed shell atoms, $\Delta n(\mathbf{r})$ is indeed very small. Then, there are two equivalent ways of calculating the polarizability—one via the total energy calculation as a function of the applied electric field $\mathbf{F}$, and the other, the usual linear response theory. There is a basic problem with the assumption concerning the electronic density made previously, which does not seem to be widely recognized, when using it in the HKS theory. HKS theory can certainly be applied to isolated atoms; for interacting atoms widely separated from each other, the ground state would be degenerate and the HKS theorem would not apply. It is implicit in the HKS theory, that the external potential preserves the spectrum characteristics of the Hamiltonian and that of the molecule is certainly very different from the isolated atoms. This is a well-known problem discussed a great deal in the literature. With this caveat in mind, we discuss the calculation of polarizabilities as a function of internuclear distances. We give a different derivation than that given by Harris et al.[108–110] as well as that given by Payne[111] whose arguments were based on HF perturbation theory. Our approach is valid for the density-functional schemes generally. The effective one-electron potential of a molecule is (cf. (2.58))

$$\begin{aligned} V_{\text{eff}}[\mathbf{r}; n(\mathbf{r}); \mathbf{F}] &= -\sum_{j=1}^{k} Z_j \frac{e^2}{|\mathbf{r}-\mathbf{R}_j|} + e^2 \int \frac{n(\mathbf{r}')}{|\mathbf{r}-\mathbf{r}'|} d\mathbf{r}' + \frac{\delta E_{xc}[n]}{\delta n(\mathbf{r})} + e\mathbf{F}\cdot\mathbf{r} \\ &\equiv V^0_{\text{eff}}[\mathbf{r}; n(\mathbf{r})] + e\mathbf{F}\cdot\mathbf{r} \end{aligned} \tag{5.29}$$

We are here dealing with a more general molecular system and we here approach it from the HKS theory. The last term describes the potential due to the static electric field $\mathbf{F}$ on the electron at $\mathbf{r}$. The total energy of the system is

$$E_{V,\mathbf{F}}[n(\mathbf{r})] = T_s[n(\mathbf{r})] + \int n(\mathbf{r}) V^0_{\text{eff}}[\mathbf{r}; n(\mathbf{r})] d\mathbf{r} + e \int \mathbf{F} \cdot \mathbf{r} n(\mathbf{r}) d\mathbf{r}$$
$$- \frac{e^2}{2} \int \int \frac{n(\mathbf{r}) n(\mathbf{r}')}{|\mathbf{r} - \mathbf{r}'|} d\mathbf{r}\, d\mathbf{r}' + \frac{1}{2} \sum_{\substack{i,j=1 \\ (i \neq j)}}^{k} \frac{Z_i Z_j e^2}{|\mathbf{R}_i - \mathbf{R}_j|}$$
$$+ E_{xc}[n(\mathbf{r})] - \int n(\mathbf{r}) \frac{\delta E_{xc}[n]}{\delta n(\mathbf{r})} d\mathbf{r} \tag{5.30}$$

where $T_s[n]$ is the kinetic energy of a noninteracting system of electrons of density $n(r)$. The effective potential V_{eff} is used to calculate the one-particle HKS states from which one can obtain the ground-state density, energy, and so on. Of course, one has the requirement that

$$\int n(\mathbf{r}) d\mathbf{r} = \text{N} \tag{5.31}$$

the total number of electrons in the system. The HKS theorem states that $E_{V,F}[n]$ is a minimum as a function of $n(\mathbf{r})$ for a given external field $\mathbf{F}$, and the potential due to the nuclei located at $\mathbf{R}_i$. The procedure of Harris et al. is to assume

$$n(\mathbf{r}) = \sum_{i=1}^{k} (n_i(\mathbf{r}) + \Delta n_i(r)) = \sum_{i=1}^{k} n_i(\mathbf{r}) + \Delta(\mathbf{r})$$

and

$$\int \Delta(\mathbf{r}) d\mathbf{r} = 0 \tag{5.32}$$

In view of the HKS theorem, it is clear that

$$E_{V,F}[n] = E_{V,F}\left[\sum_i n_i\right] + 0(\Delta^2) \tag{5.33}$$

because $E_{V,F}[n]$ is a minimum as a function of $n(\mathbf{r})$. Calculating the density $n(\mathbf{r}; \mathbf{F})$ as a function of the field $\mathbf{F}$ leads to an expression for $E_{V,F}$, which is second order in the external electric field $\mathbf{F}$ and hence the polarizability tensor $\alpha_{\lambda\mu}$ is calculated from

$$\alpha_{\lambda\mu} = -\lim_{F \to 0} \frac{\partial^2 E}{\partial F_\lambda \partial F_\mu} \tag{5.34}$$

That there is no term linear in Δ in (5.33) is called the "cancellation theorem" by Harris et al.[109] Alternatively, the dipole moment $\boldsymbol{\mu}$ can be calculated directly

$$\boldsymbol{\mu} = \int \mathbf{r} n(\mathbf{r}) d\mathbf{r} = \int \mathbf{r} \Delta(\mathbf{r}) d\mathbf{r} \tag{5.35}$$

where the second formula follows as a consequence of symmetry for the hetero-atomic case. the "cancellation theorem" is generally true in the sense that the first-order variation in density is zero because of the HKS theorem. In particular, it is also true, if one assumes the form (5.32) for $n(\mathbf{r})$ as a form for the electron density of the system. We now calculate $\Delta n(\mathbf{r})$ from our Green function approach since this method brings out the assumptions made by Harris et al. most transparently. As in Section III, we have, since there is no time dependence *in the electric field,*

$$-iG(\mathbf{r}_1t_1;\mathbf{r}_1t_1^+;\mathbf{F}) = n(\mathbf{r}_1;F) \tag{5.36}$$

and G obeys the equation:

$$\left\{i\frac{\partial}{\partial t_1}+\frac{\nabla_1^2}{2m}-V_{\text{eff}}^0[\mathbf{r}_1;n]-e\mathbf{F}\cdot\mathbf{r}_1\right\}G(\mathbf{r}_1t_1;\mathbf{r}_2t_2;\mathbf{F}) = \delta(\mathbf{r}_1-\mathbf{r}_2)\delta(t_1-t_2) \tag{5.37}$$

Going through the procedure as given in Section III.C, we obtain the change in the density induced by $V_{\text{eff}}^0(\mathbf{r}_1)$ in the lowest order of perturbation theory

$$n(\mathbf{r}_1;F) = n_0(\mathbf{r}_1;\mathbf{F}) + i\int d\bar{1}G(\mathbf{r}_1t_1;\mathbf{r}_{\bar{1}}t_{\bar{1}};\mathbf{F})G(\mathbf{r}_{\bar{1}}t_{\bar{1}};\mathbf{r}_1t_1^+;F)V_{\text{eff}}^0(r_{\bar{1}})$$

Now

$$i\int dt_{\bar{1}}G(\mathbf{r}_1t_1;\mathbf{r}_{\bar{1}}t_{\bar{1}};\mathbf{F})G(\mathbf{r}_{\bar{1}}t_{\bar{1}};\mathbf{r}_1t_1^+;\mathbf{F}) = \chi(\mathbf{r}_1;\mathbf{r}_{\bar{1}};\mathbf{F}) \tag{5.38}$$

is the response of the density to the effective potential due to all the nuclei and the electrons and hence includes all the correlations among them. Thus

$$\Delta n(\mathbf{r}_1;\mathbf{F}) = n(\mathbf{r}_1;\mathbf{F}) - n_0(\mathbf{r}_1;\mathbf{F}) = \int \chi(\mathbf{r}_1;\mathbf{r}_{\bar{1}};\mathbf{F})V_{\text{eff}}^0(r_{\bar{1}})d\mathbf{r}_{\bar{1}} \tag{5.39}$$

The additive atomic density approximation is now made in calculating χ as well as $\Delta n(\mathbf{r})$ in a non-self-consistent perturbation scheme by Harris et al. as follows. *Assume G* in (5.38) to be given by a superposition of atomic Green functions:

$$G(\mathbf{r}_1t_1;\mathbf{r}_{1'}t_{1'}) \cong \sum_{i=1}^{k} G_i(\mathbf{r}_1t_1;\mathbf{r}_{1'}t_{1'}) \tag{5.40}$$

where G_i is the Green function for the atom i with its own effective potential $V_{\text{eff}}^{(i)}[\mathbf{r};n_i(\mathbf{r})]$. In terms of the eigenvalues and eigenfunctions of the operator $(-\nabla^2/2m + V_{\text{eff}}^{(i)})$ denoted by ϵ_n^i and ϕ_n^i we have

$$G_i(\mathbf{r}_1t_1;\mathbf{r}_{1'}t_{1'}) = \sum_n \phi_n^{i*}(\mathbf{r}_1)\phi_n^i(\mathbf{r}_{1'})g_n^i(t_1t_{1'})$$

and so from (5.38) we obtain

$$\chi(\mathbf{r}_1\mathbf{r}_{\bar{1}}) \cong \sum_{i,j=1}^{k} i \int dt_{\bar{1}} G_i(\mathbf{r}_1 t_1; \mathbf{r}_{\bar{1}} t_{\bar{1}}) G_j(\mathbf{r}_{\bar{1}} t_{\bar{1}}; \mathbf{r}_1 t_1^+)$$
$$= \sum_{n,m} \sum_{i,j=1}^{k} \frac{(f_n^i - f_m^j)}{(\epsilon_n^i - \epsilon_m^j)} \phi_n^{i*}(\mathbf{r}_1)\phi_n^i(\mathbf{r}_{\bar{1}})\phi_m^{j*}(\mathbf{r}_{\bar{1}})\phi_m^j(\mathbf{r}_1) \tag{5.41}$$

where f_n^i is the function that specifies whether the state ϵ_n^i is occupied (= 1) or not (= 0). Harris et al. make a further assumption that only the terms with $i = j$ in (5.41) dominate because for closed shell systems, the $\epsilon_n^i - \epsilon_m^j$ would be large, and $\Delta n(\mathbf{r})$ genuinely very small compared to $\Sigma_i n_i(\mathbf{r})$. With this further assumption, one obtains χ to be just the sum of atomic polarizabilities given by $\chi^{(i)}(\mathbf{r}_1; \mathbf{r}_{\bar{1}})$ with $j = 1$ in (5.41). Payne[111] has allowed for the overlap as we do here, but makes other simplifying assumptions.

There are two alternative ways of examining this problem without the preceding assumptions. One is to solve the HKS equations with V_{eff} given by (5.29) and obtain the density directly for the atomic complex

$$n(\mathbf{r}; \mathbf{F}) = \sum_{n(\text{occ})} |\phi_n(\mathbf{r}; \mathbf{F})|^2 \tag{5.42}$$

and repeat the calculation for the isolated atoms with the same form for e_{xc}, and obtain

$$n_i(\mathbf{r}; \mathbf{F}) = \sum_{\lambda(\text{occ})} |\phi_\lambda^i(\mathbf{r}; \mathbf{F})|^2 \tag{5.43}$$

From this, one computes explicitly

$$\Delta n(\mathbf{r}; \mathbf{F}) = n(\mathbf{r}; \mathbf{F}) - \sum_{i=1}^{k} n_i(\mathbf{r}; \mathbf{F}) \tag{5.44}$$

and hence $\boldsymbol{\mu}$ as well as the polarizability tensor $\alpha_{\lambda\mu}$. The other alternative is to do a self-consistent linear response calculation:

$$n(\mathbf{r}; \mathbf{F}) = n(\mathbf{r}; F = 0) + \left(\frac{\delta n(\mathbf{r}; F)}{\delta \mathbf{F}}\right)_{F=0} \cdot \mathbf{F}$$

or

$$n(\mathbf{r}; \mathbf{F}) = n(\mathbf{r}; F = 0) + i\mathbf{F} \cdot \int G(\mathbf{r}t; \bar{1}) \frac{\delta G^{-1}(\bar{1}; \bar{2})}{\delta \mathbf{F}} G((\bar{2}; \mathbf{r}t^+) d\bar{1}\, d\bar{2} \tag{5.45}$$

From (5.37), we have

$$\frac{\delta G^{-1}(\bar{1}; \bar{2})}{\delta \mathbf{F}} = -\delta(\bar{1}\bar{2}) \left[+ \int \frac{\delta V_{\text{eff}}^{(0)}(\mathbf{r}_1; n; \mathbf{F})}{\delta n(\mathbf{r}_{\bar{3}})} \frac{\delta n(\mathbf{r}_{\bar{3}}; \mathbf{F})}{\delta \mathbf{F}} d\mathbf{r}_{\bar{3}}\right] \tag{5.46}$$

and from (5.29) we can obtain an equation for $\delta n/\delta \mathbf{F}$:

$$\frac{\delta n(\mathbf{r};\mathbf{F})}{\delta \mathbf{F}} = i\int G(\mathbf{r}t;\bar{1})\left[e\mathbf{r}_{\bar{1}} + \int \frac{\delta V_{\text{eff}}[\mathbf{r}_{\bar{1}};n;\mathbf{F}]}{\delta n(\mathbf{r}_{\bar{3}})}\frac{\delta n(\mathbf{r}_{\bar{3}};\mathbf{F})}{\delta \mathbf{F}}d\mathbf{r}_{\bar{3}}\right]G(\bar{1};\mathbf{r}t^{+})d\bar{1}$$

$$= -\int \tilde{\chi}(\mathbf{r};\mathbf{r}_{\bar{1}})\left[e\mathbf{r}_{\bar{1}} + \int \frac{\delta V_{\text{eff}}[\mathbf{r}_{\bar{1}};n;\mathbf{F}]}{\delta n(\mathbf{r}_{\bar{3}})}\frac{\delta n(\mathbf{r}_{\bar{3}};\mathbf{F})}{\delta \mathbf{F}}d\mathbf{r}_{\bar{3}}\right]d\mathbf{r}_{\bar{1}} \qquad (5.47)$$

where

$$\tilde{\chi}(\mathbf{r};\mathbf{r}_{\bar{1}}) = i\int G(\mathbf{r}t;\bar{1})G(\bar{1};\mathbf{r}t^{+})d\bar{1} \qquad (5.48)$$

is the density-response function of the composite system with G solved for from (5.37) with $F = 0$. This formulation generalizes the Payne calculation which was specially suited for the HF scheme. In principle, then, the two methods outlined previously can be used to compute both the $\boldsymbol{\mu}$ as well as $\alpha_{\lambda\mu}$.

Harris et al. have used their approximate scheme for calculating properties of a variety of rare-gas atomic systems, such as $(He)_2$, $(Ne)_2$, and $(Ar)_2$. We should point out that Harris and Heller[110] found a subtle difficulty in their scheme because the Hellman–Feynman theorem is violated. We showed in Section I that the Hellman–Feynman theorem is valid for the density-functional formalism. The scheme such as the one outlined by us here without the approximation of Harris et al. would be devoid of this defect. The results obtained by Harris et al. are now stated. Harris et al. used the energy method to compute the polarizabilities $\alpha_{\parallel}$ and $\alpha_{\perp}$ of $(He)_2$, $(Ne)_2$, and $(Ar)_2$ as a function of the internuclear distance, where $\parallel$ and $\perp$ imply the components parallel and perpendicular to the internuclear separation. They used the local density form for the E_{xc}, that is, electron gas version of the Gordon–Kim type. The diatomic interaction energy does not have the expected van der Waals contribution whereas the longer-range R^{-3} term was found. From $\alpha_{\parallel}$ and $\alpha_{\perp}$, they compute the second Kerr virial coefficient (B_K) and the second dielectric virial coefficient (B_ϵ). For He diatom, these density-functional results compare very well with the HF results as well as experiments, in a qualitative way. Harris et al. calculation does not and cannot give the R^{-6} coefficient in $\alpha_{\parallel,\perp}(R)$. But yet, the correct sign and order of magnitude of B_ϵ is obtained for $(Ne)_2$. No HF results exist for comparison here. Similar results were obtained for $(Ar)_2$. Thus the theory of Harris et al. seems to provide a physically sound and accurate means for calculating $\alpha_{\lambda\mu}(R)$ for a pair of interacting inert atoms. In principle, the theory outlined previously can be extended to calculate external magnetic field effects (Faraday rotation, etc.) also. Payne[111] has used the preceding type of theory to compute energy of reorganization associated with a change in electron density and has accounted theoretically for the existence of rotational barriers and the origin of closed shell repulsion.

Very recently Parr et al.[115] have proposed a density functional viewpoint of the concept of electronegativity. They show that electronegativity is the negative of the chemical potential in the HKS theory and as such is the same for all orbitals in an atom or molecule in its ground state. This demonstration does not seem to be mathematically complete for the following reason: As in Section II, these authors work with Löwdin's natural orbitals instead of the original HKS orbitals. This is not serious since they may also be chosen to be HKS orbitals as was done in Section II. In using the occupation numbers n_λ (in the notation of Section II) the authors employ the stationary property of the ground-state energy as a function of n_λ, without establishing whether the variations δn, should be of one-sign or not. Moreover, the condition of stationarity is valid only for $0 < n_\lambda < 1$ (2 if spin is included in the counting) but the end points $n_\lambda = 0$ and 1 are essential values in the theory. Even though the authors quote the constraint $0 \leqslant n_\lambda \leqslant 1$ (or 2), in their analysis of stationarity, they overlook the end points mentioned previously. It appears therefore that if we ignore the end points, then all the natural orbitals would have the same chemical potential. The discussion of the relations among ionization potential, electron affinity, and electronegativity is similar to that given by Slater concerning "transition states" in the $X\alpha$ theory and extended by Janak for the density-functional theory. We may end this discussion by pointing out that the spin-density-functional theory would be more appropriate in such discussions as the polarizability of molecular complexes as well as in the discussion of electronegativity. It appears that much can be expected in the future in the application of SDF theory to investigations of basic chemical problems.

C. Bulk Properties of Solids

A large amount of work has been done in applying the LDF and LSD theories to elucidate the bulk properties of bulk solids. These works fall into three classes: (*1*) ground-state properties of metals—cohesive energies, equilibrium lattice constants, and compressibility, magnetic susceptibility, ferromagnetic spin-splitting, magnetization, and so on; (*2*) non-ground-state properties—explicit band structure calculations; and (*3*) collective excited state properties such as plasmon and spin-wave dispersion in the long wave-length limit, frequency-dependent susceptibility calculations. Of these topics (*1*) is within the rigorous applicability of the HKS formalism whereas the topics (*2*) and (*3*) are outside the real scope of the HKS formalism but are of immense interest to see how far the formalism can be used in actuality. Recently topic (*1*) has been reviewed and collected together by Moruzzi et al.[21] and so we content ourselves with a brief summary of this important area of research. Since there is a large amount of work in topic (*2*), we confine our attention to only those metals for which the band structure has special features of interest, for example, magnetic metals. A calculation that examines the question of applicability of the HKS theory to such band calculations is briefly touched upon also. The last topic belongs in the application of the time-dependent HKS theory but, as yet, such a theory does not seem to be formally and rigorously established. It should be remarked that the ground-state properties of some interesting insulators such as diamond, boron nitride, lithium fluoride, and titanium sulfide have recently been studied by Zunger and Freeman within a LDF scheme and we begin with a description of their work.

1. Ground-State Properties

Cohen and Roy Gordon[116] were the first to apply the electron gas theory of Gordon and Kim[28] in an attempt to develop a theory of lattice energy, equilibrium structure, elastic constants, and even pressure-induced phase transitions (rocksalt to cesium chloride structures) in alkali–halide crystals. Their calculations were such that the average magnitude of the deviation between the predicted and observed bond distances is 2%, lattice energy 2%, and elastic constants 10% for lithium, sodium, potassium, and rubidium fluorides, chlorides, and iodides. The pressure induced phase changes were also predicted. They obtain a rather good overall agreement between theory and experiment. Cohen and Gordon[117] investigated similarly the properties of MgO and CaO crystals with equal success. The success of such an a priori method based on the electron gas approximation of Gordon and Kim[28] was the first such successful theory for describing ionic crystals of such a wide variety. This work preceded the HKS theory. The Gordon–Kim theory is now superceded by the more formal framework of HKS theory and it is indeed remarkable that the hint of the success of the HKS theory was already there in the pioneering work of Roy Gordon and co-workers.

We first discuss the case of insulators whose study has just begun by the LDF formalism. For all the solid-state problems, the external potential appearing in the HKS theory is the electron–nuclear and internuclear interactions. The only assumed entities are the crystal structure of the solid under consideration. Most often, the exchange-correlation potential V_{xc} is taken to be the parametrized form given by Hedin and Lundqvist[61] in a LDF scheme whereas different forms are employed (without too much differences in the final outcome) in the LSD scheme. One is then able to compute the equilibrium lattice constant, cohesive energy, compressibility, etc. associated with the ground state of the crystal. Zunger and Freeman[118] developed a self-consistent, numerical basis set of a linear-combination-of-atomic-orbitals scheme for the specific purpose of studying insulating crystals. (This should be compared with the closed-shell molecular calculations discussed earlier.) With this method, they have studied diamond[119] which has been studied by many authors both theoretically and experimentally. They found that the exchange and correlation effects on the ground-state charge density tend to increase the charge localization both in the "band" region and to an extent close to the core region. The calculated cohesive energy agrees well with experimental results and predicts an equilibrium lattice constant that is about 3% too high with exchange-only calculation whereas the inclusion of correlation brings it to about 0.5%. LiF is a prototype ionic solid and this was investigated[120] by Zunger and Freeman by their methods with LDF schemes with remarkable success, the lattice constant being 1.8% too large and cohesive energy about 7.5% too small (similar to diamond) compared to the corresponding observed quantities. The model does not include dispersion forces, it should be remarked, and its relation to the choice of E_{xc} on V_{xc} is at present unknown. If included, it is expected that LDF would give very good answers. It should be remarked that LDF does much better than the HF calculations. TiS_2 is a very interesting layered compound with unusual charge-density wave ground state. A theoretical description of the properties by conventional methods yielded contradictory results. Zunger and Freeman[121] presented the results of a LDF calculation based on their scheme[118] and could account for much of the experimental results unlike the other existing attempts. They were able to discuss the bonding properties and in particular found no metallic bonding in TiS_2, in contradiction with previous suggestions. They extrapolated from their results on TiS_2 to the case of $TiSe_2$ and were consistent with the observations. Cubic boron nitride is the simplest III–V compound, isoelectronic and isostructural with diamond and possesses extraordinary properties, which are similar to those of diamond. A comparative study of this with diamond in LDF seemed important and was reported by Zunger and Freeman.[122] They found good agreement with experimental data on the ground-state properties. They were able to reproduce remarkably well the observed trend in going from diamond to more ionic BN, that is, decrease in cohesive energy and increase in equilibrium lattice constants.

A very systemmatic account of calculations of ground-state properties of the first 50 elements of the periodic is given by Moruzzi et al.[21] They employed a local density (spin) functional scheme with "muffin-tin" approximation as a calculational method of solving the HKS equations and neglected the relativistic effects (and hence they went only up to $Z = 50$). We already discussed the results of their free atom calculations. They have computed the cohesive energy of the metal, which is defined as the total energy of the atom minus the total energy of the solid with zero-point energy included so that positive values of the cohesive energy represent the stability of the solid relative to the free atom. The same exchange-correlation approximation is used in both the atomic and the solid-state calculation, a feature which is important to cancel errors arising from misrepresentation of the exchange-correlation effects. As is well-known, the Pauli spin-susceptibility is enhanced by the exchange-correlation effects and this enhancement is an important ground-state property of the system. Gunnarsson[123] made such a calculation for investigating the magnetic properties of Pd, Mn, Ni, Co, and Cr. The calculation of the paramagnetic susceptibility enhancement in the SDF theory was formally given by Vosko and Perdew[75] and investigated thoroughly for each element (up to $Z = 50$) with a view to elucidate the magnetic properties of elemental metals by Janak.[76] The conclusion reached by this investigation was impressive in that Fe and Ni showed ferromagnetic instabilities which were borne out by the SDF calculation of their energy bands. From such a calculation, a priori values of the magnetic moment, changes in the mechanical properties due to the magnetization, and so on were obtained in general agreement with experiment.[124] It was found that Pd, which is nearly ferromagnetic experimentally, was indeed so (i.e., large susceptibility enhancement) whereas Ni and Fe were indeed ferromagnetic, although the face-centered-cubic phase of Co is almost ferromagnetic. The local density theory thus predicts (except for Co) the presence or absence of ferromagnetism in the elements through In. The enhancement factor depends on $G_{xc}^{(2)}$ $(= (\partial^2 E_{xc}/\partial m^2)_{m=0})$ of Section III. Janak[76] found that the choice of E_{xc} gives different results for the susceptibility enhancement and so this sensitivity could be used as a crucial pointer of the accuracy of the E_{xc} used. Another feature of this calculation is that the elements of the 4d series do not exhibit magnetism. In general, we may point out that, for the nonmagnetic system, the cohesive energies were within 20%, lattice constants within 0.3 Bohr radii, and bulk moduli within 10% of the experimental values. The trends across the periodic table up to $Z = 50$ were generally in accord with experiment (see Fig. 1). Note that those elements that do not fall within the theoretical curve correspond to magnetic systems and they are explained by invoking LSD with nonzero spin polarization allowed for magnetism.

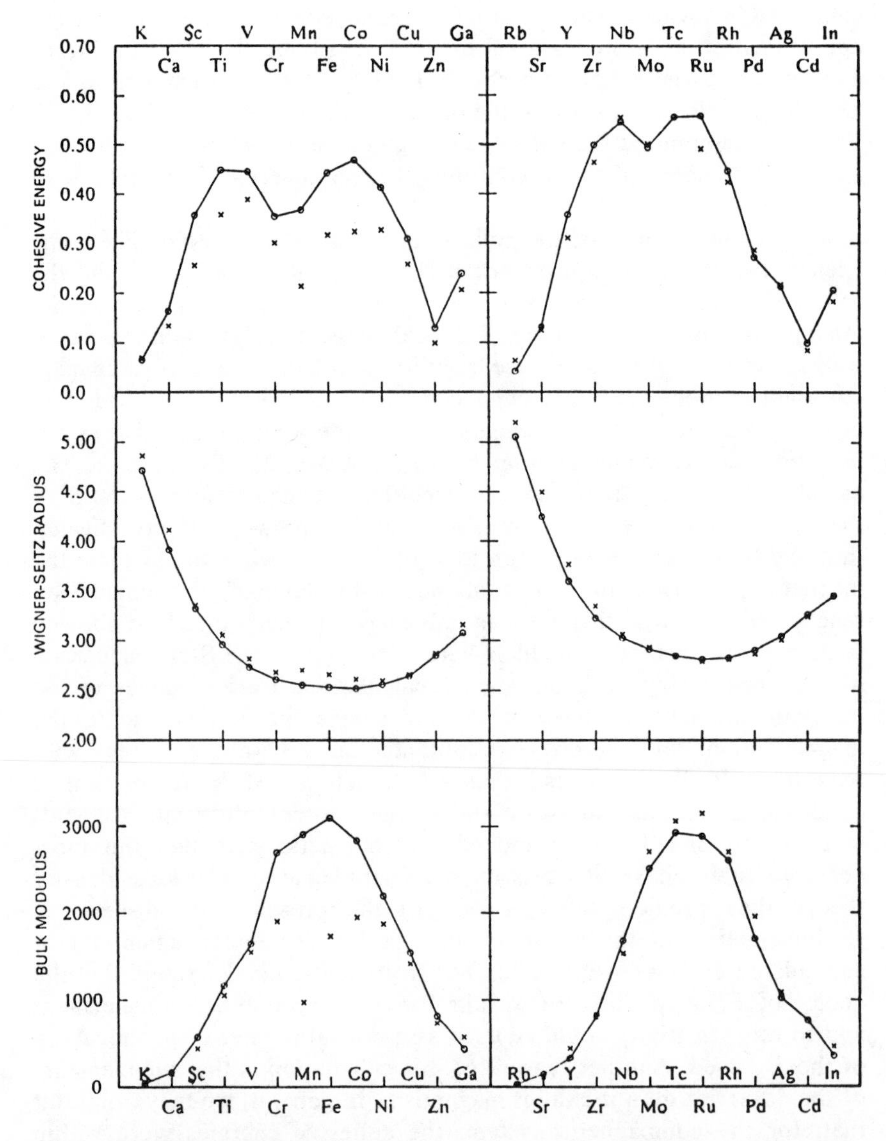

Fig. 1. Cohesive properties. Top row: cohesive energy (Ry/atom). Middle row: Wigner–Seitz radius (a.u.). Bottom row: bulk modulus (K bar). Measured values are indicated by crosses. Reproduced from Moruzzi et al.[21]

The magnetic elements had to be examined more closely. Janak and Williams[124] calculated the cohesive energy, lattice constant, bulk modulus, magnetic moment, and the pressure derivative of the magnetic moment for ferromagnetic Fe(BCC) and Ni(FCC) both for the magnetic (von Barth–Hedin) and nonmagnetic (Hedin–Lundqvist) E_{xc}. Their magnetic results of density of states of Fe and Ni are similar to those found by Callaway and co-workers.[51] The calculated results include only spin effects and almost no orbital effects. If the orbital moment is removed from experimental values, using the measured g-factors, the calculated spin moments are found to be only 1% too large for Fe and 5% too large for Ni. These calculations were all self-consistent. From Fig. 1, we see that the discrepancy occurred in the magnetic elements, which by the LSD scheme, could be explained satisfactorily.[124] On the whole, therefore, one may conclude that the fully self-consistent HKS theory with LD or LSD schemes adequately describe the trends in the ground-state properties of elements across the periodic table up to $Z = 50$. Beyond $Z = 50$, one requires a relativistic theory, which is being formulated only now,[125] some of which is outlined in Section III. We should point out that the case of antiferromagnetism of Cr has not been successfully dealt with by the SDF scheme. This is mainly because the magnetic moment in this case varies over a distance scale of a few atoms sufficient to establish a band structure. In the existing SDF scheme, the use of homogeneous electron gas results is not sufficient to describe this effect. An inherently inhomogeneous magnetic electron gas model must be the basis for parametrizing E_{xc}. For the same type of reasons, the calculation of the transition temperature exhibiting the onset of magnetism from a finite temperature theory of the LSD is not expected to be very accurate. Liu et al.[126] have reported a calculation of the temperature-dependent SDF and derived a temperature-dependent spin susceptibility. Using the $T = 0$ von Barth–Hedin E_{xc} which is expected to be not too bad an approximation, a calculation of T_c by these authors for Ni turned out to be nearly a factor 3 too large compared to the experimental value. Clearly much formal work is required before one can expect good results at finite temperatures. Liu et al.[127] have investigated the effect of nonlocality on the enhancement of the spin susceptibility. Recall, from Section III, that $G_{xc}^{(2)}$ is in general a nonlocal functional whose nature is unknown. These authors suggested some empirical forms for the nonlocality but no definite and quantitative conclusions could be reached owing to the insufficient knowledge of $G_{xc}^{(2)}$. We now turn our attention to the calculation of band structure of Ni and Fe based on LSD and briefly discuss their relationship to experiments and to the formal framework of the HKS theory.

2. Band Structure Calculation of Fe and Ni

The eigenvalues ϵ_i occurring in (2.95) and so on do not have any special significance. As stated earlier, this (and similar) equation is a mathematical artifact of the HKS formalism. These eigenvalues together with the density computed from the eigenfunctions determine the ground-state properties of the system. If one takes a pragmatic point of view and treats the ϵ_i as one-electron eigenvalues as in the one-electron theory of band structure of solids, one arrives at a "band structure" in the HKS theory. There is much controversy regarding the definition of the "Fermi surface": whether it is a ground-state property of the system or not. As shown in Section II, all the eigenvalues for which the energy is numerically equal to the chemical potential of the system define a Fermi surface. From ϵ_i therefore, one may construct such Fermi surfaces. In the magnetic systems, an important entity that one may experimentally determine is the exchange splitting of the energy band at the Fermi surfaces. As mentioned in the last subsection, fairly impressive results have been obtained for the ground-state properties of ferromagnetic Ni and Fe based on LSD. Callaway and co-workers have pioneered in the calculation of band structure and Fermi surface characteristics of Fe and Ni within a fully self-consistent LSD scheme. A summary of this work is given in Ref. 51. They used the von Barth–Hedin E_{xc} and V_{xc}, W_{xc} in the notation of Sections II and III. The experimental results of de Haas–van Alphen measurements yield various areas of the Fermi surface and an overall agreement between theory and experiment is achieved, even though there are some regions that are not at all reproduced by the theory. It is concluded that it is quite difficult to obtain reliable information about the exchange splitting from the study of the Fermi surface. Callaway also argues that there is a significant defect in the present SDF theory in that the effective potential is not sufficiently anisotropic. Recently Eastman et al.[128] using angle-resolved photoemission and synchrotron radiation, determined the bands for Ni and found the d-band width and exchange splitting. The best theoretical estimates of Callaway and Wang[51] are about a factor 2 too large compared with these experimental findings. The careful work of Eastman et al. is quite reliable and places the theory in much worse light than other experiments on similar lines.

3. Collective Excited State Properties

Callaway and Wang[51,129,130] have calculated the coefficient of the long-wavelength limit of the spin wave dispersion in ferromagnetic nickel using a straightforward generalization of the LSD method for calculating the perpendicular susceptibility, as was outlined in Section III. They obtained $\omega_{sw}/q^2 \equiv D$ for Ni to be 0.104 in atomic units which seems to compare well with the experimental value of 0.113. If one uses the assumption that $E_{xc}[n, \mathbf{s}] \cong E^h_{xc}[n, |\mathbf{s}|]$ one obtains the result of Callaway and Wang, as was shown by Rajagopal.[27] As shown generally in Ref. 27, such a calculation of the D and a similar quantity for the plasmon dispersion relation cannot be reliable as these are indeed the excited state properties of the many-electron system. The calculation of plasmon dispersion was made by Singhal and Callaway[131] for Al. Callaway and co-workers[52] have computed the optical conductivity of V within a LDF theory which also lies, in principle, outside the scope of the formalism. The experimental results do not show the sharp structure that the calculations exhibit, whereas the gross structure is similar. As realized and emphasized by Callaway and co-workers, the applicability of the HKS theory to such excited state properties is of doubtful validity. These calculations, however, do indicate that the main effect omitted in such calculations is lifetime broadening. By putting in a relaxation time phenomenologically, they obtained results in a fair agreement with experiment. It should be pointed out that theoretical work on spin wave dispersion based on Callaway–Wang type of scheme have recently been put forward independently by Liu and Vosko[132] and Edwards and Rahman.[133] All these works come under the same criticism given in Ref. 27.

In summary, we may state that the SDF theory in some ways is capable of giving a fair account of the trends in the ground-state properties of elemental metals including the magnetic properties. The calculation of excited states are not expected to be in good accord with experiment but such calculations may give an idea of where the results agree, if at all, with observed values. In the next section, we describe the theory of surface properties of metals in the LDF and LSD formalism.

D. Surface Properties—Chemisorption, Proton and Muon in Solids

We divide this subsection into four subparts: (*1*) the surface energy, work function, and so on of simple metals; (*2*) chemisorption; (*3*) motion of proton and muon in simple metals; and (*4*) application of LDF to MOSFET systems. The last one is dealt with briefly only because this subject matter is very rapidly growing. The first topic is also dealt with briefly because Lang[22] has recently written a thorough account of topic (*1*) and Langreth[134] has given a critique of the more recent work on the subject. Topics (*2*) and (*3*) are of great interest and importance, and will be reviewed in somewhat more detail here.

1. Surface Energy

The surface of a system presents a very strong inhomogeneity and as such is on a very different level than the types of inhomogeneity considered so far. Two important and basic quantities enter in the description of surface properties and they are the work function and the surface energy. The study of the surface properties have proceeded in almost the same fashion as that of the bulk properties—half-space electron gas model and the band structure of crystalline solids with a surface. These models have their usual limitations and they can, in principle, be unified by means of the density functional formalism. This formalism provides a self-consistent scheme for computing the ground-state properties of the interacting, inhomogeneous systems, as we have pointed out so often on different occasions throughout this review. Such a comprehensive theory of metal surfaces was given in two important papers by Lang and Kohn,[135,136] the first of which discussed the charge density and surface energy, and the second one, the work function. Lang[22] has written an extensive review of the subject covering the work up to 1973 and we refer the reader to this article for references and discussion of this earlier phase of the work.

The work function Φ is the minimum work that must be done to remove an electron from the metal at 0°K. The surface energy σ is the work required, per unit area of new surface formed, to split the crystal into two along a plane. These two quantities are basic to the discussion of the ground-state properties of surfaces of crystals. In the density-functional theory, the density of the system is a central quantity and its spatial behavior in the semiinfinite system near the surface is another important quantity. Lang[22] discusses several models for the computation of the properties of metal surfaces. Assuming no serious "surface reconstruction" on the surface, these models are (*a*) uniform-background model analogous to the Sommerfeld electron gas model for the bulk simple metals; (*b*) perturbative treatment of lattice effects; and (*c*) corrugated uniform-background model. The question of surface reconstruction is more frequent in semiconductors rather than in metals, as is shown by low-energy electron-diffraction experiments. Several points concerning (*a*) emerge:

(*1*) The electronic charge density $n(\mathbf{r})$ oscillates as a function of distance normal to the surface near the surface near the surface and dies exponentially outside the region where the positive charge background of the metal is absent. Deep inside, the charge density is that of the bulk; (*2*) the surface energy is negative for this model at a density corresponding to that of zinc and aluminum. This instability of the surface in this model is rectified when the lattice effects are incorporated. Moreover, the lattice effects are important because this brings out the important feature of the surface, namely, that the properties of the surface should depend on the crystallographic surface considered. This is because the positive charges are geometrically arranged on the lattice and depending on the crystal surface, one may encounter highly charged plane of lattice points. This becomes important later on when we discuss chemisorption. We may state that on the whole, when lattice effects are properly taken into consideration, the calculated work function and surface energy of simple metals are in fair agreement with experimental findings. As outlined in Section III.C, linear response of this inhomogeneous system can also be calculated and such investigations have also been carried out. Here much work has been done at various levels of sophistication and we shall briefly mention those that used the density-functional schemes. This is an important aspect of the theory because it gives information about the response of the surface electrons to perturbations due to charge distributions placed outside the surface. A point charge placed outside the metal surface experiences an image force. One may think about the image charge via the pair-correlation function for the system. This function gives the probability of finding an electron at a given position if there is another electron known to be present at some preassigned place. If one electron is outside, the image charge should be inside the metal surface and this aspect should and does emerge from the theory.[22] It should be pointed out that only the static linear-response function can be calculated in the density functional formalism. In computing the interaction energy between a charge placed outside the surface and the surface, one must include nonlinear processes which we discuss under chemisorption.

Before closing this brief account, we should mention two aspects. A controversy arose in the theory of surface energy of simple metals because the local-density-functional approach does not seem to include the surface-plasmon zero point energy. The work of Langreth and co-workers reviewed recently in Ref. 134 essentially settles this controversy. It was shown by a careful analysis of the exchange-correlation energy in terms of the collective excitations of the system, that even though the local density scheme drastically fails in incorporating the long-wavelength plasma modes, it still gives rather good answers because the weight associated with the long-wave region is relatively small. Proper treatment of the long-wave region is, of course, understandably necessary for making appropriate comparisons. Thus, it appears that local approximations employed in the calculations of the surface properties are justificable. Moreover, the lattice structure of the metal with a surface can also be handled numerically in much the same way as the bulk. The reader may consult Ref. 134 for references to recent work on the subject.

The second point is a somewhat different aspect. This concerns properties of the surface of a magnetic metal. The planar uniform background model where the electrons are spin polarized was first considered by Pant and Rajagopal[31] within LSD formalism. They found that the work function changes little when the bulk of the metal becomes a ferromagnet. The work of Pant and Rajagopal is of importance in discussing transition metal surfaces as well. Kautz and Schwartz[137] extended the work of Pant and Rajagopal.[31] They used the von Barth–Hedin form of the local spin density functional for the exchange-correlation energy. They found that the magnetization density, just as the density in the Lang–Kohn work, exhibits oscillations as a function of the distance from the surface, close to the surface. The other conclusion of Pant and Rajagopal that the work function does not change between para- and ferromagnetic states was confirmed. This area of research will become increasingly interesting in view of the experimental advances in the use of synchrotron radiation[128] in studying surface properties.

The linear response theory of surface electronic structure using the ionic lattice as a pseudopotential perturbation on a jellium slab has recently been given by Lert and Wearie.[138] They used the local density approximation and found semiquantitative agreement with previous nonlinear calculations for Na (001) surface.

2. Chemisorption

The change in the binding energy of an atomic core electron when the atom is chemisorbed on a metallic surface and the chemical bond formed between an atom and a metal surface have been studied by Lang and Williams[139,140] by a local density formalism. Their model is the self-consistent atom-jellium model and the discrete lattice substrate is treated in perturbation theory. This theory is a natural extension of the earlier work of Lang and Kohn on the theory of metallic surfaces. The importance of this work in the study of a large variety of chemical properties of surfaces is obvious. Their theory combined with the techniques of Slater's transition-state ideas has been used to explain successfully the core-level binding-energy shifts in metals for the $3d$ transition series.[141] The density-functional method affords a calculation of the energy shifts as well as the charges in the electron density which cause them in the problems mentioned previously. We first consider the calculation of core holes in chemisorbed atoms.[139] This study is motivated by the fact that the extra-atomic binding energy changes due to the interaction of the atom with the metal help one understand the results obtained from X-ray photoemission studies. The atom-jellium model is best suited for such studies because of its conceptual simplicity, both calculationally and in terms of physics, and contains within it the essential physical and chemical aspects of the problem. It is important to realize that the chemisorption geometry is lower compared with the impurity sites *inside* a metal and as such allows one to study a wider variation in the chemical environment of the excited atom than in the impurity case. The quantity of interest in this calculation is the difference between the core binding energy of the free atom and that of the chemisorbed atom. This reduction in the binding energy in going from the free to the chemisorbed atom involves *four self-consistent ground-state energy calculations* and is given by

$$\begin{aligned}-\Delta = \{&E\text{ (chemisorbed atom with core hole)} + \Phi \\ &- E\text{ (chemisorbed atom)}\} \\ &-\{E\text{ (free atom with core hole)} - E\text{ (free atom)}\}\end{aligned}$$

Here E (chemisorbed atom) is the total energy of the metal–adatom system; E (chemisorbed atom with core hole) + Φ is the total energy of the metal–adatom system with an adatom core electron removed to the vacuum level. Φ is the metal work function so that E (chemisorbed atom with core hole) is the energy of the metal–adatom system with a core electron removed and an additional electron at the Fermi level. The X-ray photoemission analysis of deep core levels probes the charge state and the chemical environment of an atom through variation in core-level binding energies, and thus the calculations of Δ are of importance. The preceding calculations automatically provide the extra-atomic screening charge distribution:

$$\begin{aligned}n(\mathbf{r};\text{ extra-atomic}) = [&n(\mathbf{r};\text{ chemisorbed atom with core}) \\ &- n(\mathbf{r};\text{ chemisorbed atom})] \\ &- [n(\mathbf{r};\text{ free atom with core hole}) - n(\mathbf{r};\text{ free atom})]\end{aligned}$$

This quantity integrates to unity and hence $n(\mathbf{r};$ extra-atomic$)$ contains one electron. Such a calculation beings out very neatly the redistribution of charge associated with a given state of the atom on chemisorption. The preceding calculation clearly goes beyond the usual linear-response theory. These results were displayed for Na and Cl as adatoms on a jellium surface whose density corresponds to a simple metal such as Al, which exhibit widely different characteristics. An important conclusion from these studies is that valence resonances and sufficiently high-lying core levels have a more readily interpretable dependence on metal–

adatom separation than do the deep core levels. In the case of Na, the simple picture of screening charge consisting of an electron supplied by the metal, but residing in the valence shell of the excited atom, seems to be accurate as the valence shell is able to accommodate the extra charge. In the case of Cl, there is insufficient "room" in the valence shell for the screening charge, in which case the environment of the atom plays a greater role. This study very naturally leads to the study of an atom bonded to the surface of a simple metal.[140] The bonding characteristics in this case is expected to be different from the bonding between two atoms in molecular formation because the metallic substrate has a continuous energy spectrum. The calculations based on atom-jellium model and HKS local density formalism contain no adjustable parameters and all the calculations are completely self-consistent. The method of attack of Lang and Williams[140] is noteworthy. They recast the HKS single-particle equations into a Lippmann–Schwinger scattering theoretic form

$$\Psi^{MA}(\mathbf{r}) = \Psi^{M}(\mathbf{r}) + \int d\mathbf{r}' G^{M}(\mathbf{r}; \mathbf{r}')\delta v_{\text{eff}}(\mathbf{r}')\Psi^{MA}(\mathbf{r}') \tag{5.49}$$

where M stands for bare metal and MA for the metal–adatom system. $\delta v_{\text{eff}}(\mathbf{r})$ is the difference between the combined metal–adatom system and the bare metal.

The Schrödinger differential equation corresponding to (5.49) is solved exactly by direct numerical integration outward from the adatom nucleus. Solutions obtained in this way, $\Psi_l(\mathbf{r})$ are characterized by their angular behavior near the nucleus but do not in general satisfy the boundary conditions contained in (5.49) because the semiinfinite metal region does not possess the same symmetry as the atom. Therefore a linear combination of the fundamental solutions,

$$\Psi^{MA}(\mathbf{r}) = \sum_l C_l \Psi_l(\mathbf{r}) \tag{5.50}$$

with the coefficients C_ls determined by substituting (5.50) into (5.49) would give us the required solution. For the mathematical details of the derivation of (5.49), the construction of the Green's function $G^M(\mathbf{r}, \mathbf{r}')$ and the solutions Ψ^M, calculation of $\Psi_l(\mathbf{r})$ and thence C_l, the reader is referred to the original paper. The local density functional scheme of Hedin–Lundqvist is employed in the calculations. One of the important quantities that is useful in the elucidation of the results obtained is the self-consistent difference in the density of the electrons of the metal–adatom system and the metal.

$$\delta n(\mathbf{r}) = n^{MA}(\mathbf{r}) - n^{M}(\mathbf{r}) \tag{5.51}$$

$n^M(\mathbf{r})$ is calculated as in the Lang–Kohn work.[135,136] $n^{MA}(\mathbf{r})$ is calculated by an iteration scheme beginning with $n^M(\mathbf{r})$. The atomic binding energy ΔE_a is the difference in total energy between the metal plus the separated atom and the metal with atom chemisorbed on its surface. The equilibrium distance d_{eq} between the adatom nucleus and the positive background edge is obtained by minimizing the calculated total energy. The calculations were made for the adatoms H, Li, O, Na, Si, and Cl which exhibit a variety of chemical behavior, chemisorbed on a high-density metallic substrate whose density corresponds roughly to that of Al. Some results for the substrate density corresponding to that of Na are also included to exhibit the dependence on this parameter. The electron density of the substrate metal (jellium) and the nuclear charge Z of the adatom are the only parameters in this calculation. From the computations, it emerges that Li and Cl represent the positive and negative ionic chemisorption, the direction of charge transfer being consistent with the magnitudes of electronegativities of Li and Cl and a high-density metallic substrate of the type of Al. The charge density contours obtained from these calculations display a spatially detailed picture of bonding. In Li, for example, charge is transferred from the vacuum side of the adatom toward the metal. For Si, the contours show a depletion of charge near the nucleus and accumulations on both the bond and the vacuum sides, showing thereby a covalent bonding character. The reader is urged to consult this work for details of many interesting results concerning the dipole moment, heat of adsorption, and so on. It should be remarked that van Himbergen and Silbey[142] have calculated the potential energy curves for helium and argon adsorbed on a metal surface based on the jellium–adatom model using Gordon–Kim procedure.

The Lang–Kohn theory has recently been applied to study the bimetallic interface energies. Swingler and Inkson[143] have derived an exact result for the interface energy of a jellium interface, assuming that the jellia are at zero separation. This should serve as a check on approximate calculations. Ferrante and Smith[144] have made a self-consistent calculation of adhesive binding energies as a function of separation based on LDF scheme. The resultant shapes of the adhesive binding energy curves is shown to vary substantially from metal to metal.

A very interesting application of the HKS theory has been made by Kalia and Vashishta[145] where they calculate the *surface structure of electron-hole drops in Ge and Si.* In indirect band-gap semiconductors such as Ge and Si excitons can be formed in large numbers by means of photoexcitation with the help of lasers. This high-density system of excitons can undergo a phase transition at liquid helium temperatures. In this connection, the formation of drops, their size, growth, etc. of such exciton liquid has become an important area of investigation. They generalize the HKS theory to a multicomponent electron system. On the basis of such a self-consistent calculation these authors predict that the electron–hole drop carries a negative charge in Ge with one conduction and one valence band. Since the surface of the electron–hole drop system is a truly inhomogeneous system of electrons and holes, it provides another testing ground for the density-functional theory. We now turn to the description of results obtained in another interesting area where a charged particle moves through a metal.

3. Motion of a Proton or Muon in Simple Metals

In contrast to chemisorption, the calculation of properties of an impurity in a metal or a jellium system is of interest in recent years because recently the study of the motion of positrons, and so on, in condensed matter system has become an important experimental tool. The problem of the screening of a point charge in an electron gas is very well studied. The density functional method enables one to compute the proper nonlinear screening of the charge in such systems. The proper nonlinear theory is essential because, in the case of attractive impurities, there is an excessive charge pile up and bound states may well occur. Among the physical systems where this seems to be important is the adsorption of hydrogen in metals such as Pd. In this subsection we describe some of the recent studies in this area.

Almbladh et al.[146] studied the screening of a proton in a jellium by employing the LDF theory. They calculated the electron density around the proton and the relaxation energy for a range of electron gas densities corresponding to $r_s = 1$ to 6 a.u. (simple metallic densities). These are compared with linear screening theory and are used to estimate the hydrogen heats of solution for jellium. This self-consistent calculation shows that the electron density at the proton varies very little with mean electron gas density over the metallic region ($r_s = 1$ to 6 a.u.). This shows that the polarization cloud around the proton is rigid. But the electron density at the proton is greatly enhanced over the mean density and the linear response theory grossly underestimates the charge pile up particularly for the lower mean densities. Moreover, the so-called Friedel oscillations in the density away from the proton is very different from those in the linear theory. For $r_s > 1.9$ a.u. a bound state is obtained by these authors. This is an H^- state and remains quite shallow, being deepest at $r_s \simeq 4$ a.u. for which the binding energy is 0.3 eV. From these calculations, it was found that simple metals dissolve only very small amounts of hydrogen in conformity with experimental results.

Jena et al.[147] calculated the electron spin-density around a positive muon using the LSD formalism in a self-consistent calculation. These results were used to estimate the hyperfine field at interstitial positive-muon sites in ferromagnetic Fe, Co, Ni, and Gd. In view of much experimental work using the μ^+ in matter, this work is bound to spark further investigations. Jean and Singwi[148] calculated the electron density distribution around a proton in a jellium with the density range in the metallic region ($r_s \sim 1$ to 6) including the first gradient correction, unlike Ambladh et al.[146] Very shallow bound states continued to exist between $r_s = 2.07$ and 5 (compare Ambladh et al.'s $r_s = 1.9$ and above). It is concluded that the HKS formalism even with local density approximation provides reliable quantitative results. Jena and Singwi[148] also used the LDF to calculate the Knight shift in paramagnetic metals at the positive muon site. These results are not in accord with experimental results and reasons for this are given.

4. Application of LDF to MOSFET Systems

When a strong electric field is applied perpendicular to the surface of a semiconductor, the electronic states form two-dimensional energy bands. The electrons are essentially confined to a two-dimensional region parallel to the surface of the semiconductor. This system is a different kind of an inhomogeneous electron system in that it is a two-dimensional inhomogeneity in an otherwise three-dimensional system. The density of these electrons can essentially be varied over a wide range over three decades from 10^9 to $10^{12}/\text{cm}^2$. This system has many practical applications and is very interesting theoretically. Ando[149] has applied the HKS theory to calculate the g factor and the effective mass of these electrons with results that are in satisfactory agreement with experiments, in contrast to other theories that were not successful. Ando, Fowler, and Stern[150] are presently writing a comprehensive review article on the subject and the reader may refer to this for more details. We describe here Ando's theory. In the HKS formalism, the electronic density $n(\mathbf{r})$ is a crucial element, all other entities being functionals of $n(\mathbf{r})$, for a description of the ground state of any inhomogeneous system of electrons. What Ando did was to note that $n(\mathbf{r})$ in this problem is essentially inhomogeneous in the z direction, perpendicular to the surface of the semiconductor and so is a function of z only. In other words, he assumed the density distribution to be given in terms of a set of one-dimensional wave function, $\xi_n(z)$, which describes the electrons moving in a two-dimensional (xy) plane at a distance z from the surface. These functions obey HKS one-particle equations as before which are appropriately constructed for this case. He combined the effective mass equation with the HKS equations so that the kinetic energy part, T_s, now mimics the effective mass Hamiltonian, being replaced by the operator

$$-\frac{\hbar^2}{2m_t}\left(\frac{\partial^2}{\partial x^2}+\frac{\partial^2}{\partial y^2}\right)-\frac{\hbar^2}{2m_l}\frac{\partial^2}{\partial z^2}$$

in contrast to the usual $-(\hbar^2/2m)\nabla^2$, where m_l, m_t are the longitudinal and transverse effective band masses of the electrons appropriate to the surface of the semiconductor. Ando then expressed the quasiparticle effective mass m^* and the effective g factor g^* in terms of $n(z)$. Using the three-dimensional electron gas results as input, Ando computed these quantities. The many-body contributions are thus properly taken into account in this theory and the results obtained are satisfactory. Since only LDF was employed, this work can be perhaps criticized on that account. In the application of straightforward many-body theory to this system, one assumed essentially a two-dimensional character to the many-electron system[151] and computed a variety of experimentally relevant quantities that were not as successful as Ando's. Certainly the reality is somewhere between the two and further work remains to be done in this area. For a more detailed account of various aspects of this system one may consult the review article of Ando, Fowler, and Stern.[150]

E. Liquid-state Properties

This subject matter is vast and it is not our intention to review even in barest detail the various aspects of properties of liquids and their theoretical explanations. We should mention that the functional methods have been in use in this field for a long time. The equilibrium properties of uniform liquids described in terms of Percus–Yevik equation and Ornstein–Zernicke relations are examples of such a functional theory. The reader may refer for a description of these, to recent books on the subject by Balescu[152] and Croxton.[153] We review here those aspects of the work which make use of the Hohenberg–Kohn theorem in much the same way as in the electronic properties of metals described so far. It should be pointed out that the HKS theorems are all valid for classical as well as quantum mechanical Fermi or Bose systems. The use of the HKS theorems for inhomogeneous liquid systems such as liquids with free surface, droplets, and vortex line structure, has appeared recently and we review these aspects here. We must mention at the outset that such a formulation has been quite successful and incorporates into the density-functional formalism for liquids these new aspects of the inhomogeneities in the system in a unified theoretical framework.

The first application of the HKS theorem to the free surface of liquid helium and the electron bubble was made by Padmore and Cole.[154] They used a phenomenological expression for the free energy functional and used the variational principle. They obtained physically reasonable answers from their simple theory and suggested that this approach lends itself neatly to the study of other restricted geometries such as pores and channels. Ebner and Saam[155] placed this phenomenological theory on a sounder footing. They were able to include in their formalism the effects of zero-point vibrations of both the bulk and surface excitations, and studied liquid helium at zero temperature. In particular, they obtained a surface tension of 0.384 erg/cm^2 (0.578 erg/cm^2 expt.) for liquid helium at $T = 0$°K. They also were able to study various aspects of the planar free-surface, vortex-line structure, droplets, and so on. Ebner, Saam, and Stroud[156] put forward a density-functional theory appropriate to nonuniform simple classical fluids which is exact in the linear-response regime. Using only the Lennard-Jones 6–12 potential as input, this theory was found to yield surface tensions and density profiles in very good agreement with results of Monte Carlo computations. More recently, Saam and Ebner[157] compared the results of their functional theory with different approximate theories for dealing with nonuniform classical fluids. In particular, they show that the density-functional theory produced an unsaturated liquid film near the fluid–substrate interface whereas the other theory yields no such film. In another formal paper, Saam and Ebner[158] developed a formally exact density-functional theory for nonuniform classical systems by expressing the free energy in terms of direct density-correlation functions. This is similar to the corresponding work of Almbladh,[43] Gunnarsson and Lundqvist,[44] and Rajagopal and von Barth[45] for electronic systems. We must point out that Chihara[159] has derived the quantum hypernetted chain equation from the HKS theory. Manninen and Hautojärvi[160] have the density-functional formalism to study the clustering of atoms around the positron and positive ions in gaseous He, Ne, and Ar. We discuss their results at the end of this section.

We begin by outlining the Saam–Ebner theory. The steps are essentially the same as in Section II but we outline here the theorem again because, in the theory of liquids, some of the definitions have a different connotation. We explicitly use here the classical formulation even though much of what is being proved is valid for quantum systems.

Following Saam and Ebner,[158] consider a classical system of particles with fixed chemical potential μ at a temperature T in the presence of an external potential $v(\mathbf{r})$ which couples to the particle number density $n(\mathbf{r})$. Using Mermin's[25] formulation of the HK theorem, we note that there exists a functional $\tilde{\Omega}[n]$ of $n(\mathbf{r})$ such that the minimum value of

$$\Omega \equiv \tilde{\Omega}[n] + \int d\mathbf{r} v(\mathbf{r}) n(\mathbf{r}) \tag{5.52}$$

w.r.t. variations of $n(\mathbf{r})$ at constant μ, T, $v(\mathbf{r})$, and volume V, is the equilibrium grand free energy of the system. At the minimum, $n(\mathbf{r})$ is the equilibrium number density. The minimum is determined by

$$\left.\frac{\delta\Omega}{\delta n(\mathbf{r})}\right|_{T,\mu,v(\mathbf{r}),V} = 0 = -v_n(\mathbf{r}) + v(\mathbf{r}) \tag{5.53}$$

where

$$v_n(\mathbf{r}) \equiv -\left.\frac{\delta\tilde{\Omega}[n]}{\delta n(\mathbf{r})}\right|_{T,\mu,v} \tag{5.54}$$

As before, for any given $n(\mathbf{r})$, $v_n(\mathbf{r})$ is the external potential which would produce this $n(\mathbf{r})$.

The direct correlation function $C(\mathbf{r}, \mathbf{r}')$ is defined by[152]

$$\frac{\delta v_n(\mathbf{r})}{\delta n(\mathbf{r}')} = \beta^{-1}\left(\frac{\delta(\mathbf{r}-\mathbf{r}')}{n(\mathbf{r})} - C(\mathbf{r}, \mathbf{r}')\right) \tag{5.55}$$

where $\beta^{-1} = k_B T$ as before. See Section II.B. The pair correlation function $h(\mathbf{r}, \mathbf{r}')$ is defined by

$$\frac{\delta n(\mathbf{r})}{\delta v_n(\mathbf{r}')} = -\beta n(\mathbf{r})[\delta(\mathbf{r}-\mathbf{r}') + n(\mathbf{r}')h(\mathbf{r}, \mathbf{r}')] \tag{5.56}$$

This should be contrasted with (3.38). Now, one has the general identity, in the theory of functional differentiation,

$$\int d\mathbf{r}'' \frac{\delta n(\mathbf{r})}{\delta v_n(\mathbf{r}'')} \frac{\delta v_n(\mathbf{r}'')}{\delta n(\mathbf{r}')} = \delta(\mathbf{r}-\mathbf{r}') \tag{5.57}$$

Using the definitions (5.55) and (5.56) in (5.57), we obtain the Ornstein–Zernicke relation[152] for nonuniform systems

$$h(\mathbf{r}, \mathbf{r}') = C(\mathbf{r}, \mathbf{r}') + \int d\mathbf{r}'' h(\mathbf{r}, \mathbf{r}'') n(\mathbf{r}'') C(\mathbf{r}'', \mathbf{r}') \tag{5.58}$$

The next set of steps are very similar to the coupling constant integration trick used in Section II in obtaining (2.112), except that the arguments proceed differently. Suppose one has some means of obtaining $C(\mathbf{r}, \mathbf{r}')$ for a given $n(\mathbf{r})$ which is arbitrary but physically reasonable (nowhere negative, and integrates to the total number of particles and represents a physical system). Then (5.55) can be functionally integrated w.r.t. $n(\mathbf{r})$ from some initial density $n_0(\mathbf{r})$ to a final density given by $n(\mathbf{r})$.

To do this, we choose a path[161] in the space of density functions characterized by a single parameter θ varying from 0 to 1. Then the integration gives[161]

$$\beta[v_n(\mathbf{r}) - v_{n_0}(\mathbf{r})] = -\ln\left(\frac{n(\mathbf{r})}{n_0(\mathbf{r})}\right) + \int_0^1 d\theta \int d\mathbf{r}' \frac{\partial n(\mathbf{r}', \theta)}{\partial \theta} C(\mathbf{r}, \mathbf{r}'; \theta) \tag{5.59}$$

Choose

$$n(\mathbf{r}; \theta) = n_0(\mathbf{r}) + \theta[n(\mathbf{r}) - n_0(\mathbf{r})] \tag{5.60}$$

so that (5.59) becomes

$$\beta[v_n(\mathbf{r}) - v_{n_0}(\mathbf{r})] = -\ln\left(\frac{n(\mathbf{r})}{n_0(\mathbf{r})}\right) + \int_0^1 d\theta \int d\mathbf{r}' C(\mathbf{r}, \mathbf{r}'; \theta)[n(\mathbf{r}') - n_0(\mathbf{r}')] \tag{5.61}$$

The existence of $\tilde{\Omega}[n]$ guarantees that the result, (5.61), is independent of the path of integration which implies that, as a function of θ, the functions behave mathematically properly which precludes application of this formalism to the critical region of a continuous phase transition. Let us choose

$$\left.\begin{aligned} n_0(\mathbf{r}) &= n_0 = \text{constant} \\ &\text{and} \\ v_{n_0}(\mathbf{r}) &= 0 \end{aligned}\right\} \tag{5.62}$$

Compare Section II, (2.104). In Section II, we used a potential that maintained the density $n(\mathbf{r})$ when the system went from noninteracting to interacting situation as the coupling parameter θ varied from 0 to 1. Then, (5.61) becomes

$$\begin{aligned} \beta v_n(\mathbf{r}) &= -\ln\frac{n(\mathbf{r})}{n_0} + \int_0^1 d\theta \int d\mathbf{r}' C(\mathbf{r}, \mathbf{r}'; \theta)[n(\mathbf{r}') - n_0] \\ &= -\beta \left.\frac{\delta\tilde{\Omega}[n]}{\delta n(\mathbf{r})}\right|_{T,\mu,v} \end{aligned} \tag{5.63}$$

using (5.54). Integrating this once again along the same path, we obtain

$$\begin{aligned} \beta[\Omega - \Omega_0] = \beta \int d\mathbf{r} v(\mathbf{r})[n(\mathbf{r}) - n_0] + \int d\mathbf{r} n(\mathbf{r}) \ln \frac{n(\mathbf{r})}{n_0} - \int d\mathbf{r}(n(\mathbf{r}) - n_0) \\ - \int_0^1 d\theta \int_0^\theta d\theta' \iint d\mathbf{r}\, d\mathbf{r}' C(\mathbf{r}, \mathbf{r}'; \theta')[n(\mathbf{r}) - n_0][n(\mathbf{r}') - n_0] \end{aligned} \tag{5.64}$$

where Ω_0 is the grand free energy of the reference frame. The Mermin theorem guarantees that Ω is unique, that is, path independent. If one can calculate $C(\mathbf{r}, \mathbf{r}'; \theta)$ for any density along the path, (5.64) provides a useful way of finding Ω and $n(\mathbf{r})$ for a given external potential $v(\mathbf{r})$. The final density can be parametrized and R.H.S. of (5.64) minimized w.r.t. the parameters. This then gives a scheme for modeling the inhomogeneous system. For instance, from the Percus–Yevick theory of uniform liquids, one can obtain $C(\mathbf{r}, \mathbf{r}'; \theta)$ with the Lennard-Jones potential as input. In the work of Ebner, Saam, and Stroud,[156] the formalism used was patterned after the theory described in Section II applied to an inhomogeneous Bose system of particles. Then, one obtains a result similar to (5.64), with $C(\mathbf{r}, \mathbf{r}'; \theta)$ replaced by $V(\mathbf{r}-\mathbf{r}')g(\mathbf{r}\mathbf{r}'; \theta)$ where $V(\mathbf{r}-\mathbf{r}')$ is the Lennard-Jones potential of interaction among the particles and $g(\mathbf{r}\mathbf{r}'; \theta)$ is related to the density–density response function of the system. The Percus–Yevick equation is

$$C(\mathbf{r}, \mathbf{r}') = (1 + h(\mathbf{r}, \mathbf{r}'))(1 - e^{\beta V(\mathbf{r}-\mathbf{r}')}) \tag{5.65}$$

where $V(\mathbf{r}-\mathbf{r}')$ is the interatomic potential, usually taken to be the Lennard-Jones potential. Equations 5.65 and 5.58 must be solved together with that for $n(\mathbf{r})$ given by (5.63) and (5.53). Chihara[159] suggests the use of the Schrödinger-like one-particle equation of the HKS type instead of (5.53) for the quantum description of the liquid and obtain $n(\mathbf{r})$ from the one-particle solutions as in Section II. This approach seems to be attractive but has not been applied so far in practical calculations.

From the preceding general formulations we can obtain the previous models as special cases. Before doing that, we can make a few general remarks. The work of Ebner and Saam[155] is similar to the HKS theory applied to a system with slowly varying density of a Bose system and the purpose was to deal with liquid helium at $T = 0°\mathrm{K}$ with a variety of nonuniformities in it. In that case, the ground-state energy $E[n]$ is a functional of density $n(\mathbf{r})$ and when $n(\mathbf{r}) = n_0 + \delta n(\mathbf{r})$ with $|\delta n(\mathbf{r})/n_0| \ll 1$ and $\delta n(\mathbf{r})$ slowly varying in space, the second-order perturbation theory gives us an accurate result

$$E[n] = E_0 + \tfrac{1}{2}\int\int d\mathbf{r}\, d\mathbf{r}'\, W[\mathbf{r}-\mathbf{r}'; n_0]\delta n(\mathbf{r})\delta n(\mathbf{r}') \tag{5.66}$$

where

$$W[\mathbf{r}-\mathbf{r}'; n_0] = -\tfrac{1}{4}\int \frac{d^3\mathbf{q}}{(2\pi)} \frac{e^{i\mathbf{q}\cdot(\mathbf{r}-\mathbf{r}')}}{\chi_q(n_0)} \tag{5.67}$$

Here $\chi_q(n_0)$ is the retarded density-density response function at a wave number $\mathbf{q}$ for the uniform system, with density n_0. The basic assumption here is that the uniform system of superfluid He at zero temperature may be uniquely described by a functional of the number density $n(\mathbf{r})$. We should point out that this precludes rotational motions of the fluid and is applicable to irrotational velocity flows of the superfluid system.[34,35] In the limit $q \to \infty$, χ_q approaches the free-particle result

$$\chi_q^{0^{-1}}(n_0) = \frac{\hbar^2 q^2}{4mn} \tag{5.68}$$

where m is the mass of the ^{4}He atom, and separating this out in (5.66) and after some rearrangement of terms, one obtains

$$E[n] = \int d\mathbf{r}\epsilon(n(\mathbf{r})) + \frac{\hbar^2}{2m}\int d\mathbf{r}(\nabla n^{1/2}(\mathbf{r}))^2 - \tfrac{1}{4}\int\int d\mathbf{r}\, d\mathbf{r}' \times \left\{\int \frac{d^3q}{(2\pi)^3} e^{i\mathbf{q}\cdot(\mathbf{r}-\mathbf{r}')}[\chi_q^{-1}(n) - \chi_q^{0^{-1}}(n_0)]\right\}(n(\mathbf{r}) - n(\mathbf{r}'))^2 \tag{5.69}$$

The second term in the preceding is the usual quantum pressure term,[34,35] and here it is derived purely from an appeal to the density fluctuations for $q \to \infty$ of the uniform system. Ebner and Saam[155] then used for $\chi_q^{-1}(n_0)$, the Feynman result that

$$\chi_q^{-1}(n_0) = \frac{\hbar^2 q^2}{4mn_0 S_q^2} \tag{5.70}$$

where S_q is the liquid structure factor. A further approximation used is to take $n_0 = [n(\mathbf{r}) + n(\mathbf{r}')]/2$ and using (5.69), with the help of the variational principle an integro-differential equation for $n(\mathbf{r})$ is obtained. They also included the surface mode contributions to $E[n]$ so that the zero point contributions from the low-lying collective modes at long wavelengths make appropriate contributions in determining the density profile of the inhomogeneous system. In this way, they extended the work of Padmore and Cole[154] and obtained a reasonable description of the liquid helium system with planar free surface, vortex line, and droplets of ^{4}He.

The work of Ebner, Saam, and Stroud[156] employed the classical density functional theory, proceeding as in the liquid helium work.[155] The approximations used in this work can be made more precise using the general theory of Saam and Ebner[158] outlined previously. Define

$$\tilde{C}(\mathbf{r}, \mathbf{r}') \equiv 2\int_0^1 d\theta \int_0^\theta d\theta' C(\mathbf{r}, \mathbf{r}'; \theta') \tag{5.71}$$

and

$$\tilde{C}_0(\mathbf{r}) \equiv \int d\mathbf{r}' \tilde{C}(\mathbf{r}, \mathbf{r}') \tag{5.72}$$

Then (5.64) may be rewritten in the form

$$\Omega - \Omega_0 = \int d\mathbf{r}\,\omega(\mathbf{r}) + \int d\mathbf{r}\,V(\mathbf{r})[n(\mathbf{r}) - n_0] + \frac{k_B T}{4} \int\int d\mathbf{r}\,d\mathbf{r}'\,\tilde{C}(\mathbf{r},\mathbf{r}')[n(\mathbf{r}) - n(\mathbf{r}')]^2 \tag{5.73}$$

with

$$\omega(\mathbf{r}) = k_B T\left[n(\mathbf{r}) \ln \frac{n(\mathbf{r})}{n_0} - (n(\mathbf{r}) - n_0) - \tfrac{1}{2}\tilde{C}_0(r)(n(\mathbf{r}) - n_0)^2\right] \tag{5.74}$$

The approximate theory of Ref. 156 essentially employs a local density scheme, where $\omega(\mathbf{r})$ and $\tilde{C}(\mathbf{r},\mathbf{r}')$ in (5.73) are replaced by the corresponding expressions for a *uniform* system with density $n(\mathbf{r})$ with $\tilde{C}(\mathbf{r},\mathbf{r}')$, as the direct correlation function of a uniform system at a density $(n(\mathbf{r}) + n(\mathbf{r}'))/2$, that is,

$$\tilde{C}(\mathbf{r},\mathbf{r}') \cong C^h\left(\mathbf{r} - \mathbf{r}'; \frac{n(\mathbf{r}) + n(\mathbf{r}')}{2}\right)$$

This approximation is perhaps reasonable for the regime within which the linear response theory is valid. The work of Ref. 156 reported reasonable numerical results. In Ref. 158, it is concluded that this theory guessed the correlation functions (5.71), (5.72) correctly and hence was able to obtain good results. In Ref. 157 Saam and Ebner compared the results of their theory[158] with several existing theoretical models of inhomogeneous liquids. The reader is referred to this paper for detailed comparisons. The most important point that emerges from this study is that the density functional scheme is capable of correctly handling large nonlinear perturbations such as those leading to formation and growth of unsaturated films on solid substrates, where the other theories have not succeeded.

We conclude this section with a brief description of the work of Manninen and Hautojärvi.[160] This work is of interest because it combines the density-functional theory in an interesting way with the general variational principle for free energy for a system of a positive charge in a rare gas system. They deal with the rare gas system by means of a density-functional of the Saam–Ebner type, and add to it a term that couples the positive charge to the gas density using an optical psuedo-potential. The wave function for the positive charge then obeys the usual Schrödinger equation with the density of the gas as input which in turn is determined in terms of the positive charge interaction with the gas via the HKS theorem. Using a local density functional model with the van der Waals equation of state and the optical potential for the interaction between the gas atoms and the positive charge, these authors explain abnormal behavior of the positron at low temperatures in rare gases.

A nonlocal scheme such as that given in Section III.B can be given for the Bose gas system following the same kind of procedures.[162] The density–density correlation function for the Boson system at finite temperatures can be defined as

$$n(\mathbf{r})n(\mathbf{r}')g(\mathbf{r},\mathbf{r}') = \mathrm{Tr}\{P\Phi^{+}(\mathbf{r})\Phi^{+}(\mathbf{r}')\Phi(\mathbf{r}')\Phi(\mathbf{r})\} \tag{5.75}$$

where P is the statistical operator for the Boson system and Φ, Φ^{+} are the annihilation and creation operators for the bosons. Then we can prove the following properties:

A. $g(\mathbf{r},\mathbf{r}')$ is nonnegative

B.
$$\int d\mathbf{r}'(g(\mathbf{r},\mathbf{r}') - 1)n(\mathbf{r}') = -1 + \frac{1}{\beta}\frac{\partial}{\partial\mu}(\ln n(\mathbf{r})) \tag{(5.76a)}$$

and

$$\int d\mathbf{r} n(\mathbf{r})(g(\mathbf{r},\mathbf{r}') - 1) = -1 + \frac{1}{\beta}\frac{\partial}{\partial\mu}(\ln n(\mathbf{r}')) \tag{5.76b}$$

where μ is the chemical potential of the system.

C. $g(\mathbf{r},\mathbf{r}')$ is symmetric in $\mathbf{r}, \mathbf{r}'$.

One can then express the free energy of the system in terms of $g(\mathbf{r},\mathbf{r}')$ and obtain the one-particle potential, $V_{xc}(\mathbf{r})$, as before by functional differentiation once. The procedure can then be extended to compute the density-response functions also, which will involve the second derivative. The functional derivatives of g w.r.t. $n(\mathbf{r})$ can then be determined using the sum rules (5.76) and an ansatz of the form (3.29) and (3.49). Using the results of homogeneous systems as input, these equations can be solved and thus a closed set of equations obeying the basic requirements of symmetry and sum rules can be developed. This will enable one to go beyond the approximation used in Ref. 156. Combined with the work of Saam and Ebner, such a scheme may provide a sound basis for further work in the theory of inhomogeneous liquids, both classical and quantum systems.

VI. CONCLUDING REMARKS

In this final section, we touch upon various attempts to improve upon the LDF scheme, criticisms of LDF, some new work on the calculation of phonon frequencies of covalent semiconductors, progress in relativistic density functional theory, and some work we missed earlier under Section V.A.

The self-consistent LDF scheme has been applied, as we have seen in Section V, to a wide variety of inhomogeneous systems with surprising success. Whenever the inhomogeneity could be represented as an external potential irrespective of the extent of inhomogeneity, be it a pointlike situation as in atoms or many-point inhomogeneities as in molecules and bulk solids, or two-dimensional inhomogeneity as in surface problems and MOSFET systems, or even two solid surfaces as in the investigation of adhesion between solids, LDF scheme has been applied. The LDF scheme is valid theoretically as long as the scale of inhomogeneity is small compared to the reciprocal of the Fermi momentum. On the whole, in all the applications made so far this seems to have been violated and yet the results obtained are in surprisingly good accord with experimental ground-state properties. One of the ways of examining this question has been to numerically compute properties of the system where the nonlocality of the interparticle potential may be significant and see if significantly different results are obtained compared to those of LDF. Sham and Kohn[20] suggested a scheme for computing the nonlocal self-energy of an inhomogeneous system which goes beyond the LDF theory. This has now been numerically tested in two situations—in band structure and Fermi surface calculations[163,164] and in the investigation of light impurities in simple metals.[165] Nickerson and Vosko[163] calculated the Fermi surface of alkali metals and found that they are more sensitive to the approximation made to the self-energy than are the other properties like the density and chemical potential. The latter are adequately described by the local scheme. The Fermi surface of copper was investigated similarly by Wang and Rasolt[164] who observed that nonlocal contributions to the Fermi surface are relatively small compared to the local contributions resulting from the nonlocal theory. They also performed the calculation in which only the s, p bands are dealt with using a nonlocal scheme for the self-energy so that the experimental results of Fermi surface of Cu was reproduced. The work of Vinter is more transparent regarding the effect of the nature of nonlocality.[165] He considered the problem of Amladh et al.[146] where a charged impurity is imbedded in an electron gas. He used the Sham–Kohn[20] scheme to incorporate the energy dependence and nonlocality of the self-energy. By varying the charge on the impurity, he can go from a system with a strongly localized bound state to a completely delocalized system. He found that the bound state associated with the impurity is not shifted appreciably when the nonlocal, energy-dependent exchange and correlation is used instead of the local, energy-independent potential as long as the bound state is above the plasmaron band. This is mainly because of large cancellations of contributions. Another point that emerges from all these works is that even though the eigenvalues appearing in the HKS equation do not have any direct significance, the use of the eigenvalues as the actual one-particle excitations of the system has *some* physical meaning. In view of Janak's recent work,[24] it appears that, with his reinterpretation, these eigenvalues do have some significance. It appears that due to subtle cancellations of contributions, much of the LDF results seem to require little correction. It should be pointed out that there has been no formal proof of these cancellations except in special models. Overhauser has attempted to disprove the preceding conclusions based on a theory of nonuniform electron gas that he has proposed.[166,167] Overhauser raises many basic questions concerning the HKS development. The most important objection raised is that the HKS scheme involves no velocity dependence in its self-consistent potential. We pointed out on two occasions in this chapter that the general structure of the many-particle Hamiltonian involves current operators in general and when one has irrotational flow, one can use a density-functional theory, at least for Bose systems.[34,35] Overhauser considers a model system that is inhomogeneous in that he considers a simple cosine function as the external potential. He then goes on to construct the exchange and corelation potentials using the one-particle states generated by this external potential, and shows that this potential is nonlocal and does not go over to the HKS potential in the limit of long wavelengths of the modulations of the density. This work is of importance because it calls into question in a definite way the correctness of the HKS formalism. One criticism of the Overhauser calculation seems to us to be worth pointing out. In his calculations, only the exchange-correlation energies were computed using the solutions for the cosine potential. This implies that he has treated the interaction effects in perturbation theory whereas in the HKS development, the entire many-particle system was considered placing the kinetic and interaction parts on an equal footing. It appears, therefore, that the proper calculation ought to proceed by constructing explicitly the many-electron solution for a model inhomogeneous problem. Put another way, there is a kinetic-energy contribution to the exchange and correlation potential that Overhauser has not computed. This remains to be incorporated before one can conclusively rule out the HKS formalism as being a fundamentally unjustified theory. We should point out that the use of the coupling-constant integration method overcomes the question of dealing with the kinetic-energy contributions to the exchange-correlation part. But this makes the theory much more complicated. As far as we know, no counter example has been constructed to disprove the HK theorem. The objections of Overhauser concerning the HKS theory are based on physical grounds and indeed seem reasonable, but no satisfactory and conclusive proof has so far been put forward. We should remind the reader that the HKS development concerns the ground state of the many-body system. The calculations of the excited states within LDF are mostly ad hoc and do not have a theoretical basis. The work of Peuckert[26] on the HKS theorem for time-dependent inhomogeneous perturbations is promising even though it lacks the full mathematical treatment.

The calculation of phonon frequencies based on HKS formalism involves two ground-state calculations corresponding to two configurations of the lattice and is similar to the molecular vibration work. Wendel and Martin[168] have recently derived the structural properties of covalent semiconductors, in particular Si, based on LDF theory. Since two self-consistent calculations are made, this method is not restricted to small perturbations. These authors consider displacements of atoms in periodic patterns so that both the undistorted and distorted systems are crystalline and compute the difference in their total energies. In this fashion, they derive structural properties directly from the electron Hamiltonian. Their calculations were not self-consistent. The bulk modulus calculated with no adjustable parameters is found to be 28% lower than the observed value. For distortions lowering the crystal symmetry, they choose patterns corresponding to a transverse optic (TO) phonon at the zone center, a (C_{11}–C_{12}) elastic deformation, and a transverse acoustic (TA) phonon at the X-point of the Brillouin zone. The latter two distortions require directional covalent forces in phenomenological models. The TO frequency calculated in this way is found to be in good agreement with experiment. The results for (C_{11}–C_{12}) and ω_{TA} are not as good and it is hoped a self-consistent calculation would lead to better answers. The basic structural properties of Si such as the elastic constants and optic mode frequencies are obtained in this way in reasonable agreement with experiment. The calculations do predict the softening of $\omega_{TA}(X)$ and its pressure dependence. There are convergence difficulties in actual computations because of large cancellations of contributions thus requiring more complete self-consistent calculations before one can obtain definitive numerical results. We should point out that two ground-state energy calculations are in themselves variational in character but their difference is not. This point may be of subtle importance in such calculations.

Recently two interesting self-consistent relativistic band structure calculations based on nonrelativistic LSD for E_{xc} but Dirac-like HKS equation have appeared. Glötzel[169] calculated the equilibrium distance, bulk modulus, spin susceptibility of lanthanum, cerium, and theorium. It is found that the $l = 3$ partial waves make a significant contribution to crystal bonding in all the three cases considered. Also, at the atomic volume of the γ-phase of Ce, a ferromagnetic instability is found. The equilibrium constants are about 5% smaller than the measured radii, bulk moduli are 20% lower than the experiment. Skriver et al.[170] have performed similar self-consistent relativistic calculations for seven actinides (Ac–Am). This work concerns again only the bulk properties of these metals and it is found that the variation of the atomic volume and bulk modulus through the $5f$ series can be explained in terms of an increasing $5f$ binding up to plutonium followed by a sudden localization in americium via a complete spin polarization. These two papers are of interest partly because they show that one now requires an appropriate LSD formulation for the relativistic case, an outline of which was given in Section II. We[171] have now constructed a relativistic exchange potential for a local density scheme. This has also been independently done by MacDonald and Vosko.[172] The Slater $X\alpha$-type calculation of the exchange was also recently published by Ellis.[173] The most important point made in Refs. 171 and 172 and missed in Ref. 173 is that the transverse photon contributions do make significant contribution to the total exchange potential and changes its sign for $\hbar k_F/mc \simeq 1.95$. The spin generalization of these results are not straightforward because, unlike in the nonrelativistic theory, the spin is coupled intimately with the linear momentum of the particle. We are presently investigating this important problem within a relativistic formulation.

The LDF theory has been applied to study the structure of finite nuclei by Negele.[174] The results of nuclear matter theory was the only input into the theory in much the same way as the homogeneous electron gas theory was in the electronic calculations described previously. The theory for O^{16}, Ca^{40}, Ca^{48}, Zr^{90}, and Pr^{208} are shown to yield very satisfactory agreement with experimental binding energy, single-particle energy, and electron scattering cross section. This work is similar to that reported in Refs. 90 and 91.

Recently Janak[175] calculated the ground-state energy of cobalt as a function of magnetization and found two minima, one of which is lower than the other. The interesting feature of this work is that here one has a case with multiple minima and so an incomplete calculation would have given one or the other minimum. Another upshot of this work is that by applying pressure one may be able to induce the system to go to the other state. This is a definite prediction of the theory.

The improvement in the results can come only by better suggestions for the potential, which more satisfactorily incorporate the exchange-correlation effects. In Section III.C we outlined a scheme that is based on pair correlation functions and the exact sum rules they obey. Such a scheme holds promise because several physical aspects of correlations are properly taken into account in that formulation. The question of nonlocality can thus be handled. A better formal understanding of the time-dependent theory may shed light on the questions of the calculation of band structure, excited states, collective modes, and so on. One possible way is to incorporate the conservation approximations of many-body theory into a numerically amenable functional theory. Another approach is to follow Overhauser's suggestion and build a parametrized model inhomogeneous system for which the energy functional can be determined accurately in terms of the parameters. Using this as an input instead of the results of the homogeneous gas one can then go beyond the LD theory. Also, as we noted in Section III, the LSD theory uses an E_{xc} that depends only on the magnitude of the spin-density vector since the homogeneous electron gas has this dependence. To go outside this approximation, one requires a model magnetic system for which the ground-state energy is known accurately as a function of spin-density vector. Another area for future work is the relativistic density functional theory. From the experience gained in the non-relativistic calculation it appears that one should carefully set up an appropriate local potential for this problem. Such a model ought to be very useful in describing the magnetic states of heavy rare earths without too many ad hoc parameters just as the magnetism of $3d$ elements were dealt with by means of LSD theory. Also, highly stripped atoms with large Z, which occur in laser fusion work where relativistic effects are of paramount importance, can be treated by such a relativistic theory. Another interesting area of research involves extending the DF theory to finite temperatures in setting up the appropriate E_{xc}, and so on, and applying it to study effects in laser-excited plasmas. An extension of the HK theorem for a system containing both Bose and Fermi particles would be of interest as one can then deal with another important quantum system, that is, He_3–He_4 mixtures. This would be a more subtle extension of the HK theorem than that given by Kalia and Vashishta,[145] for a multicomponent plasma with the same statistics for the constituents.

The LSD or LD theory has made possible the calculation of a wide variety of properties of systems in a relatively easy way (computationally!). Many of the questions posed and answered so far would not have been possible before the LDF theory was made practical and workable. With the relativistic generalization, its scope will be further increased.

There remain many basic and fundamental questions concerning the formalism. Progress is bound to be made here also just as the formal work of Lieb and Simon[11] put the Thomas–Fermi theory on a footing not recognized before. It would be interesting to see if the density functional theory is asymptotically exact in the same sense as the TF theory is.

Since the manuscript was completed in October 1978, we have become aware of some work which must be read as part of the review.

A. Professor Elliott Lieb kindly supplied me with the following comments for which I am grateful.

1. The TFD theory has not been discussed by Simon and Lieb.

2. A rigorous proof that energy minimizing solutions to the HFS equations actually exist can be found in Lieb and Simon, *Comm. Math. Phys.* **53**, 185 (1977).

3. The TF approximation to the kinetic energy, given on page 7 of the text is, with a modified constant, actually a lower bound, as shown by Lieb and Thirring, *Phys. Rev. Letts.* **35**, 687 (1975).

4. For a review of item (3) above, as well as a short summary of the rigorous results in TF theory, see Lieb, *Rev. Mod. Phys.* **48**, 553 (1976).

5. The sign of the many body (i.e., intermolecular) potentials in TF theory is found in R. Benguria and Lieb, *Ann. Phys. (N.Y.)* **110**, 34 (1978). This is relevant to Chapter 5B of our review.

6. In R. Benguria and Lieb, *Comm. Math. Phys.* **63**, 193 (1978), it is proved that Teller's theorem (p. 64) can be strengthened (in the neutral case) to the statement that the energy is monotone decreasing under dilation of a molecule. This is much stronger than "no binding." It should also be noted that Teller's theorem holds even for the subneutral case.

7. In H. Brezis and Lieb, "Long Range Atomic Potentials in TF theory," *Comm. Math. Phys.* (to appear 1979), it is proved that in the neutral case the long-range interatomic potential falls off as R^{-7} for large separation R. Some people had thought it went as R^{-6}.

8. Lieb has proved (preprint 1978) that the exchange energy in the $\rho^{4/3}$-form is actually a universal lower bound for a suitable constant, C, premultiplying it; that is, $E_x \simeq -C\, e^{2/3} \int \rho^{4/3}(x) d^3x$. The best C has not yet been found.

B. C. Ebner and C. Punyanitya, *Phys. Rev. A* **19** 856 (1979) have applied the density-functional theory of simple classical fluids to the problem of localized excess electron states in rare-gas liquids. Their results are in agreement with previous theories—no localized states in dilute gases of helium, neon, argon, krypton, and xenon while they exist in dense helium gas and in liquid neon near the liquid-gas coexistence wave. This theory also predicts that the localized states in neon will become unstable at sufficiently large liquid densities below those at which liquid-solid transition occurs. Another feature of this work is that no oscillatory behavior of the density profile is found.

C. A. Ghazali and P. Leroux Hugon, *Phys. Rev. Letts.* **41**, 1569 (1978) have proposed a density-functional theory of the metal-insulator transition in doped semiconductors. H. B. Shore, E. Zaremba, J. H. Rose, and L. Sander, *Phys. Rev.* **B18**, 6506 (1978) have used a density-functional theory of Wigner-crystallization in very dilute electron gas systems. We should caution the reader that while such approaches are of immense interest, in developing such theories of phase-changes using the density-functional formalism, one ought to be careful in not violating the basic *assumptions* of the formalism such as nondegeneracy of the ground-state, local-density approximation, and so forth.

D. In collaboration with Mr. M. V. Ramana, we have now developed a functional theory of inhomogeneous relativistic spin-polarized electron system. This enables us to set up a scheme for the band-structure calculations of heavy elements in a consistent formalism.

E. In collaboration with Mr. Uday Gupta, based on the temperature-dependent density-functional formalism, we have investigated the effects of intermediate-density on the screening of a point charge in high-density plasmas relevant to the laser-induced plasma situation. This work clearly brings out the effects of spatial dependence as well as the temperature dependence of the impurity potential, a subject of tremendous current interest. We are presently working on further improvements on the temperature-dependent density-functional theory.

References

1. D. R. Hartree, *Proc. Comb. Phil. Soc.*, **24**, 89 (1928).
2. L. H. Thomas, *Proc. Comb. Phil. Soc.*, **23**, 542 (1972).
3. E. Fermi, *Rend. Acad. Naz. Lincei*, **6**, 602 (1927).
4. P. A. M. Dirac, *Proc. Comb. Phil. Soc.*, **26**, 376 (1930).
5. J. C. Slater, *Phys. Rev.*, **35**, 210 (1930).
6. V. A. Fock, *Z. Physik*, **61**, 126 (1930).
7. W. Lenz, *Z. Physik*, **77**, 713 (1932).
8. N. H. March, *Self-Consistent Fields in Atoms*, Pergamon, New York, 1975.
9. D. A. Kirzhnitz, Yu. E. Lozovik, and G. V. Shpatakovskaya, *Sov. Phys. Usp.*, **18**, 649 (1976).
10. E. Teller, *Rev. Mod. Phys.*, **34**, 627 (1962).
11. E. H. Lieb and B. Simon, *Adv. Math.*, **23**, 22 (1977).
12. J. C. Slater, *Phys. Rev.*, **81**, 385 (1951); **82**, 538 (1951).
13. J. C. Slater, in *Advances in Quantum Chemistry* (Ed. P. O. Lowdin), Vol. 6, Academic, New York, 1972, p. 1.

14. J. W. D. Connolly, in *Semiempirical Methods of Electronic Structure Calculations, Part A: Techniques* (Ed. G. A. Segal), Vol. VI, Chapter 4, Plenum, New York, 1977, p. 105.
15. J. C. Slater, *Int. J. Qu. Chem. Symposium*, **9**, 7 (1975).
16. R. Gaspar, *Acta Physica Hungaria*, **3**, 263 (1954).
17. T. Koopmans, *Physica*, **1**, 104 (1933).
18. P. Hohenberg and W. Kohn, *Phys. Rev.*, **136**, B864 (1964).
19. W. Kohn and L. J. Sham, *Phys. Rev.*, **140**, A1133 (1965).
20. L. J. Sham and W. Kohn, *Phys. Rev.*, **145**, 561 (1966).
21. V. L. Moruzzi, J. F. Janak, and A. R. Williams, *Calculated Electronic Properties of Metals*, Pergamon, New York, 1978.
22. N. D. Lang, in *Advances in Solid State Physics* (Eds. Ehrenreich, Seitz, and Turnbull) Vol. 28, Academic, New York, 1973, p. 225.
23. J. F. Janak, *Phys. Rev.*, **B9**, 3985 (1974).
24. J. F. Janak, *Phys. Rev.* **B18**, 7165 (1978).
25. N. D. Mermin, *Phys. Rev.*, **137**, A1441 (1965).
26. V. Peuckert, *J. Phys.* **C11**, 4945 (1978).
27. A. K. Rajagopal, *Phys. Rev.*, **B17**, 2980 (1978).
28. R. Gordon and Y. S. Kim, *J. Chem. Phys.*, **56**, 3122 (1972); and Y. S. Kim and R. Gordon, *ibid.*, **60**, 1842 (1974).
29. W. Münch, J. Reiss, and H. Primas, private communication, 1978.
30. L. P. Kadanoff and G. Baym, *Quantum Statistical Mechanics*, W. Benjamin, New York, 1962.
31. M. M. Pant and A. K. Rajagopal, *Solid State Comm.*, **10**, 1157 (1972).
32. U. von Barth and L. Hedin, *J. Phys.*, **C5**, 1629 (1972).
33. A. K. Rajagopal and J. Callaway, *Phys. Rev.*, **B7**, 1912 (1973).
34. R. F. Dashen and D. H. Sharp, *Phys. Rev.*, **165**, 1867 (1968).
35. G. S. Grest and A. K. Rajagopal, *Phys. Rev.*, **A10**, 1395 (1974).
36. T. L. Gilbert, *Phys. Rev.*, **B12**, 2111 (1975).
37. G. Baym, *Phys. Rev.*, **127**, 1391 (1962).
38. J. M. Luttinger and J. C. Ward, *Phys. Rev.*, **118**, 1417 (1960).
39. J. D. Björken and S. D. Drell, *Relativistic Quantum Mechanics*, McGraw-Hill, New York, 1964; *Relativistic Quantum Fields*, McGraw-Hill, New York, 1965.
40. B. Swirles, *Proc. Roy. Soc. (London)*, **A152**, 625 (1935); **A157**, 680 (1936).
41. I. Lindgren and A. Rosen, *Case Studies in Atomic Physics*, **4**, 93 (1974).
42. Y. K. Kim, *Phys. Rev.*, **154**, 17 (1967); **159**, 190 (1967).
43. O. Almbladh, private communication via U. von Barth.
44. O. Gunnarsson and B. I. Lundqvist, *Phys. Rev.*, **B13**, 4274 (1976).
45. A. K. Rajagopal and U. von Barth, unpublished notes, 1977.
46. A. K. Rajagopal, J. Kimball, and M. Banerjee, *Phys. Rev.*, **B18**, 2339 (1978).
47. Yngve Öhrn, in *The New World of Quantum Chemistry* (Eds. B. Pullman and R. Parr), D. Reidel, Boston, 1976, pp. 57–78.
48. A. K. Rajagopal and E. C. G. Sudarshan, *Phys. Rev.*, **A10**, 1852 (1974); see also U. M. Titulaer, ibid., **A11**, 2204 (1975).
49. S. Srebrenik, *Int. J. Qu. Chem. Symposium*, **9**, 375 (1975); see also S. Srebrenik, R. F. W. Bader, and T. Tung Nguyen-Dang, *J. Chem. Phys.*, **68**, 3667 (1978) and R. F. W. Bader, S. Svebrenik, and T. Tung Nguyen-Dang, ibid., **68**, 3680 (1978).
50. J. H. Wood, *Adv. Quantum Chem.*, **7**, 143 (1973).
51. J. Callaway and C. S. Wang, *Physica*, **91B**, 337 (1977).
52. J. Callaway, D. Laurent, and C. S. Wang, *Inst. Phys. Conf. Ser.*, **39**, 41 (1978).

53. O. K. Andersen and O. Jepsen, *Physica*, **31B**, 317 (1977); see also O. K. Andersen, *Inst. Phys. Conf. Ser.*, **39**, 1 (1978).
54. A. K. Rajagopal, *Physica*, **91B**, 24 (1977).
55. R. McWeeny, in *The New World of Quantum Chemistry* (Eds. B. Pullman and R. Parr), D. Reiderl, Boston, 1976, pp. 3–31.
56. B. Y. Tong, *Phys. Rev.*, **A4**, 1375 (1971).
57. M. Gell-Mann and K. A. Brueckner, *Phys. Rev.*, **106**, 364 (1957).
58. E. P. Wigner, *Phys. Rev.*, **46**, 1002 (1934).
59. K. S. Singwi, M. P. Tosi, R. H. Land, and A. Sjölander, *Phys. Rev.*, **176**, 589 (1968); *Phys. Rev.*, **B1**, 1044 (1970); the final paper in the series which is most satisfactory is that of P. Vashishta and K. S. Singwi, *Phys. Rev.*, **B6**, 875 (1972).
60. A. K. Rajagopal, unpublished notes, 1968, 1975.
61. L. Hedin and B. I. Lundqvist, *J. Phys.*, **C4**, 2064 (1971).
62. A. K. Rajagopal, S. P. Singhal, M. Banerjee, and J. C. Kimball, *Phys. Rev.*, **B17**, 2262 (1978).
63. A. K. Rajagopal, S. P. Singhal, and J. C. Kimball, unpublished work, 1977.
64. J. F. Janak, V. L. Moruzzi, and A. R. Williams, *Phys. Rev.*, **B12**, 1257 (1975).
65. P. Rennert, *Acta Phys. Hungarica*, **37**, 219 (1974).
66. U. von Barth, unpublished notes, 1977.
67. O. Gunnarsson, M. Jonson, and B. Lundqvist, *Phys. Letts.*, **A59**, 177 (1976).
68. O. Gunnarsson, M. Jonson, and B. Lundqvist, *Solid State Comm.*, **24**, 765 (1977).
69. J. A. Alanso and L. A. Girifalco, *Solid State Comm.*, **24**, 135 (1977).
70. J. A. Alanso and L. A. Girifalco, *Phys. Rev.*, **B17**, 3735 (1978).
71. L. J. Sham, *Phys. Rev.*, **B7**, 4357 (1973).
72. J. C. Stoddart and P. Hanks, *J. Phys.*, **C10**, 3167 (1977).
73. Y. Kawazoe, H. Yashuhara, and M. Watabe, *J. Phys.*, **C10**, 3293 (1977).
74. G. Niklasson, A. Sjölander, and K. S. Singwi, *Phys. Rev.*, **B11**, 113 (1975).
75. S. H. Vosko and J. P. Perdew, *Can. J. Phys.*, **53**, 1385 (1975).
76. J. F. Janak, *Phys. Rev.*, **B16**, 255 (1977).
77. (*a*) A. K. Rajagopal, *Phys. Rev.*, **142**, 152 (1965); (*b*) A. K. Rajagopal, J. Rath, and J. C. Kimball, *Phys. Rev.*, **B7**, 2657 (1973); (*c*) A. K. Rajagopal, *Pramana*, **4**, 140 (1975).
78. A. K. Rajagopal, H. Brooks, and N. R. Ranganathan, *Nuovo Cim. Suppl.*, **5**, 807 (1967).
79. F. Bloch, *Z. Phys.*, **81**, 363 (1933); *Helv. Phys. Acta*, **7**, 385 (1934).
80. H. Jensen, *Z. Phys.*, **106**, 620 (1937).
81. E. C. G. Sudarshan and N. Mukunda, *Classical Mechanics, A. Modern Perspective*, Wiley, New York, 1974.
82. D. H. Kobe and G. C. Coomer, *Phys. Rev.*, **A7**, 1312 (1973) and references therein.
83. H. Primas and M. Schleicher, *Int. J. Qu. Chem.*, **9**, 855 and 871 (1975).
84. S. C. Ying, *Il Nuovo Cim.*, **23B**, 270 (1974).
85. S. Lundqvist, *Int. J. Qu. Chem. Symposium*, **9**, 23 (1975): See also G. Mukhopadhyay and S. Lundqvist, *Il Nuovo Cim.*, **27B**, 1 (1975).
86. A. K. Kerman and S. E. Koonin, *Ann. Phys.* (*N.Y.*), **100**, 332 (1976).
87. F. Villars, *Nuclear Phys.*, **A285**, 269 (1977).
88. L. V. Keldysh, *Sov. Phys. JETP*, **20**, 1018 (1965).
89. J. C. Slater, *The Self-Consistent Field for Molecules and Solids*, Vol. 4, McGraw-Hill, New York, 1974.
90. K. A. Brueckner, J. R. Buchler, S. Jorna, and R. J. Lombard, *Phys. Rev.*, **171**, 1188 (1968).

91. K. A. Brueckner, J. R. Buchler, R. C. Clark, and R. J. Lombard, *Phys. Rev.*, **181**, 1543 (1969).
92. S. E. Koonin and J. Randup, *Nucl. Phys.*, **A289**, 475 (1977).
93. C. Y. Wong and H. H. K. Tang, *Phys. Rev. Letts.*, **40**, 1070 (1978).
94. B. Y. Tong and L. J. Sham. *Phys. Rev.*, **144**, 1 (1966).
95. H. P. Kelly, *Phys. Rev.*, **131**, 684 (1963); See also the review article of H. P. Kelly, *Correlation Effects in Atoms and Molecules*, Vol. 14 of *Advances in Chemical Physics* (Eds. G. LeFebve and C. Moser), Interscience, New York, 1969.
96. B. Y. Tong, *Phys. Rev.*, **B6**, 1189 (1972).
97. O. Gunnarsson, B. I. Lundqvist, and J. W. Wilkins, *Phys. Rev.*, **B10**, 1319 (1974).
98. U. von Barth, *IBM Report*, to be published (1978).
99. J. C. Slater, *Quantum Theory of Atomic Structure*, Vol. II, McGraw-Hill, New York, 1960.
100. H. B. Shore, J. H. Rose, and E. Zaremba, *Phys. Rev.*, **B15**, 2858 (1977).
101. K. Schwarz, *J. Phys.*, **B11**, 1339 (1978).
102. C. L. Pekeris, *Phys. Rev.*, **126**, 1470 (1962).
103. O. Gunnarsson and P. Johansson, *Int. J. Qu. Chem.*, **X**, 307 (1976).
104. O. Gunnarsson, J. Harris, and R. O. Jones, *Int. J. Qu. Chem. Symposium*, **11**, 71 (1977).
105. O. Gunnarsson, J. Harris, and R. O. Jones, *Phys. Rev.*, **B15**, 3027 (1977).
106. O. Gunnarsson, J. Harris, and R. O. Jones, *J. Chem. Phys.*, **67**, 3970 (1977).
107. J. Harris and R. O. Jones, *J. Chem. Phys.*, **68**, 1190 (1978).
108. R. A. Harris, D. F. Heller, and W. M. Gelbart, *J. Chem. Phys.*, **61**, 3854 (1974).
109. D. F. Heller, R. A. Harris, and W. M. Gelbart, ibid., **62**, 1947 (1975).
110. R. A. Harris and D. F. Heller, ibid., **62**, 3601 (1975).
111. P. W. Payne, ibid., **68**, 1242 (1978).
112. J. C. Slater, *Quantum Theory of Molecules and Solids*, Vol. I, McGraw-Hill, New York, 1963.
113. W. Kołos and L. J. Wolniewicz, *J. Chem. Phys.*, **43**, 2439 (1965).
114. W. Kołos and C. J. Roothaan, *Rev. Mod. Phys.*, **32**, 219 (1960).
115. R. G. Parr, R. A. Donnelly, Mel Levy, and W. E. Palke, *J. Chem. Phys.*, **68**, 3801 (1978).
116. A. J. Cohen and R. G. Gordon, *Phys. Rev.*, **B12**, 3228 (1975).
117. A. J. Cohen and R. G. Gordon, *Phys. Rev.*, **B14**, 4593 (1976).
118. A. Zunger and A. J. Freeman, *Phys. Rev.*, **B15**, 4716 (1977).
119. A. Zunger and A. J. Freeman, *Phys. Rev.*, **B15**, 5049 (1977).
120. A. Zunger and A. J. Freeman, *Phys. Rev.*, **B16**, 2901 (1977).
121. A. Zunger and A. J. Freeman, *Phys. Rev.*, **B16**, 906 (1977).
122. A. Zunger and A. J. Freeman, *Phys. Rev.*, **B17**, 2030 (1978).
123. O. Gunnarsson, *J. Phys.*, **F6**, 587 (1976).
124. J. F. Janak and A. R. Williams, *Phys. Rev.*, **B14**, 4199 (1976).
125. A. K. Rajagopal, Unpublished notes, 1978. A preliminary report of the relativistic density-functional theory is given in A. K. Rajagopal, *J. Phys. C* (*Letts.*), **11**, L943 (1978).
126. K. L. Liu, A. H. MacDonald, and S. H. Vosko, *Inst. Phys. Conf. Ser.*, **39**, Chapter 7, 557, 1978.
127. K. L. Liu, A. H. MacDonald, and S. H. Vosko, *Can. J. Phys.*, **55**, 1991 (1977).
128. D. E. Eastman, F. J. Himpell, and J. A. Knapp, *Phys. Rev. Letts.*, **40**, 1514 (1978).
129. J. Callaway and C. S. Wang, *J. Phys. F: Metal Phys.*, **5**, 2119 (1975).

130. C. S. Wang and J. Callaway, *Solid State Comm.*, **20**, 255 (1976).
131. S. P. Singhal and J. Callaway, *Phys. Rev.*, **B4**, 2347 (1976).
132. K. L. Liu and S. H. Vosko, *J. Phys. F: Metal Phys.*, **8**, 1539 (1978).
133. D. M. Edwards and M. A. Rahman, *J. Phys. F: Metal Phys.*, **8**, 1501 (1978).
134. D. C. Langreth, *Comments Solid State Phys.*, **8**, 129 (1978).
135. N. D. Lang and W. Kohn, *Phys. Rev.*, **B1**, 4555 (1970).
136. N. D. Lang and W. Kohn, *Phys. Rev.*, **B3**, 1215 (1970).
137. R. L. Kautz and B. B. Schwartz, *Phys. Rev.*, **B14**, 2017 (1976).
138. P. W. Lert and J. H. Weare, *J. Phys.*, **C11**, 1865 (1978).
139. N. D. Lang and A. R. Williams, *Phys. Rev.*, **B16**, 2408 (1977).
140. N. D. Lang and A. R. Williams, *Phys. Rev.*, **B18**, 616 (1978).
141. A. R. Williams and N. D. Lang, *Phys. Rev. Letts.*, **40**, 954 (1978).
142. J. E. Van Himbergen and R. Silbey, *Solid State Comm.*, **23**, 623 (1977).
143. J. N. Swingler and J. C. Inkson, *Solid State Comm.*, **24**, 305 (1977).
144. J. Ferrante and J. R. Smith, *Solid State Comm.*, **23**, 527 (1977).
145. R. K. Kalia and P. Vashishta, *Phys. Rev.*, **B17**, 2655 (1978). For a review of the properties of electron–hole liquids, see T. M. Rice, in *Adv. in Solid State Phys.* **32**, 1 (1978) (Eds. Ehrenreich, Seitz, & Turnbull).
146. C. O. Almbladh, U. von Barth, Z. D. Popovic, and J. J. Stott, *Phys. Rev.*, **B14**, 2250 (1976).
147. P. Jena, K. S. Singwi, and R. M. Nieminen, *Phys. Rev.*, **B17**, 301 (1978).
148. P. Jena and K. S. Singwi, *Phys. Rev.*, **B17**, 3518 (1978).
149. T. Ando, *Phys. Rev.*, **B13**, 3468 (1976).
150. T. Ando, A. B. Fowler, and F. Stern, review article in preparation for *Rev. Mod. Phys.* (1980).
151. See, for instance, the work of A. K. Rajagopal and J. C. Kimball, *Phys. Rev.*, **B15**, 2819 (1977) and also Ref. 62.
152. R. Balescu, *Equilibrium and Nonequilibrium Statistical Mechanics*, Wiley, New York, 1975, Chap. 8.
153. C. A. Croxton, *Liquid State Physics—A statistical Mechanical Introduction*, Cambridge Univ. Press, United Kingdom, 1974.
154. T. C. Padmore and M. W. Cole, *Phys. Rev.*, **A9**, 802 (1974).
155. C. Ebner and W. F. Saam, *Phys. Rev.*, **B12**, 923 (1975).
156. C. Ebner, W. F. Saam, and D. Stroud, *Phys. Rev.*, **A14**, 2264 (1976).
157. W. F. Saam and C. Ebner, *Phys. Rev.*, **A17**, 1768 (1978).
158. W. F. Saam and C. Ebner, *Phys. Rev.*, **A15**, 2566 (1977).
159. J. Chihara, *Prog. Theor. Phys.*, **59**, 76 (1978).
160. M. Manninen and P. Hautojärvi, *Phys. Rev.*, **B17**, 2129 (1978).
161. J. K. Percus and G. J. Yevick, *Phys. Rev.*, **110**, 1 (1958); and J. L. Lebowitz and J. K. Percus, *J. Math. Phys.*, **4**, 116 (1963) are the original papers where the functional methods were employed.
162. A. K. Rajagopal, Unpublished notes, 1978.
163. S. B. Nickerson and S. H. Vosko, *Phys. Rev.*, **B14**, 4399 (1976).
164. J. Shy-Yih Wang and M. Rasolt, *Phys. Rev.*, **B15**, 3714 (1977).
165. B. Vinter, *Phys. Rev.*, **B17**, 2429 (1978).
166. K. J. Duff and A. W. Overhauser, *Phys. Rev.*, **B5**, 2799 (1972).
167. A. W. Overhauser, *Phys. Rev.*, **B10**, 4918 (1974).
168. H. Wendel and R. M. Martin, *Phys. Rev. Letts.*, **40**, 950 (1978) and a preprint.
169. D. Glötzel, *J. Phys. F. Metal Phys.*, **8**, L163 (1978).

170. H. L. Skriver, O. K. Andersen, and B. Johansson, *Phys. Rev. Letts.*, **41**, 42 (1978).
171. A. K. Rajagopal, *J. Phys. C.* **11**, L943 (1978).
172. A. H. MacDonald and S. H. Vosko, Preprint (1978). I thank Dr. MacDonald for sending a preprint of this paper.
173. D. E. Ellis, *J. Phys. B, Atomic Phys.*, **10**, 1 (1977).
174. J. W. Negele, *Phys. Rev.*, **C1**, 1260 (1970).
175. J. F. Janak, *Solid State Comm.*, **25**, 53 (1978).

THEORIES OF LIPID MONOLAYERS

F. W. WIEGEL

Department of Applied Physics,
Twente University of Technology,
P.O. Box 217, 7500 AE Enschede, The Netherlands

AND

A. J. KOX

Institute of Theoretical Physics,
University of Amsterdam,
Valckenierstraat 65, 1018 XE Amsterdam, The Netherlands.

CONTENTS

I. INTRODUCTION

Any substance consisting of chain molecules with a polar hydrophilic head group and a hydrophobic hydrocarbon tail (so-called amphipathic or amphiphilic chain molecules), will form a monomolecular film when

This review is based in part on some lectures presented by one of us (F.W.W.) at the Los Alamos Scientific Laboratory. This author thanks the members of the Group in Theoretical Biology and Biophysics, and especially Dr. George I. Bell, for their hospitality during the late fall of 1977. A.J.K. acknowledges useful discussions with Dr. Norman L. Gershfeld.

spread on water. Although such monolayers have been widely studied experimentally,[1,2] it is only recently that the first studies have appeared in which attempts were made to explain and predict the properties of monolayers with the help of statistical physics. It is the aim of this paper to review these studies. First, however, we give a general description of the structure and properties of monolayers, and a brief sketch of the relevance of their study for biophysics.

In Fig. 1 the structure formulae are given for two examples of amphipathic molecules: a fatty acid (*n*-tetradecanoic acid or myristic acid) and a phospholipid (dipalmitoyl phosphatidylcholine or DPPC, which is also known as dipalmitoyl lecithin). In general, fatty acids have a relatively simple head group and only one tail; phospholipids have a more complicated structure and have two or even more tails. The cross section of the head group of DPPC is of the order of[3] 40 Å^2; the tails have an approximate length of 20 Å; the molecular weight is 733. The stereochemical structure of the hydrocarbon tails is such that each carbon–carbon bond may assume three states: one *trans* state, with zero energy; and two *gauche* states, with energy ϵ ($\epsilon \simeq 0.5$ kcal/mol $= 0.35 \times 10^{-13}$ erg[4]). In its *gauche* states, a bond is rotated through an angle of 120° or 240° around the previous bond. The situation is illustrated in Fig. 2, where we also give the potential energy of a bond as a function of the

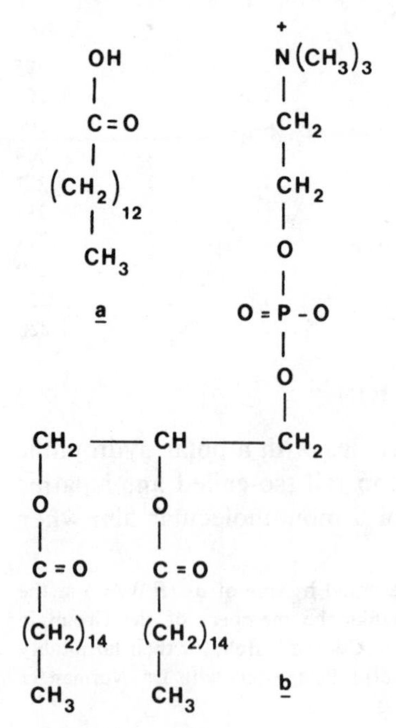

Fig. 1. The structure formulae of (*a*) myristic acid and (*b*) DPPC.

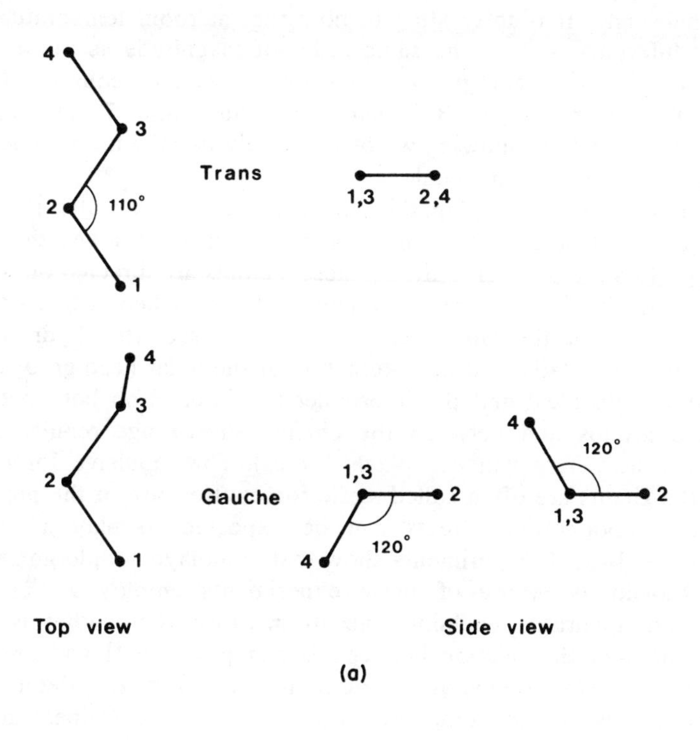

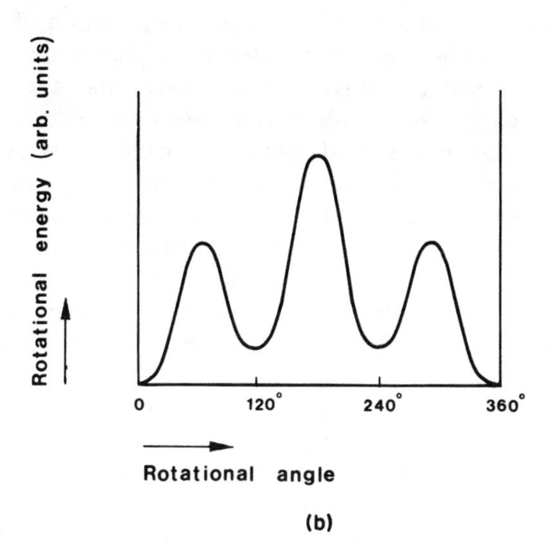

Fig. 2. Stereochemical structure of *trans* and *gauche* bonds, and the angular dependence of the rotational energy.

rotational angle. It is interesting to note that at room temperature the energy difference ϵ is of the same order of magnitude as the average rotational kinetic energy of a bond, which equals $\frac{1}{2}k_B T = 0.69 \times 10^{-16} \times T$ erg (k_B is Boltzmann's constant and T denotes the temperature). In the following we often loosely use the term "lipid" for any amphipathic chain molecule.

In a monolayer of amphipathic chain molecules on water the head groups are in close contact with the substrate, in such a way that their dipole vectors are parallel to the surface; the tails are directed out of the surface. Two kinds of interactions may be distinguished in this system: the interaction of the molecules with the substrate (the hydrophobic forces due to the tails, and the interaction of the polar head groups with the water molecules) and the intermolecular interaction both between the head groups and between the chains (short-range repulsive and long-range attractive van der Waals forces). The repulsive forces between the chains are often called steric forces. Because of the presence of forces, cooperative effects can be expected to play a role in monolayers. In fact, experiments show that monolayers undergo several phase transitions. Many of these experiments employ a Langmuir trough, an apparatus enabling one to measure the isotherms of a monolayer, i.e., the relation between lateral pressure Π and area per chain A, at constant temperature. We do not go into further detail as far as experiments are concerned (reviews are given by Gaines[1] and by Gershfeld[2]), but only summarize their results.

Although there is still ambiguity in the interpretation of some of the data, most authors now agree that three main phases may be found in monolayers: a crystalline phase, a fluid phase, and a gaseous phase, separated from each other by phase transitions (see Fig. 3).

In the *crystalline* phase (with area per chain below $A \simeq 25$ Å^2 and lateral pressure above $\Pi \simeq 10$ to 30 dyne/cm) the system is highly ordered: The head groups of the molecules are arranged in a regular (probably hexagonal) array and the C–C bonds in the tails are all in the *trans* state. This phase is usually denoted as the "liquid condensed phase". The direction of the tails is either perpendicular to the surface (at high densities) or tilted. In some substances a phase transition (probably of second order) occurs between these two states.[5]

At higher values of the area per chain and lower pressures ($A \simeq 40$ Å^2/chain, $\Pi \simeq 10$ dyne/cm) the monolayer may be compared to a *fluid*; this state is called the "liquid expanded state" or "liquid crystal state." Now the head groups are disordered and the C–C bonds are free to assume the *gauche* states. For that reason the phase transition between the liquid condensed and liquid expanded states, the "main transition",

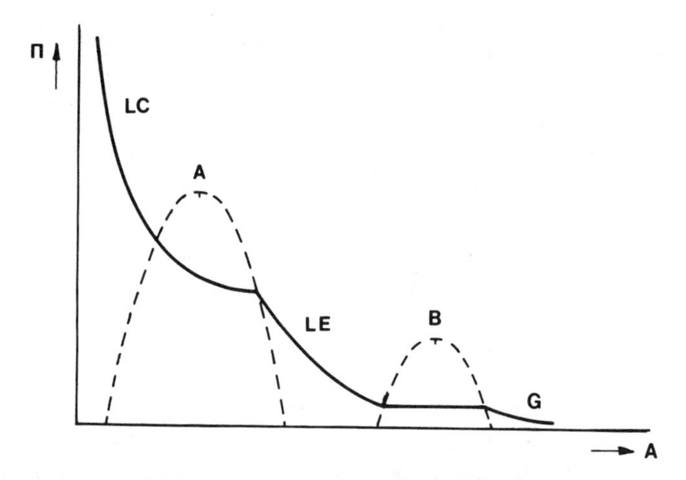

Fig. 3. Qualitative representation of the Π(A) isotherm. The different phases are denoted by LC (liquid condensed), LE (liquid expanded), and G (gaseous). A and B represent the critical points for the phase transitions between LC and LE, and LE and G, respectively. The coexistence curves, drawn as dashed lines, in reality show considerable asymmetry. Note that the figure is not drawn to scale: The phase transition between LE and G occurs at much lower values of Π and much higher values of A.

is usually associated with the "melting" of the tails (cf. Ref. 6). Initially, at low values of A, in spite of the high mobility of the molecules, the liquid expanded phase shows some anisotropy in the orientation of the tails, to be compared with the anisotropy in liquid crystals.[7] At higher values of A, however, this phenomenon disappears.

Finally, at much lower pressures and much higher areas ($\Pi \simeq 0.15$ dyne/cm, $A \simeq 10^3$ Å^2/chain), the monolayer is in its *gaseous* phase. It now is completely isotropic, with some of the tails almost flat on the surface.

We must emphasize that the picture given previously is a very general one. Whether all phases are to be observed depends on the specific substance studied. Also, since a critical point is associated with each phase transition (with critical temperatures ranging from 30 to 60°C), the number and nature of the phases through which an isotherm will pass will depend on the temperature.

It is important to note that Fig. 3 suggests that the phase transitions between the three main phases are both similar to first-order transitions. For the transition between the gaseous and the liquid expanded states this is generally recognized to be correct: The isotherms show a horizontal part. For the main transition the situation is more complicated as this transition is characterized by isotherms that are nonhorizontal but do show a discontinuity in slope. This has led some authors to believe that the main transition is of second-order and that between the liquid condensed and the liquid expanded states an intermediate (mesomorphous) phase exists.[8,9] If one assumes the main transition to be of first order, no such intermediate phase exists and an explanation must be found for the finite slope of the experimental isotherms in the coexistence region. Several causes have been suggested: impurities in the monolayer,[10,11] the finite size of the Langmuir trough,[12] the interaction of the polar head groups with the water molecules,[13,14] and the occurrence of metastability.[2,15] The first possibility is made unlikely by recent experiments[5]; for the last one experimental evidence exists.[15] One should also realize that experimentally pure monomolecular layers do not exist: There is always a certain amount of lipid in solution in the substrate. We take the point of view that (at least as a first approximation) a theoretical study of monomolecular layers is relevant and that the main transition may be treated as a first-order transition.

A statistical mechanical theory of lipid monolayers should explain the form of the observed isotherms and the nature of the different phases, and of the phase transitions, in terms of the intermolecular interactions and the structure of the molecules. The two main elements such a theory should contain are: (*1*) a calculation of the total number of allowed configurations of the hydrocarbon chains in the monolayers; (*2*) an incorporation of the long-range attractive intermolecular forces. In the following sections several theories are reviewed that incorporate these elements in different ways. (We have included most of the papers that appeared through June 1978.)

As has been mentioned earlier, the study of lipid monolayers is of relevance for biophysics. This relevance stems from the fact that the structural framework of biological membranes is formed by a lipid bilayer[16]: two parallel monolayers with their head groups on the outside surfaces and the tails pointing inward. Furthermore, it turns out that thermodynamically biomembranes and lipid bilayers resemble each other closely, in spite of the fact that biomembranes contain a large amount of other compounds, such as cholesterol and proteins. These matters are reviewed in detail by Chapman,[17] by Melchior and Steim,[18] and by Tredgold.[19] Also of interest in this connection is a somewhat less technical paper by Nagle and Scott.[20] The connection between lipid monolayers and bilayers is made by noting that the similarity of their structures implies that a bilayer may be compared with a monolayer at a constant lateral pressure (50 dyne/cm according to some authors,[12,21,22] 12.5 dyne/cm according to others[5]). The analogy is not complete: The coupling between the two layers of the bilayer produces a phase transition (the "pretransition"), which is accompanied by the formation of ripples,[23] and which has no counterpart in monolayers. The main transition of the monolayer, however, corresponds to a similar one in both lipid bilayers and biomembranes. This justifies the conclusion that the study of lipid monolayers may lead to a better understanding of the structure and function of biological membranes.

II. TWO-DIMENSIONAL MODELS

It is of interest to study the combinatorial problems that arise in membrane statistics first in the case of two dimensions. The advantage of lowering the dimension is that some of the ensuing models can be solved exactly; a disadvantage is the very drastic and "uncontrolled" simplification of the model. However, it can be expected that the exact solution to a two-dimensional model will often provide clues to the behavior of its three-dimensional counterpart.

The use of two-dimensional models for membrane statistics was initiated by Nagle.[24–27] This author introduced a pair of lattice models (called model A and model B) which could be reduced to dimer models for which the statistical mechanics can be solved exactly using the method of Pfaffians. A still simpler, and less realistic, model was considered by Wiegel[28]; this model can be reduced to the free-fermion model which can be solved with the combinatorial method.[29] The models in this section are especially designed to gain insight into the main transition of monolayers.

Consider Fig. 4, an appropriate exemplification of Nagle's model B and of the model of Ref. 28. The head groups are constrained to the vertices of a square lattice and the hydrocarbon chains are solid lines covering some of the edges of the lattice. In this model chains are not allowed to fold back: Per CH_2 group one *trans* and only one *gauche* conformation are permitted. The *trans* conformation is represented by an angle of 90° between two successive bonds, and the permitted *gauche* conformation by an angle of 180°. Nagle[24] has also introduced an

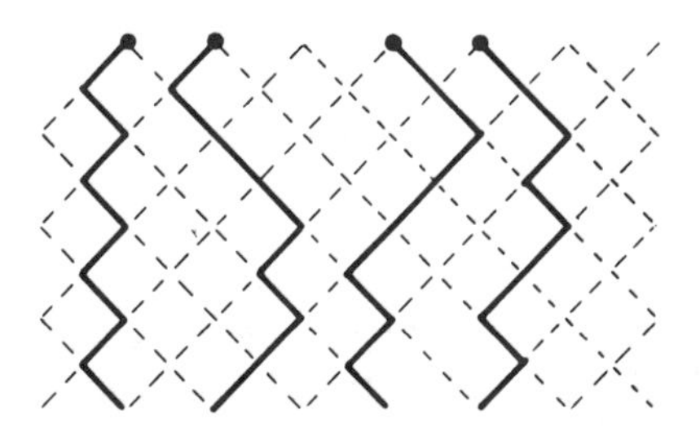

Fig. 4. A two-dimensional model for a lipid monolayer. The dashed lines indicate the underlying square lattice, the dots the head groups of the lipids, the heavy lines the hydrocarbon chains of the lipids.

alternative model (model A) on a triangular lattice in which two *gauche* conformations are allowed, but in which backfolding is still forbidden. The condition that no backfolding occurs has to be imposed to obtain an exactly solvable model. This condition is realistic at fairly high densities, but not at low densities. As has already been mentioned in the introduction, a phase transition from the liquid expanded to the gaseous state occurs at low densities. Since at these densities backfolding will be common, this phase transition cannot be studied in the models that form the subject of this section.

The transformation of models A and B to dimer models and their subsequent solution can be found in detail in Refs. 24 and 25. We summarize the results of Ref. 28 for model B in which the energy difference between a *trans* and a *gauche* conformation has been neglected. In the absence of attractive interactions the lateral pressure Π_0 for chains consisting of N repeating units is in the grand canonical ensemble given by:

$$\Pi_0(z, T) = \frac{k_B T N}{8\pi^2 \sqrt{2} l} \int_0^{2\pi} d\theta \int_0^{2\pi} d\phi \ln[1 + 2z^2 + 2z(\cos\theta + \cos\phi) + 2z^2 \cos(\theta - \phi)] \tag{2.1}$$

where l denotes the length of the bonds of the underlying square lattice and z stands for the fugacity. The other thermodynamic functions can be found from this exact result by differentiation with respect to the appropriate parameters.

Rather than repeating the derivation of (2.1) we use this section to solve model B in a continuum approximation that was developed in Ref. 28 and which uses an analogy with the theory of random walks.[30,31] Consider an x-axis parallel to the surface of the monolayer and project the end points of the bonds of a specific chain onto the x-axis. For a chain of N repeating units this projection will form a one-dimensional random walk of N steps. If no other chain molecules are present one may proceed in the following way. On a coarse-grained scale the probability density $F(x; N)$ to find the end of the Nth bond at position x is the solution of a diffusion equation:

$$\left[\frac{\partial}{\partial N} - D\frac{\partial^2}{\partial x^2}\right] F(x; N) = \delta(x)\delta(N) \tag{2.2}$$

with

$$D = \tfrac{1}{4} l^2 \tag{2.3}$$

As the total number of configurations equals 2^N, the configuration sum $Q(x; N)$ over those configurations that have the starting point of the first repeating unit fixed in the origin $x = 0$ and the end of the Nth repeating unit at x can be written as:

$$Q(x; N) = 2^N F(x; N) \tag{2.4}$$

For a system of M chain molecules, with initial positions $x_1^{(0)} < x_2^{(0)} < \cdots < x_M^{(0)}$, the configuration sum $Q(x_1, x_2, \ldots, x_M; N)$ over those configurations for which the chains end at positions $x_1, x_2, \ldots, x_M$ can similarly be written in the form:

$$Q(x_1, x_2, \ldots, x_M; N) = 2^{NM} F(x_1, x_2, \ldots, x_M; N) \tag{2.5}$$

where F is the solution of the diffusion equation:

$$\left[\frac{\partial}{\partial N} - D \sum_{i=1}^{M} \frac{\partial^2}{\partial x_i^2}\right] F(x_1, x_2, \ldots, x_M; N) = \delta(N) \prod_{i=1}^{M} \delta(x_i - x_i^{(0)}) \tag{2.6}$$

The constraint which forbids the chains of different molecules to intersect now gives the boundary conditions:

$$F(x_1, x_2, \ldots, x_M; N) = 0 \qquad \text{if} \qquad x_j = x_i \quad (j > i) \tag{2.7}$$

The solution of (2.6) can be written in terms of the orthonormal eigenfunctions of the M-dimensional Laplacian:

$$F = \sum_k \phi_k(x_1, x_2, \ldots, x_M) \phi_k^*(x_1^{(0)}, x_2^{(0)}, \ldots, x_M^{(0)}) \exp(-\lambda_k N) \tag{2.8}$$

where the λ_k are the eigenvalues:

$$-D \sum_{i=1}^{M} \frac{\partial^2 \phi_k}{\partial x_i^2} = \lambda_k \phi_k \tag{2.9}$$

The boundary conditions on ϕ_k are the "hard point" boundary conditions of (2.7). In addition, one should impose hard wall or periodic boundary conditions on a very large interval $[0, L]$, with L the length of the monolayer. In the thermodynamic limit this quantity will tend to infinity. It is convenient to impose periodic boundary conditions in the direction perpendicular to the "surface" of the monolayer by identifying x_i with $x_i^{(0)}$. The total configuration sum $Q_M(N)$ is then found by substituting (2.8) into (2.5) and integrating over all space coordinates:

$$Q_M(N) = 2^{NM} \sum_k \exp(-\lambda_k N) \tag{2.10}$$

In the limit of infinitely long chains (which is always implicitly taken in Refs. 24–28) the ground state λ_0 dominates the sum in the right-hand side and one finds for the configurational entropy per lattice vertex s_0:

$$\frac{s_0}{k_B} = \rho \ln 2 - \lim_{M_0 \to \infty} \left(\frac{\lambda_0}{M_0}\right) \tag{2.11}$$

In this formula M_0 denotes the number of lattice vertices divided by N and $\rho \equiv M/M_0$ stands for the fraction of lattice vertices occupied by CH_2 groups.

It is of interest to note that λ_0 equals the ground-state energy (with the replacement $\hbar^2/2m \to D$) of a one-dimensional quantum gas of hard point particles. As the ground state is symmetrical in the particle coordinates, λ_0 also equals the ground-state energy of the hard-point-Bose gas. This system has been considered by Girardeau[32] and other authors.[33] They find

$$\lim_{M_0 \to \infty} \frac{\lambda_0}{M_0} = \frac{\pi^2}{6} \frac{D}{l^2} \rho^3 \tag{2.12}$$

Substitution of this result together with (2.3) into (2.11) leads to

$$\frac{s_0}{k_B} = \rho \ln 2 - \frac{\pi^2}{24} \rho^3 \tag{2.13}$$

For small densities the entropy increases linearly; it reaches a maximum at intermediate densities. For low densities these results are in quantitative agreement with the exact solution of model B in Refs. 24–28. For high densities the replacement of a lattice model by a continuum model is only qualitatively correct. For example, the exact entropy vanishes at close packing density with an infinite negative slope. The lateral pressure per repeating unit Π_0/N follows from

$$\beta \frac{\Pi_0}{N} \equiv \frac{\partial}{\partial L}\left(M_0 \frac{s_0}{k_B}\right) = \frac{1}{l\sqrt{2}} \frac{\pi^2}{12} \rho^3 \tag{2.14}$$

where $\beta = (k_B T)^{-1}$. This expression is identical to the low-density limit of the pressure in the exact solution of model B given by (2.1); at high densities the exact pressure diverges because of the hard core of the chains.

Let us now consider the effect on the thermodynamic functions of two possible refinements of model B. Inclusion of the energy difference (ϵ) between a *gauche* and a *trans* conformation has essentially two effects: (*1*) The configuration sum of a single chain becomes $(1 + e^{-\beta\epsilon})^N$. (*2*) The effective diffusion coefficient decreases. Its temperature dependence is roughly given by

$$D \cong \tfrac{1}{2} l^2 (1 + e^{\beta\epsilon})^{-1} \tag{2.15}$$

The configurational entropy per lattice vertex thus equals

$$\frac{s_0}{k_B} = \rho \ln(1 + e^{-\beta\epsilon}) - \frac{\pi^2}{12} (1 + e^{\beta\epsilon})^{-1} \rho^3 \tag{2.16}$$

and the lateral pressure per repeating unit follows from

$$\beta \frac{\Pi_0}{N} = \frac{1}{l\sqrt{2}} \frac{\pi^2}{6} (1 + e^{\beta\epsilon})^{-1} \rho^3 \tag{2.17}$$

Both entropy and pressure vanish in the limit $T \to 0$ because in this limit the chains are frozen in their all-*trans* conformations.

The second refinement consists of incorporating the long-range attractive forces which hold the monolayer together. As was pointed out in Section I these attractive forces are a superposition of polar head group interactions, hydrophobic interactions and van der Waals forces. If one assumes that their range is very large as compared to the bond length, these forces can be taken into account as follows: (*1*) For the case of an exponentially decaying attraction a rigorous treatment can be given (cf. Ref. 28). (*2*) For an arbitrary long range interaction the mean-field approximation can be used.[34] In both cases the lateral pressure per repeating unit of the system with attractive forces is given by

$$\frac{\Pi}{N} = \frac{\Pi_0}{N} - \frac{w_0}{4l^2} \rho^2 \qquad (\rho \geq \rho_0) \tag{2.18}$$

$$\Pi = 0 \qquad (\rho \leq \rho_0) \tag{2.19}$$

where w_0 is defined as

$$w_0 \equiv \frac{1}{N} \int_{-\infty}^{+\infty} |V(x)| dx \tag{2.20}$$

A phase transition develops at the density ρ_0 for which the right-hand side of (2.18) vanishes. Using (2.17) one finds:

$$\rho_0(\beta) = \frac{3}{\pi^2\sqrt{2}} \frac{\beta w_0}{l} (1 + e^{\beta\epsilon}) \tag{2.21}$$

A typical isotherm of this model with attractive forces is drawn in Fig. 5. The isotherm shows only a vague resemblance to experimental isotherms and at this point a comparison between theory and experiment is not warranted.

We finish our discussion of the two-dimensional models by considering the flat part (2.19) of the isotherm in more detail. The analogy between this flat part and the flat section of a gas–liquid isotherm in the coexistence region suggests that for $\rho \leq \rho_0$ the monolayer consists of two coexisting phases. The nature of these two phases has been elucidated by Wiegel[35,36] for model B. It turns out that the monolayer consists of stretches of relatively densely packed lipids separated by pores in which the lipid density vanishes. This spontaneous division of the membrane is the direct outcome of the process of competition between

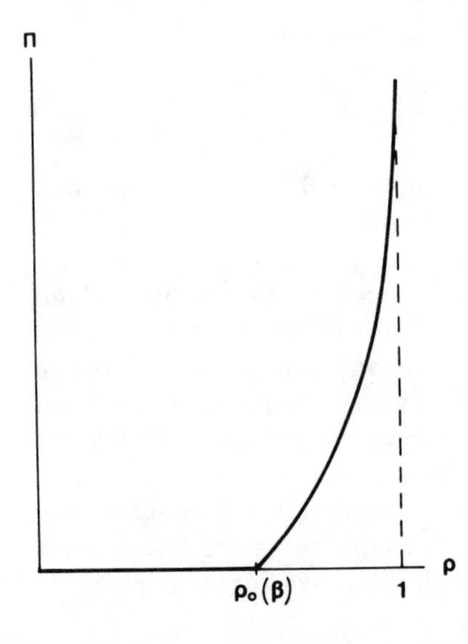

Fig. 5. Qualitative representation of an isotherm (2.18)–(2.19) for a two-dimensional monolayer model with attractive forces.

the long-range attractive forces (which tend to pull the lipids close together) and the "entropic forces" (which push the lipids apart). This is an example of spontaneous breaking of symmetry: Whereas the model itself is translationally invariant, the statistically favored configurations of the monolayer correspond to alternating regions of high and vanishing lipid density. The statistical properties of these pores have been analyzed in Refs. 35 and 36. It is found that they are small and occur infrequently for densities larger than ρ_0. When the density approaches ρ_0 from above the pore size increases rapidly and the pores also become more numerous. Indications for the existence of regions of low density also show up in the computer simulation of membrane dynamics by Cotterill[37] (see also Section IV). A recent phenomenological theory due to Stoeckly,[38] which leads to fair agreement with experimental data, also assumes clustering of the lipids in the liquid expanded phase, similar to the cluster formation found in this two-dimensional model. This theory is discussed in the next section.

III. PHENOMENOLOGICAL THEORIES

The exactly solvable models of the previous section have two serious shortcomings: They apply only to a two-dimensional world and are essentially high-density theories. Early attempts to derive an equation of state for a three-dimensional monolayer[39–42] are all based on modifications of the van der Waals equation. Stoeckly[38] has developed a theory—also along phenomenological lines—which applies especially to the low-density regime of a three-dimensional monolayer. In view of its direct applicability to some recent experiments Stoeckly's theory is outlined in this section.

As was mentioned in the introduction, at very low densities and pressures a phase transition between the liquid expanded and gaseous states occurs in some monolayers. Experiments by Hawkins and Benedek[43] and by Kim and Cannell[44] on monolayers of pentadecanoic acid $CH_3(CH_2)_{13}COOH$ show the existence of a phase transition with a critical point at a critical pressure $\Pi_c \cong 0.174$ dyne/cm, a critical area $A_c \cong 240$ Å^2/chain and a critical temperature $T_c \cong 26.27°C$.

The low-density phase transition is remarkable because of the fact that the experimentally observed critical exponents are those of the mean-field theory. Renormalization group calculations of Fisher, Ma, and Nickel[45] show that the potential energy of the dipole–dipole head group attraction, which decays as r^{-3}, is just sufficient to produce mean-field critical exponents. This corroborates the importance of the long-range dipole–dipole interactions.

In Stoeckly's theory it is assumed that the lipids form tightly bound clumps even at low densities. The number of clumps of each size is found from the sequence of reactions:

$$C_1 + C_j \rightleftarrows C_{j+1} \tag{3.1}$$

where C_j denotes a clump of j molecules. If M_j denotes the number of clumps of j molecules the equilibrium condition reads

$$\frac{M_1 M_j}{M_{j+1}} = \frac{Z_1 Z_j}{Z_{j+1}} \tag{3.2}$$

where Z_j is the partition function of a clump of j molecules. Assuming a constant binding energy E_0 per molecule in a clump, the partition functions Z_j can be approximated by the expression:

$$Z_j \cong \Omega \frac{2\pi m k_B T j}{h^2} \exp\left\{\frac{(j-1)E_0}{k_B T}\right\} \tag{3.3}$$

where m is the mass of a lipid, h Planck's constant, and Ω the total area of the monolayer. The number densities $\rho_j \equiv M_j/\Omega$ are found from (3.2) and (3.3) as

$$\rho_j = j\rho_1^j r^{j-1} \tag{3.4}$$

$$r = \frac{h^2}{2\pi m k_B T} \exp\left(\frac{E_0}{k_B T}\right) \tag{3.5}$$

The number density of clumps is given by

$$\bar{\rho} = \sum_{j=1}^{\infty} j\rho_1^j r^{j-1} \tag{3.6}$$

and the number density of lipid molecules by

$$\rho = \sum_{j=1}^{\infty} j^2 \rho_1^j r^{j-1} \tag{3.7}$$

The series in the last two equations can be summed explicitly by differentiating a geometric series.

If the clumps behaved as a two-dimensional ideal gas the equation of state would be given by $\Pi = \bar{\rho} k_B T$. Stoeckly uses two *ad hoc* modifications of this equation of state to take into account the interactions between the clumps: (*1*) It is assumed that each lipid occupies an area a_0 which is excluded for the other lipids; this leads to the equation of state $\Pi = \bar{\rho} k_B T(1 - \rho a_0)^{-1}$. (*2*) The long range attraction between the clumps is incorporated by multiplying with a Dieterici factor. This gives the phenomenological equation of state:

$$\Pi = \frac{\bar{\rho} k_B T}{1 - \rho a_0} \exp\left(\frac{-\alpha\rho}{k_B T}\right) \tag{3.8}$$

The critical point is found by setting

$$\left(\frac{\partial \Pi}{\partial \Omega}\right)_c = 0 \tag{3.9}$$

$$\left(\frac{\partial^2 \Pi}{\partial \Omega^2}\right)_c = 0 \tag{3.10}$$

With three adjustable parameters E_0, a_0, and α, the critical point can be fitted to the critical point as determined experimentally by Kim and Cannell.[44] This procedure leads to a calculated critical isotherm with the correct qualitative features.

At present Stoeckly's theory is one of the few existing attempts to derive the low-density phase transition from some simple assumptions (see also Ref. 46, discussed in Section V). The clump formation on which this theory relies is similar to the clump formation that is found in the exactly solvable models in Section II. In spirit this model is akin to the droplet model of condensation.[47,48] It is also of interest to note that the existence of clumps is in agreement with the liquid crystalline character of the liquid expanded phase if one assumes an anisotropy in the tails within each clump.

IV. NUMERICAL METHODS

Numerical simulation is often a useful tool to gain insight into the behavior of many-particle systems. Both major methods in this field, the Monte Carlo method and molecular dynamics, have been applied to models of monolayers. In most cases these models were rather simple, since the amount of computer time needed grows sharply with increasing complexity (number of degrees of freedom) of the model. Nevertheless, some interesting results have emerged from these studies.

So far only Cotterill[37] and Toxvaerd[49] have performed molecular dynamics calculations on monolayers. They both consider the case in which the individual molecules are reduced to purely two-dimensional objects, moving in the plane of the monolayer. The properties of these systems are determined by numerical integration of the equations of motion of the particles. Owing to the two-dimensional character of the models, the influence of the steric interactions of the chains and of their large number of possible configurations is neglected. As a consequence, the calculations at best simulate the crystalline state of a monolayer, with the chains parallel and in the all-*trans* state. Also, the mechanism responsible for the main transition, the melting of the tails, is absent here. These are rather severe limitations.

Cotterill's model[37] consists of 240 rodlike molecules. The interactions are a 12–6 Lennard-Jones potential acting between the end points of the rods (which correspond to the hydrocarbon chains), and a Coulomb potential acting between the midpoints. In this way, both long-range attractive and short-range repulsive forces, and polar forces are treated. It turns out that this system undergoes a phase transition from an ordered state to an isotropic fluidlike one.

Toxvaerd[49] considers 256 particles which are point-centers of force. Their interaction potential is a generalization of one proposed by Salem:

$$\phi(r) = \frac{\epsilon}{q-1}\left[\left(\frac{\sigma}{r}\right)^{5q} - q\left(\frac{\sigma}{r}\right)^{5}\right] \tag{4.1}$$

Here ϵ and σ are parameters with the dimension of energy and length, respectively. This potential was originally proposed[50] to describe the interaction between two straight hydrocarbon chains. In Salem's version, the parameter q took the value $q = 5$; Toxvaerd treats the cases $q = 5$ and $q = 2$. In our opinion, it is not likely that the difference between (4.1) and a 12–6 Lennard-Jones potential will have a significant effect on the outcome of the calculations. In Toxvaerd's model polar forces are accounted for in an approximate way: They are added as a perturbation to ϕ. This author has calculated the pressure for a number of values of the density and the temperature, using the virial theorem. A phase transition was found, the critical point of which could not be determined. The connection with experiments was made by a calculation of the derivatives of the pressure, in particular the isothermal compressibility, and a comparison of these with experimentally found results. The calculated values are at least one order of magnitude too small, probably due to the shortcomings of the model.

A second numerical method that has been applied to monolayers is the Monte Carlo method. It consists of generating a random sequence of allowed states of one or more molecules, which makes it especially well suited for studying the conformational properties of hydrocarbon chains. Whittington and Chapman[51] have applied this method to a system of three chains with up to 10 bonds restricted to a two-dimensional hexagonal lattice. (Note that here the concept "two-dimensional" is used in the sense of Section II, in contract with the discussion in the preceding paragraphs.)

An elaborate calculation, using the Monte Carlo method, was performed by Scott.[52] He treated (three-dimensional) systems of up to 10 chains comprising up to 15 bonds each. All bonds were allowed to assume three states. Results are given for a very important series of parameters: the order parameters S_n, the values of which are a measure for the degree of order in the monolayer. Denoting an average over all chains by $\langle\rangle$, their definitions read

$$S_n \equiv \tfrac{3}{2}\langle\cos^2\theta_n\rangle - \tfrac{1}{2} \tag{4.2}$$

with θ_n the angle between the plane of the nth CH_2 group and the plane of the monolayer, or, alternatively, the azimuth angle of the nth bond, minus its value in the all-*trans* state, measured with respect to the chain axis in this latter state. Thus, in the ordered all-*trans* state we have $S_n = 1$, whereas total disorder corresponds to $S_n = 0$. The importance of these parameters stems from the fact that they can be measured experimentally, with the help of ESR or NMR techniques.[53,54] In Scott's paper the order parameters are determined for a series of densities and temperatures. Although all CH_2 groups as well as all head groups are represented by hard spheres, so that attractive forces are missing, agreement with experiment is fair.

For systems of one or two chains it is possible to generate all allowed chain configurations, instead of merely a random sequence. This is done by Belle and Bothorel[55] and by Bothorel, Belle, and Lemaire[56] for two chains, of five bonds each, enclosed in a hexagonal box. For the intermolecular potential the Lennard-Jones form is chosen. Thermodynamic quantities, as well as order parameters, are calculated. The short lengths of the chains preclude a detailed comparison with experiment.

An interesting combination of analytical and numerical methods is given by Marčelja.[57-59] This author applies the Maier–Saupe theory for liquid crystals to monolayers. He considers a single molecule with a 10-bond chain and replaces the attractive potential due to all other molecules by an effective field Φ, proportional to the order parameters:

$$\Phi = V_0 \left\langle \left(\frac{n_{tr}}{n_{tot}}\right) \sum_n \left(\tfrac{3}{2}\cos^2\theta_n - \tfrac{1}{2}\right) \right\rangle \tag{4.3}$$

In this equation V_0 denotes a proportionality constant that can be determined experimentally; n_{tr} is the number density of bonds in the *trans* state; and n_{tot} the total number density of bonds. The total energy of the molecule is given by

$$E(\Phi, \Pi) = -\left(\frac{n_{tr}}{n_{tot}}\right)\Phi \sum_n \left(\tfrac{3}{2}\cos^2\theta_n - \tfrac{1}{2}\right) + E_{int} + \Pi A \tag{4.4}$$

the first term of which results from the effective field. The second term accounts for the internal energy of the chain and the third one is due to the steric repulsive forces between the chains. (Π is the lateral pressure; A the area per chain.) A self-consistent equation for Φ is found by explicitly writing out the average in (4.3):

$$\Phi = V_0 Z^{-1} \sum_{\substack{\text{all}\\ \text{config.}}} \left(\frac{n_{tr}}{n_{tot}}\right) \sum_n \left(\tfrac{3}{2}\cos^2\theta_n - \tfrac{1}{2}\right) \exp[-\beta E(\Phi, \Pi)] \tag{4.5}$$

The partition function Z is defined as

$$Z = \sum_{\substack{\text{all}\\ \text{config.}}} \exp[-\beta E(\Phi, \Pi)] \tag{4.6}$$

A solution for Φ is obtained by numerically generating all configurations of the chain and iterating equation (4.5). Once Φ is known, the partition function may be calculated and, hence, the thermodynamic functions. In this way the equation of state of a monolayer is found. The isotherms show a first-order phase transition. Moreover, they are in satisfactory agreement with the behavior of the experimentally measured isotherms near the main transition. Its inclusion of a treatment of the chain packing problem makes Marčelja's theory more realistic than those mentioned earlier in this section. There is one shortcoming, however: the absence of steric repulsion. A consideration of this interaction would have made the model even more valuable.

V. LATTICE GAS THEORIES

In this section we review some theories based on lattice gas models. The calculations are sometimes very crude, although they can often be considerably refined.

To our knowledge the first attempt to derive the isotherm of a monolayer of lipids from first principles was by Ohki.[60] He studied especially the main phase transition between the liquid expanded state and the liquid condensed state. The flexibility of the hydrocarbon chains was neglected and the long-range attractive interaction replaced by a nearest-neighbor interaction. Thus, in this model the lipids are hard rods (perpendicular to the surface for high densities) with short-range attractions. This system is now mapped onto the two-dimensional lattice gas: a square lattice in which each lattice vertex is either empty or occupied by a lipid. The lattice gas is equivalent to the two-dimensional Ising model[61] and some of its thermodynamic functions can be calculated rigorously.[62] In this way Ohki obtained the $\Pi(A)$ curve at high densities.

The most serious approximation in Ohki's theory is the complete neglect of the degrees of freedom of the hydrocarbon chains. As can be seen from the exactly solvable models discussed in Section II and from the numerical calculations of the previous section these degrees of freedom dominate the main transition. Hence, the fair agreement with experimental data which this author claims must be considered to be somewhat spurious.

Ohki's model has been refined by Firpo et al.[46] These authors studied the projections of the lipids on the surface of the monolayer and used a lattice gas model in which each molecule corresponds to p neighboring aligned lattice cells. Attractive interactions are taken into account by assuming an interaction energy between two rectangles which is proportional to the length of common sides. Hence, for $p = 1$, Ohki's model is recovered; for $p \geqslant 2$, an exact solution is not possible but a mean-field theory leads to some interesting conclusions. For $p = 4$, the model of Firpo et al. leads to fair agreement with the experiments of Kim and Cannell[44] on the phase transition between the gaseous and the liquid expanded phases. For $p = 4$ this model also shows a second phase transition at higher densities which these authors identify with the main transition. However, as the model does not take into account the variety of chain configurations it seems to us that its usefulness is restricted to the phase transition which occurs at low density.

The model of Firpo et al. is very similar to models treated by Caillé and Ågren[63] and by Caillé et al.[64] However, the theory of Ref. 63 does not include attractive forces. In Ref. 64 these forces are taken into account in the mean-field approximation. These authors compare their results with the main transition and find qualitative agreement. As the abundance of chain configurations is not incorporated in these models either, the critique of the last paragraph applies to this work as well.

In a series of papers, Scott has presented a considerable improvement of Ohki's lattice gas theory. In Scott's first paper[65] Ohki's assumption of rigid molecules is slightly relaxed by permitting three states for one chain: one *trans* and two *gauche* states. This model shows some rotational disordering at high temperatures but is nevertheless unrealistic because not just one but each bond in a chain has three states. The treatment of the packing problem is, therefore, inadequate.

In Scott's second paper[11] the lattice gas model is generalized in two respects: (*1*) Instead of a single two-dimensional layer of cells he considers several layers of cells, stacked on top of each other. Each cell is large enough to contain a CH_2 group but small enough that no more than one CH_2 group will fit into it. Scott studies models with up to five layers of cells. (*2*) Each chain is represented by many configurational states. The maximum number of states considered is 14. Having calculated the configurational entropy Scott includes a correction term due to the energy of the attractive forces of the form:

$$E_{\text{attr.}} \sim -\rho^{3/2} \tag{5.1}$$

where ρ is the lipid number density. It should be remarked here that at the present time no statistical mechanical derivation exists of a density dependence of the van der Waals interaction of the form (5.1). This term was first introduced by Nagle[24] using qualitative estimates. A derivation from first principles does exist for a term of the form

$$E_{\text{attr.}} \sim -\rho^2 \tag{5.2}$$

but only for the case in which the range of the attraction is very large as compared to the size of the molecules (see Ref. 34). Using (5.1) Scott finds isotherms that are qualitatively in agreement with experiments on the main transition in dipalmitoyl lecithin monolayers.

Models, similar to the one in Scott's second paper, form the basis of work by Caillé[66] and by Priest.[67] The theory of Priest forms a three dimensional extension of the work by Nagle discussed in Section II.

The isotherms derived in Ref. 11 are flat in the coexistence region, in contrast with the rounded isotherms observed experimentally (see also Section I). In Scott's third paper[13] the rounding of the experimental isotherms is ascribed to the surface pressure of the polar head groups located in the water surface. In this paper the lateral pressure Π_{chains} due to the chains is calculated from a model with 15 layers and 21 states and the lateral pressure Π_{heads} due to the head groups is taken to be

$$\Pi_{\text{heads}} = \sigma_0 - \sigma \tag{5.3}$$

where σ_0 is the surface tension of pure water and σ the surface tension of the monolayer. The total pressure is given by

$$\Pi = \Pi_{\text{chains}} + \Pi_{\text{heads}} \tag{5.4}$$

In this way the theory predicts isotherms that are rounded in the coexistence region. A further refinement of this model has been presented by Scott and Cheng.[14]

VI. ONE-CHAIN MODELS

Although the theories to be discussed in this section were originally devised to describe lipid bilayers, the methods employed justify their inclusion in this review. In all cases, the partition function of a system of lipid molecules is found by a semiepirical calculation of the partition function of one chain, and a mean-field treatment of the intermolecular interactions. Thus, the many-chain excluded volume problem is simplified to a one-chain problem. The actual calculation of the partition function is performed with the help of assumptions concerning the relative importance of the possible chain configurations. Since, furthermore, no attempt is made to account for the interaction between the two halves of a bilayer, or for its specific geometrical structure, these theories are of relevance to monolayers as well as to bilayers.

In a paper by Marsh[68] it is assumed that configurations containing two adjacent *gauche* bonds are forbidden. In this way the effect of steric repulsions between chains is approximated. Attractive forces are neglected altogether. The admitted sequences of three bonds are now restricted to three types: *ttt*, *tgt* and *gtg*, with *t* denoting the *trans* bond, and *g* the *gauche* bonds. The last type is often called a "kink." Its net effect is a lateral displacement of the part of the chain beyond it. For the calculation of the partition function standard matrix methods from polymer statistics are used. From the partition function thermodynamic quantities, such as the configurational entropy of a chain, are calculated. Agreement with bilayer experimental data is fair. A similar model has been studied by Caillé and Rapini[69] with the help of generating functions rather than matrix methods.

McCammon and Deutch[70] improve on the preceding models in several ways. They separately consider the liquid condensed and liquid expanded states. Furthermore, both attractive and repulsive intermolecular forces are included. These authors assume that in the liquid condensed phase the chains are in their all-*trans* states. Denoting the interaction energy of two such chains by E_{tt}, the partition function for one molecule (i.e., two chains) reads

$$Z_0(T) = \exp\left(\frac{-E_{tt}}{k_B T}\right) \tag{6.1}$$

It is postulated that the chains are hexagonally ordered. This implies that each pair of chains has 10 immediate neighbors. The partition function for a system of N molecules now takes the form

$$Z(N, T) = \exp\left(\frac{-6NE_{tt}}{k_B T}\right) \tag{6.2}$$

To find the partition function for the liquid expanded state the following assumptions are made: (*1*) Depending on the model chosen, the terminal two or four bonds of any chain may assume any combination of *trans* and *gauche* states, except two successive opposite *gauche* ones (this restriction represents the effect of intramolecular steric repulsions). (*2*) The remaining parts of the chains are either all-*trans* or kinked in one or more places. (*3*) The head groups have an energy E_h, as compared to the liquid condensed phase. (*4*) Molecules with the same number of kinks form clusters which do not interact with each other. For a molecule in an l-kink cluster the partition function may now be written

$$Z_0^{(l)}(T) = F(T) \exp\left[\frac{-E_h - E_{kk} - 4l\epsilon}{k_B T}\right] \tag{6.3}$$

with E_{kk} (independent of l) the interaction energy between two kinked chains and ϵ the energy difference between a *gauche* and a *trans* state. The function $F(T)$ stands for the contribution of the disordered end bonds; its precise form depends on the model chosen. Using (6.3) as a starting point, McCammon and Deutch calculate the partition function for the whole system, summing over all partitions in clusters. Transition entropies and energies for the main transition are calculated by subtracting the corresponding values for the two different phases found from the two partition functions. The parameters E_{tt}, E_{kk}, and E_h are fixed by equating theoretical and experimental bilayer transition temperatures. Although the theoretical chain-length dependence of this temperature agrees with experiment, a difference of 30 to 40% is found for the transition entropy and energy.

The theory of Jacobs et al.[71] has several features in common with the previous ones. The main difference is that in this case the steric forces are not dealt with by excluding certain chain configurations, but are taken into account in the form of an effective pressure P, which results when a chain bends and thus increases its effective diameter (cf. Marčelja's theory, discussed in Section IV). The energy per chain associated with this pressure is

$$E(\alpha) = P\Delta a(\alpha) \tag{6.4}$$

where $\Delta a(\alpha)$ is the increase in diameter, measured with respect to the all-*trans* state. The argument α represents the particular configuration of the chain. In the calculation of the partition function a summation over all possible chain states should be performed. As a simplification, it is postulated that $\Delta a(\alpha)$ depends only on the location of the *gauche* bond closest to the head group. Furthermore, the head groups are treated as hard disks, hence, their contribution to the partition function is the hard disk partition function. Numerical results for this quantity, as well as for the pressure P, which is taken to be the hard disk pressure, are provided by molecular dynamics calculations. Attractive forces are treated in a mean-field approximation. The partition function found with the help of the preceding approximations shows a first-order phase transition. For a comparison with experiment three parameters are fixed at their experimental values; for two other parameters the values are chosen that give optimal agreement between theory and experiment for the chain-length dependence of the transition temperature of bilayers. Transition enthalpies are also calculated. Although they are of the right order of magnitude, it is unfortunate that the values of the adjusted parameters are only supported by order-of-magnitude estimates.

In Jackson's theory[72] the packing problem is tackled in a different way. This author observes that formation of a kink creates a "vacancy" into which neighboring chains may kink, forming so-called "falling" kinks. In this way sequences of kinks in adjacent chains arise. It is assumed that contributions from such sequences dominate in the liquid expanded state. The energy of a system of lipids is now found in the following way: A sequence starting at bond n (i.e., the first *gauche* bond of the first kink is the nth bond of the chain, counted from the end of the chain) contains at most $n-1$ falling kinks. The average number is set equal to $(n-1)g$ where g is a function of the temperature, the form of which may be deduced from a qualitative argument. Each falling kink contributes 2ϵ to the energy; the first one, the formation of which requires more energy, gives a contribution of $2\epsilon + \epsilon_n$. For ϵ_n the estimate $\epsilon_n = n\omega/(A - A_0)$ is made, with A_0 the area per chain in the liquid condensed phase, and ω a parameter. The number of sequences starting at n is denoted by $J(n)$. The energy now reads

$$E=\sum_{n=1}^{N-2} J(n)\left\{\frac{n\omega}{A-A_0}+2\epsilon(n-1)g+2\epsilon\right\} \tag{6.5}$$

where N is the total number of bonds in a chain. A similar expression for the entropy in combination with (6.5) leads to the free energy. The unknown quantities $J(i)$ are eliminated by treating them as parameters and minimizing the free energy with respect to them.

Part of Jackson's paper is devoted to a careful treatment of the intermolecular forces. He separately considers the hydrophobic forces, the hydrocarbon van der Waals forces, and the polar group repulsions. Experimental data are used for their precise forms. Adding their contributions to the free energy found earlier, the total free energy is obtained, from which a first-order phase transition follows. To compare with experiment the parameter ω is adjusted in such a way that for one chain length experimental (bilayer main transition) and theoretical transition temperatures coincide. Agreement of calculated transition enthalpies with experimental ones is satisfactory but the predicted chain-length dependence of the transition temperature is less good.

Summarizing this section, we remark that the relative success of the theories presented here once more bears out the importance of a careful consideration of the packing problem of the hydrocarbon chains.

VII. THE RANDOM WALK ANALOGY

The continuum approximation which was used in Section II to solve a two-dimensional model can also be applied to three-dimensional models. In this connection some recent work by Kox and Wiegel[73,74] is outlined.

Choose a Cartesian system of coordinates in such a way that the plane $z=0$ coincides with the surface of the monolayer and project the end points of all repeating units onto the x, y-plane. The projections of a chain consisting of N repeating units will give rise to a random walk of N steps. If the monolayer consists of M chains, these correspond to M points in the x, y-plane which perform a random walk in such a way that no two points can approach each other within a distance equal to the diameter of the hydrocarbon chains.

Using this random walk analogy one finds for the configuration sum of M lipids without attractive forces the expression:

$$Q_M(N)=\frac{q^{NM}}{M!}\sum_k \exp(-\lambda_k N) \tag{7.1}$$

where q denotes a positive constant. This formula is analogous to (2.10).

The λ_k are the eigenvalues of the differential equation:

$$-D\sum_{i=1}^{M}\left(\frac{\partial^2}{\partial x_i^2}+\frac{\partial^2}{\partial y_i^2}\right)\phi_k(x_1, y_1, \ldots, x_M, y_M) = \lambda_k\phi_k(x_1, y_1, \ldots, x_M, y_M) \tag{7.2}$$

with boundary condition:

$$\phi_k = 0 \quad \text{if} \quad \{(x_i - x_j)^2 + (y_i - y_j)^2\}^{1/2} \leqslant a \qquad (i \neq j) \tag{7.3}$$

The last two equations are analogous to (2.9) and (2.7). The effective diffusion coefficient D, like the constant q in (7.1), has a temperature dependence which follows from the details of the underlying model.

The quantities λ_k correspond to the energy eigenvalues of a two-dimensional quantum gas of hard disks of diameter a (with Boltzmann statistics) with $\hbar^2/2m$ replaced by D. The ground-state energy $\lambda_0(\rho)$ has been calculated for low densities by Schick (75):

$$\lambda_0(\rho) \cong \Omega\,\frac{4\pi D\rho^2}{\ln(\rho_0/\rho)} \qquad (\rho \ll \rho_0) \tag{7.4}$$

with Ω the area of the monolayer, ρ the number density and ρ_0 the close-packing density. Bruch[76] has shown that the first correction term to (7.4) is smaller by a factor of the order of magnitude $\{\ln(\rho_0/\rho)\}^{-1}$.

The sum in (7.1) can now be evaluated using the approximation

$$\lambda_k \cong \Omega\,\frac{4\pi D\rho^2}{\ln(\rho_0/\rho)} + \Delta\lambda_k^{(0)} \qquad (\rho \ll \rho_0) \tag{7.5}$$

where $\Delta\lambda_k^{(0)}$ equals the free-particle spectrum. This leads in a straightforward way to the equation of state of the system without attractive forces:

$$\beta\Pi_0(\rho, \beta) \cong \rho + 4\pi ND\,\frac{\rho^2}{\ln(\rho_0/\rho)} \qquad (\rho \ll \rho_0) \tag{7.6}$$

The effect of a long-range attraction between the lipids is to replace this equation of state by

$$\beta\Pi(\rho, \beta) = \rho + 4\pi ND\,\frac{\rho^2}{\ln(\rho_0/\rho)} - \tfrac{1}{2}\beta w\rho^2 \tag{7.7}$$

(+ Maxwell's equal area construction)

where w denotes the absolute value of the integral of the attractive potential over the surface (cf. (2.20)). The details of this derivation can be found in Ref. 34 or in Section 4 of Ref. 74.

The isotherm (7.7) is in qualitative agreement with experiments[77,78,12]; quantitative agreement, however, is poor. This was to be expected for two reasons: (*1*) The effects of backfolding were neglected at all densities, instead of at high densities only. (*2*) The expression (7.4) for the ground-state energy of the hard disks quantum gas, which is correct at low densities, has been used at intermediate densities. Hence the theory only serves to show the connection between the ground-state energy of the two-dimensional quantum gas and the isotherm of the lipid monolayer. It is to be hoped that, maybe through a numerical approach, the function $\lambda_0(\rho)$ will become available for intermediate and high densities in the near future. Such a result would give an accurate equation of state of the monolayer for intermediate densities.

VIII. SELF-CONSISTENT FIELD THEORY

In this section we present a theory for a three-dimensional model of a monolayer which is specifically designed to account for the effects of backfolding.[79] The theory is of the mean-field type and leads to a very simple formalism, yet its conclusions are in qualitive agreement with experiment.

In this model the monolayer consists of identical chain-molecules with their head groups confined to the x, y-plane of a Cartesian coordinate frame; the chains are restricted to the half-space $z \geq 0$. A chain is represented by a random walk of N steps in this half-space, each of length l. These random walks are treated in the continuum approximation which has especially been used by Edwards. Review of this method can be found in papers by Freed[80] and Wiegel[81] On a coarse-grained scale a chain can be described by a continuous trajectory $\mathbf{r}(\nu) \equiv (x(\nu), y(\nu), z(\nu))$ where the label ν of the repeating units of the chain is treated as a continuous variable: $0 \leq \nu \leq N$. The number density $P_0[\mathbf{r}(\nu)]$ of chain configurations of the type $\mathbf{r}(\nu')$ is proportional to

$$P_0[\mathbf{r}(\nu)] \sim \exp\left\{-\frac{3}{2l^2}\int_0^N \left(\frac{d\mathbf{r}}{d\nu'}\right)^2 d\nu'\right\} \tag{8.1}$$

This is the continuous limit of a product of Gaussian transition probabilities.[80,81] Note that the energy difference between a *trans* and a *gauche* conformation has not been taken into account in an explicit way. This situation can be remedied by interpreting the length l as an effective step length $l(T)$ which depends on the temperature. In the presence of a gravitational field of acceleration g, directed along the negative z-axis, the Boltzmann factor $B[\mathbf{r}(\nu)]$ corresponding to the configuration $\mathbf{r}(\nu)$ equals

$$B[\mathbf{r}(\nu)] = \exp\left\{-\beta m_0 g \int_0^N z(\nu')d\nu'\right\} \tag{8.2}$$

where m_0 denotes the mass of a repeating unit.

For the time being, we consider only the hard-core steric repulsion between the chains; a long-range attraction is added shortly. Suppose a fraction $f(z)$ of space is on the average occupied by polymeric material, that is, per unit volume only a volume $1-f(z)$ is available for the end points of the steps. The probability that a chain with end points at $\mathbf{r}_0, \mathbf{r}_1, \ldots, \mathbf{r}_N$ will nowhere overlap with other chains is thus approximately

$$W \cong \prod_{\nu=1}^{N} (1-f(z_\nu)) \tag{8.3}$$

In the continuous limit this becomes

$$W[\mathbf{r}(\nu)] \cong \exp\left\{\int_0^N \ln[1-f(z(\nu'))]d\nu'\right\} \tag{8.4}$$

Combination of (8.1), (8.2), and (8.4) shows that the weight functional for a chain of configuration $\mathbf{r}(\nu)$, $0 \leqslant \nu \leqslant N$ is proportional to

$$P[\mathbf{r}(\nu)] \sim \exp\left\{-\int_0^N \left[\frac{3}{2l^2}\left(\frac{d\mathbf{r}}{d\nu'}\right)^2 - \ln(1-f(z(\nu'))) + \beta m_0 g z(\nu')\right]d\nu'\right\} \tag{8.5}$$

At this point some order of magnitude estimations are useful. The thickness of a monolayer is about 35 Å, hence $z(\nu)$ is at most of order 10^{-8} m. With $g \cong 10\ \text{ms}^{-2}$, $m_0 \cong 10^{-25}$ kg, and $k_B T \cong 4\times 10^{-21}$ J one finds that $\beta m_0 g z(\nu)$ is at most of order 10^{-12}. As the fraction f of space occupied by polymeric material is of the order 1 to 10^{-2} the last term in the exponential on the right-hand side of (8.5) can be neglected with respect to the second term. This gives for the weight functional the form:

$$P[\mathbf{r}(\nu)] \sim \exp\left\{-\int_0^N L[\mathbf{r}(\nu')]d\nu'\right\} \tag{8.6}$$

$$L[\mathbf{r}(\nu)] = \frac{3}{2l^2}\left(\frac{d\mathbf{r}}{d\nu}\right)^2 - \ln(1-f(z(\nu))) \tag{8.7}$$

For reasons of symmetry the most probably configuration of the chain, which is denoted by $\mathbf{r}^*(\nu) \equiv (x^*(\nu), y^*(\nu), z^*(\nu))$, is a configuration in which the chain trajectory moves perpendicularly to the surface of the substrate. Hence $x^*(\nu) = \text{constant}$, $y^*(\nu) = \text{constant}$, and $z^*(\nu)$ is the solution of the Euler–Lagrange equation

$$\frac{3}{l^2}\frac{d^2 z^*}{d\nu^2} = (1-f)^{-1}\frac{df}{dz^*} \tag{8.8}$$

which obeys the boundary condition $z^*(0) = 0$. This is the equation of motion of a fictitious classical particle with coordinate z^*, mass $3/l^2$, and time ν, moving in an external potential $\ln(1-f(z^*))$. The total "energy" E is a constant of the motion, hence

$$\frac{3}{2l^2}\left(\frac{dz^*}{d\nu}\right)^2 = E - \ln(1-f) \tag{8.9}$$

As $dz^*/d\nu$ should vanish in those parts of space in which no polymeric material is present ($f = 0$) one has $E = 0$ and consequently

$$dz^* = \left\{-\tfrac{2}{3}l^2 \ln(1-f)\right\}^{1/2} d\nu \tag{8.10}$$

The function f can now be determined in a self-consistent way by the following argument, which is similar to an argument used by de Gennes[82] in the theory of self-avoiding random walks. Suppose the monolayer has area Ω and consists of M different hydrocarbon chains. Consider a slice of space with z-coordinates between z^* and $z^* + dz^*$ and with volume $\Omega\, dz^*$. If every configuration is approximated by the most probable configuration one finds that $M\, d\nu$ repeating units are contained in a volume $\Omega\, dz^*$. If every repeating unit excludes an effective volume γ for occupancy by the center of any other repeating unit, the excluded fraction of space equals

$$f = \gamma \frac{M}{\Omega}\frac{d\nu}{dz^*} = \gamma\rho\{-\tfrac{2}{3}l^2 \ln(1-f)\}^{-1/2} \tag{8.11}$$

Equation 8.11 is an implicit equation which uniquely defines f. Since z does not appear f is constant throughout the monolayer.

The value of the Lagrangian (8.7) for the most probable configuration is:

$$L[\mathbf{r}^*(\nu)] = -2\ln(1-f) \equiv \chi \tag{8.12}$$

Hence, in this approximation the configuration sum of M molecules without attractive forces is given by

$$Q_0(M, \Omega, \beta) = \frac{1}{M!}\left(\frac{\Omega}{\Omega_0}\right)^M \exp(-\chi N M) \tag{8.13}$$

where Ω_0 is some small volume the explicit value of which will not be needed here (cf. Section 3 of Ref. 74). The Helmholtz free energy F_0 equals

$$F_0 = k_B T \Omega\{\rho \ln(\rho\Omega_0) - \rho + N\rho\chi(\rho)\} \tag{8.14}$$

and the lateral pressure is

$$\Pi_0 = -\left(\frac{\partial F_0}{\partial \Omega}\right)_{M,T} = k_B T\left(\rho + N\rho^2 \frac{d\chi}{d\rho}\right) \tag{8.15}$$

Note that for long chains the second term on the right-hand side dominates this expression.

Long-range attractive forces can now be included in the mean-field approximation (see the discussion in Section VII). The result is the equation of state:

$$\beta\Pi = \rho + N\rho^2 \frac{d\chi}{d\rho} - \tfrac{1}{2}\beta w \rho^2 \tag{8.16}$$

(+ Maxwell's equal area construction)

where w has the same meaning as in (7.7). The equation of state can be brought in a dimensionless form by introduction of the dimensionless variables:

$$\tilde{\rho} = \rho \frac{\gamma}{l} \sqrt{3/2} \tag{8.17}$$

$$\tilde{\Pi} = \beta\Pi \frac{\gamma}{l} \sqrt{3/2} \tag{8.18}$$

$$\tilde{w} = \beta w \left(\frac{\gamma}{l} \sqrt{3/2} \right)^{-1} \tag{8.19}$$

Expressed in these variables the equation of state is determined uniquely by the set of equations:

$$\tilde{\Pi} = \tilde{\rho} + N\tilde{\rho}^2 \frac{d\chi}{d\tilde{\rho}} - \tfrac{1}{2}\tilde{w}\tilde{\rho}^2 \tag{8.20}$$

(+ Maxwell's equal area construction)

$$\chi(\tilde{\rho}) = -2 \ln(1 - f) \tag{8.21}$$

$$\tilde{\rho}^2 = -f^2 \ln(1 - f) \tag{8.22}$$

In this form the isotherms can be compared with the experimentally determined isotherms.

The last two equations give the low-temperature limit of the function $\chi(\tilde{\rho})$:

$$\chi(\tilde{\rho}) \cong 2\tilde{\rho}^{2/3} \qquad (\tilde{\rho} \ll 1) \tag{8.23}$$

The thickness of the monolayer is found from (8.10) and (8.21):

$$\text{Thickness} = Nl\sqrt{\chi/3} \tag{8.24}$$

$$\cong Nl\sqrt{2/3}\,\tilde{\rho}^{1/3} \qquad (\tilde{\rho} \ll 1) \tag{8.25}$$

As the thickness of the monolayer cannot exceed Nl the present theory is restricted to densities less than the density for which the right-hand side of (8.24) equals Nl. This gives the condition $\tilde{\rho} \leq \sqrt{3/2}\,(1 - e^{-3/2}) \cong$

TABLE I.
The Values of the Dimensionless Density $\tilde{\rho}_c$. Pressure $\tilde{\Pi}_c$, and Interaction Energy $\tilde{w}_c$ at the Critical Point, as a Function of the Number of Repeating Units N.

N	$\tilde{\rho}_c$	$\tilde{\Pi}_c$	$\tilde{w}_c$
2	0.426	0.196	13.7
3	0.384	0.192	19.3
4	0.360	0.194	24.9
5	0.343	0.198	30.4
6	0.331	0.205	35.9
7	0.322	0.211	41.4
8	0.315	0.219	46.8
9	0.309	0.227	52.3
10	0.305	0.235	57.7
11	0.301	0.243	63.1
12	0.297	0.252	68.6
13	0.294	0.260	74.0
14	0.292	0.269	79.4
15	0.289	0.280	84.9
16	0.287	0.286	90.3
17	0.286	0.296	95.7
18	0.284	0.305	101
19	0.283	0.314	107
20	0.281	0.323	112
21	0.280	0.332	117
22	0.279	0.341	123
23	0.278	0.350	128
24	0.277	0.359	134
25	0.276	0.368	139
26	0.275	0.380	145
27	0.274	0.387	150
28	0.274	0.397	155
29	0.273	0.406	161
30	0.272	0.415	166
31	0.272	0.424	172
32	0.271	0.433	177
33	0.271	0.442	182
34	0.270	0.453	188
35	0.270	0.462	193
36	0.269	0.471	199
37	0.269	0.480	204
38	0.268	0.490	209
39	0.268	0.500	215
40	0.268	0.508	220
41	0.267	0.518	226
42	0.267	0.526	231
43	0.267	0.537	237
44	0.266	0.545	242
45	0.266	0.555	247

0.951. At this density a fraction of space $f_0 = (1 - e^{-3/2}) \cong 0.777$ is blocked by polymeric material.

The equation of state, (8.20) to (8.22), predicts a critical temperature below which coexisting phases appear. Because of the mean-field approximation made while incorporating the attractive forces the critical exponents have the classical values. The critical point occurs for that dimensionless density $\tilde{\rho}_c$ for which $d\tilde{\Pi}/d\tilde{\rho} = 0$ and $d^2\tilde{\Pi}/d\tilde{\rho}^2 = 0$. For a given value of N this implies a specific value $\tilde{w}_c$ of the dimensionless integral of the attractive potential. In Table I we have numerically calculated $\tilde{\rho}_c$, $\tilde{\Pi}_c$, and $\tilde{w}_c$ as a function of N.

As was mentioned in the introduction, the experiments do not show flat isotherms in the coexistence region. This might be due to the metastable nature of the high-density phase.[15] Nagle[12] has estimated, through a process of extrapolation, a critical point for dipalmitoyl phosphatidylcholine (DPPC):

$$T_c \cong 305\ \text{K} \tag{8.26}$$

$$\Pi_c \cong 25\ \text{dyne/cm} \tag{8.27}$$

$$\rho_c^{-1} \cong 26.5 \times 10^{-16}\ \text{cm}^2/\text{chain} \tag{8.28}$$

We can use these figures to estimate the order of magnitude of the various other parameters in the theory. First one calculates

$$\frac{\tilde{\Pi}_c}{\tilde{\rho}_c} = \frac{\Pi_c}{k_B T_c \rho_c} \cong 1.57 \tag{8.29}$$

Table I gives $N \cong 31$. The number of residues in a chain of DPPC equals 15, so the effective number of repeating units is of the right order of magnitude. As we have $\tilde{\rho}_c \cong 0.27$, (8.22) shows that at the critical point about 39% of space is blocked by polymeric material. Equation 8.17 gives $(\gamma/l)\sqrt{3/2} \cong 7.2 \times 10^{-16}\ \text{cm}^2$. In the present model the repeating units are small spheres of diameter l, hence γ is given by $\gamma = \frac{4}{3}\pi l^3$. Solving l from the last two equations one finds $l \cong 1.2$ Å. This effective size of the repeating units is of the right order of magnitude. Using Table I again we find $\tilde{w}_c \cong 172$; with (8.19) one finds $w \cong 0.52 \times 10^{-26}\ \text{cm}^2\ \text{erg}$. This value of the integral of the attractive potential between two chains is also of the right order of magnitude.[24]

IX. CONCLUSIONS AND OUTLOOK

The statistical mechanics of lipid monolayers does not yet form a unified theory. The approaches discussed in Sections II through VIII use different models and concepts, and their interrelation is often poorly understood. Although most of these theories are in qualitative agreement with some aspect of the experiments, quantitative agreement is nearly always poor unless the theory contains a large number of adjustable parameters. In future extensions of the theory the following points deserve special attention:

1. The interactions between the lipids, and especially the interactions between the polar head groups of the lipids, should be identified with the effective interaction between lipids in an aqueous medium and not with the interaction of lipids in vacuum. Actually, as the head groups often carry charges or electric dipoles, the bare lipid–lipid interaction will be "renormalized" to a considerable extent by the ionic composition of the aqueous medium. At present the theory does not take this effect into account. In the case of biological membranes it is conceivable that a change in the metabolism of the cell will lead to such a change in the ionic composition of the medium, and hence of the effective lipid–lipid interaction, that the membrane will make a transition to another phase. As the permeability of the membrane for the cell fluid and the substances dissolved in it is very different in the different phases, this could provide a tentative mechanism for some of the regulatory functions of biomembranes.
2. The presence of the head groups with their electric dipole moments will also change the structure of water at the surface. The resulting change in the surface tension of water could lead to an explanation of the curved form of the $\Pi(A)$ isotherm in the coexistence region of the main transition, as suggested by Scott.[13]
3. To date no theory exists for the order–disorder transition which might occur in the hexagonal packing of the head groups.
4. In some substances a transition seems to occur in the crystalline phase between a state in which the tails are perpendicular to the surface and a state in which the tails are tilted with respect to the surface.[5] No model calculations have been made for this transition.
5. The nature of the liquid expanded phase has not been clarified in any of the theories discussed previously. A certain amount of anisotropy can be present in this phase, hence the monolayer seems to be in a liquid–crystalline mesophase. Actually, in the fluid regime the monolayer might be in one of several mesophases,[5] in which case the full theory of liquid crystals would have to be used.[7]
6. Not too much importance should be assigned to the studies of exactly solvable models in two dimensions which were reviewed in Section II. These models are so highly simplified that most features of a three-dimensional monolayer are lost.

7. The numerical approach of Section IV is very promising. At present the authors are developing a molecular dynamics simulation of a fairly realistic model of a lipid monolayer.
8. A combination of a lattice gas theory (of the type discussed in Section V) with a more precise counting of the configurations of a single chain (as in Section VI) might lead to considerable improvements.
9. A calculation of the ground state of the quantum gas of hard disks for intermediate densities could lead, through the random walk analogy of Section VII, to a more reliable $\Pi(A)$ isotherm. To date no such results are known to us. However, it seems feasible to make progress towards a solution of this problem using Monte Carlo techniques similar to the ones employed in the analogous three-dimensional case.
10. It is of considerable interest to study the effects of a small concentration of impurities (such as proteins) on the thermodynamic functions of the lipid monolayer. Some steps in this direction have already been taken by Marčelja,[59] Schröder,[83] and Pink and Carroll.[84] It is here that the theory of lipid monolayers merges with the theory of biological membranes.

References

1. G. L. Gaines, *Insoluble Monolayers at Liquid-Gas Interfaces*, Wiley-Interscience, New York, 1966.
2. N. L. Gershfeld, *Ann. Rev. Phys. Chem.*, **27**, 349 (1976).
3. G. D. Fasman, Ed., *Handbook of Biochemistry and Molecular Biology. Lipids, Carbohydrates and Steroids*, CRC Press, Cleveland, 1975.
4. P. J. Flory, *Statistical Mechanics of Chain Molecules*, Interscience, New York, 1969.
5. O. Albrecht, H. Gruler, and E. Sackmann, *J. de Phys.*, **39**, 301 (1978).
6. J. E. Rothman, *J. Theor. Biol.*, **38**, 1 (1973).
7. P. G. de Gennes, *The Physics of Liquid Crystals*, Clarendon, Oxford, 1974.
8. D. G. Dervichian, *J. Chem. Phys.*, **7**, 931 (1939).
9. W. D. Harkins, T. F. Young, and E. Boyd, *J. Chem. Phys.*, **8**, 954 (1940).
10. S. Marčelja, *Biochim. Biophys. Acta*, **367**, 165 (1974).
11. H. L. Scott, *J. Chem. Phys.*, **62**, 1347 (1975).
12. J. F. Nagle, *J. Membrane Biol.*, **27**, 233 (1976).
13. H. L. Scott, *Biochim. Biophys. Acta*, **406**, 329 (1975).
14. H. L. Scott and Wood-Hi Cheng, *J. Coll. Interface Sci.*, **62**, 125 (1977).
15. L. W. Horn and N. L. Gershfeld, *Biophys. J.*, **18**, 301 (1977).
16. R. A. Nystrom, *Membrane Physiology*, Prentice-Hall, Englewood Cliffs, New Jersey, 1973.
17. D. Chapman, *Q. Rev. Biophys.*, **8**, 185 (1975).
18. D. L. Melchior and J. M. Steim, *Ann. Rev. Biophys. Bioengng.*, **5**, 205 (1976).
19. R. H. Tredgold, *Adv. Phys.*, **26**, 79 (1977).
20. J. F. Nagle and H. L. Scott. *Phys. Today*, Febr. 1978, p. 38

21. M. C. Phillips, R. M. Williams, and D. Chapman, *Chem. Phys. Lipids*, **3**, 234 (1969).
22. S. W. Hui, M. Cowden, D. Papahadjopoulos, and D. F. Parsons, *Biochim. Biophys. Acta*, **382**, 265 (1975).
23. M. J. Janiak, D. M. Small, and G. G. Shipley, *Biochemistry*, **15**, 4575 (1976).
24. J. F. Nagle, *J. Chem. Phys.*, **58**, 252 (1973).
25. J. F. Nagle, *Proc. R. Soc.*, **A337**, 569 (1974).
26. J. F. Nagle, *Phys. Rev. Lett.*, **34**, 1150 (1975).
27. J. F. Nagle, *J. Chem. Phys.*, **63**, 1255 (1975).
28. F. W. Wiegel, *J. Stat. Phys.*, **13**, 515 (1975).
29. F. W. Wiegel, *Can. J. Phys.*, **53**, 1148 (1975).
30. S. Chandrasekhar, *Rev. Mod. Phys.*, **15**, 1 (1943).
31. P. G. de Gennes, *J. Chem. Phys.*, **48**, 2257 (1968).
32. M. Girardeau, *J. Math. Phys.*, **1**, 516 (1960).
33. E. H. Lieb and D. C. Mattis, *Mathematical Physics in One Dimension*, Academic, London, 1966, Chapter 5.
34. P. C. Hemmer and J. L. Lebowitz, in C. Domb and M. S. Green, Eds., *Phase Transitions and Critical Phenomena*, Vol. 5b, Academic, London, 1976, p. 108.
35. F. W. Wiegel, *Phys. Lett.*, **57A**, 393 (1976).
36. F. W. Wiegel, *Physica*, **89A**, 397 (1977).
37. R. M. J. Cotterill, *Biochim. Biophys. Acta*, **433**, 264 (1976).
38. B. Stoeckly, *Phys. Rev.*, **A15**, 2558 (1977).
39. J. N. Phillips and E. Rideal, *Proc. R. Soc.*, **A232**, 149 (1955).
40. D. G. Hedge, *J. Coll. Sci.*, **12**, 417 (1957).
41. M. Nakagaki, *Bull. Chem. Soc. Japan*, **32**, 1232 (1959).
42. T. Smith, *J. Coll. Interface Sci.*, **23**, 27 (1967).
43. G. A. Hawkins and G. B. Benedek, *Phys. Rev. Lett.*, **32**, 524 (1974).
44. M. W. Kim and D. S. Cannell, *Phys. Rev.*, **A13**, 411 (1976).
45. M. E. Fisher, S. K. Ma, and B. G. Nickel, *Phys. Rev. Lett.*, **29**, 917 (1972).
46. J. L. Firpo, J. J. Dupin, G. Albinet, A. Bois, L. Casalta, and J. F. Baret, *J. Chem. Phys.*, **64**, 1369 (1978).
47. M. E. Fisher, *Physics*, **3**, 255 (1967).
48. M. E. Fisher, *Rep. Prog. Phys.*, **30**, 731 (1967).
49. S. Toxvaerd, *J. Chem. Phys.*, **67**, 2056 (1977).
50. L. Salem, *J. Chem. Phys.*, **37**, 2100 (1962).
51. S. G. Whittington and D. Chapman, *Trans. Faraday Soc.*, **62**, 3319 (1966).
52. H. L. Scott, *Biochim. Biophys. Acta*, **469**, 264 (1977).
53. W. L. Hubbell and H. M. McConnell, *J. Am. Chem. Soc.*, **93**, 314 (1971).
54. J. Seelig and W. Niederberger, *Biochemistry*, **13**, 1585 (1974).
55. J. Belle and P. Bothorel, *Biochem. Biophys. Res. Comm.*, **58**, 433 (1974).
56. P. Bothorel, J. Belle, and B. Lemaire, *Chem. Phys. Lipids*, **12**, 96 (1974).
57. S. Marčelja, *J. Chem. Phys.*, **60**, 3599 (1974).
58. S. Marčelja, *Biochim. Biophys. Acta*, **367**, 165 (1974).
59. S. Marčelja, *Biochim. Biophys. Acta*, **455**, 1 (1976).
60. S. Ohki, *J. Theor. Biol.*, **15**, 346 (1967).
61. S. G. Brush, *Rev. Mod. Phys.*, **39**, 883 (1967).
62. B. M. McCoy and T. T. Wu, *The Two-Dimensional Ising Model*, Harvard Univ. Press, Cambridge, Mass., 1973.
63. A. Caillé and G. Ågren, *Can. J. Phys.* **53**, 2369 (1975).
64. A. Caillé, A. Rapini, M. J. Zuckermann, A. Cros, and S. Doniach, *Can. J. Phys.*, **56**, 348 (1978).

65. H. L. Scott, *J. Theor. Biol.*, **46**, 241 (1974).
66. A. Caillé, *Can. J. Phys.*, **52**, 839 (1974).
67. R. G. Priest, *J. Chem. Phys.*, **66**, 722 (1977).
68. D. Marsh, *J. Membrane Biol.*, **18**, 145 (1974).
69. A. Caillé and A. Rapini, *Phys. Lett.*, **58A**, 357 (1976).
70. J. A. McCammon and J. M. Deutch, *J. Am. Chem. Soc.*, **97**, 6675 (1975).
71. R. E. Jacobs, B. Hudson, and H. C. Andersen, *Proc. Nat. Acad. Sci. USA*, **72**, 3993 (1975).
72. M. B. Jackson, *Biochemistry*, **15**, 2555 (1976).
73. F. W. Wiegel and A. J. Kox, *Annal. Israel Phys. Soc.*, **2**, 966 (1978).
74. A. J. Kox and F. W. Wiegel, *Physica*, **92A**, 466 (1978).
75. M. Schick, *Phys. Rev.*, **A3**, 1067 (1971).
76. L. W. Bruch, *Physica*, **93A**, 95 (1978).
77. F. Villalonga, *Biochim. Biophys. Acta*, **163**, 290 (1968).
78. M. C. Phillips and D. Chapman, *Biochim. Biophys. Acta*, **163**, 301 (1968).
79. F. W. Wiegel and A. J. Kox, *Phys. Lett.* **68**A, 286 (1978).
80. K. Freed, *Adv. Chem. Phys.*, **22**, 1 (1972).
81. F. W. Wiegel, *Phys. Rep.*, **16**, 57 (1975).
82. P. G. de Gennes, *Rep. Prog. Phys.*, **32**, 187 (1969).
83. H. Schröder, *J. Chem. Phys.* **67**, 1617 (1977).
84. D. A. Pink and C. E. Carroll, *Phys. Lett.*, **66A**, 157 (1978).

THE REACTION $F + H_2 \rightarrow HF + H$

JAMES B. ANDERSON

Department of Chemistry
The Pennsylvania State University
University Park, Pennsylvania 16802

CONTENTS

I. INTRODUCTION

If the reaction $H + H_2 \rightarrow H_2 + H$ is ranked first on the list of reactions of fundamental interest in chemistry, the reaction $F + H_2 \rightarrow HF + H$ must be ranked a very close second. For the $H + H_2$ system we have accurate potential energies,[1,2] accurate thermal rate measurements,[3,4] and accurate classical[4,5] and quantum[4,6] calculations of reaction rate. The success in theory and experiment for $H + H_2 \rightarrow H_2 + H$ is a milestone in the development of chemical kinetics and molecular dynamics. Looking ahead, we should expect the reaction $F + H_2 \rightarrow HF + H$ to provide the next test of our capabilities in theory and experiment.

The reaction $F + H_2 \rightarrow HF + H$ has a great deal more "character" than the reaction $H + H_2 \rightarrow H_2 + H$. It is highly exoergic (by ~32 kcal/mole) and produces vibrationally excited HF. It is the energy source for powerful lasers. The dynamics of the reaction can be studied by a variety of experimental techniques—infrared chemiluminescence, chemical laser, molecular beam, hot atom—which give reaction rates, product velocity distributions, product state distributions, product angular distributions, effects of reactant states and translational energies. Together these provide a wealth of experimental detail unavailable for the simpler reaction of H with H_2.

The reaction $F + H_2 \rightarrow HF + H$ presents greater theoretical difficulties and thus a greater challenge than the reaction $H + H_2 \rightarrow H_2 + H$. With eleven electrons, compared to three, the FHH system requires much more elaborate calculations in the determination of potential energies. In addition, the $F + H_2$ reaction has multiple potential energy surfaces accessible at ordinary temperatures. Calculations of molecular dynamics are more difficult because of the lower symmetry of the system. Nevertheless, it seems likely that within a decade we will see not only potential energy surfaces accurate to 0.1 kcal/mole for the $F + H_2$ system but also full quantum calculations of the reaction dynamics using these surfaces in the prediction of reaction attributes that may be compared in detail with experimental measurements.

Theoretical work to date for the $F + H_2$ system includes the determination of semiempirical and *ab initio* potential energy surfaces. Whereas these surfaces may not be sufficiently accurate to allow a fair test of the available methods of treating the reaction dynamics by comparisons with experiment, three-dimensional classical and semiclassical trajectory calculations with these surfaces have reproduced qualitatively the major reaction characteristics. Full quantum calculations have been made for collinear systems. Preliminary three-dimensional quantum calculations of the molecular dynamics have been completed. Further progress will undoubtedly occur by the use of increasingly larger basis sets in the calculation of the potential energy surfaces by variational methods and the step-by-step elimination of approximations required in the quantum treatments of the molecular dynamics in three dimensions.

Both theoretical and experimental studies of the $F + H_2$ reaction are complicated by the possibility of reaction of excited states of the F atom. The excited state $F(^2P_{1/2})$ lies only 1.16 kcal above the ground state $F(^2P_{3/2})$. For equilibrium at 300°K about 7% of the F atoms exist in the higher state. It has not been possible to distinguish between reactions of the two species in most experimental studies. The multiple states and multiple potential energy surfaces may lead to electronically nonadiabatic reactions.

We may expect new types of experiments and refinements of older experimental methods to eliminate certain questions presented by data available at present. There exist uncertainties and apparent ambiguities in the relations between threshold energy and activation energy which may be resolved by more accurate and detailed measurements of reaction cross sections and thermal rates. The reactions of the two low-lying states of the F atom may soon be distinguished experimentally in measurements of rates and product state distributions.

The most recent reviews of the subject reaction and related reactions are those by Polanyi and Schreiber,[7] "The Reaction of $F+H_2 \rightarrow HF+H$, A Case Study in Reaction Dynamics"; by Cohen and Bott,[8] "A Review of Rate Coefficients in the H_2–F_2 Chemical Laser System"; by Jones and Skolnik,[9] "Reaction of Fluorine Atoms"; and by Foon and Kaufmann,[10] "Kinetics of Gaseous Fluorine Reactions."

In succeeding sections experimental and theoretical work to date is reviewed. The experiments are discussed in detail with emphasis on the limitations and achievements of the most important, most accurate experiments. The theory is discussed in detail with emphasis on the most important, most accurate treatments. The areas of agreement and disagreement are evaluated and the need for further work is examined.

II. EXPERIMENTAL WORK

A. General Considerations

The extreme reactivity of both reactants and products of the reaction $F+H_2 \rightarrow HF+H$ presents the experimenter with a number of difficult problems. Most of these are associated with the production of F atoms, measurement of F atom concentration, and the elimination of secondary reactions. The reaction is rapid, occurring with a probability of 0.01 to 0.1 per collision at room temperature for typical reaction times of a few microseconds in thermal systems. Except for HF none of the reactant or product species can be measured with accuracy in such a short time scale. The experimenter is thus limited to (*a*) steady-state measurements of one or more species in flow systems, (*b*) transient measurements of HF production by chemiluminescence in static systems, or (*c*) competitive rate measurements. In flow systems the problem of providing rapid mixing of reagents may occur. In static systems the F atoms must be generated in the presence of H_2.

Atomic fluorine may be produced by a variety of methods: thermal dissociation of F_2, electrical discharges through gases containing fluorine compounds, photolytic dissociation of fluorine compounds, chemical reactions producing F atoms as products, nuclear recoil reactions yielding ^{18}F. All these methods have been used to produce F atoms for investigation of reactions with H_2.

The detection of atomic fluorine and measurement of its concentration may be accomplished by: mass spectrometry, EPR spectroscopy, titration by secondary reagents, resonance absorption spectroscopy, resonance fluorescence spectroscopy, and rate of corrosion of metals. Of these the first three have been employed in investigations of the $F + H_2$ reaction.

The experimenter's task is to bring an F atom into a collision with an H_2 molecule and determine what happens. In the ideal experiment the states (electronic, rotation, vibration) of the two species are specified exactly as are the relative velocity, the center-of-mass velocity, the vibrational and rotational phases and axis of rotation of the diatomic molecule, and even the impact parameter. The identical properties are then determined for the species departing the collision either as unreacted species or as reaction products. Of course, no such ideal experiment has been performed for $F + H_2$ or any other reaction, but a number of these properties have been specified for reactants and measured for products. The experimenter is limited to examining large numbers of individual events collectively in which the distributions of reactant species and collision properties are known.

The experiments fall into two classes: thermal and crossed-beam. In thermal systems reactant and collision properties are known in many cases to be those corresponding to Boltzmann distributions. In crossed-beam systems it is, in general, possible to specify or limit reactant states and velocities and to measure these properties for the collision products. Specification of states of rotation is not too far-fetched for $F + H_2$, but selection of phases and impact parameters seems nearly beyond imagination.

The reaction properties measured to date include for thermal systems: reaction rate and the distribution of rotational–vibrational states of the HF product. For crossed-beam systems the angular distribution of the HF (DF) product has been measured for several relative velocities of the collision partners.

B. Thermal Reaction Rate—Early Estimates

The first information about the reaction $F+H_2 \rightarrow HF+H$ came from experiments with mixtures of H_2 and F_2 which are now known to undergo reaction by the chain mechanism involving H and F atoms:

$$F+H_2 \rightarrow HF+H \quad (1)$$

$$H+F_2 \rightarrow HF+F \quad (2)$$

$$\overline{H_2+F_2 \rightarrow 2HF}$$

Moissan and Dewar[11] had reported about 1900 that the contact of fluorine with liquid hydrogen resulted in immediate explosions. In 1930 Wartenberg and Taylor[12] reported little or no reaction for mixtures of H_2 and F_2 even with ultraviolet irradiation. From his early calculations of potential energy surfaces for hydrogen–halogen reactions Eyring[13] predicted in 1931 activation energies of 5 to 15 kcal/mole for $F+H_2$ (1), 0 kcal/mole for $H+F_2$ (2), and 50 to 71 kcal/mole for the direct four-center reaction H_2+F_2. In subsequent experiments Eyring and Kassel[14] also found little or no reaction at room temperature for mixtures of H_2 and F_2 in the absence of a catalyst or other means of initiating an explosion.

In 1937 Bodenstein, Jockusch, and Chong[15] reported measurements of the slow reaction of F_2 and H_2 at room temperature with and without ultraviolet light to dissociate F_2. From these data Schumacher[16] estimated an activation energy of 7.5 kcal/mole for the reaction $F+H_2$ (1) in agreement with Eyring's prediction of 5 to 15 kcal/mole. Later experimental and theoretical values of the activation energy are substantially lower, but these first estimates must be judged as remarkably accurate.

C. Direct Thermal Rate Measurements

There are eleven studies that yield directly (or nearly directly) rate constants for the reaction $F+H_2 \rightarrow HF+H$. As shown in Table I nine of these are restricted to single temperatures near 300°K whereas two span temperature ranges of 100°K near 300°K. The experiments have incorporated a variety of F atom sources and F and/or HF detection methods.

The most accurate measurement is probably that of Clyne, McKenney, and Walker[20] at 298°K yielding a rate constant of 1.5×10^{13} cm^3/mole-sec. Many of the other results lie within a factor of about two higher or lower for 298°K.

Clyne, McKenney and Walker[20] used a flow system in which a stream containing F atoms was passed through a tube provided with H_2 inlet

TABLE I.
Direct Thermal Rate Measurements

Investigators	Method	T (°K)	k_3(298°K) (cm^3/mol-sec)	A (cm^3/mol-sec)	E_a (cal/mole)
Homann et al.[17] (1970)	flow F from $N + NF_2$ reaction HF by mass spec. F by ClNO titration	300–400	1.2×10^{13}	$10^{14.2}$	1600
Dodonov et al.[18] (1971)	flow F from discharge in F_2 HF by mass spec.	293	1.8×10^{13}	—	—
Kompa and Wanner[19] (1972)	WF_6 photolysis in H_2 HF by chemiluminescence	298	3.8×10^{13}	—	—
Clyne, McKenney, and Walker[20] (1973)	flow F from discharge in CF_4 HF by mass spec. F by Cl_2 titration	298	1.5×10^{13}	—	—
Igoshin, Kulakov, and Nikitin[21] (1973)	F from discharge in NF_3 HF by chemiluminescence	195–294	1.5×10^{13}	$10^{14.0}$	1080
Pearson et al.[22] (1973)	F from discharge in NF_3 HF by chemiluminescence	(~500)	(2.4×10^{13})	—	—
Hon, Axworthy, and Schneider[23] (1973)	flow F from thermal F_2 F by ESR	298	1.0×10^{13}	—	—
Lam, Peyron, and Puget[24] (1974)	flow system F from discharge in SF_6 HF by chemiluminescence	~300	0.14×10^{13}	—	—
Goldberg and Schneider[25] (1976)	flow F from discharge in CF_4 or F_2 F by ESR	300	1.0×10^{13}	—	—
Quick and Wittig[26] (1977)	F from SF_6 photolysis	~300	$(1.6–2.2) \times 10^{13}$	—	—
Flynn et al.[27] (1977)	F from SF_6 photolysis HF by chemiluminescence	~300	2.0×10^{13}	—	—

jets at several positions down the tube. An on-line mass spectrometer was placed at the tube exit. The F atoms were generated with a microwave discharge in a stream of CF_4 in Ar. The H_2 concentration was determined from the mass spectrometer data whereas the concentration of F atoms was determined by titration with Cl_2 using the extremely rapid reaction

$$F+Cl_2 \rightarrow ClF+Cl$$

with the end point determined with the aid of the mass spectrometer. The reaction was carried out with the F-atom concentration considerably in excess of the H_2 concentration. The scatter in the rate constants from seven runs was approximately 50%.

In the same set of experiments Clyne, McKenney, and Walker[20] determined rate constants for a number of different reactions of F atoms. Of particular interest are the results for $F+CH_4 \rightarrow HF+CH_3$ since rates for this reaction have been obtained recently by Clyne and Nip[28] using absorption spectrometry for F-atom detection and since the relative rates of reaction of F with H_2 and CH_4 have been determined by competitive rate measurements. In Ref. 20 the reported rate for $F+CH_4$ is $3.6\ (\pm 100\%) \times 10^{13}\ cm^3/mole\text{-}sec$ whereas in Ref. 28 the reported rate is $4.5(\pm 20\%) \times 10^{13}\ cm^3/mole\text{-}sec$. Clyne and Nip[28] suggest the earlier error limit is probably too conservative. In view of this and with allowance for the reduced error for the combined results of the seven runs for $F+H_2$ in Ref. 20 the reported uncertainty of 50% in the rate constant for $F+H_2$ by Clyne, McKenney, and Walker is also conservative.

The first direct measurements leading to an activation energy were carried out by Homann et al.[17] using a flow system. A mixture of He and N_2 was passed through a microwave discharge to produce N atoms to react with NF_2 introduced downstream to release F atoms in the reaction $N+NF_2 \rightarrow N_2+2F$. Further downstream H_2 was introduced and the mixture was passed through a reaction tube to an exit port provided with a mass spectrometer sampling system. The F-atom concentrations were determined by titration with added ClNO reacting with F according to

$$F+ClNO \rightarrow NO+FCl$$

The extent of the reaction $F+H_2 \rightarrow HF+H$ was determined from either F-atom loss or HF production as given by the mass spectrometer.

Repeated measurements were made at three temperatures spanning the range 300 to 400°K. The results were fit to an Arrhenius expression $k = A \exp(-E_a/RT)$ with $A = 1.6 \times 10^{14}\ cm^3/mole\text{-}sec$ and $E_a = 1600$ cal/mole and a rate at 300°K of $1.2 \times 10^{13}\ cm^3/mole\text{-}sec$.

As pointed out by Clyne et al.[20] and by Cohen and Bott[8] there are several possible problems with the experimental method used by Homann et al.[17] With F formed by the reaction $N + NF_2 \rightarrow N_2 + 2F$ an excess of either N atoms or unreacted NF_2 can occur unless the N and NF_2 are introduced in equal amounts. With excess NF_2 reaction with H atoms might increase the yield of HF by a factor of 2. In the titration of F with ClNO the highly reactive species FCl may cause difficulties in accurate calibrations and F atoms may be recreated by the reaction $F_2 + NO \rightarrow FNO + F$. The effects of these processes resulting from use of ClNO on the calculated rate constants are uncertain. Since the response of any one of the problems to changes in temperature is unknown the combined effect of these problems is an uncertainty of at least 50% in the activation energy determined.

Dodonov et al.[18] carried out a measurement at 293°K similar to that of Clyne, McKenney, and Walker.[20] In this case F atoms were generated in a discharge through F_2 and He. The extent of reaction was determined from changes in H_2, H, F, and HF concentrations as followed with a mass spectrometer. The mass spectra indicated the absence of F_2 and complications that would arise from its reactions, but SiF_4 from wall reactions was found in concentrations similar to those of F. The reaction rate obtained, on the assumption of a negligible interference of the reaction of H atoms with SiF_4, was 1.8×10^{13} cm^3/mole-sec with an uncertainty of 33%. Goldberg and Schneider[25] used a fast flow system with ESR measurements of F-atom concentrations. Atomic F was produced by a microwave discharge of a mixture of either CF_4 or F_2. Dissociation of CF_4 to CF_3 and F was 10 to 20% and that of F_2 to F was 94 to 100%. The rate constant determined for reaction with H_2 was $1.00 \pm 0.08 \times 10^{13}$ cm^3/mole-sec at the single temperature of 298°K. In a similar experiment with ESR detection but with F generated by thermal dissociation Hon, Axworthy, and Schneider[23] obtained a rate constant of 1×10^{13} cm^3/mole-sec for reaction with H_2 at 298°K.

In another class of experiments, HF concentrations were followed by measurements of the laser gain or the rate of chemiluminescence of vibrationally excited HF formed after photolysis of fluorine-containing compounds in mixtures with H_2. At least three possible difficulties may occur: collisional relaxation of vibrationally excited HF, secondary reactions of photolysis fragments with H or H_2, and the reaction of translationally energetic F atoms from photolysis. Avoiding collisional relaxation of HF requires low pressure, but promoting collisional relaxation of F-atom translation requires high pressures. Little is known about the rates of reaction of most of the photolysis fragments although upper limits to these rates may be estimated.

Kompa and Wanner[19] used WF_6 as the source of F atoms in photolysis in the presence of H_2. The reaction was carried out in a laser cavity, and the laser emission from excited HF was monitored as a function of time. The system was operated at low pressures so that collisional deactivation of excited HF could be made small in comparison with the stimulated emission. Kompa and Wanner[19] cite inaccuracy in H_2 pressure measurement and translational excitation of F atoms from photodissociation as sources of error in the thermal rate constant determined. Cohen and Bott[8] suggest the secondary reactions $H + WF_6$ or $H + WF_5$ could result in a rate coefficient too high by a factor of 2. The rate constant reported is 3.8×10^{13} cm^3/mole-sec for 298°K.

Igoshin, Kulakov, and Nikitin[21] conducted laser-gain experiments in the temperature range 195 to 294°K using a pulsed discharge to generate F atoms in mixtures of NF_3 with H_2 (and D_2). Interpretation of results is complicated by the production of excited H_2 and other species in the discharge as well as the necessity to take into account collisional deexcitation of the HF formed. The computed rate constants are thus dependent on the values assumed for other processes. The reported rate constants are: 1.5×10^{13} cm^3/mole-sec (extrapolated) for $F + H_2$ at 298°K with an activation energy of 1080 ± 170 cal/mole, and 1.3×10^{13} cm^3/mole-sec (extrapolated) for $F + D_2$ at 298°K with an activation energy of 790 ± 180 cal/mole. Because of the uncertainties in the experiment itself and the discrepancies between the ratio k_{H_2}/k_{D_2}, from these experiments and those from competitive rate measurements (see text following) these rate constants and the associated activation energies are uncertain by a factor of at least two.

Lam, Peyron, and Puget[24] used a continuous HF laser of the nozzle type operating with F atoms from a discharge in SF_6 entering an H_2 stream at 300°K. The rate constant deduced, 0.14×10^{13} cm^3/mole-sec, is a factor of 10 lower than those from most other studies.

Experiments with a pin-discharge HF laser using NF_3 as the F-atom source for reaction with H_2 were carried out by Pearson et al.[22] Calculations based on pulse-shape data yield a rate constant of 2.4×10^{13} cm^3/mole-sec at 500°K.

Most recently, Quick and Wittig[26] and Preses, Weston, and Flynn[27] have investigated the rates of reaction of F atoms generated by multiphoton dissociation of SF_6 in the presence of H_2. The output of a pulsed TEA CO_2 laser was focussed in a cell containing an SF_6/H_2 mixture at low pressure. The time dependence of the chemiluminescence was measured to determine rate constants at about 300°K. Since the rate of collisional deexcitation of HF was low the rise time of the spontaneous emission established the rate constants. The rate constants determined,

(1.6 to 2.2) $\times 10^{13}$ cm^3/mole-sec by Quick and Wittig[26] and (2.0 ± 0.3) $\times$ 10^{13} cm^3/mole-sec by Preses et al.,[27] either indicate by their agreement with other values that the dissociation of SF_6 gave thermal F atoms or add to the list of similar values if the F atoms are assumed near thermal as indicated in measurements of SF_6 dissociation by Coggiola et al.[29] Repeated experiments with a Ne moderator to assure thermal equilibration of the F atoms would be valuable.

D. Rates Derived from Measurements in Complex Systems

Several studies of the reactions occurring in H_2,F_2 systems have been carried out without isolation of the single reaction of F with H_2. In these the deduced rate constant for the reaction of F with H_2 depends strongly on the values of rate constants for the other reactions occurring in the system. On the basis of reversibility considerations applied to measurements of HF decomposition in shock tubes[30,31] rough estimates of high-temperature rates may be obtained. From studies of H_2–F_2 flames with mass spectrometric determination of the species concentrations Homann and MacLean[32] and MacLean[33] estimated a rate constant of 9×10^{13} cm^3/mole-sec at 2000°K. Levy and Copeland[34] measured rates in a flow system and examined the inhibiting effect of added oxygen. The complexity of possible oxygen effects prevents a reliable determination of rate for the reaction of F with H_2 from the data gathered.

Rabideau, Hecht, and Lewis[35] conducted ESR measurements in a flow system in which H_2 dissociated in a microwave discharge was added to a stream of F_2. Modeling of the system to reproduce the measured atom concentration profiles led to an estimate of (4 ± 1) $\times 10^{12}$ cm^3/mole-sec for $F + H_2$ at 300°K. This result must be considered uncertain since it depends on a number of other rate constants.

E. Rates Derived from Competitive Rate Measurements

Competitive rate experiments in which two or more species compete in reaction with F atoms offer a means for determining relative rates without measurement of F atom concentrations or reaction times. Thus, relative rates can be determined with considerably greater accuracy than absolute rates. The rate of reaction of F with H_2 relative to other species may be used to determine absolute rates for $F + H_2$ reaction provided the absolute rate is known for the other species. Unfortunately, the measurements of absolute rates for F atoms with other species are just as difficult as they are for F atoms with H_2.

Of most interest in establishing rate constants for $F + H_2$ are experiments involving the competition of H_2 and CH_4 (Table II) and the direct

TABLE II.
Competitive Rate Measurements: H_2 vs. CH_4

Investigators	T(°K)	$k_{H_2}/k_{CH_4}{}^a$	k_{H_2}/k_{CH_4}(300°K)
Mercer and Pritchard[36] (1959)	298–423	$0.95 \exp[-(500 \pm 200)/RT]$	0.41
Foon and Reid[37] (1971)	253–348	$1.22 \exp[-(630 \pm 300)/RT]$	0.42
Williams and Rowland[38,39] (1971, 1973)	283	0.33, 0.34	—
Jonathan et al.[40] (1971)	300	0.74 ± 0.07	0.74
Lam, Peyron, and Puget[24] (1974)	~300	0.38	0.38
Manning et al.[41] (1975)	303	0.39 ± 0.01	—
Kapralova, Margolin, and Chaikin[42] (1970)	77–353	$0.41 \exp[-935/RT]$	0.09

^{a}Activation energies in calories/mole.

measurements of rate for reaction of F with CH_4. For the temperature range 298 to 423°K Mercer and Pritchard[36] found $k_{H_2}/k_{CH_4} = 0.95 \exp[-(500 \pm 200)/RT]$, E_a in calories. Experiments by Foon and Reid[37] with common competitors for H_2 and CH_4 yield $k_{H_2}/k_{CH_4} = (1.22 \pm 0.01) \exp[-(630 \pm 300)/RT)$ in the temperature range 253 to 348°K. Other experiments have given the ratio of rate constants at single temperatures: Williams and Rowland,[38,39] $k_{H_2}/k_{CH_4} = 0.34$ at 283°K; Jonathan et al.,[40] 0.74 at 300°K; Lam et al.,[24] 0.38 at 300°K; Manning et al.,[41] 0.39 at 303°K. A less reliable result may be inferred from experiments by Kaprolova, Margolin, and Chaikin[42] separately measuring the competitions H_2/D_2 and D_2/CH_4 to give $k_{H_2}/k_{CH_4} = 0.41 \exp(-935/RT)$ in the range 77 to 353°K.

A single measurement of the temperature dependence of the $F + CH_4$ reaction has been made. Wagner, Warnatz, and Zetzsch,[43] in an experiment similar to that by Homann et al.,[17] obtained a rate constant 4.8×10^{13} cm^3/mole-sec at 298°K with an activation energy of 1100 cal/mole. Measurements by Clyne, McKenney, and Walker[20] gave a constant of 3.6 ($\pm$ 100%) $\times 10^{13}$ cm^3/mole-sec at 298°K. From flash photolysis of WF_6 Kompa and Wanner[19] obtained 4.3×10^{13} cm^3/mole-sec. Most recently, Clyne and Nip,[28] using atomic resonance absorption of F in a flow system, obtained 4.5×10^{13} cm^3/mole-sec and recommended a value $(4.3 \pm 1.0) \times 10^{13}$ cm^3/mole-sec.

In reviewing the relative rate data available as of 1964 for competitions between H_2, CH_4, C_2H_6 and higher hydrocarbons Fettis and Knox[44] arrived at an activation energy of 1710 cal/mole for the reaction $F + H_2$. This was based on the assumption of zero activation energy for the higher hydrocarbons and activation energies for the intermediate competitions as follows: H_2/CH_4, 500 ± 200 cal/mole; CH_4/C_2H_6, 930 ± 80 cal/mole; C_2H_6/C_3H_8, 280 ± 25 cal/mole. The combined uncertainty of 305 cal/mole in addition to the uncertainty of zero activation energy for C_3H_8 leads to an overall uncertainty for $F + H_2$ in excess of 500 cal/mole for the activation energy of $F + H_2$.

F. Thermal Rate Constants—Discussion

As shown in Table I direct measurements of the rate of reaction of F with H_2 at 300°K give a range of rate constants centered about a value 1.5×10^{13} cm^3/mole-sec which is that given by the experiments of Clyne, McKenney, and Walker[20] which are judged to be the most accurate. A rate constant in the range $(1.0–2.0) \times 10^{13}$ cm^3/mole sec is compatible with nearly all the experiments listed. This range is also compatible with the rate of 4.3×10^{13} cm^3/mole-sec recommended by Clyne and Nip[28] for $F + CH_4$ together with a relative rate of 0.3 to 0.5 for the H_2/CH_4 competition as listed in Table II.

The most reliable estimates of the activation energy must come directly from the experiments of Homann et al.[17] yielding 1600 cal/mole. The experiments by Wagner et al.[43] giving 1100 cal/mole for $F + CH_4$, together the values of 500 cal/mole by Mercer and Pritchard[36] and 630 cal/mole by Foon and Reid[37] for H_2/CH_4 competition, lead to an activation energy of 1600 to 1730 cal/mole. These values are consistent with the value of 1710 cal/mole deduced by Fettis and Knox.[44] It seems highly unlikely that the activation energy could lie outside the range 1300 to 1900 cal/mole.

G. Competitive Thermal Rate Measurements: $H_2/D_2/HD$

The relative rates of the reactions of F with H_2, D_2, and HD are of interest in testing theoretical predictions of reaction rate since these reactions should show differing degrees of quantum effects. As indicated, relative rates are more easily obtained experimentally than absolute rates and are thus more reliable.

The results of available measurements for the H_2/D_2 competition are listed in Table III. The experiments include direct competitive studies as well as separate experiments for the two systems in the same apparatus. The rate constants for 298°K are in remarkably good agreement giving a value of 1.6 to 2.0 for the ratio k_{H_2}/k_{D_2} in eight of the nine studies.

Two studies with entirely different methods, those by Persky[46] and Grant and Root,[49] give essentially exact agreement for the relative rate constant and its temperature dependence. Persky[46] used a flow system with F atoms produced by a microwave discharge of CF_4 or SF_6 in

TABLE III.
Competitive Rate Measurements: H_2 vs. D_2

Investigators	Method	T (°K)	k_{H_2}/k_{D_2} [a]	k_{H_2}/k_{D_2} (298°K)
Kapralova, Margolin, and Chaikin[42] (1970)	thermal reactor F from F_2 HF, DF by ESR	77–293	1.48 exp(45/RT)	1.6
Williams and Rowland[38] (1971)	near thermal F from nuclear recoil gas chromatography	283	1.8	—
Foon, Reid, and Tait[45] (1972)	thermal reactor common competitors mass spec	250–300	1.5 exp(130/RT)	1.9
Persky[46] (1973)	flow F from discharge in SF_6 or CF_4 HF, DF by mass spec	163–417	1.04 exp(370/RT)	1.94
Berry[47] (1973)	CF_3I photolysis in H_2, D_2 HF, DF chemiluminescence	297	1.8	1.8
Igoshin, Kulakov, and Nikitin[21] (1973)	F from discharge in NF_3 HF, DF chemiluminescence separate experiments	195–294	1.9 exp(−290/RT)	1.1
Grant and Root[48] (1974)	moderated F from nuclear recoil common competitor chromatography	303–457	1.11 exp(356/RT)	2.0
Grant and Root[49] (1975)	moderated F from nuclear recoil common competitor chromatography	273–457	1.04 exp(382/RT)	1.98
Preses, Weston, and Flynn[27] (1977)	F from SF_6 photolysis HF, DF by laser gain separate experiments	~300		1.65

[a] Activation energies in calories/mole.

helium added to a mixture of H_2/D_2 in large excess. The ratio of rates was determined as the ratio of products HF/DF detected with a mass spectrometer. The ratio of rate constants found was $k_{H_2}/k_{D_2} = (1.04 \pm 0.02) \exp[370 \pm 10)/RT]$ for the temperature range 163 to 417°K. Grant and Root[49] measured the rates of $F+H_2$ and $F+D_2$ relative to the rate of the addition $F+C_3F_6$ in nuclear recoil experiments. These gave a ratio $k_{H_2}/k_{D_2} = (1.04 \pm 0.06) \exp[(382 \pm 35)/RT]$ for the temperature range 303 to 475°K.

The uncertainty in the results of any of the other experiments listed in Table III is sufficient for overlap with the Persky[46] and Grant–Root[49] results. Kapralova et al.[42] used a reaction mixture containing molecular fluorine and analysis of collected HF and DF was made by EPR of H and D atoms generated in a high-frequency discharge of the collected products. The ratio obtained by Foon, Reid, and Tait[45] is in agreement with the Persky and Grant–Root results at room temperature. At 130 ± 300 cal/mole the activation energy difference obtained is also in agreement. Results of room temperature measurements by Williams and Rowland,[38] Berry[47] and Preses, Weston, and Flynn[27] are reasonably close. The one set of measurements giving a much lower value for the room temperature ratio and a different activation energy is that by Igoshin et al.[21] in which a discharge through the reaction mixture to generate F atoms complicates the analysis.

It is thus reasonable to accept the Persky[46] and Grant–Root[49] results within the accuracies claimed by these investigators.

In an experiment similar to that of Persky,[46] Klein and Persky[50] measured the relative rates for the competition of normal (25% para, 75% ortho) and para-enriched (75% para, 25% ortho) H_2 with D_2. The calculated values of the relative rates k_{H_2} (75% para)/k_{H_2} (25% para) were 1.02 ± 0.01 (175°K), 0.98 ± 0.01 (237°K), 1.00 ± 0.01 (298°K).

The reaction of F with HD may proceed in two ways

$$F + HD \rightarrow HF + D$$

$$F + DH \rightarrow DF + H$$

Two sets of experiments using different methods have given rate constant ratios for these reactions. In an experiment run in the same way as his experiments for H_2/D_2 competitions Persky[51] obtained the ratio $k_{F+HD}/k_{F+DH} = (1.26 \pm 0.02) \exp[70 \pm 6)/RT]$ for the temperature range 159 to 413°K. Berry,[47] by measuring relative gain coefficients for laser transitions of HF and DF produced after photolysis of CF_3I in the presence of HD, obtained a ratio $k_{F+HD}/k_{F+DH} = 1.42 \pm 0.10$ at room temperature in nearly exact agreement with Persky's results.

H. Measurements of Chemiluminescence (nonlaser)

The reaction of F with H_2 is exothermic by approximately 31.9 kcal/mole and releases about 65% of the energy available to products HF and H as vibrational excitation of HF. Figure 1 shows the vibrational energy levels accessible to HF and to DF formed in reaction. Measurements of the emission of infrared radiation of newly formed HF have been carried out by a number of workers as listed in Table IV.

Measurements of the infrared chemiluminescence of the type pioneered by Polanyi and his co-workers[52–57] have yielded detailed rate constants for the production of specific rotation–vibration states of HF produced in reaction of F with H_2. In these experiments the reactants are introduced through separate tubes into a chamber where they meet

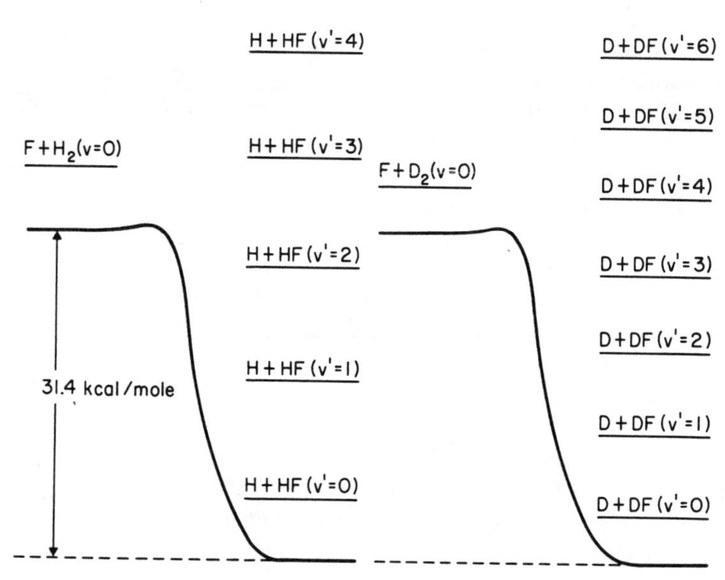

Fig. 1. Energetics of the reactions $F + H_2 \rightarrow HF + H$ and $F + D_2 \rightarrow DF + D$.

TABLE IV.
Measurements of Chemiluminescence (nonlaser)

Investigators	Reactants	Notes
Polanyi and Tardy[52] (1969)	$F + H_2$	HF vib., rot., trans. distribution
Anlauf et al.[53] (1970)	$F + H_2$, $F + D_2$	translational energy by difference
Jonathan et al.[59] (1971)	$F + H_2$	product disttributions
Jonathan et al.[40] (1971)	$F + H_2$	some relative rate constants
Polanyi and Woodall[54] (1972)	$F + H_2$, $F + D_2$	higher accuracy, less relaxation
Chang and Setser[58] (1973)	$F + H_2$	product distributions
Perry and Polanyi[55] (1976)	$F + HD$	branching ratios
Perry and Polanyi[56] (1976)	$F + H_2$, $F + D_2$, $F + HD$	reactant temperature variation
Douglas and Polanyi[57] (1976)	$F + H_2$ ortho/para	effect of H_2 rotation
Quick and Wittig[26] (1977)	$F + H_2$	rate constant
Preses, Weston, and Flynn[27] (1977)	$F + H_2$, $F + D_2$	rate constants

as uncollimated crossed molecular beams or jets. The emission of excited HF in the region of interaction is detected with a spectrometer. The observed infrared intensities are converted to relative populations of the rotation–vibration states produced in reaction.

The experimental techniques have been refined over a period of years.

For the $F + H_2$ reaction initial experiments were reported by Polanyi and Tardy[52] in 1971. A particular problem is the collisional relaxation of the excited HF by reactants and products within the chamber. Pressures have been lowered sufficiently to make negligible the extent of vibrational relaxation, but even in the most recent experiments rotational relaxation is appreciable. A means for correcting the observed rotational distributions to give the initial rotational distributions has been devised and found adequate. The most refined studies for $F + H_2$ and $F + D_2$ are those by Polanyi and Woodall.[54] Following this have come investigations of the reaction of F with HD by Perry and Polanyi[55,56] and of the effects of reactant rotational and translational energy variations on product distributions for reaction of F with H_2 by Douglas and Polanyi.[57]

The measurements of Polanyi and Woodall[54] gave the following values for relative rate constants for reaction of F with H_2 at approximately 300°K to produce HF in vibrational state v': $k(v' = 1) = 0.31$, $k(v' = 2) = 1.0$, $k(v' = 3) = 0.47$. Figure 2 shows a triangular plot giving the distribution of vibrational and rotational energies with contour lines joining points of equal rate constants $k(v', J')$. Since the energy available to products is approximately constant at 34.7 kcal/mole, the dashed

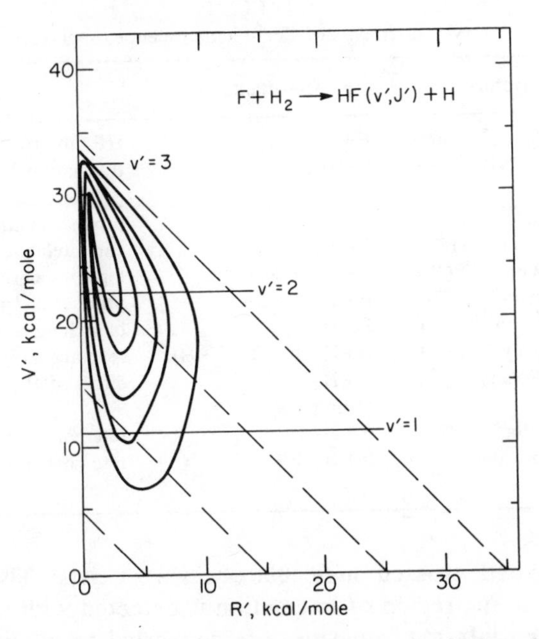

Fig. 2. Triangular plot for the $F + H_2$ reaction showing the distribution of product translational, rotational, and vibrational energy as measured by Polanyi and Woodall[54] for reaction at ~300° K. Contours join points of equal detailed rate constant ignoring quantization.

diagonal lines indicate the translational energy of H and HF. Polanyi and Woodall[54] obtained results similar in appearance for the reaction of F with D_2.

The chemiluminescence measurements of Polanyi and Woodall[54] are quantitatively confirmed by experiments in a similar system at somewhat higher background pressures by Chang and Setser.[58] Measurements in a flow system with an Ar buffer gas by Jonathan et al.[40,59] are more difficult to interpret because of faster HF relaxation. With measurements of HF emission at several locations down the flow tube extrapolation could be made to zero relaxation time. The relative rate constants for $HF(v')$, as revised by Berry[47] with new values for the transition probabilities, are very close to those of the low pressure systems. Preses, Weston and Flynn[27] have also measured chemiluminescence but not rate constants for specific states $HF(v')$.

The more recent experiments by Perry and Polanyi[55,56] provide detailed rate constants for the production of $HF(v', J')$ and $DF(v', J')$ in the reactions of F atoms at ~300°K with HD, H_2, and D_2 at 77 to 1315°K. The mean fractions of the available energy entering vibration f'_v and rotation f'_R of HF and DF were found for room temperature reactants to be: $F+HD \rightarrow HF+D$, 0.59 and 0.125; $F+DH \rightarrow DF+H$, 0.63 and 0.066; $F+H_2 \rightarrow HF+H$, 0.66 and 0.083; $F+D_2 \rightarrow DF+D$, 0.67 and 0.076. Increasing the temperature of the reactants HD, H_2, or D_2 was found to increase the translational and rotational energy of the products.

An additional study of the effects of changing reactant energy was carried out by Douglas and Polanyi[57] with the use of para-enriched and normal hydrogen at inlet temperatures of 77 and 290°K. The effective temperatures for collisions with F atoms from a source at 290°K were 97 and 290°K. The para-enriched H_2 was 90 to 95% para ($J=0, 2, \ldots$) whereas the normal H_2 was 75% ortho ($J=1, 3 \ldots$). Product HF rotational excitation was found unaffected by the H_2 rotational state J whereas the fraction of energy released to vibration of HF decreased slightly in going from $J=0$ to 1 and increased slightly in going from $J=1$ to 2.

I. Measurements of Chemiluminescence (laser)

Following the discovery of chemical laser action in the H_2–Cl_2 system by Kasper and Pimentel,[60] laser emission was observed by Deutsch[61,62] in fluorocarbon–hydrogen discharges and by Kompa and Pimentel[63,64] for flash photolyzed mixtures of UF_6 and H_2. Table V lists experiments with laser systems yielding information on the reaction of F with H_2. Studies by Kompa, Parker, and Pimentel[65] identified the important processes in the UF_6–H_2 and UF_6–D_2 systems and provided preliminary estimates of

TABLE V.
Measurements of Chemiluminescence (laser systems)[a]

Investigators	Reactants	Notes
Kompa and Pimentel[63,64] (1967)	$F+H_2$, $F+D_2$	UF_6 photolysis, HF vib. populations
Kompa, Parker, and Pimentel[65] (1968)	$F+H_2$, $F+D_2$, $F+HD$	UF_6 photolysis, HF and DF vibrational populations
Parker and Pimentel[66] (1969)	$F+H_2$	equal gain, vib. population ratios
Kompa and Wanner[19] (1972)	$F+H_2$	pulse shape for rate constant
Coombe and Pimentel[68] (1973)	$F+H_2$, $F+D_2$	equal gain, zero gain; temp. dependence of vib. distributions
Coombe and Pimentel[69] (1973)	$F+H_2$	equal gain, zero gain; effect of H_2 rotation (ortho/para)
Berry[47] (1973)	$F+H_2$, $F+D_2$, $F+HD$	grating selection; relative rate coefficients
Igoshin et al.[21] (1973)	$F+H_2$	pulse shape for rate constant
Pearson et al.[22] (1973)	$F+H_2$	pulse shape for rate constant

[a]List excludes a large number of experiments important for laser development but not providing details of the reaction of F with H_2. See, for example, Refs. 61, 62, 67.

the rate constant ratios for the production of excited HF (or DF) in several vibrational levels by reaction of F with H_2 (or D_2). Conditions were found for which collisional deexcitation of HF vibration was minimized but rotational equilibration was nearly complete.

Parker and Pimentel[66] made more accurate measurements of the rate constant ratios for production of HF($v'=1,2$) and DF($v'=2,3$) by use of the "equal-gain temperature" method. Mixtures of UF_6–H_2 and UF_6–D_2 were used with Ar or SF_6 added to control temperature rise and promote rotational relaxation of the reaction products. With variation of temperature of the system, equal to the rotational temperature of the rotationally equilibrated HF, two vibration–rotation transitions of HF could be caused to initiate simultaneously with the same gain. The complexity of multiple transitions and cascade effects was avoided by considering only the first transition observed. From the relationship between gain and the vibrational populations, together with the known rotational temperatures, the ratio of rate constants $k(v'=2)/k(v'=1)$ could be determined. With a correction for vibrational deexcitation the estimated ratio was $k(v'=2)/k(v'=1)\sim 5.5$ at 498°K. Similarly, for $F+D_2 \rightarrow DF+D$, $k(v'=3)/k(v'=2)\sim 1.6$.

Improved experiments were reported by Coombe and Pimentel[68] who used the equal-gain temperature and zero-gain temperature methods. In the latter method a separate driver laser is used to provide emission of a single transition and the gain for that transition in a second optical cavity containing the photolyzed mixture of interest is measured. The rotational temperature of the vibrationally excited but rotationally relaxed HF is varied by controlling the system temperature. For zero gain the relative populations of the vibrational levels may then be calculated using the equation for laser gain. Coombe and Pimentel[68] found for $F + H_2 \rightarrow HF + H$ an Arrhenius type of temperature dependence: $k(v' = 2)/k(v' = 1) = 2.14 \exp(+254/RT)$ and $k(v' = 3)/k(v' = 2) = 0.39 \exp(+117/RT)$. For $F + D_2 \rightarrow DF + D$ an increase with temperature was found for the ratio $k(v' = 3)/k(v' = 2)$. Using normal H_2 and para-enriched H_2 Coombe and Pimentel[69] were able to isolate the effects of rotational and translational excitation of the reactants $F + H_2$ and found that high vibrational states of product HF were (most likely) favored by lower J states of reactant H_2.

Berry[47] determined the vibrational populations of HF and DF in reactions of F with H_2, D_2, and HD using a somewhat different approach. Relative gain was measured for single transitions and all other transitions were eliminated by use of a grating replacing one end mirror. This allowed ratios between $v' = 0, 1, 2, 3$ for HF and $v' = 0, 1, 2, 3, 4$ for DF to be determined. Reactions were carried out at room temperature with CF_3I as the fluorine source and added Ar as the buffer gas. Berry also determined from the data estimated values of the relative thermal rate constants at room temperature for $F + H_2 \rightarrow HF + H$, $F + D_2 \rightarrow DF + D$, $F + HD \rightarrow HF + D$, $F + DH \rightarrow DF + H$. These ratios are in good agreement with the (presumed) more accurate values from more direct thermal rate measurements.

J. Product State Distributions—Discussion

Results of the chemiluminescence and chemical laser experiments giving vibrational state distributions of HF and DF for reactions of F with H_2, D_2, and HD at room temperature are listed for comparison in Table VI. This table is a revision of that given by Berry[47] in 1973. The relative rate constants $k(v' + 1)/k(v')$ may be seen to be in excellent agreement except for a few ratios involving the less-populated states. Krogh, Stone, and Pimentel[70] have discussed the possible effects of differing energy transfer rates on the distributions reported. On the basis of the reproducibility obtained for the later experiments of any one type and the agreement between the refined experiments of different types, the uncertainties listed in the table appear entirely reasonable in most cases. One may simply accept the values obtained by Perry and Polanyi[56] and Berry[47] as accurate within the error limits indicated.

Only the low-pressure chemiluminescence experiments yield rota-

TABLE VI.
Product Vibrational State Ratios for ~300°K

	N_4/N_3	N_3/N_2	N_2/N_1	N_1/N_0	Method	Investigators
$F + H_2 \rightarrow HF + H$						
	—	—	~5.5	—	cl	Parker and Pimentel[66]
	—	0.58 ± 0.12	3.6 ± 0.2	—	ir	Jonathan et al.[40]
	—	0.53 ± 0.10	3.4 ± 0.7	—	ir	Polanyi and Woodall[54]
	—	0.56 ± 0.06	3.4 ± 0.3	—	ir	Chang and Setser[58]
	—	0.48 ± 0.01	3.3	—	cl	Coombe and Pimentel[68]
	—	0.63 ± 0.04	3.40 ± 0.10	5.2 ± 0.4	cl	Berry[47]
	—	0.55	3.6	~7	ir	Perry and Polanyi[56]
$F + D_2 \rightarrow DF + D$						
	—	~1.6	—	—	cl	Parker and Pimentel[66]
	0.66 ± 0.13	1.5 ± 0.3	2.3	—	ir	Polanyi and Woodall[54]
	—	1.5 ± 0.1	—	—	cl	Coombe and Pimentel[68]
	0.4 ± 0.2	1.80 ± 0.10	2.35 ± 0.10	2.3 ± 0.2	cl	Berry[47]
	0.59	1.92	3.5	~4	ir	Perry and Polanyi[56]
$F + HD \rightarrow HF + D$						
	—	0.15 ± 0.05	3.1 ± 0.2	5.2 ± 0.4	cl	Berry[47]
		0.14	3.3	~5	ir	Perry and Polanyi[56]
$F + DH \rightarrow DF + H$						
	0.3 ± 0.2	1.60 ± 0.10	2.25 ± 0.10	2.2 ± 0.2	cl	Berry[47]
	0.61	1.85	3.0	~3	ir	Perry and Polanyi[56]

tional distributions. In these there is a greater uncertainty due to the partial relaxation by collisions with background gases. The results of Polanyi and Woodall[54] and Perry and Polanyi[56] as obtained with a "truncation" correction must be considered uncertain in a quantitative sense but there is little question that the general features are correct.

K. Crossed-Beam Experiments

A crossed molecular beam study of the $F+D_2$ reaction was carried out by Lee and co-workers[71,72] in 1970. Fluorine atoms were generated in a nickel oven by thermal dissociation at 675°C and velocity selected with a rotating slotted disk (velocity spread 20% fwhm). The D_2 beam (velocity spread $\approx 8\%$ fwhm) was generated by a nozzle source of variable temperature. The beams intersected at right angles. The angular distribution of product DF was measured with a rotatable mass spectrometer. Distributions have been reported for five collision energies in the range 0.80 to 4.20 kcal/mole (center-of-mass). Figure 3 shows the measured angular distribution and the Newton diagram for 1.68 kcal/mole.

At 0.80 kcal/mole the product DF is scattered almost exclusively in the backward direction (with respect to the attacking F atom). As energy is increased an increasing fraction of DF is observed scattered in the

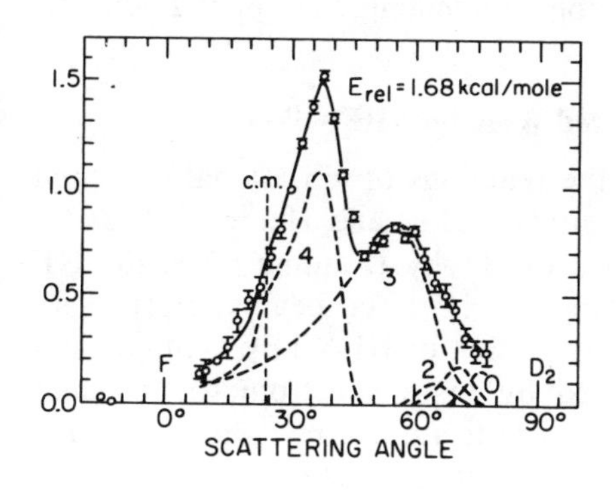

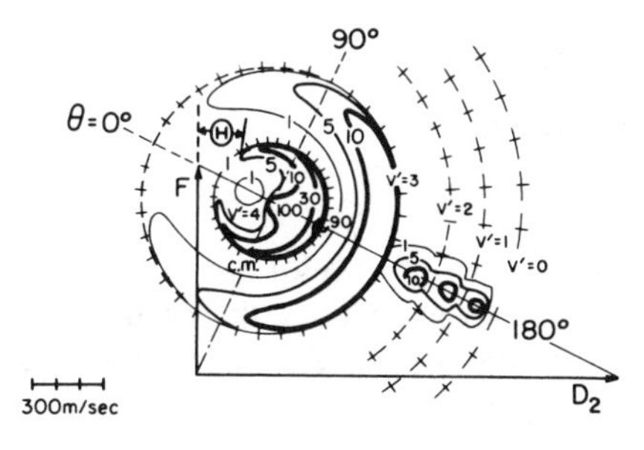

Fig. 3. Angular distribution of DF intensity (density) as measured by Lee and co-workers[71,72] for the reaction of F with D_2 at an average relative kinetic energy of 1.68 kcal/mole.

forward direction. At the lower collision energies the favorable kinematics give separate peaks in the angular distribution for DF($v' = 3$) and DF($v' = 4$). With more energy in vibration for $v' = 4$ there is less available kinetic energy for separation of products D and DF and the velocity of back-scattered DF in the center-of-mass system is lower. As indicated by the Newton diagram this results in HF($v' = 4$) appearing at an angle close to that of the center-of-mass. The ratio of products DF($v' = 4$) to DF($v' = 3$) was found to be about 0.75 for an energy of 0.80 kcal/mole. This is in reasonable agreement with the room temperature results for chemiluminescence listed in Table VI.

Additional experiments with velocity analysis of the reaction products to resolve the product distributions in angle and velocity space will yield more detailed information.

L. Hot Atom Experiments

As indicated in Section II.G Grant and Root[48,49] carried out several competitive rate studies with thermalized F atoms. They have also investigated the competition between reactions of F with H_2/C_3F_6 and D_2/C_3F_6. Relative yields of products show a temperature dependence. The interpretation is complex and uncertain but the results may be explained by a temperature effect in reaction with C_3F_6 and a ratio of reaction probabilities of 1.3 for the non-Boltzmann reaction of F with H_2 relative to D_2.

M. Studies of the Related Reactions HF + D →

It has been possible to examine the reactions of vibrationally excited HF with D atoms in two types of studies. Bott and Heidner[73] found a large increase in the rate of removal of HF by D atoms when the HF was excited to $v' = 3$ by absorption of HF laser beams. Bartoszek, Manos, and Polanyi[74] measured the depletion of HF(v') by D atoms in a chemiluminescence experiment aided by mass spectrometric product detection. The results of these studies indicate a threshold vibrational energy requirement of $v' = 3$ for the abstraction reaction

$$\mathrm{HF}(v' \geq 3) + \mathrm{D} \rightarrow \mathrm{F} + \mathrm{HD}$$

and a threshold $v' = 5$ for the exchange reaction

$$\mathrm{HF}(v' \geq 5) + \mathrm{D} \rightarrow \mathrm{H} + \mathrm{DF}$$

For the abstraction reaction the observation of a vibrational energy requirement of $v' = 3$ for reaction of HF is consistent with the production of HF($v' = 3$) in the reaction of F with HD. For the exchange reaction the observation of the requirement of HF($v' \geq 5$) is consistent with a barrier height of 45 to 50 kcal/mole for exchange.

III. THEORETICAL WORK

A. Potential Energy Surfaces

The attempts to estimate and to determine the potential energy surface(s) for the reaction of F with H_2 have a long history beginning with Eyring's[13] use of the London–Eyring–Polanyi (LEP) method in 1931. Nearly all available empirical, semiempirical, and approximate methods as well as modern variational methods with large basis sets have been used since then.

Three potential energy surfaces may be involved in reaction at thermal energies. A correlation diagram is shown in Fig. 4. The ground-state surface $1^2A'$ ($^2\Sigma$ for collinear configurations) correlates with ground-state reactants $F(^2P_{3/2}) + H_2(^1\Sigma_g^+)$ and with ground-state products $HF(^1\Sigma^+) + H(^2S)$. A second surface $^2A''$ ($^2\pi$ for collinear) correlates with ground-state reactants but with excited-state products $HF(^3\pi) + H(^2S)$. The third surface $2^2A'$ ($^2\pi$ for collinear) correlates with excited state reactants $F(^2P_{1/2}) + H_2(^1\Sigma_g^+)$ and with excited state products $HF(^3\pi) + H(^2S)$. The energy separation between the two accessible F atom states, $^2P_{1/2}$ and $^2P_{3/2}$, is only 1.16 kcal/mole which is of the same magnitude as estimated barrier heights. The product $HF(^3\pi)$ lies about 30 kcal above $HF(^1\Sigma^+)$ and may be inaccessible. The possibility of reaction on excited surfaces by nonadiabatic processes with surface hopping exists and has been investigated in several studies (see Section III.E).

The several surfaces proposed for the F–H–H system are listed in Table VII. Following Eyring's LEP surface are several modifications of the LEP method including that by Sato[76] to give the convenient LEPS (London–Eyring–Polanyi–Sato) surfaces used for a number of studies of

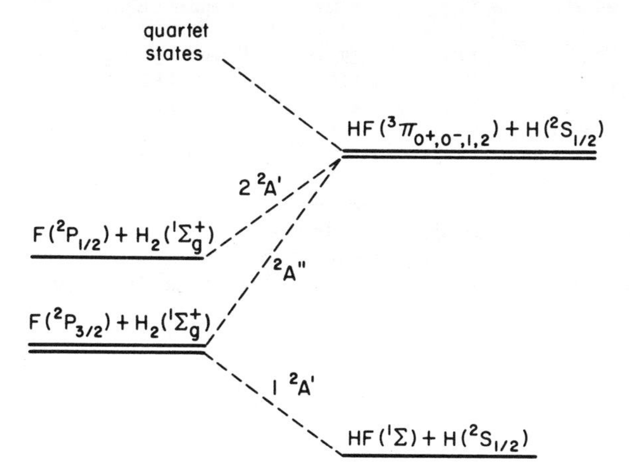

Fig. 4. Correlation diagram for the reaction $F + H_2 \rightarrow HF + H$.

TABLE VII.
F–H–H Potential Energy Surfaces

Authors	Acronym	Barrier[a] (kcal/mole)	Type[b]
Eyring[13] (1931)	—	5	LEP
Hirschfelder[75] (1941)	—	6.0	Modified LEP
Sato[76] (1955)	—	0.5	LEPS
Johnston and Parr[77] (1963)	—	2	Bond energy–bond order
Ellison and Patel[78] (1964)	—	5	Diatomics in molecules
Anderson[79] (1970)	JA	1.68	LEPS
Muckerman[80] (1971)	M1	0.90	LEPS
Muckerman[81] (1972)	M2	0.89	LEPS
Muckerman[81] (1972)	M3	1.43	LEPS
Muckerman[81] (1972)	M4	2.55	LEPS
Muckerman[82] (1973)	M5	1.06	LEPS
Wilkins[83] (1972)	W	0.98	LEPS
Thompson[84] (1972)	T	—	LEPS
Newton[85] (1973)	—	—	SCF
Bender et al.[86] (1973)	—	5.72	CI—214 configurations
Bender et al.[86] (1973)	BOPS	1.66	CI—338 configurations
Blais and Truhlar[87] (1973)	BT	1.24	Semiempirical valence bond
Tully[88] (1973)	T1	15.5	Diatomics-in-molecules ($1^2A'$, $2^2A'$, $^2A''$)
	T2	1.1	Diatomics-in-molecules ($1^2A'$, $2^2A'$, $^2A''$)
Polanyi and Schreiber[7,89] (1973)	SE1	2.16	Extended LEPS
Polanyi and Schreiber[90] (1974)	BOPS-fit	1.26	LEPS fit to CI of Ref. 8
	SE1	2.16	Extended LEPS
Bender et al.[91] (1975)	—	—	CI, H–F–H barrier
Jaffe et al.[92] (1975)	—	—	GRHF ($^2\Sigma$, $^2\pi$)
	—	—	GRHF – CI ($^2\Sigma$, $^2\pi$)
Rebentrost and Lester[93] (1975)	—	—	SCF, ($1^2A'$, $2^2A'$, $^2A''$)
Feng, Grant, and Root[94] (1976)	FR	2.50	LEPS fit to CI of Ref. 8 with raised barrier
Polanyi and Schreiber[7] (1977)	SE2	2.27	Extended LEPS
	SE3	2.02	Extended LEPS
	SE4	2.25	Extended LEPS
	SE5	2.63	Extended LEPS
Ungemach et al.[95] (1977)	—	~3.35	CI > 338 configurations
Botschwina and Meyer[96]	—	, H–F–H	CI, H–F–H barrier
Wadt and Winter[97] (1977)	—	—	CI, H–F–H barrier

[a] Barrier for the $1^2A'$ ($^2\Sigma$) reaction of F with H_2.

[b] LEP—London, Eyring, Polanyi; LEPS—London, Eyring, Polanyi, Sato; CI—configuration interaction; GRHF—generalized restricted Hartree Fock; SCF—self consistent field.

the reaction dynamics. The semiempirical valence bond surface by Blais and Truhlar[87] and the diatomics-in-molecules surface *T*2 by Tully[88] have much stronger bases in theory.

Ab initio variational calculations for collinear F–H–H have been made by Bender, O'Neill, Pearson, and Schaefer;[86] by Jaffe, Morokuma, and George;[92] by Rebentrost and Lester;[93] and by Ungemach, Schaefer, and Liu.[95] The initial calculations by Bender et al.[86] for the ground-state surface gave a barrier height of 1.66 kcal/mole (by difference of calculated barrier and calculated reactant energy). Jaffe et al.[92] made an equivalent calculation for the ground state surface and a similar calculation for the $^2\pi$ surface. Rebentrost and Lester[93] made calculations at the SCF level for all three surfaces. The most recent calculations by Ungemach et al.[95] yield barrier heights of 3.93 and 3.99 kcal/mole with the largest basis sets used. By considering the changes in barrier height with basis set these authors predict a barrier height of 3.35 kcal/mole and judge it improbable this value is more than 1 kcal/mole higher than the exact value. The evolution of these surfaces has been described by Bender and Schaefer.[98]

Calculations of the barrier for the exchange reaction $H + FH \rightarrow HF + H$ have been made by Bender, Garrison, and Schaefer,[91] by Botschwina and Meyer[96] and by Wadt and Winter.[97] These all yield barrier heights of 40 to 50 kcal/mole for this reaction. As noted by Bender et al.[91] this barrier in the H–F–H configuration provides for semiempirical potential energy surfaces a critical test that is failed by most of the LEPS surfaces.

In treatments of the dynamics of the $F + H_2$ reactions the LEP (and LEPS) semiempirical method has provided simple, convenient formulas with a questionable basis in theory. Coolidge and James[99] have discussed the weaknesses in detail. Truhlar and Wyatt[100] have recently reviewed the LEP method as applied to the $H + H_2$ system. Whether the LEP and LEPS formulas are semiempirical or empirical is of little importance in regard to the $F + H_2$ system at present. The LEP and LEPS surfaces listed in Table VII have been adjusted to agree in one way or another with either experimental evidence or the surfaces generated by other theoretical means.

All the ground state surfaces of Table VII are qualitatively similar for collinear F–H–H. The BOPS surface is shown in Fig. 5. The barrier to reaction is located well out in the entrance valley which intersects the side of the much deeper exit valley. The barrier is higher for configurations deviating from collinear. The surface characteristics are such that the dynamics of reaction yield vibrationally excited HF for reaction through the favored collinear configuration.

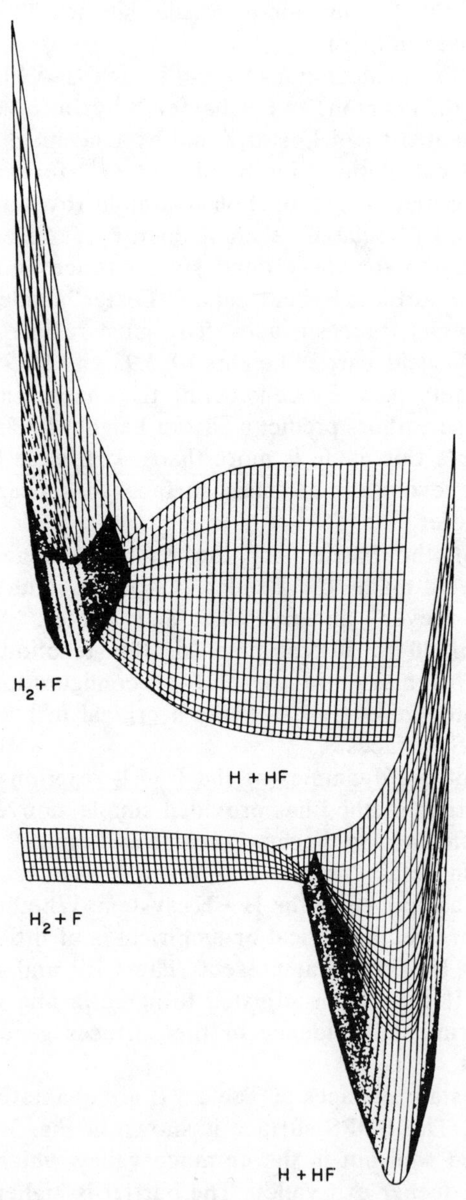

Fig. 5. The BOPS (Bender, O'Neill, Pearson, and Schaefer[86]) potential energy surface for collinear $F + H_2$.

The surface used most often for dynamics calculations is the M5 surface of Muckerman[82] which was obtained as the LEPS surface giving the best agreement with experiment for quasiclassical trajectory calculations. A second such surface is Polanyi and Schreiber's[7,89] SE-1 which gives a satisfactory representation of product energy distributions for quasiclassical trajectories. A third surface to receive attention in dynamics studies is the BOPS *ab initio* surface which has been fit by extended-LEPS and other representations.

There is general agreement between the Muckerman M5, Polanyi and Schreiber SE-1, and the *ab initio* BOPS surfaces in regard to the location of the saddle point in the barrier to reaction. However, the M5, SE-1, and BOPS-fit LEPS surfaces have an attractive well of depth ~0.4 kcal/mole in the entrance valley which does not appear in the *ab initio* calculation. Also, the barrier for the exchange reaction $H + FH \rightarrow HF + H$ is much lower for the M5 (1.2 kcal/mole) and SE-1 (3.5 kcal/mole) than is indicated by the *ab initio* calculations of Bender, Garrison, and Schaefer[91] (49.0 and 67.8 kcal/mole).

The magnitude of the problem faced in *ab initio* calculations is shown by comparison of the total electronic energy of the FHH system (~ 100 hartrees or 62,700 kcal/mole) with the barrier height of ~1 kcal/mole. A total accuracy of 1 part in 62,700 seems completely out of the question but accuracies of 1 kcal/mole in the difference between similar configurations seem entirely possible. Here, one depends on a cancellation of similar errors or omissions for the same basis set. The exothermicity of reaction, 31.3 kcal/mole, predicted by Ungemach et al.[95] from their calculations for $F + H_2$ and $FH + H$ is in agreement with the experimental value, 31.5 ± 0.5 kcal/mole. This gives a degree of confidence in the energies of intermediate configurations such as the saddle point relative to the energy of reactants. Of course, the passing of the exothermicity test is necessary but not sufficient to ensure an accurate potential energy surface. The ultimate test of a given *ab initio* calculation is a better *ab initio* calculation with a larger basis set.

B. Classical Trajectory Calculations

The molecular dynamics of the FHH system has been subjected to a large number of studies using classical mechanics. These are listed in Table VIII. The first studies (1970–1971) were undertaken to determine whether classical dynamics with assumed potential energy surfaces could reproduce the major characteristics observed experimentally for reaction of F with H_2. The results were found in substantial agreement with experiment.

Since then additional studies have been undertaken to investigate the

TABLE VIII.
Classical Trajectory Studies of the FHH System

Authors	Surface	Notes
Anderson[79] (1970)	JA	1D; $H+HF \rightarrow H_2+F$
Jaffe and Anderson[101] (1971)	JA	3D; $F+H_2$; $F+D_2$
Muckerman[80] (1971)	M1	3D; $F+H_2$; $F+HD$; $F+D_2$
Muckerman[81] (1972)	M1–M4	3D; $F+H_2$; $F+HD$
Muckerman[102] (1972)	M3	3D; $F+HD$; hot atom
Wilkins[83] (1972)	W	3D; $F+H_2$
Thompson[84] (1972)	T	3D; HF relaxation with H
Blais and Truhlar[87] (1973)	BT	3D; $F+D_2$
Ding et al.[89] (1973)	SE1	3D; $F+D_2$
Jaffe et al.[103] (1973)	JA	3D; $F+H_2$ $HF+H$; combined phase-space trajectory method
Wilkins[104] (1973)	W	3D; $F+HD$
Wilkins[105] (1973)	W	3D; $H+HF \rightarrow H_2+F$
Schatz et al.[106] (1973)	M5	1D; $F+H_2$
Polanyi and Schreiber[90] (1974)	SE1 BOPS–fit	3D; $F+H_2$
Whitlock and Muckerman[107] (1974)	M5	1D; $F+D_2$
Bowman et al.[108] (1974)	M5	1D; $F+H_2$, $F+D_2$, $H+HF$, $D+DF$
Wilkins[109] (1974)	W	3D; $F+D_2$
Wilkins[110] (1975)	W	3D; $H+HF$, etc.
Schatz et al.[111] (1975)	M5	1D; $F+H_2$
Schatz et al.[112] (1975)	M5	1D; $F+D_2$
Feng et al.[94] (1976)	FR	3D; $F+H_2$, $F+D_2$, hot atom
Polanyi and Schreiber[7] (1977)	SE2–SE5	3D; $F+H_2$

relations between: experimental observations, classical predictions, semiclassical predictions, full quantum predictions, details of potential energy surfaces, and isotope substitution. These relations have been examined for collinear and three-dimensional systems. Not all comparisons furnish information that is decisive. For example, a classical trajectory study with a questionable potential energy surface may disagree with certain experimentally measured details. One is then unsure whether classical mechanics, the potential energy surface, or both are at fault. If agreement is obtained there may be compensating errors. However, a comparison of classical and quantum calculations for the same potential energy surface may reveal difficulties inherent in classical mechanics.

The Muckerman[82] M5, Polanyi and Schreiber[7,89] SE-1, and fits to the BOPS[86] collinear surface have been used extensively for collinear classical trajectory studies, especially for comparison with full-quantum and semiclassical treatments. Of these surfaces only the surface SE-1 has been used for three-dimensional classical trajectories.

Polanyi and Schreiber[7,90] have described in detail their results for the reaction of F with H_2 in three dimensions on surface SE-1. The computed thermal rate constant at 300°K is 0.06×10^{13} cm^3/mole-sec which may be compared with the value 1.5×10^{13} cm^3/mole-sec from experiment. The computed activation energy is 1.937 kcal/mole which may be compared with the estimate of 1.7 kcal/mole from experiment. The computed distributions of energy among reaction products are compared in Table IX with those measured by Polanyi and Woodall.[54] Figure 6 shows the calculated "triangle plot" for comparison with the experimentally determined plot of Fig. 2. The agreement may be seen to be extremely good. However, the calculated vibrational distribution is too broad and the calculated rotational distribution too narrow.

The angular distribution of HF products found by Polanyi and Schreiber[7] for SE-1 is similar in character to that observed in crossed-beam experiments by Schafer et al.[71,72] for $F+D_2$ in crossed-beam experiments. The HF is scattered primarily backwards at low translational energies and shifts forward with increasing translational energy of reactants.

TABLE IX.
Comparison of Quasiclassical Trajectory Results for Surface SE-1 with Experimental Measurements[a]

	Computed[b]	Experimental[c]
$\langle f'_v \rangle$[d]	0.665	0.66
$\langle f'_R \rangle$	0.095	0.08
$\langle f'_T \rangle$	0.239	0.26
$k(v')/k_{max}$		
$v'=1$	0.22	0.31
$v'=2$	[1.00]	[1.00]
$v'=3$	0.26	0.47
$\langle J' \rangle$		
$v'=1$	9.98	9.35
$v'=2$	6.51	7.06
$v'=3$	3.39	2.40

[a] Reaction $F+H_2 \rightarrow HF(v', J')+H$, 300°K.
[b] Polanyi and Schreiber.[7,90]
[c] Polanyi and Woodall.[54]
[d] Average fraction of energy to vibration, rotation, translation.

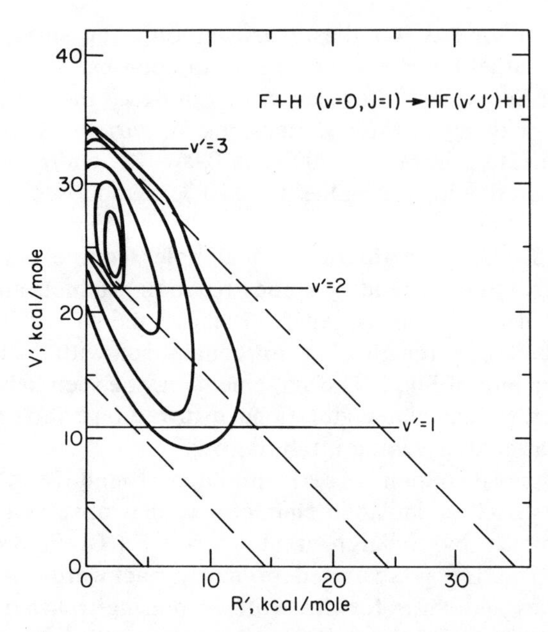

Fig. 6. Triangular plot of the product energy distribution for the reaction $F + H_2(v = 0, J = 1)$ at 300° K as determined by Polanyi and Schreiber[7] from classical trajectories with surface SE-1.

Additional details examined by Polanyi and Schreiber[7] include the effects of enhanced reagent translation, vibration, and rotation on the energy distribution among products. Enhanced H_2 vibration leads to enhanced HF vibration. Enhanced reagent translation produces enhanced translation and rotation of products. Enhanced H_2 rotation produces enhanced HF rotation. The decrease in vibrational excitation of HF with increasing rotation from $J = 0$ to $J = 1$ for H_2, observed experimentally by Coombe and Pimentel,[69] is not observed.

It is clear that by optimization of the surface characteristics one can reproduce with classical mechanics most of the reaction characteristics observed experimentally. It is encouraging that the surface SE-1 producing agreement in this way is similar to the BOPS *ab initio* surface in its collinear region. However, one may not conclude from the observed agreement that either classical mechanics or the surface SE-1 is adequate except in a qualitative way.

Classical calculations with the next generation of *ab initio* surfaces or comparisons with the next generation of three-dimensional quantum calculations of dynamics using current surfaces will probably allow definite conclusions regarding the adequacy of classical trajectory calculations for quantitative predictions.

C. Quantum Treatments—Collinear

For three-particle collinear reaction systems it has been possible to make nearly exact full-quantum calculations of molecular dynamics. The general effects of varied characteristics of potential energy surfaces had been investigated in full-quantum collinear calculations by Diestler et al.[113,114] for a series of diagnostic surfaces. These showed behavior in general accord with classical trajectory results. For the $F + H_2$ system the first calculations were reported by Wu, Johnson, and Levine[115] and Schatz, Bowman, and Kupperman[106] who used close-coupling methods with LEPS surfaces. Subsequent calculations have been carried out to determine reaction characteristics for other $F + H_2$ surfaces and to investigate alternative calculation methods. Table X lists the several investigations.

The reaction probabilities determined by Schatz, Bowman, and Kupperman[106,108,111,112] for $F + H_2$ on surface M5 with a close coupling method have been reproduced reasonably well by Adams, Smith, and Hayes[116] with an integral equation method, by Light and Walker[119] with an *R*-matrix method, and by Connor, Jakubetz, and Manz[120] using the

TABLE X.
Quantum Treatments of the Reaction $F + H_2 \rightarrow HF + H$

Authors	Notes
Wu, Johnson, and Levine[115] (1973)	1D, close-coupling, several surfaces
Schatz, Bowman, and Kupperman[106] (1973)	1D, close-coupling, M5
Bowman, Schatz, and Kupperman[108] (1974)	1D, close-coupling, M5, also $F + D_2$
Adams, Smith and Hayes[116] (1974)	1D, integral equation, M5
Connor, Jakubetz, and Manz[117] (1975)	1D, state path sum, fit to BOPS
Schatz, Bowman, and Kupperman[111,112] (1975)	1D, close-coupling, M5, also $F + D_2$
Redmon and Wyatt[118] (1975)	3D, close-coupling, M5
Light and Walker[119] (1976)	1D, *R*-matrix approach, M5
Connor, Jakubetz, and Manz[120] (1978)	1D, state path sum, M1, M5, fit to BOPS
Shan, Choi, Poe, and Tang[121] (1978)	3D, distorted-wave approximation, M5

state-path-sum method. For surface M1 the results obtained by Wu et al.[215] are in good agreement with the more recent results of Connor et al.[120] with minor exceptions which have been discussed by Connor et al.[120] The several methods give essentially the same results for a given surface. There remain a few discrepancies which might be examined in checks of the numerical procedures used.

Figures 7–8 show the reaction probabilities for the collinear reaction $F + H_2(v = 0) \rightarrow FH(v') + H$ as a function of the initial relative translational energy of reactants as determined by Schatz et al.[111] Reaction occurs primarily to $v' = 2$ and $v' = 3$. For $v' = 0$ to 1 the translational threshold energy is well below the classical barrier height of 1.06 kcal/mole. Reaction probability peaks at translation energies of about 0.3 kcal/mole and decreases thereafter for these product states. For $v' = 3$ the behavior is remarkably different. The threshold is about 1.0 kcal/mole with reaction probability a maximum at about 2.4 kcal/mole.

A comparison of reaction probabilities for producing $v' = 2$ and $v' = 3$ as given by the quantum calculations and by quasiclassical (quantized reactants) calculations of Schatz et al. is given in Fig. 7. Large differences, expecially in the behavior near threshold, may be seen. Both sets of results have been treated to give properties of reaction in (collinear) thermal systems. The rate constants for the quantum results give Arrhenius activation energies ($T = 200$–$400°K$) of 0.411 kcal/mole for $v' = 2$ and 2.279 kcal/mole for $v' = 3$. The total rate constant does not

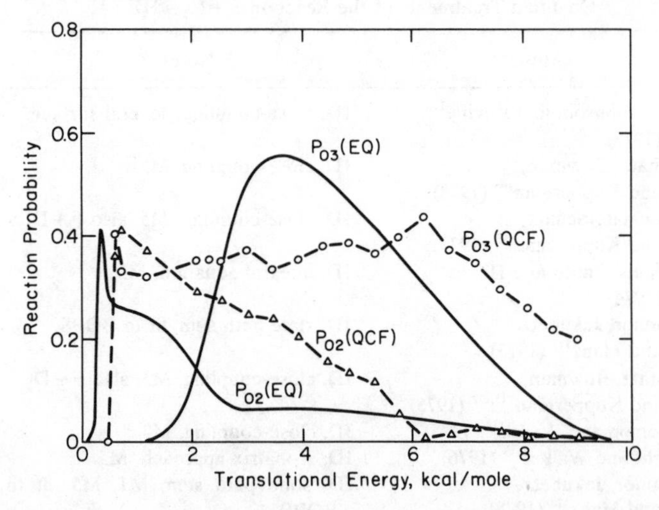

Fig. 7. Exact quantum and quasiclassical reaction probabilities for the collinear reaction $F + H_2(v = 0) \rightarrow HF(v') + H$ as determined by Schatz, Bowman, and Kupperman.[111] Reaction probability P_{02} for $v' = 2$, P_{03} for $v' = 3$. Exact quantum: EQ. Quasiclassical: QCF.

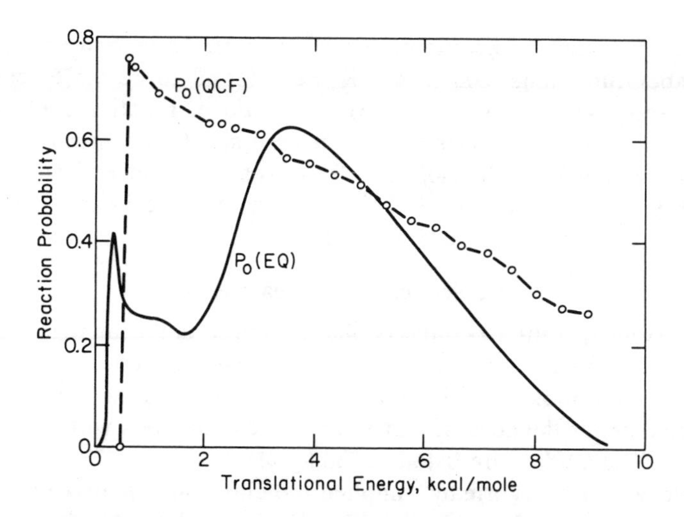

Fig. 8. Exact quantum and quasiclassical total reaction probabilities for the collinear reaction $F+H_2(v=0) \rightarrow HF(v')+H$ as determined by Schatz, Bowman, and Kupperman.[111] See Fig. 7.

show simple Arrhenius behavior. For the quasiclassical system the activation energies ($T=200$–400°K) are 0.791 kcal/mole for $v'=2$ and 0.853 kcal/mole for $v'=3$. The large differences in the energy dependences of reaction probabilities are reflected in the product distributions. The rate constant ratio $k(v'=3)/k(v'=2)$ increases from near zero at 200°K to 2.7 at 1200°K for the quantum system but varies from 0.9 at 200°K to about 1.2 at 1200°K for the quasiclassical system.

Effects of varying the potential energy surface have been revealed by the state-path-sum calculations of Connor, Jakubetz, and Manz.[120] Surfaces investigated were the Muckerman M1 and M5 and a surface based on the BOPS *ab initio* calculation. Despite the similarity of the surfaces the results are qualitatively and quantitatively different. For the BOPS surface reaction to HF ($v'=3$) peaks at a translational energy of about 1.4 kcal/mole—half that for surface M5. The peak near threshold for $v'=2$ is absent. Connor et al.[120] indicate that the diatomic molecule states, resonance scattering regions, and corner and saddle point regions need to be known very accurately for meaningful quantitative results to be obtained.

The isotopic variants $F+D_2$, $F+HD$, and $F+DH$ were also examined by Schatz et al.[112] in close-coupled calculations for surface M5. For $F+D_2 \rightarrow FD+D$ the dominant product states are $DF(v'=3)$ and $DF(v'=4)$ which show an energy dependence similar to that for $HF(v'=2)$ and $HF(v'=3)$. The threshold translational energies are about 0.3 kcal/mole for DF($v' = 3$) and 1.1 kcal/mole for DF($v' = 4$). Activation energies for the temperature range 200 to 400°K were found to be 0.676 kcal/mole ($v' = 3$) and 2.167 kcal/mole ($v' = 4$). The ratio of reaction probabilities $P(v' = 4)/P(v' = 3)$ increases from near zero at a translational energy of zero to about 4.3 at 2.8 kcal/mole. The crossed beam experiments for $F + D_2$ by Lee and co-workers[71,72] show an increase in this ratio with translational energy from 0.75 at 0.8 kcal/mole to 3.5 at 2.7 kcal/mole.

D. Semiclassical Treatments

With exact quantum results available for the collinear reaction of F with H_2 the use of semiclassical methods can add little to an understanding of the molecular dynamics of the system. However, much can be learned about the accuracy of semiclassical methods with $F + H_2$ as a test case. The studies are listed in Table XI.

Whitlock and Muckerman[107] applied the classical S-matrix or uniform semiclassical method developed by Miller[125] and by Marcus[126] to the collinear reaction of F with D_2 on surface M5. Schatz et al.[111–112] obtained similar results in an analogous calculation and also treated the reaction of F with H_2. In general, the reaction probabilities given by the semiclassical procedures are in good qualitative agreement with those of the exact quantum treatments. Differences appear in behavior near threshold energies as well as in an oscillatory behavior of the semiclassical probabilities at higher energies. The semiclassical results are significantly closer to the exact quantum results than are the quasiclassical results.

Extensive use of semiclassical methods has also been made for investigation of the effects of the multiple electronic states accessible to the $F + H_2$ system. These studies are described in the next section.

TABLE XI.
Semiclassical Treatment of the Reaction $F + H_2 \rightarrow HF + H$

Authors	Notes
Tully[122] (1974)	3D, approx. semiclassical, nonadiabatic
Whitlock and Muckerman[107] (1974)	1D, surface M5, S-matrix
Schatz et al.[111] (1975)	1D, $F + H_2$, surface M5, also reverse
Schatz et al.[112] (1975)	1D, $F + D_2$, surface M5, also reverse
Laing et al.[123] (1975)	1D, two-state semiclassical, nonadiabatic
Komornicki, George, and Morokuma[124] (1976)	1D, nonadiabatic, 3 surfaces

E. Nonadiabatic Behavior

The general problem of electronic degeneracy and possible nonadiabatic behavior for collisions of F with H_2 has been outlined by Truhlar[127] and by Muckerman and Newton.[128] As indicated in Section II.A there are two accessible states of F atoms near room temperature: $^2P_{1/2}$ and $^2P_{3/2}$ separated by an energy $\Delta = 1.16$ kcal/mole. The correlation diagram is shown in Fig. 4. For electronically adiabatic collisions the fraction of $F+H_2$ encounters which take place on the ground-state $1^2A'$ surface is between $\frac{1}{3}$ and $\frac{1}{2}$ depending on temperature according to $[\exp(-\Delta/kT)+2]^{-1}$. This suggests adjustment of rate coefficients calculated for the ground state surface by a factor $\frac{1}{3}$ to $\frac{1}{2}$ before comparison with experimental rate measurements.

For slightly separated states the orbital and rotational angular momentum of the nuclei as well as spin–orbit interaction may lead to important coupling between the states. In this case reaction does not occur on a simple ground-state potential energy surface. The energy separation of states determines the importance of the effects.

Both SCF[93] and GRHF-CI[92] calculations show a rapidly increasing potential for the $2^2A'$ surface as the F atom approaches H_2 within an F–H distance of 2 Å. The diatomics-in-molecules surface T2 by Tully[88] shows similar behavior. The GRHF-CI calculations of Jaffe, Morokuma, and George[92] indicate a well in the entrance valley for the $2^2A'$ ($^2\pi$ collinear) surface having a depth of 2.5 kcal. This well is absent in the DIM surfaces.

Using an approximate semiclassical method with the DIM surface T2[88] Tully[122] estimated the reaction rate of $F(^2P_{1/2})$ with H_2 to be an order of magnitude less than that of $F(^2P_{3/2})$. Zimmerman and George[129] and Laing, George, Zimmerman, and Lin[123] considered transitions between states in the nonreactive collinear DIM system by full-quantum and semiclassical methods. More recently, Komornicki, George, and Morokuma[124] investigated transitions between the $1^2A'$ and $2^2A'$ surfaces in classical trajectory calculations in three dimensions with transitions allowed between the two surfaces. The $1^2A'$ surface was taken as Muckerman's M5 and the $2^2A'$ surface was a modification of the valence-bond surface of Blais and Truhlar.[87] For collisions of $F(^2P_{1/2})$ starting on the $2^2A'$ surface switching to the $1^2A'$ surface was observed to result in a significant reaction cross section for production of HF + H. For collisions of $F(^2P_{3/2})$ beginning on the $1^2A'$ surface the reaction cross section was significantly lowered at translational energies above 2 to 3 kcal/mole when switching to the $2^2A'$ surface was allowed.

All of these calculations suggest important effects arise from the interactions of the multiple surfaces. The reaction does not take place on a single surface within the spirit of the Born–Oppenheimer approximation. The character of the effective "ground-state" surface may be significantly altered by the presence of the upper electronic states.

F. Quantum Treatments—Three Dimensional

Two preliminary three-dimensional quantum calculations have been reported. Redmon and Wyatt[118] used a coupled-channel method with surface M5. Estimates of the product vibrational distribution for $F + H_2$ $(v = 0,\ J = 0) \rightarrow HF(v') + H$ were obtained with use of a rotationally near-adiabatic (total angular momentum of zero) but vibrationally nonadiabatic model. The calculated reaction probabilities for producing $HF(v')$ were found in qualitative agreement with those given by exact quantum treatments of the collinear reaction on the same surface.

Shan, Choi, Poe, and Tang[121] used a method based on the distorted-wave Born approximation to obtain an approximate distribution of product states for $F + H_2(v = 0, J = 0) \rightarrow HF(v', J') + H$ on the M5 surface. For a translational energy of 2.3 kcal/mole the product ratio $HF(v' = 1):HF(v' = 2):HF(v' = 3)$ was 3:17:80. The ratio of $HF(v' = 3)$ to $HF(v' = 2)$ is much larger than that indicated by the calculations of Redmon and Wyatt[118] at the same energy and that given by chemiluminescence experiments.

Both sets of calculations may be described as preliminary. Both sets of authors call for more accurate calculations. Since only a short time has passed since the much simpler system $H + H_2 \rightarrow H_2 + H$ has yielded to full-quantum three-dimensional treatment, one should not expect immediate success for $F + H_2$. Nevertheless, one may expect increasingly accurate calculations to appear over the next few years.

G. Discussion

Comparisons of exact quantum and classical (quantized reactants) treatments for the collinear reaction on a single potential energy surface reveal large quantum effects. Preliminary quantum results for three dimensions show these effects remain for the three-dimensional case. It is clear that the classical treatments can give at best a qualitative prediction of the true behavior of the system. Although this is perhaps disappointing, it should be recognized that a correct qualitative prediction is extremely valuable. For many practical applications a qualitative prediction is all that is required.

Since the reaction rates and product distributions predicted by classical dynamics are significantly different from those of exact quantum treatments, a potential energy surface optimized to reproduce experimental observations in a classical trajectory study cannot be the exact surface. However, without accurate and economical methods available for optimization with a three-dimensional quantum treatment of the dynamics a surface optimized for classical mechanics may be the best that is possible at present.

As long as we may question the accuracy of the potential energy surface and the accuracy of the method of treating the dynamics on the surface as well as the effects of nearby electronic states, disagreement with experiment cannot reveal exactly which part of the whole is in error. Thus, advances in two of the three areas—three-dimensional potential energy surfaces, three-dimensional quantum dynamics, three-dimensional nonadiabatic behavior—are required before detailed comparison with experimental measurements can be used to judge success in the remaining area.

Although it has been shown by Connor, Jakubetz, and Manz[120] that the full-quantum molecular dynamics of the collinear system is highly sensitive to certain critical areas of the potential energy surface it is not at all certain that such a sensitivity extends to the three-dimensional system. We may hope that it does not and that surfaces more accurate than 0.1 kcal/mole are not required for quantitative reproduction of results of experiments in the three-dimensional world.

We can look forward to increasingly accurate potential energy surfaces for the collinear system in the next few years. At the same time we can expect crossed-beam experiments to yield the variation of reaction cross-section with collision energy. Combined, these advances will undoubtedly resolve questions of barrier height, threshold energy, and activation energy. We may also look forward to improved chemiluminescence experiments with narrower velocity distributions in the colliding streams. These will give product state distributions more readily compared with dynamics calculations. The problem of the quantum molecular dynamics of the three particles in three dimensions with nonadiabatic behavior will ultimately yield. The goal is in sight and is probably within our grasp.

References

1. B. Liu, *J. Chem. Phys.* **58**, 1925 (1973).
2. Review: D. G. Truhlar and R. E. Wyatt, *Adv. Chem. Phys.*, **36**, 141 (1977).
3. D. N. Mitchell and D. J. LeRoy, *J. Chem. Phys.*, **58**, 3449 (1973).
4. Review: D. G. Truhlar and R. E. Wyatt, *Ann. Rev. Phys. Chem.*, **27**, 1 (1976).
5. M. Karplus, R. N. Porter, and R. D. Sharma, *J. Chem. Phys.*, **43**, 3259 (1965).
6. G. C. Schatz and A. Kupperman, *J. Chem. Phys.*, **65**, 4668 (1976).
7. J. C. Polanyi and J. L. Schreiber, *Faraday Disc. Chem. Soc.*, **62**, 267 (1977).

8. N. Cohen and J. F. Bott, "A Review of Rate Coefficients in the H_2–F_2 Chemical Laser System," SAMSO-TR-76-82, Aerospace Corporation, El Segundo, California (1976).
9. W. E. Jones and E. G. Skolnik, *Chem. Rev.*, **76**, 563 (1976).
10. R. Foon and M. Kaufman, *Prog. Reaction Kinet.* **8**, 81 (1975).
11. H. Moissan and J. Dewar, *Compt. Rend.*, **124**, 1202 (1894); **136**, 641, 785 (1903).
12. H. N. Wartenburg and J. Taylor, *Nachr. Ges. Wiss. Göttingen. Geschäft. Mitt. Math.-physik Klasse*, **1**, 119 (1930).
13. H. Eyring, *J. Am. Chem. Soc.*, **53**, 2537 (1931).
14. H. Eyring and L. S. Kassel, *J. Am. Chem. Soc.*, **55**, 2796 (1933).
15. M. Bodenstein, H. Jockusch, and S.-H. Chong, *Anorg. Allgem. Chem.*, **231**, 24 (1937).
16. H. J. Schumacher, *Chemische Gasreaktionen*, Theodor Steinkopff, Leipzig, 1938, p. 443.
17. K. H. Homann, W. C. Solomon, J. Warnatz, H. G. Wagner, and C. Zetzsch, *Ber. Bunsenges. Phys. Chem.*, **74**, 585 (1970).
18. A. F. Dodonov, G. K. Lavrovskaya, I. I. Morozov, and V. L. Talroze, *Dokl. Akad. Nauk. SSSR*, **198**, 622 (1971).
19. K. L. Kompa and J. Wanner, *Chem. Phys. Lett.*, **12**, 560 (1972).
20. M. A. A. Clyne, D. J. McKenney, and R. F. Walker, *Can. J. Chem.*, **51**, 3596 (1973).
21. V. I. Igoshin, L. V. Kulakov, and A. I. Nikitin, *Kratk. Soobshich. Fiz.*, **1**, 3 (1973); *Sov. J. Quant. Electron.*, **3**, 306 (1974).
22. R. K. Pearson, J. O. Cowles, G. L. Hermann, D. W. Gregg, and J. R. Creighton, *IEEE J. Quant. Electron.*, **9**, 879 (1973).
23. J. F. Hon, A. Axworthy, and G. Schneider, *Interim Report*, R-9297, Rocketdyne Division, Rockwell International (1973).
24. T. M. Lam, M. Peyron, and P. Puget, *J. Chim. Phys.*, **71**, 377 (1974).
25. I. B. Goldberg and G. R. Schneider, *J. Chem. Phys.*, **65**, 147 (1976).
26. C. R. Quick and C. Wittig, *Chem. Phys. Lett.*, **48**, 420 (1977).
27. J. M. Preses, R. E. Weston, and G. W. Flynn, *Chem. Phys. Lett.*, **48**, 425 (1977).
28. M. A. A. Clyne and W. S. Nip, *Int. J. Chem. Kin.*, **10**, 367 (1978).
29. M. J. Coggiola, P. A. Schulz, Y. T. Lee, and Y. R. Shen, *Phys. Rev. Lett.*, **38**, 17 (1977).
30. T. A. Jacobs, R. R. Geidt, and N. Cohen, *J. Chem. Phys.*, **43**, 3688 (1965).
31. J. A. Blauer, *J. Phys. Chem.*, **72**, 79 (1968).
32. K. H. Homann and D. I. MacLean, *Ber. Bunsenges. Phys. Chem.*, **75**, 945 (1971).
33. D. I. MacLean, *Air Force Report*, AD-760770 (1973).
34. J. B. Levy and B. K. W. Copeland, *J. Phys. Chem.*, **67**, 2156 (1963); **69**, 408 (1965); **72**, 3168 (1968).
35. S. W. Rabideau, H. G. Hecht, and W. B. Lewis, *J. Mag. Resonance*, **6**, 384 (1972).
36. P. D. Mercer and H. O. Pritchard, *J. Phys. Chem.*, **63**, 1468 (1959).
37. R. Foon and G. P. Reid, *Trans. Faraday Soc.*, **67**, 3513 (1971).
38. R. L. Williams and F. S. Rowland, *J. Phys. Chem.*, **75**, 2709 (1971).
39. R. L. Williams and F. S. Rowland, *J. Phys. Chem.*, **77**, 301 (1973).
40. N. Jonathan, C. M. Melliar-Smith, S. Okuda, D. H. Slater, and D. Timlin, *Mol. Phys.*, **22**, 561 (1971).
41. R. G. Manning, E. R. Grant, J. C. Merrill, N. J. Parks, and J. W. Root, *Int. J. Chem. Kin.*, **7**, 39 (1975).
42. G. A. Kapralova, A. L. Margolin, and A. M. Chaikin, *Kinet. Katal.*, **11**, 810 (1970).
43. H. G. Wagner, J. Warnatz, and C. Zetzsch, *Ann. Assoc. Quim. Argent.* **59**, 169 (1971).
44. G. C. Fettis and J. H. Knox, *Prog. Reaction Kin.*, **2**, 1 (1964). See also G. C. Fettis, J. H. Knox, and A. F. Trotman-Dickenson, *Can. J. Chem.*, **38**, 1643 (1960).

45. R. Foon, G. P. Reid, and K. B. Tait, *Trans. Faraday Soc.*, **68**, 1131 (1972).
46. A. Persky, *J. Chem. Phys.*, **59**, 3612 (1973).
47. M. J. Berry, *J. Chem. Phys.*, **59**, 6229 (1973).
48. E. R. Grant and J. W. Root, *Chem. Phys. Lett.*, **27**, 484 (1974).
49. E. R. Grant and J. W. Root, *J. Chem. Phys.*, **63**, 2970 (1975).
50. F. S. Klein and A. Persky, *J. Chem. Phys.*, **61**, 2472 (1974).
51. A. Persky, *J. Chem. Phys.*, **59**, 5578 (1973).
52. J. C. Polanyi and D. C. Tardy, *J. Chem. Phys.*, **51**, 5117 (1969).
53. K. G. Anlauf, P. E. Charters, D. S. Horne, R. GL MacDonald, D. W. Maylotte, J. C. Polanyi, W. J. Skrlac, D. C. Tardy, and K. B. Woodall, *J. Chem. Phys.*, **53**, 4091 (1970).
54. J. C. Polanyi and K. B. Woodall, *J. Chem. Phys.*, **57**, 1574 (1972).
55. D. S. Perry and J. C. Polanyi, *Chem. Phys.*, **12**, 37 (1976).
56. D. S. Perry and J. C. Polanyi, *Chem. Phys.*, **12**, 419 (1976).
57. D. J. Douglas and J. C. Polanyi, *Chem. Phys.*, **16**, 1 (1976).
58. H. W. Chang and D. W. Setser, *J. Chem. Phys.*, **58**, 2298 (1973).
59. N. Jonathan, C. M. Melliar-Smith, and D. W. Slater, *Mol. Phys.*, **20**, 93 (1971).
60. J. V. V. Kasper and G. C. Pimentel, *Phys. Rev. Lett.*, **14**, 352 (1965).
61. T. F. Deutsch, *Appl. Phys. Lett.*, **10**, 234 (1967).
62. T. F. Deutsch, *Appl. Phys. Lett.*, **11**, 18 (1967).
63. K. L. Kompa and G. C. Pimentel, *J. Chem. Phys.*, **47**, 857 (1967).
64. K. L. Kompa and G. C. Pimentel, *Ber. Bunsenges. Phys. Chem.*, **72**, 1067 (1968).
65. K. L. Kompa, J. H. Parker, and G. C. Pimentel, *J. Chem. Phys.*, **49**, 4257 (1968).
66. J. H. Parker and G. C. Pimentel, *J. Chem. Phys.*, **51**, 91 (1969).
67. D. J. Spencer, T. A. Jacobs, H. Mirels, and R. W. F. Gross, *Int. J. Chem. Kin.*, **1**, 493 (1969).
68. R. D. Coombe and G. C. Pimentel, *J. Chem. Phys.*, **59**, 251 (1973).
69. R. D. Coombe and G. C. Pimentel, *J. Chem. Phys.*, **59**, 1535 (1973).
70. O. D. Krogh, D. K. Stone, and G. C. Pimentel, *J. Chem. Phys.*, **66**, 368 (1977).
71. T. D. Schaefer, P. E. Siska, J. M. Parson, F. P. Tully, Y. C. Wong, and Y. T. Lee, *J. Chem. Phys.*, **53**, 3385 (1970).
72. Y. T. Lee, in *The Physics of Electronic and Atomic Collisions* (Eds. T. R. Govers and F. J. DeHeer), North-Holland, Amsterdam, 1971, p. 357.
73. J. F. Bott and R. F. Heidner, *J. Chem. Phys.*, **68**, 1708 (1978).
74. F. E. Bartoszek, D. M. Manos, and J. C. Polanyi, *J. Chem. Phys.*, **69**, 933 (1978).
75. J. O. Hirschfelder, *J. Chem. Phys.*, **9**, 645 (1941).
76. S. Sato, *J. Chem. Phys.*, **23**, 2465 (1955).
77. H. S. Johnston and C. Parr, *J. Am. Chem. Soc.*, **85**, 2544 (1963).
78. F. O. Ellison and J. C. Patel, *J. Am. Chem. Soc.*, **86**, 2615 (1964).
79. J. B. Anderson, *J. Chem. Phys.*, **52**, 3849 (1970).
80. J. T. Muckerman, *J. Chem. Phys.*, **54**, 1155 (1971).
81. J. T. Muckerman, *J. Chem. Phys.*, **56**, 2997 (1972).
82. J. T. Muckerman, unpublished (1973).
83. R. L. Wilkins, *J. Chem. Phys.*, **57**, 912 (1972).
84. D. L. Thompson, *J. Chem. Phys.*, **57**, 4170 (1972).
85. M. D. Newton, Brookhaven National Laboratory, unpublished work (1973).
86. C. F. Bender, S. V. O'Neill, P. K. Pearson, and H. F. Schaefer, *Science*, **176**, 1412 (1972). See also C. F. Bender, P. K. Pearson, S. V. O'Neill, and H. F. Schaefer, *J. Chem. Phys.*, **56**, 4626 (1972).
87. N. C. Blais and D. G. Truhlar, *J. Chem. Phys.*, **58**, 1090 (1973).

88. J. C. Tully, *J. Chem. Phys.*, **58**, 1396 (1973).
89. A. M. G. Ding, L. J. Kirsch, D. S. Perry, J. C. Polanyi, and J. L. Schreiber, *Farad. Disc. Chem. Soc.*, **55**, 252 (1973).
90. J. C. Polanyi and J. L. Schreiber, *Chem. Phys. Lett.*, **29**, 319 (1974).
91. C. F. Bender, B. J. Garrison, and H. F. Schaefer III. *J. Chem. Phys.*, **62**, 1188 (1975).
92. R. L. Jaffe, K. Morokuma, and T. F. George, *J. Chem. Phys.*, **63**, 3417 (1975).
93. F. Rebentrost and W. A. Lester, Jr., *J. Chem. Phys.*, **63**, 3737 (1975).
94. D. Feng, E. R. Grant, and J. W. Root, *J. Chem. Phys.*, **64**, 3450 (1976).
95. S. R. Ungemach, H. F. Schaefer, and B. Liu, *Disc. Faraday Soc.*, **62**, 330 (1977).
96. P. Botschwina and W. Meyer, *Chem. Phys.*, **20**, 48 (1977).
97. W. R. Wadt and N. W. Winter, *J. Chem. Phys.*, **67**, 3068 (1977).
98. C. F. Bender and H. F. Schaefer, in *Fluorine-Containing Free Radicals*, ACS Symposium Series, Vol. 66 (J. W. Root, ed.), p. 283 (1978).
99. A. S. Coolidge and H. M. James, *J. Chem. Phys.*, **2**, 811 (1934).
100. D. G. Truhlar and R. E. Wyatt, *Advan. Chem. Phys.*, **36**, 141 (1977).
101. R. L. Jaffe and J. B. Anderson, *J. Chem. Phys.*, **54**, 2224 (1971).
102. J. T. Muckerman, *J. Chem. Phys.*, **57**, 3388 (1972).
103. R. L. Jaffe, J. M. Henry, and J. B. Anderson, *J. Chem. Phys.*, **59**, 1128 (1973).
104. R. L. Wilkins, *J. Phys. Chem.*, **77**, 3081 (1973).
105. R. L. Wilkins, *J. Chem. Phys.*, **58**, 3038 (1973).
106. G. C. Schatz, J. M. Bowman, and A. Kupperman, *J. Chem. Phys.*, **58**, 4023 (1973).
107. P. A. Whitlock and J. T. Muckerman, *J. Chem. Phys.*, **61**, 4618 (1974).
108. J. M. Bowman, G. C. Schatz, and A. Kupperman, *Chem. Phys. Lett.*, **24**, 378 (1974).
109. R. L. Wilkins, *Mol. Phys.*, **28**, 21 (1974).
110. R. L. Wilkins, *Mol. Phys.*, **29**, 555 (1975).
111. G. C. Schatz, J. M. Bowman, and A. Kupperman, *J. Chem. Phys.*, **63**, 674 (1975).
112. G. C. Schatz, J. M. Bowman, and A. Kupperman, *J. Chem. Phys.*, **63**, 685 (1975).
113. D. J. Diestler, *J. Chem. Phys.*, **56**, 2092 (1972).
114. K. P. Fong and D. J. Diestler, *J. Chem. Phys.*, **56**, 3200 (1972).
115. S. Wu, B. R. Johnson, and R. D. Levine, *Mol. Phys.*, **25**, 839 (1973).
116. J. T. Adams, R. L. Smith, and E. F. Hayes, *J. Chem. Phys.*, **61**, 2193 (1974).
117. J. N. L. Connor, W. Jakubetz, and J. Manz, *Molec. Phys.*, **29**, 347 (1975).
118. M. J. Redmon and R. E. Wyatt, *Int. J. Quantum Chem. Symposium*, **9**, 403 (1975).
119. J. C. Light and R. B. Walker, *J. Chem. Phys.*, **65**, 4272 (1976).
120. J. N. L. Connor, W. Jakubetz, and J. Manz, *Molec. Phys.*, **35**, 1301 (1978).
121. Y. Shan, B. H. Choi, R. T. Poe, and K. T. Tang, *Chem. Phys. Lett.*, **57**, 379 (1978).
122. J. C. Tully, *J. Chem. Phys.*, **60**, 3042 (1974).
123. J. R. Laing, T. F. George, I. H. Zimmerman, and Y. W. Lin, *J. Chem. Phys.*, **63**, 842 (1975).
124. A. Komornicki, T. F. George, and K. Morokuma, *J. Chem. Phys.*, **65**, 48, 4312 (1976).
125. W. H. Miller, *J. Chem. Phys.*, **53**, 1949, 3578 (1970).
126. R. A. Marcus, *J. Chem. Phys.*, **54**, 3965 (1971).
127. D. G. Truhlar, *J. Chem. Phys.*, **56**, 3189 (1972).
128. J. T. Muckerman and M. D. Newton, *J. Chem. Phys.*, **56**, 3191 (1972).
129. I. H. Zimmerman and T. F. George, *Chem. Phys.*, **7**, 323 (1975).

NONRADIATIVE PROCESSES IN SMALL MOLECULES IN LOW-TEMPERATURE SOLIDS

V. E. BONDYBEY

and

L. E. BRUS

Bell Laboratories
Murray Hill, New Jersey 07974

CONTENTS

I. INTRODUCTION

Studies of molecular guests in low-temperature condensed phases began in the early 1920s with the experiments of Vegard, who observed emission from a variety of low temperature molecular and rare gas solids.[1-3] A large increase of interest in the field came with the realization that low-temperature solids can be used to isolate and study spectroscopically free radicals and other transient molecules. This discovery occured independently and almost simultaneously in the laboratories of Porter and Pimentel. While Porter[4,5] observed the optical spectra of several free radicals in organic solvent, Pimentel's group pioneered the use of rare gas hosts in infrared studies[6-8] and also coined the phrase "matrix isolation." Today substantial progress has occured in the field of transient spectroscopy via the widespread use of matrix isolation techniques.[9-11]

In the last few years attention has turned away from pure spectroscopic structural studies and toward understanding of the internal dynamics of isolated molecules. Against the prevailing intuitive feeling that relaxation processes in the solid should be very efficient, there was an increasing body of evidence suggesting remarkably slow vibrational relaxation. Actually Vegard[12] recognized in the early 1930s that vibrationally unrelaxed emission from a metastable excited state of N_2 (labeled today $A^3\Sigma_u^+$) occurs in solid Ar host. This result implied that the phosphorescence radiative rate of $A^3\Sigma_u^+$ is of the same order of magnitude, or faster than, the vibrational relaxation rate. Using an elementary rotating blade phosphoroscope, he determined that the phosphorescence delays on $a \approx 1$ sec time scale! Additional observations of slow relaxation of this type occured in the laboratories of Broida[13-15], Robinson,[16-18] and other investigators. In 1965 Sun and Rice[19] also predicted on theoretical grounds that vibrational relaxation should be slow.

A great surge of interest in this field has been brought about by the availability of intense, monochromatic laser sources.[20,21] Tunable lasers, in the first place, permit selective excitation of discreet vibrational levels in the molecular guests and provide thus an invaluable tool in relaxation studies. In the second place, the feasibility of selective excitation also raises the possibility of inducing a selective chemical reaction. Clearly one can envision exciting selectively and reacting one component of a mixture in the collision free low-temperature environment. Even more ambitiously one might contemplate exciting and selectively reacting or isomerizing a functional group in a complex molecule. The possibilities of such "molecular engineering" or "molecular surgery" have been widely discussed. Obviously in this context the rates and pathways of energy flow within the molecule and from the molecule into the environment must be known.

A similarity exists between the relaxation in a large molecule and relaxation of a small molecule in the low-temperature solid,[22–24] and this analogy has, in fact, been used in some of the early theories of the guest molecule relaxation. The dense vibrational manifold accepts the energy during relaxation between the electronic states of a large molecule, whereas multiquantum phonon states of the host lattice accept the energy in the relaxation process between the internal vibronic states of the small guest.

Conversely, a small molecule in a low-temperature solid is a simple model system for studies of various types of radiationless transitions, and the results obtained provide often insight into relaxation processes in other environments. Thus, to give an example, matrix studies have shown that vibrational relaxation in excited electronic states of diatomic molecules often occurs via a cascading process involving nearby electronic states.[25,26] Similar relaxation mechanisms were recently established for the collisional relaxation of several diatomic species by gas phase rare gas atoms.[27,28]

A similar analogy seems to exist for polyatomic molecules, where intramolecular mode-to-mode vibrational energy transfer processes seem to depend strongly both in gas and condensed phases on the anharmonic intermode coupling elements. In cases where such coupling is enhanced, for instance by Fermi resonance, efficient transfer takes place.[29–31]

Unimolecular nonradiative relaxation cannot, because of energy conservation considerations, take place in an isolated small molecule[32] (excluding predissociative processes for levels immersed in dissociative continua). When relaxation does occur, it is due to the interaction of the molecule with its environment. The appropriate molecule–rare gas atom interaction potentials therefore control the relaxation, whether in the gas or in the condensed phase, leading to the similar relaxation behavior described previously.

Usually the detailed properties of the multidimensional potential describing the guest–host interaction are not known. Fortunately, useful information is provided by the spectroscopy of the guest molecule and, in particular, by the study of the lineshapes associated with the internal guest transitions. Although the primary purpose of the present article is to discuss guest relaxation processes, we review in Section II the spectroscopy of matrix isolated molecules and the factors controlling the observed lineshapes. In the following sections we then consider various types of radiationless transitions occuring in the low-temperature solids.

In Section III we give a brief review of the experimental techniques used in studies of nonradiative relaxation processes occuring in the matrix. The simplest case of such a process is direct vibrational relaxation, where the vibrational quantum of the guest is converted into the delocalized lattice phonons, and this is discussed in Section IV. If the guest is also excited electronically, it can undergo internal conversion or intersystem crossing into a different electronic state, and various examples of electronic relaxation, both in atomic and molecular guests, are discussed in Section V. Finally, vibrational energy transfer is discussed in Section VI.

II. MATRIX-ISOLATED MOLECULES AND THEIR SPECTRA

A. Physical Origin of Molecular Lineshapes

The atoms of the rare gas lattice are bound by very weak van der Waals forces. Thus the pairwise interaction energy is only $\approx 60\ cm^{-1}$ for two Ne atoms and $\approx 400\ cm^{-1}$ for the more polarizable Xe atoms.[33] The acoustical lattice phonons are therefore of low frequency, and the Debye cutoff for the rare gas solids is near $65\ cm^{-1}$. The energies associated with the internal bonds in the covalently bound molecules are, on the other hand, of the order of several electron volts, and their vibrational frequencies are correspondingly higher. Rebane[34] has therefore proposed that to a good approximation a Born–Oppenheimer type separation of the internal molecular vibrations from the low-frequency lattice modes exists.

Under these circumstances there will exist, for each internal vibronic state of the guest, a separate multidimensional potential surface describing the equilibrium positions and motion of the solvent atoms. These potential surfaces will then determine the spectral lineshapes associated with the vibronic transitions in the guest molecule. The origin of the lineshapes is schematically shown for typical potential curves in single dimension in Fig. 1. If there is no change in the shape or equilibrium position of the potential surface between the lower and upper guest internal states, then processes conserving the number of phonons will be favored. In the low-temperature $\approx 4°K$ solid only a single sharp line will be observed for each vibronic band. Under these circumstances the spectrum of the guest will resemble that of a nonrotating gas phase molecule. When, on the other hand, a displacement along some normal coordinate accompanies the absorption or emission process, Franck–Condon type considerations will dictate the appearance of a progression in that particular phonon normal mode as shown in Fig. 1*b*. Since,

LINESHAPES IN OPTICAL SPECTRA
OF MATRIX ISOLATED SPECIES

a) No change in guest-host potential : weak coupling

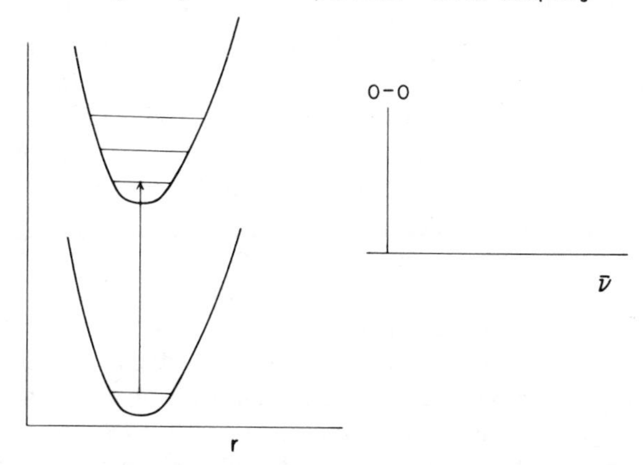

b) Large change in guest host potential: strong coupling

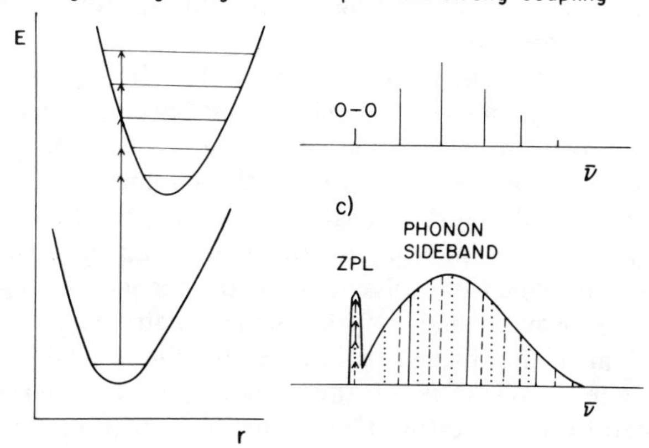

Fig. 1. Figure explaining schematically the origin of matrix lineshapes. (*a*) If the guest–host potential remains unchanged, transitions conserving the number of phonons are favoured and single line results. (*b*) If shift in the lattice atom positions occurs, a Franck–Condon progression will appear. Convolution of many such progressions gives a sharp ZPL containing their common 0–0 origins, and a broad phonon sideband (see text).

however, the problem is not restricted to one dimension and a large number of phonon modes may couple to the electronic transition, there will be a large number of such progressions. This is depicted for several lattice modes in Fig. 1*b*. The (0, 0) origins of all these progressions are coincident and give rise to a sharp line in the spectrum generally referred to as the zero phonon line (ZPL). This corresponds physically to the adiabatic process connecting the fully relaxed equilibrium geometry in the lower state with the new equilibrium around the upper level. The higher members of these progressions will, in general, not overlap exactly due to the varying magnitudes of the individual lattice vibrations. The envelope of all these lines will give rise to a relatively broad and usually continuous wing which is called the phonon sideband. Physically, the maximum of the phonon sideband corresponds to the vertical Franck–Condon process which leaves the geometry around the initial state unchanged. This is then followed by fast, $\leq 10^{-12}$ sec time scale resolvation of the final state. It should be emphasized that a particular lineshape is not a property of a particular guest state but instead of a transition between the initial and the final state. It should also be noted that the appearance of sharp zero phonon lines and absence of appreciable phonon sidebands does not necessarily imply a weak interaction between the guest and the solvent. It simply means that no change in this interaction occurs during the transition. Thus numerous molecular cations exhibit sharp, zero-phonon type spectra, with little evidence for phonon wings, although ion solvation energies[35] are often in excess of 1 eV.

Experimentally one finds that only the ZPLs generally appear in infrared and Raman spectra. Under normal circumstances the guest–host interaction is insensitive to the vibrational state of the guest, and so the vibrational transitions are in the weak coupling limit. Electronic transitions are usually more strongly coupled to the phonons as evidenced by the more or less intense phonon wings in the guest spectra, reflecting the extent to which the solvent geometry changes during the transition. Different vibronic bands of the same electronic transition generally display similar lineshapes, again reflecting the insensitivity of potentials to the vibrational state of the guest. The guest–host interaction is controlled by the electron density distribution rather than by the bond lengths and nuclear geometry of the guest. Thus in the $X^1\Sigma_g^+ \rightarrow B^3\Pi\, 0_u^+$ transition in Cl_2 the bond length increases by ≈ 0.4 Å, and yet there is insignificant intensity in the phonon sidebands.[36,37] Conversely, the bands of the $X^1\Sigma_g^+ \rightarrow A^1\Pi_u$ Phillips transition of C_2 show very intense phonon wings,[38] although there is negligible change in the r_e. The $a^3\Pi_u \rightarrow d^3\Pi_g$ transition of the same molecule, on the other hand, exhibits entirely different lineshapes with mainly ZPLs.

As we have discussed previously, the phonon sideband is the result of coupling of the transition to the lattice phonons and has generally the appearance of a broad continuum. The transition will actually couple preferentially to those modes having a large amplitude in the neighborhood of the guest. In particular, the introduction of the molecular impurity itself introduces several low-frequency vibrations associated with its rotational and translational degrees of freedom. Preferential coupling to these pseudolocalized modes is particularly apparent in the spectra of some diatomic hydrides, which exhibit a discrete structure in their phonon sidebands resembling the gas phase rotational spectrum.[39–41] We show later that these localized modes are of key importance for the relaxation process.

B. Strongly Interacting Guests and Adiabatic Potential Changes

For most transitions between valence states of neutral molecules the bond energies within the guest are large with respect to the guest–host interactions. Under these circumstances the guest potential function, including the anharmonic parts, is not significantly perturbed by the solid. In special cases where either the guest–host interactions are particularly large, or where the guest itself is only weakly bound, this may not be true and some interesting spectroscopy and dynamics may be observed in these cases. For instance, much stronger interactions between the matrix and the guest can be expected for molecules in Rydberg states, and also, in general, for ionic guests. Similarly, very strongly polar molecules, such as alkali halides, are known to interact rather strongly with the lattice.[42] Conversely, even relatively weak interaction with the host can lead to rather strong perturbations of the guest spectrum if the guest itself is a weakly bound dimer or a van der Waals complex. We consider each of these examples in the following sections.

1. Solvation of Rydberg States

Electronic wave functions of Rydberg states are diffuse, and highly polarizable. It is therefore of interest to explore, how these states will be affected by the solid environment. Rydberg states in solid rare gases were first studied via VUV absorption spectroscopy. It was generally found that the absorption bands into the excited Rydberg states were rather broad.[43,44] The observations were interpreted using a modified Wannier model.[44] This model physically corresponds to an electron orbitting an ionic core; the binding energy between the ion and the electron is decreased due to dielectric shielding by the host atoms.

For higher Rydberg states near the ionization limit, the Wannier model is undoubtedly correct. However, an alternative model was recently advanced for the low-lying $A^2\Sigma^+$ state of NO in rare gas solids, which presents a particularly interesting example. In the gas phase the NO spectra are known to show very strong perturbations between several excited valence and Rydberg states. In rare gas solids these perturbations of the excited valence states practically disappear, and no distinct absorption due to the Rydberg states is seen.[45–47] It was concluded that the Rydberg absorptions have broadened into an apparent continuum. This broadening was recently studied using the laser-induced fluorescence technique.[48,49] It was concluded that the quantum yield of the Rydberg fluorescence in the rare gas solids is close to unity; furthermore an interesting asymmetry between the emission and excitation spectrum was seen, with the bands being considerably broader in excitation than in fluorescence. The broadening in both fluorescence and absorption must represent an extensive phonon sideband. This asymmetric broadening, coupled with a measurement of the purely radiative lifetimes in the solid environment, implies that the excited Rydberg state creates a small "bubble" for itself. There is only a relatively minor Wannier-like delocalization of the wave-function tail into the solvent. In this situation the guest–host radial interaction potential is shifted outward in the Rydberg state as shown in Fig. 2. One thus probes in absorption the steeply rising hard sphere repulsion inner limb, whereas the shallow outer limb of the ground electronic state potential controls the fluorescence. The ZPL, which should occur between the origin of the absorption and the origin of the fluorescence, is strongly forbidden by the Franck–Condon factors and is not observed experimentally.

NO is interesting in that one can simultaneously observe $3s\sigma$ "bub-

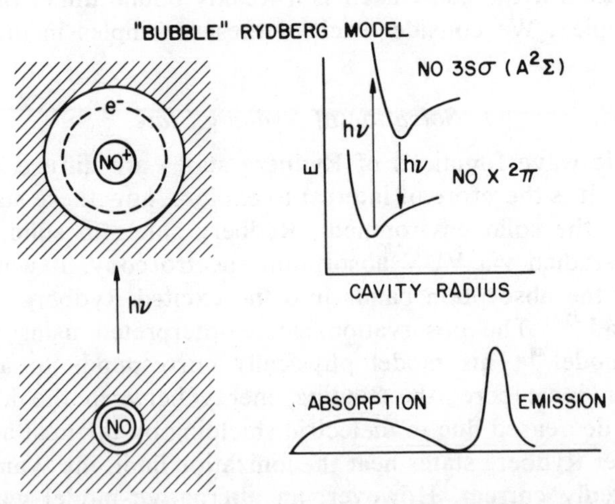

Fig. 2. Schematic diagram illustrating the small "bubble" Rydberg model. The Franck–Condon principle, when applied to the NO-solvent radial potential curve, predicts that the phonon sideband will be wider in absorption than in fluorescence. The shallow outer limb of the ground state represents the relatively small amount of energy necessary to create a small cavity.

ble" fluorescence and nearly isoenergetic $B^2\Pi$ valence fluorescence. The initial distribution of sites around the $X^2\Pi$ ground state produces an inhomogeneous width of $\approx 30\ cm^{-1}$ on the $B^2\Pi$ fluorescence ZPL. This same distribution of environments produces $3s\sigma$ "bubble" emission with site shifts of several thousand cm^{-1}! Obviously bubble formation, and penetration of the wavefunction tail into the solvent, are extremely sensitive to the exact local structure. The metastable sites in the bubble emission may correspond to vacancies in the second coordination sphere.

2. *Ion Spectroscopy in Rare Gas Hosts*

Another interesting example of strongly interacting guests are molecular cations. Their interaction energies with rare gas solids are large and in the heavier rare gases—Ar, Kr, and Xe—exceed 1 eV. Nevertheless, numerous ions exhibit, both in absorption and in fluorescence, extremely sharp ZPL type spectra.[50,51] This is exemplified in Fig. 3 by a section of the excitation spectrum of the hexafluorobenzene radical cation $C_6F_6^+$ in solid Ne. The ion spectra often also show surprisingly little perturbation of the molecular potential function by the solid. For instance in the cases of CS_2^+ and diacetylene[52], $C_4H_2^+$, where the gas phase spectra have been studied at high resolution[53,54] and comparisons can be made, there are less than 0.2% shifts in the T_e values, and also the vibrational frequencies change by less than a few wave numbers. In the $C_6F_6^+$ and the sym—$C_6H_3F_3^+$ species which exhibit Jahn–Teller splitting in their doubly degenerate ground states, the vibrational structure in the matrix is also unperturbed.

When the ionization energy of the host substantially exceeds the electron affinity of the guest ion, one can expect little charge transfer interaction. The solvation energies, although still rather large, will mainly consist in the polarization of the host atoms by the charged guest and will not be strongly dependent on the particular vibronic state of the guest. As a result very little resolvation may accompany the transition, and principally ZPLs will be seen in the electronic spectrum. In view of the high, $\approx$19-eV, ionization potential of Ne atoms, the preceding condition is fulfilled for most cations, and accordingly sharp and unperturbed ion spectra are observed in Ne host. The ionization potentials of Ar and the heavier rare gases are, on the other hand, much lower and are often commensurate with the electron affinities of organic ions. Much stronger charge-transfer interactions will therefore occur in this case. Even stronger interactions can be expected for the excited states of the cations, whose electron affinities will be higher by their respective excitation energies.

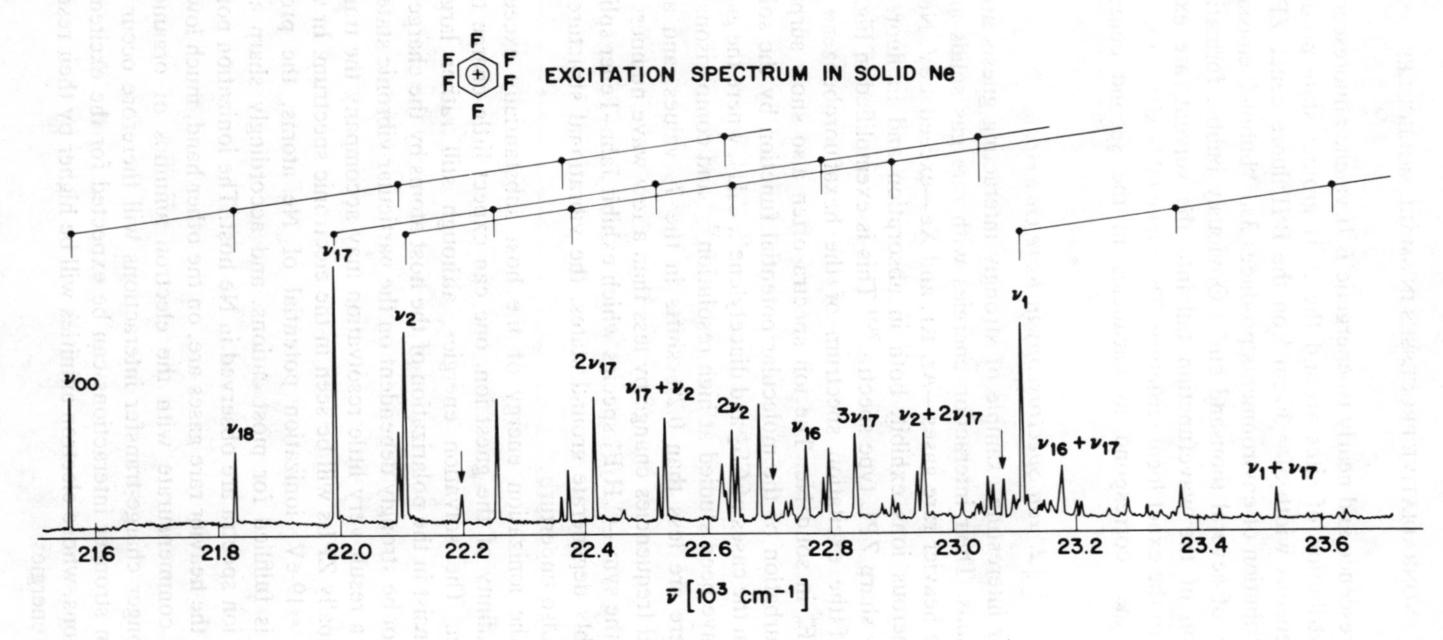

Fig. 3. Part of the excitation spectrum of the $C_6F_6^+$ radical cation fluorescence in solid Ne. The vibrational mode numbering shown is based on a D_{6h} symmetry.

An interesting example has been observed for $C_6H_6^+$, dimethyldiacetylene radical cation.[55] This species exhibits in solid Ne sharp, well resolved spectra. The (0–0) band occurs both in emission and absorption at 20499 cm^{-1}, only negligibly shifted from its gas phase position, and the vibrational structure is, within the experimental uncertainty, identical to the gas phase spectrum. The spectra observed in solid Ar are quite different and are shown in Fig. 4. The bands, both in excitation and emission spectra are broad, and the vibrational structure is rather strongly perturbed. The (0–0) phonon sideband maximum in emission is shifted over 440 cm^{-1} from its position in the excitation spectrum. An interesting asymmetry can be seen, with the emission bands being a factor of 5 to 6 broader than the excitation bands. The Franck–Condon argument advanced to explain these observations is very similar to that used to explain the Rydberg spectra of NO. A strong charge transfer interaction between the cation and the lattice atoms apparently takes place. This interaction is stronger in the excited state with its higher electron affinity and results in a decreased guest–host equilibrium distance. Consequently, the steep, hard sphere repulsion part of the potential controls the fluorescence, and gives rise to the extremely broad

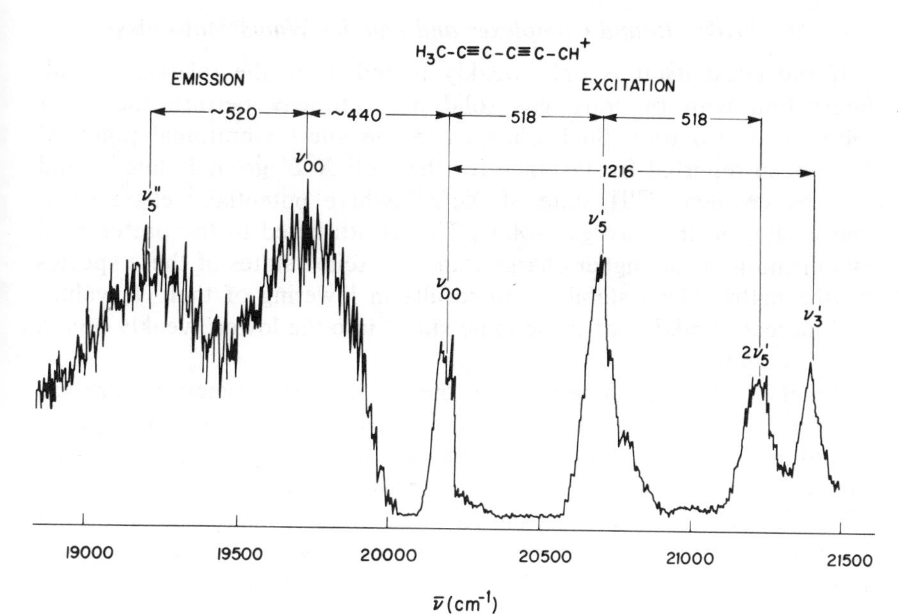

Fig. 4. Portions of the (left) emission and (right) excitation spectrum of the dimethyldiacetylene radical cation, $C_6H_6^+$, in solid Ne. The spectra show the asymmetry in band shapes between emission and absorption, and the apparent noncoincidence of the (0–0) origin.

phonon sidebands. The absorption process probes the shallow outer limb and results in the relatively sharper bands. As in the case of the Rydberg transitions in NO, the ZPL is strongly Franck–Condon forbidden and has not been observed either in absorption or in fluorescence.

The appearance of sharp vibronic structure in the spectra of a variety of molecular cations in solid Ne, and the absence of appreciable perturbation of the guest indicate that electronic spectroscopy of matrix isolated cations should in the future provide extensive information about these elusive species.

It should be noted that, although the rather strong interaction of the ions with the solid does not perturb their vibronic structure, it seems to lead to a very efficient vibrational relaxation. Despite the very short radiative lifetimes of most of the ionic transitions studied, no vibrationally unrelaxed emission has been observed. Similarly, in the NO small "bubble" $3s\sigma$ Rydberg state, vibrational relaxation is fast[49] despite the fact that the NO^+ core is isoelectronic with N_2 and has a similarly large $\approx 2300\ cm^{-1}$ vibrational quantum. The strong interaction in the "bubble" state leads to many orders of magnitude faster relaxation than occurs in the ground electronic state of N_2.

3. *Weakly Bound Complexes and van der Waals Molecules*

If the guest itself is only weakly bound, even the relatively weak interaction with the rare gas solid may strongly perturb the guest vibrational structure. Such changes in the guest vibrational potential have been reported for instance for the XeF $X^2\Sigma^+$ ground state[56,57] and for the covalent $C^1\Pi$ state of XeO,[58] where potential wells deepen appreciably in the rare gas solids. This is attributed to the preferential stabilization of the higher charge transfer excited states of these species by the matrix. This stabilization results in lowering of their T_e values and increased mixing of these ionic states into the lower, weakly bound covalent states.

Another interesting example are the Ca_2 and Mg_2 dimers in rare gas matrices.[59–62] The potential well of these species, which are only weakly bound in the gas phase, appears to deepen appreciably in the matrix, and the ω_e values increase substantially. It is interesting to note that similar diatomic species, which are known to be more strongly bound, do not show this perturbation by the host. For example, diatomic Pb_2 shows negligible shifts of its molecular constants by the rare gas solids.[63]

4. *Hydrogen Bonding to Rare Gas Atoms*

Even strongly bound covalent species can show an appreciable change in adiabatic potential curves if the interaction with the solvent is stronger than simple dispersion and polarization. OH radical presents an interesting example.[64] In neon host, simple ZPL $X^2\Pi \rightarrow A^2\Sigma^+$ spectra are observed, without change in potential curve shape from the gas phase. The phonon sideband fine structure indicates that slightly perturbed free rotation occurs. In solid argon, the vibronic bands are broad phonon wings that are asymmetric in absorption and fluorescence. A fine structure observed in absorption can be interpreted in terms of the low-frequency ν_3 OH–Ar vibrational mode of a linear hydrogen bonded complex. The hydrogen bond is stronger with a shorter r_e in the excited state, as shown in Fig. 5. The complex binding energy is $\approx$675 cm^{-1} in the $A^2\Sigma^+$ state and over 1000 cm^{-1} for this excited state in OH–Kr. In fluorescence, the hard sphere repulsive wall of the weakly bound ground electronic state complex is probed above its dissociation limit. ν_3 structure is not observed. The majority of the spectroscopic and dynamical properties observed in argon are reproduced for isolated OHAr in solid

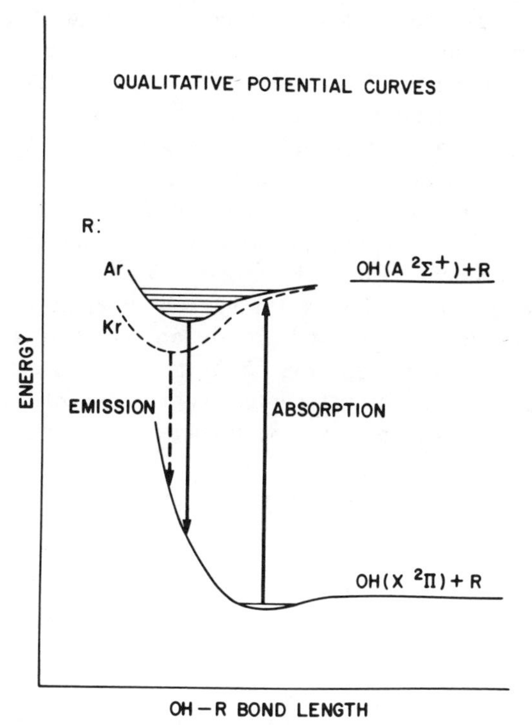

Fig. 5. Qualitative excited and ground state potential curves for the hydrogen bond between a rare gas atom and OH. The solid lines refer to OH–Ar. The dashed line refers to OH–Kr and shows how the phonon sideband maximum in fluorescence is strongly red shifted for OH–Xe and OH–Kr.

Ne, thus showing the specific nature of the interaction. OH, and to a lesser extent NH, are strongly acidic in their excited states and may hydrogen bond to the heavier rare gas atoms.[65] Charge transfer from the rare gas appears to contribute appreciably to these hydrogen bonds, just as it does in the case of the XeF ($X^2\Sigma^+$) "complex."[56]

C. Cage Effect and Molecular Dissociation

As previously mentioned, nonradiative transitions cannot occur in isolated small molecules without interaction with the surrounding solvent. One exception involves molecules excited into dissociative or predissociative states. This leads to the interesting question of the effect of the solid environment on molecular dissociation, and on the shape of the potential curves near the dissociation limit. We have seen that for low and moderate levels of excitation, the guest–host interaction is insensitive to the vibrational state of the guest. This is due to the fact that the valence electrons which determine the interaction potential are normally insensitive to nuclear motion. As the guest excitation approaches the molecular dissociation limit, the amplitude of the vibrational motion increases, and eventually must exceed the physical dimension of the available "cavity" in the host lattice. This leads to some interesting spectroscopic effects, which were studied in some detail in several halogen molecules and alkyl iodides.[66,67]

Cl_2 represents a particularly clear example. Figure 6 shows the excitation spectrum of Cl_2 fluorescence in solid Ar, in the region just below the dissociation limit[37] near 4750 Å. Here the ground state $r_e = 2.0$ Å, and the highest $^3\Pi_0$ level observed ($v' = 15$) has an outer turning point near ≈ 3.5 Å. Despite this large bond expansion, there is no perceptible change in the anharmonic spacing of the last few observed ZPL. The effect of the solvent is to transfer intensity from the ZPL to the phonon sideband as one goes from low v' to high v'. For high v' the Ar nuclei

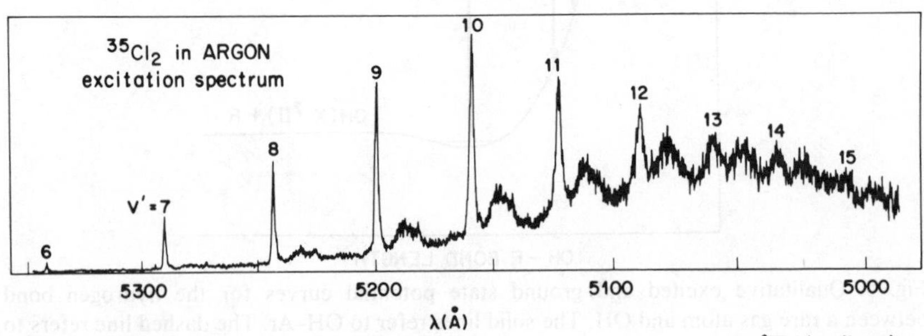

Fig. 6. Excitation spectrum of Cl_2 fluorescence in Ar host. The $B\ ^3\Pi\, 0_u^+$ vibrational assignments are shown.

must undergo greater resolvation around the excited state. However, this resolvation motion, which displaces Ar atoms as the Cl_2 bond expands, experimentally requires little energy as the adiabatic potential curve is not changed in shape from the gas phase. For excitation at still higher energies, the host cage effect prevents dissociation. Even for excitation 2 eV above the convergence limit one sees fully relaxed emission from the $v' = 0$ level of the low-lying $^3\Pi_2$ state. The quantum yield for permanent dissociation appears to be close to zero.

The interhalogen ICl represents another interesting case.[68-70] Figure 7 shows that in the gas phase two excited 0^+ curves exhibit an avoided crossing, such that the lower one exhibits four bound vibrational levels below the barrier to dissociation formed by the avoided crossing. Experimentally in rare gas matrices it is observed that excitation as much as 0.8 eV above this dissociation barrier produces $v' = 0$ fluorescence from the lower 0^+ potential and from the still lower A $^3\Pi_1$ bound state, with subnanosecond rise times. A complete cage effect again occurs with the quantum yield of permanent dissociation being $\leq 10^{-4}$. Nevertheless, the solvent does not induce quantized vibrational levels in the dissociative region, as the fluorescence excitation spectra in the dissociative region are perfectly smooth. In the 0^+ bound region below

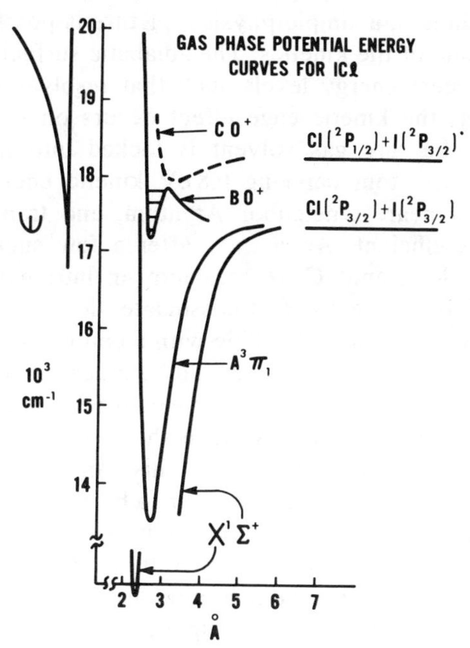

Fig. 7. Gas phase potential energy curves of ICl.

the dissociation barrier, five quantized levels are observed in the condensed phase experiments. This appears to occur because there is a differential shift in the T_e values of the zero-order bound and repulsive 0^+ states; such that the repulsive 0^+ crosses the bound state at a higher energy in the solid than in the gas phase.

The technique of photoselection[71] in rare gas matrices was applied to the question of whether transient local melting occurs during the ICl dissociation–recombination sequence. It was found that the fluorescence remains completely polarized even for excitation 0.8 eV above the dissociation barrier, implying that the axis of the emitting molecule has the same absolute direction in space as it does in the ground electronic state before excitation. This result indicates that the specific local environment (i.e., the equilibrium positions of the nearby rare gas atoms) remains unaltered through the transient cage process. If a local melting did occur, the cavity around ICl would reform with a random orientation with respect to the ICl axis before dissociation.

Two apparently contradictory facts emerge: (*a*) There is a strong *kinetic* cage effect with the yield of permanent dissociation being $\leqslant 10^{-4}$ without local melting. (*b*) There is a negligible *spectroscopic* cage effect, in the sense that there are no bound levels in the dissociative region and the gas phase anharmonicity, just below the dissociation limit, is unchanged. Nevertheless, a simple physical picture is possible if we recall the short time scale of the kinetics. The adiabatic surface only describes the position of guest energy levels such that resolvation continuously occurs. However, the kinetic cage effect occurs on the 10^{-14} sec time scale such that the rare gas solvent is locked into its ground state configuration. A Cl atom carrying 0.8 eV kinetic energy collides impulsively with its nearest neighbor Ar atom, and transfer of nuclear kinetic energy is efficient. Apparently after a few such collisions the excess energy is lost, and ICl relaxes into an intramolecularly bound excited state. If ICl were to photodissociate slowly in a "gedanken" experiment, the Cl atom would collide with a coupled group (the lattice) of Ar atoms, whose effective mass would be much higher than that of a single Ar atom. Nuclear energy transfer would be much less efficient.

It is striking that the adiabatic surface shows such little hindrance of the large-amplitude vibrational motion. This can be rationalized if we realize that rare gas solids at 4.2°K appear to be as elastic as any known solid material. The macroscopic bulk modulus of solid Ne at 4.2°K is essentially equal to that of room temperature liquids such as ethanol. Let us consider the solvent to be a continuous, elastic material, and ask how much energy is required to open up a cavity of ≈ 4 Å diameter that would accommodate molecular motion. We find that a negligible amount is required if the resulting solvent compression is allowed to propagate to regions far removed from the cavity. In other words, if we wait for solvent relaxation the energy required to open a small cavity is negligible.[66]

This simple picture ignores possible barriers to viscous flow or activation energies for solvent motion. Such effects, if important, could create host-induced negative anharmonicities and barriers to dissociation.[72] To our knowledge such phenomena have not been observed in rare gas solids. Apparently the isotropic nature of the van der Waals forces allows elastic motion without appreciable activation energies.

We may imagine the following concerning solvent involvement in molecular reactions and isomerizations: Any molecular state, which in the gas phase would isomerize or react, should not be appreciably hindered from doing so in rare gas solids, *if* this state lives for a time long with respect to resolvation times. These ideas are consistent with the fact that diverse organic and inorganic synthetic reactions can be carried out in solid Ar at 4.2°K.[73–76] For example, the predissociation of *s*-tetrazine, which is a ring fragmentation producing $N_2 + (HCN)_2$ dimer, occurs apparently unhindered in Ar matrices.[77]

The shapes of fluorescence excitation spectra in photodissocation regions, even though they contain no vibrational structure, do yield information about ultrafast radiationless transitions occuring before arrival of the population in the bound fluorescent $v' = 0$ state. In the ICl case, the normalized fluorescence excitation curve dropped sharply when crossing above the gas phase dissociation barrier, and it was concluded that the recombination quantum yield into the fluorescent $v' = 0$ $B0^+$ dropped from $Q \approx 1.0$ below the barrier to $Q \approx 0.01$ above the barrier. Above the barrier, the remaining population explores longer ICl bond lengths and then recombines along other bound potential curves.

In the case of Br_2 in solid Ar[78] the dissociative region excitation spectrum of $v' = 0$ $B0^+$ emission showed only the $B0^+$ dissociation continuum, and not the stronger $^1\Pi_1$ continuum. The repulsive $^1\Pi_1$ curve, which correlates with ground-state $^2P_{1/2}$ Br atoms, does not radiationlessly feed the $B0^+$ state, which correlates with the excited dissociation limit $^2P_{1/2} + {}^2P_{3/2}$.

The quantitative theory of predissociation in solids was formulated[58] for XeO in solid Ar. As observed for ICl previously, a differential shift in energy between the zero-order bound and repulsive curves can markedly influence the predissociation rate of a given vibronic level. However, for a given set of potential curves, the transient cage phenomena do not appear to influence the vibronic coupling between the bound and free wave functions.

III. EXPERIMENTAL TECHNIQUES

A. Preparation of Matrix Samples

Matrix deposition techniques and equipment have been extensively reviewed,[9-11] and we do not repeat these discussions here. We point out, however, that in many experiments finding the correct combination of deposition conditions and guest concentrations is the most time-consuming part of the experiment. In situations where the lifetimes are very long, extensive energy transfer could occur and dominate the relaxation process. Under these circumstances, the level of impurities in the matrix (especially H_2O) is critical. The vacuum integrity of the apparatus is essential.

To cleanly excite specific vibronic states and to observe the vibrational structure of the guest, the ability to generate sharp spectra is important. In this regard one can sometimes exploit the difference between the homogeneous (phonon contour) lineshape discussed in the lineshape section, and the inhomogeneous linewidth representing the distribution of local environments in the solid.[79-81] If the inhomogeneous linewidth is larger, then excitation with a sharp-frequency laser source will produce sharp luminescence from the small subset of guests which is in exact resonance with the laser. The resulting luminescence spectrum will show only the homogeneous linewidth. Excitation spectra showing only the homogeneous lineshape can also be obtained by allowing the emission spectrometer to observe only this small subset, and then scanning the laser excitation wave length.[70]

B. Electronic Spectroscopy and Data Processing Techniques

Time resolved spectra in the 10^{-9} sec and longer time scales can be obtained with the apparatus in Fig. 8. In this particular arrangement a small computer controls the functions of the hard-wired signal averager, and the wave length drives of the emission spectrometer and dye laser. At the pulsed laser repetition rate of ≈ 10 Hz, it is convenient to use a transient digitizer to record the complete time dependence at one emission wave length. Ultraviolet excitation is possible by using a servomechanism to adjust an angle tuned doubling crystal as the dye laser wave length is scanned. This apparatus, and variations employing boxcars, vidicons and so on can be used to follow excited state relaxation processes in real time.

C. Ground Electronic State Dynamics and Infrared Excitation

Direct excitation of infrared active vibrations can be achieved with either fixed frequency of tunable IR lasers. CO, CO_2, and tellurium

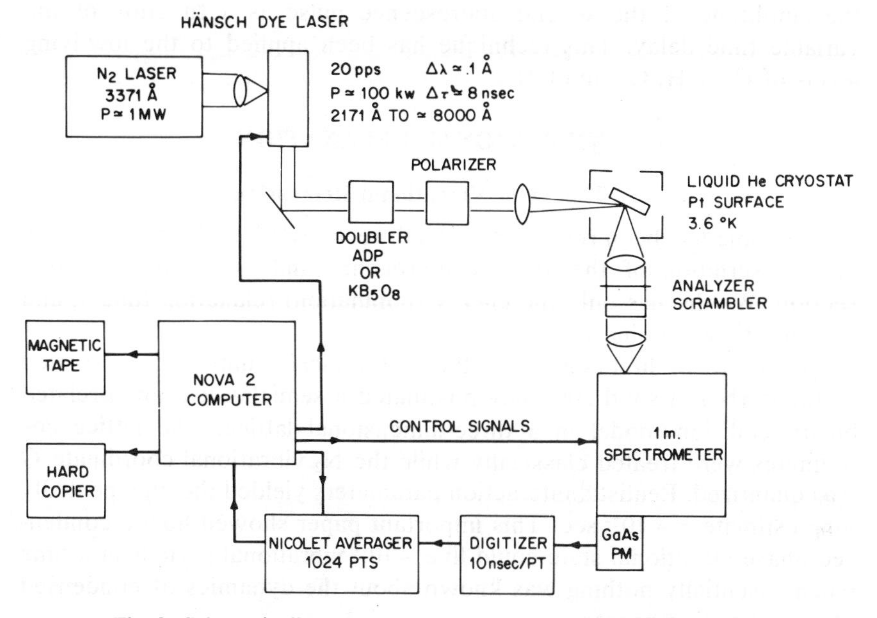

Fig. 8. Schematic diagram of a laser-excited luminescence apparatus.

crystal doubled CO_2 fixed frequency lasers, as well as tunable spin-flip Raman and OPO sources have all been employed. In the case of $v = 1 \rightarrow v = 0$ diatomic transitions, one has the problem of strong scattered laser light occuring at the same wave length as the resulting infrared fluorescence. The dynamics can be observed by simply monitoring the infrared emission, or by a double resonance experiment. For example, a CW infrared laser can monitor the recovery of the ground-state population, if the transition is initially saturated. If a convenient, fluorescent excited electronic state is available, the $v = 1$ population could be monitored by exciting this population to the excited electronic state after a variable time delay. Infrared excitation experiments will become more important with the development of high power OPOs and F-center lasers.

If there are convenient, fluorescent excited electronic states (as normally occurs in free radicals), then optical–optical double resonance experiments can probe ground-state relaxation dynamics. With favorable Franck–Condon factors and short excited-state lifetimes, one can produce population in excited ground-state levels by a laser-induced absorption–fluorescence sequence, and then reexcite this population into the upper electronic state after a variable time delay. One then monitors the amplitude of the second fluorescence pulse as a function of the variable time delay. This technique has been applied to the low-lying levels of C_2^-, NH, C_2, and CN.

IV. VIBRATIONAL RELAXATION

A. Theory of Vibrational Relaxation

This topic has been reviewed in depth recently[82–84] and we give only a brief description of the various approaches and assumptions in this section. We discuss only the energy (population) relaxation time τ, and not vibrational dephasing.

Sun and Rice[19] first considered the prototypical situation of an isolated diatomic (N_2) in solid Ar. They postulated a semiclassical uncorrelated binary collision model in a three-dimensional lattice. The lattice coordinates were treated classically while the N_2 vibrational coordinate Q was quantized. Realistic interaction parameters yielded the unexpectedly long estimate $\tau \approx 10^{-2}$ sec! This important paper showed how a condensed phase vibrational state could live $\approx 10^{11}$ vibrational periods at a time when essentially nothing was known about the dynamics of condensed phase relaxation process.

Quantum mechanical modeling of the relaxation mechanism began with the ultra simple picture of a two-level system ($v = 1$ and $v = 0$) weakly coupled to a harmonic heat bath.[85–89] These theories generally predicted a very strong temperature dependence due to stimulated phonon emission, and an exponential decrease in the relaxation rate with ω_e (the energy gap law). Diestler[90] incorporated a shift in the lattice equilibrium positions from $v = 1$ to $v = 0$, and Lin[91] introduced an exponential dependence on lattice coordinates in the perturbation Hamiltonian. Finally Lin[92] has reformulated the theory in terms of a Born–Oppenheimer type separation between the high-frequency Q motion and the low-frequency motions of the lattice.

Following proposals by experimentalists that rotation accepts energy during diatomic hydride relaxation,[93,94] the rotational and center-of-mass translational local modes have been incorporated into these models. Gerber and Berkowitz[95] considered a three-dimensional model coupling vibration with rotation, and were able to quantitatively calculate the experimental relaxation rate of NH ($X^3\Sigma$) in Ar by assuming the applicability of a published HCl–Ar interaction potential. They found that the highest available J level of $v = 0$ ($J_m = 13$) is directly populated, with the slight energy mismatch being taken up by the center of mass local mode. More recently, the model has been extended[96,97] to include the competition between these two local motions and the delocalized lattice phonons. They predict an intermediate mass range exists where neither local modes nor lattice phonons can produce an appreciable relaxation rate. Freed and Metiu[98] and Freed et al.[99] have considered a simplified two-dimensional model of coupling to rotation. They find that rates should decrease exponentially with $\sqrt{\omega/B}$ with only a weak temperature dependence. Finally Diestler has attempted to numerically apply a two-dimensional model to NH, OH, and HCl. He finds rotation is important in most but not all of these situations.

Experimental evidence exists that complexes may form between diatomics and rare gas atoms; that is, the potentials are not pairwise additive[64,65]. Berkowitz and Gerber[96] have attempted to model complexes by shortening the distance to one nearest neighbor atom; they find rate increases of 10 to 10^3. For comparison, a single nearest-neighbor vacancy enhances the rate by only a factor of ≈ 3. Also, Diestler et al.[100] have explicitly considered the normal modes of a diatomic coupled to two atoms with the remaining atoms taken as a heat bath.

Instead of postulating analytical models, one could in principle simply solve the equations of motion for all the nuclei. This is not practical for a realistic three-dimensional crystal. A simplification occurs in the stochastic trajectory method, in which one solves these equations for the diatomic and a few nearby atoms. The remaining crystal is simulated by random forces and damping kernals. This approach has been applied by Shugard et al.[101,102] to the case of Cl_2 in argon. The results show sensitivity to the exact shape of the phonon spectrum assumed for the remainder of the crystal. The calculated relaxation rates were in reasonable agreement with the values deduced from experimental spectroscopic observations.

B. Experimental Studies of Vibrational Relaxation

The pioneering work of Vegard[12] almost 50 years ago demonstrated that a metastable excited state of N_2, now known as $A^3\Sigma_u^+$, had vibrational relaxation times of ≈ 1 sec in solid Ar host at liquid He temperatures. In 1968 Tinti and Robinson[17] confirmed this result by directly time resolving the N_2 phosphorescence following X-ray excitation of the matrix. This discovery, that condensed phase relaxation times can be "long" and are not uniformly $\approx 10^{-13}$ sec, suggested that direct time resolved mechanism studies could be performed. Bondybey and Nibler[21] have used a fixed-frequency CW Ar^+ laser in a two-photon sequential excitation experiment to observe extremely slow relaxation in the ground state of C_2^-.

The advent of tunable visible and infrared laser sources opened the possibility of pulsed excitation of specific vibronic states. In 1975 frequency doubled tunable dye lasers were used to directly time resolve vibrational relaxation in the excited electronic states OH ($A^2\Sigma^+$)[93] and NH ($A^3\Pi$)[94]. Relaxation rates were deduced from time resolved emission of the directly populated $v \geqslant 1$ level, and lower levels populated via vibrational relaxation. The hydride rates were an order of magnitude or more faster than the deuteride rates, and these rates were independent of temperature up to the softening points of the matrices. In these studies, the electronic spectra fine structure shows that these diatomic hydrides undergo slightly perturbed free rotation at 4.2°K in the solid environment. These results did not agree with the energy gap law predictions of models in which delocalized lattice phonons directly accepted energy, and it was proposed that the rotational degree of freedom actually accepts the energy. It was argued that since the rotational constant B scales with μ^{-1} and the frequency ω_e scales with $\mu^{-1/2}$, relaxation in the hydride is a lower-order process in the sense that a smaller J is populated than in the deuteride. This reasoning qualitatively suggests that hydride relaxation should be faster. Additionally, only a mild temperature dependence should be observed since the stimulated emission properties of harmonic oscillator phonon functions are not involved.

These results were extended to the NH $X^3\Sigma^-$ ground electronic state using two independent lasers.[103] The $v = 1$ level of the ground state was populated by laser-induced fluorescence, and a second laser then probed this population with variable time delay after the first (pump) laser. The rates were a factor of $\approx$200 slower than in the excited state. Nevertheless, a similar absence of temperature dependence and dominance of the NH rate over the ND rate were observed. It seemed that, although the interaction with the surrounding solvent was considerably weaker in the ground electronic state, the mechanism remained unchanged.

Obviously, a truly free rotation must conserve angular momentum, and thus the deviation from spherical symmetry must be the key factor in these rates. This dependence upon local symmetry was investigated for the case of NH ($A^3\Pi$) in various Ar/Kr alloys.[65] The pure Ar and pure Kr rates were compared with the rates for 11 Ar and 1 Kr nearest neighbors, 10 Ar and 2 Kr neighbors, and so on. These results demonstrated that vibrational relaxation is caused by short-range forces, in the sense that nearby Kr atoms (or H_2 and CO) that are not nearest neighbors do not affect the rate. If the Kr atom is a nearest neighbor, then a saturation effect occurs as the relaxation rate for one Kr nearest neighbor is essentially the same as in pure Kr. It was concluded that a weak donor–acceptor, or hydrogen-bonded NHKr complex forms. The relaxation mechanism would then best be modeled as internal vibrational redistribution within the triatomic complex, whose ν_2 mode corresponds to the NH rotational motion. A stronger OH–Ar complex has been subsequently observed.[66] The OH spectra are entirely different in solid Ar than in solid neon. In neon almost free rotation fine structure occurs, whereas in argon v_3 OH–Ar vibrational progressions are observed. Moreover, the relaxation rate of OH $A^2\Sigma^+$ in neon is increased by more than a factor of 10^3 by introducing just one nearest-neighbor Ar atom. The complex model certainly appears justified for these hydrides, which have strongly acidic protons in their excited electronic states.

Studies of ground electronic state relaxation via infrared laser techniques have been pioneered by Legay and co-workers.[83,104] A frequency doubled CO_2 laser was first used to pump CO ($v = 1$) in neon and argon matrices. It was discovered that the ensuing emission is extremely sensitive to small traces of impurities, and to the concentration of CO. This occurs as the intrinsic vibrational relaxation time is very long, enabling exactly resonant (dipole–dipole) transfer of the excitation among the COs to occur many times during its lifetime. The excitation is ultimately deactivated at a CO having a nearby impurity, or appears as infrared emission. In the limit of high dilution in impurity free matrices, CO ($v = 1$) decays entirely via infrared emission.[105]

Abouaf-Marguin et al.[106] have used two CO_2 lasers to study the relaxation of the ν_3 C–F stretching mode of CH_3F and CD_3F in solid Kr. An intense pulsed CO_2 laser populates this state, and the ground-state recovery is monitored by the transmitted power of a weak, continuous probe laser. Again, an absence of temperature dependence and a tenfold longer CD_3F lifetime as compared with the CH_3F lifetime are observed. It is concluded that methyl group rotation about the symmetric top axis accepts the CF stretch energy. These authors also observed that NH_3 relaxes with an order of magnitude faster rate in rare gas solids than in solid N_2. This was attributed to the fact that NH_3 rotation about the symmetry axis is blocked in solid N_2, but occurs freely in rare gas solids.

Careful and detailed infrared emission studies of HCl vibrational relaxation in solid Ar have been performed by Wiesenfeld and Moore[107] using a tunable infrared OPO to directly excite the $v = 1$ and $v = 2$ levels. The effects of impurities have been convincingly eliminated. Energy transfer becomes negligible at dilution ratios higher than 1:2000, and both HCl and DCl are observed to decay principally nonradiatively. The relaxation rate of HCl ($v = 2$) is 32 times faster than that of DCl ($v = 2$), with a very mild (less than a factor of 2) temperature dependence in the range of 9 to 20°K. Infrared emission from isolated $(HCl)_2$ dimers was not observed, and orders of magnitude faster relaxation via the dimer bending modes is implied.

These diatomic and polyatomic studies suggest the generality of the rotation accepting mechanism for hydrides. With some success, Legay[83] has in fact correlated the available rate data as an exponentially decreasing function of $J_m \approx \sqrt{\omega_e/B}$. J_m is the largest rotational quantum number that can be populated. Does this same mechanism apply to nonhydride species, with higher barriers to free rotation and moments of inertia?

An isotopic study of O_2 ($b^1\Sigma_g^+$) vibrational relaxation within isolated $(O_2)_2$ dimers in solid neon has been carried out.[108,109] A laser-induced intersystem crossing process produced a vibrational cascade in the dimer state $X^3\Sigma(v=0)+b^1\Sigma(v=n)$. The cascade occurs within the $b^1\Sigma_g^+$ state; the neighboring O_2 is in its electronic and vibrational ground state. The energy gap law was systematically violated, with the $(^{18}O_2)_2$ rates being a factor of ≈ 40 slower than those of $(^{16}O_2)_2$. The rates for $^{16}O_2(^3\Sigma, v=0)+{}^{18}O_2(^1\Sigma, v=3)$ are a factor of two higher than those of $^{18}O_2(^3\Sigma, v=0)+{}^{18}O_2(^1\Sigma, v=3)$. That is, the rate also depends upon the isotopic identity of the unexcited nearest neighbor in the dimer. These results are consistent with the somewhat correlated rotational manifolds of the two O_2 molecules accepting the energy.

Very recently, the first mechanistic study of an isolated, nonhydride molecule has been performed.[110] The risetime on NO ($a^4\Pi$, $v=0$) phosphorescence in solid Ar was time resolved following ArF excimer laser excitation at 1933 Å. $^{14}N^{16}O$, $^{15}N^{16}O$, and $^{14}N^{18}O$ isotopic species were employed in an attempt to separate two properties which might influence relaxation: (*1*) the reduced mass μ, and (*2*) the asymmetry of vibration. The dramatic isotope effects seen in the $O_2(b^1\Sigma_g^+)$ dimer studies were absent. The small $\approx$30% isotopic rate differences observed did not correlate with either the reduced mass or the asymmetry of vibration. However, the two heavier isotopes did relax slightly faster than $^{14}N^{16}O$. This effect is in the direction of a conventional energy gap law. It is possible that for this symmetrical situation of NO in argon, the lattice phonons directly accept energy. In the $(O_2)_2$ dimer experiments, the excited $O_2(b^1\Sigma_g^+, v=n)$ state sees a strongly asymmetric local environment which could create the vibration–rotation coupling necessary for the rotation accepting mechanism.

Vibrational relaxation in the homonuclear diatomic C_2^- anion has been studied extensively by Nibler and co-workers.[111,112] They generate the C_2^- molecular ion by VUV photolysis of acetylene and measure its lifetimes by an optical–optical double resonance technique. They observe that, even at very high dilutions, the lifetimes are strongly dependent on whether C_2H_2 or deuterated acetylene, C_2D_2 was the parent compound. They conclude that energy transfer from C_2^- to an impurity, possibly the parent C_2H_2, controls the relaxation process.

One can observe ultrafast vibrational relaxation as a lifetime broadening in the spectra of weakly bound Ca_2 dimers in solid Kr and Ar.[61] The ground-state vibrational frequency is very low ($\approx 72\,cm^{-1}$ in Ar), close to the Debye frequency of the rare gas solids. While the lowest v'' levels in the emission spectrum appear sharp, the higher levels exhibit distinct broadening. If energy relaxation is the prevailing broadening mechanism, lifetimes of <1 psec are implied for levels above $v'' > 8$. The excited electronic state is more strongly bound and the ω_e' is considerably larger ($\approx 120\,cm^{-1}$ in Ar). As a result the relaxation is slower and particularly in solid Kr, extensive vibrationally unrelaxed fluorescence is observed. One can deduce from the relative intensities of the different progressions a $\tau \approx 3$ nsec for $v' = 1$ with progressively decreasing values for the higher levels.

C. Vibrational Relaxation in Polyatomic Molecules

Vibrational relaxation of the lowest frequency mode in polyatomic molecules does not differ qualitatively from relaxation in diatomics. The previously discussed studies of relaxation of ν_2 of NH_3 and of ν_3 of CH_3F fall into this category. For higher vibrational levels, however, one has to contend with mode-to-mode conversion of vibrational energy, which may compete with direct vibrational relaxation. This competition makes relaxation in polyatomics particularly interesting. Energy gap law considerations favor crossing to the closest lower-lying vibrational level, regardless of mode type, while the general weakness of anharmonic intermode coupling elements tends to favor relaxation within one normal mode. Several studies of intermode relaxation pathways have been carried out by Raman techniques[113,114] in room-temperature liquids, where the lifetimes are in the low picosecond range.

The weakly interacting low-temperature matrices often give well-resolved vibronic spectra, and would appear to be an ideal medium for this type of study. Unfortunately, electronic emission spectra of polyatomic molecules in matrices are in most cases completely vibrationally relaxed. This fast relaxation reflects a high density of vibrational states, and slower relaxation would be expected for strongly bound smaller polyatomics having only a few modes of high frequency.

The first clear observation of vibrationally unrelaxed emission was reported in the CF_2 radical.[115] Emission from the $(0,1,0)$ and $(0,2,0)$ levels in solid Ar was observed and lifetimes of 10 and 5 nsec, respectively, were measured for the relaxation controlled lifetimes of these levels. The first information about the relaxation pathway mechanism in a polyatomic molecule was provided by two independent studies of CNN.[116,117] In this molecule all the vibrational modes were observed in the electronic spectrum and could be independently excited by the laser. It was found that excitation of the intense (1, 0, 0) vibronic band or of any of the higher-lying levels of the upper electronic state produced vibrationally unrelaxed emission from the (0, 2, 0) and (0, 1, 0) levels in addition to the vibrationally relaxed (0, 0, 0) fluorescence. The relative intensities of emission from these three levels were $\approx 1:10:90$, independent of which of the higher-lying levels is excited. These observations imply that the stretching vibrational modes relax via intermode conversions into the overtones of the low-frequency bending vibration ν_2. The relaxation in this molecule would thus seem to follow the energy gap law.

The importance of intermode coupling elements was demonstrated in recent time-resolved emission studies of ClCF and NCO. The laser induced fluorescence of matrix isolated ClCF comes mainly from the vibrationless (0, 0, 0) level.[30,118] In solid Ar a small fraction of the emission intensity originates from the (0, 0, 1) state, containing one quantum of the C–Cl stretch. The excitation spectrum of the relaxed fluorescence in Fig. 9*a* can be analyzed in terms of all three upper state modes. The excitation spectrum of the unrelaxed fluorescence in Fig. 9*b* on the other hand, contains only vibrational levels which involve excitation of the ν_3 mode. The states corresponding to excitation of the pure ν_2 vibrational progression, whose expected positions are denoted by arrows in Fig. 9*b*, are absent. This implies that the overtone states of this progression do not undergo single phonon relaxation into quasiresonant levels involving the stretching vibrations. They relax by multiphonon vibrational relaxation processes within the ν_2 manifold.

NCO presents a particularly interesting example.[31] This linear free radical has a rather complex spectroscopy[119,120] with Renner splitting in the ground electronic state and numerous Fermi resonances in the excited electronic state. These resonances occur between the symmetric stretching mode ν_1 and overtones of the bending mode. The electronic spectrum shows remarkably little perturbation of the vibronic structure by the solid medium. Excitation of the higher vibrational levels results in extensive vibrationally unrelaxed emission, indicating that vibrational relaxation is slow compared with the 170-nsec radiative lifetime. The observed relaxation processes are denoted by the wiggly arrows in the energy level diagram in Fig. 10. Selective excitation of individual vibronic levels reveals fast relaxation within groups of levels which are, from the spectroscopic studies, known to be in Fermi resonance. However, no relaxation involving other levels is observed. For instance, the (0, 0, 2) state relaxes completely into the (0, 4, 0) state 200 cm^{-1} below in less than about 5 nsec. On the other hand, this state does not further relax into the (1, 0, 0) state which lies 220 cm^{-1} lower in energy; the

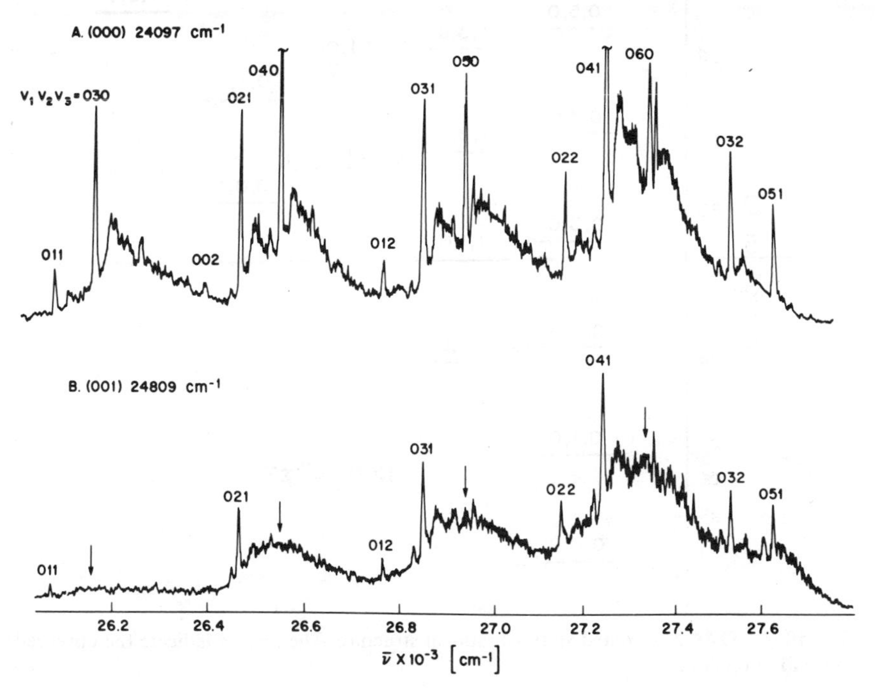

Fig. 9. Excitation spectra of ClCF fluorescence in solid Argon; (*a*) relaxed fluorescence exhibiting both bending and stretching modes; (*b*) unrelaxed (0, 0, 1) fluorescence containing only levels involving ν_3 excitation. Expected positions of the missing members of the pure ν_2 bending progression are denoted by arrows.

corresponding relaxation rate is slower by at least four orders of magnitude.

Even when no unrelaxed spectra are observed, analysis of emission spectra can sometimes yield information about ground-state vibrational relaxation.[121] In the C_3 radical, whose stretching vibrations both occur at relatively high frequencies,[122] the ν_2 bending mode is uncommonly low ($\approx 65\ cm^{-1}$) and lies actually below the Debye cutoff of the host lattice spectrum. The emission progressions involving the low-frequency bending mode appear to be severely broadened, and the corresponding levels relax on subpicosecond timescale. Levels involving the high-frequency modes are, on the other hand, sharp (FWHM $< 1\ cm^{-1}$) and unperturbed, and clearly relax considerably slower.

The results described previously clearly indicate that vibrational relaxation processes in polyatomic molecules cannot be explained by an

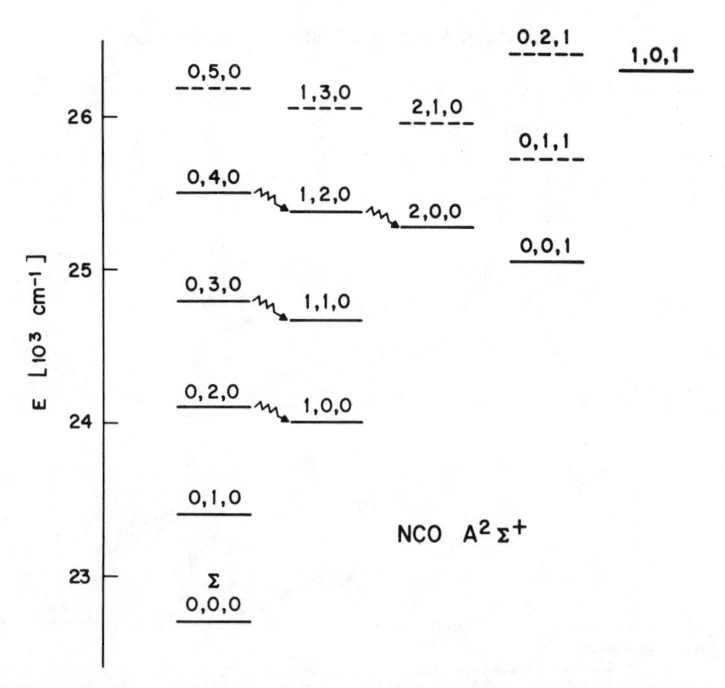

Fig. 10. NCO ($A^2\Sigma^+$) excited state vibrational structure. The arrows indicate the observed relaxation processes.

energy gap law alone. The presently unknown intermode anharmonic coupling elements must also be taken into consideration. Effects that enhance intermode coupling, such as Fermi resonance, enhance intermode transitions in the solid phase just as they do in gas phase collisions.[29]

There are as yet no similar studies of vibrational relaxation in molecules larger than triatomic. One can obtain information about the relaxation processes from a laser fluorescence experiment of the preceding type only if the radiative and nonradiative relaxation rates are of comparable magnitude. In large polyatomic molecules relaxation is generally too fast, and only emission from the vibrationless level is observed. Different techniques will clearly be needed in this case.

Recently, various line narrowing and hole burning methods have been used successfully to measure the lifetimes of the pure electronic transitions in porphyrins and other large molecules in organic glasses.[123–125] Typically, in these experiments a narrow-bandwidth laser is used to bleach a hole in the inhomogeneously broadened spectral line. The width of the resulting notch in the inhomogeneous site distribution is then indicative of the homogeneous linewidth and, provided lifetime is the dominating broadening mechanism, of the lifetime. Similar techniques should be applicable to the matrix relaxation studies. In a recent investigation[126] using an unnarrowed, $0.5\,cm^{-1}$ dye laser, a distinct broadening of this kind was observed for the high vibrational levels in the excitation spectrum of H_2-Phtalocyanine in solid Ar. Studies with more monochromatic sources in molecules of intermediate size should provide invaluable information about vibrational relaxation mechanisms and pathways in polyatomic molecules.

We should mention in this connection the pioneering experiments of Turner et al. in infrared laser-induced photochemistry.[127,128] They observe that excitation of the CO stretching vibrations of various isotopic varieties of $Fe(CO)_4$ enriched in $^{13}C^{18}O$ using a CW CO laser leads to intramolecular rearrangement of the ligands. The energy of the IR laser photon is insufficient to cause isomerization via the dissociation–recombination sequence. The authors conclude that the process again involves conversion of the CO stretching vibrational energy into the lower-frequency Fe–CO stretching and bending modes, with the resulting high-amplitude distortions leading to the observed rearrangements.

An interesting example of polyatomic vibrational relaxation, related to large amplitude molecular motion previously discussed in connection with photodissociation, occurs when relaxation requires a large change in geometry along a very low frequency normal mode. Biphenyl, for example, is twisted in the ground electronic state and nearly planar in an excited Π–Π* state. If twisted Π–Π* biphenyl is produced by vertical Franck–Condon excitation, can relaxation to the nearly planar form occur? Phenyl ring twisting would require extensive motion by nearby rare gas atoms. In neon host, analysis of a strong absorption-fluorescence asymmetry in the torsion-phonon contours of high-frequency vibronic lines shows that the Π–Π* excited state rotates substantially toward the planar conformation before fluorescence occurs.[129] Some elastic energy of compression must be stored in the surrounding environment in this case.

V. ELECTRONIC RELAXATION PROCESSES

A. Vibrational Relaxation via Real Intermediate States

In the early classical studies of large organic molecules in solid solution at low temperatures, only totally relaxed emission from the vibrationless level was observed. Furthermore, it was often found that triplet states decayed by a very long-lived phosphorescence, with the nonradiative electronic relaxation being inefficient.[9] It is therefore not surprising that in the early theoretical works it was assumed that intrastate vibrational relaxation is much faster than the interstate electronic transition. Although this picture is still valid for most larger molecules, especially at higher temperatures, recent studies described in the previous section indicate that diatomic vibrational relaxation is often very inefficient. In many cases it is slower than competing processes like radiative relaxation or intermolecular energy transfer. We now show that host-induced interelectronic transitions can be much faster than host-induced vibrational relaxation.

Robinson[16] predicted that most radiationless electronic transitions in the condensed phases will occur when a strong interstate coupling is present in the isolated molecule. In particular, one can expect efficient electronic relaxation in those cases, where strong rotational perturbations exist in the spectrum of the gas phase molecule. A particularly interesting example of interstate relaxation processes occurs in CN and has been studied in some detail.[26,130] The most straightforward technique for probing vibrational relaxation in an excited electronic state involves selective excitation of a given vibrational level, and observing the time-resolved behavior of the emission of both the level initially populated and of the lower levels populated by relaxation. A coupled two-level system can be described by the well-known first-order equations:

$$\frac{d[A]}{dt} = -k_A[A] \tag{5.1a}$$

$$\frac{d[B]}{dt} = k_A[A] - k_B[B] \tag{5.1b}$$

where $[A]$ and $[B]$ are populations of the two levels and k_A and k_B are their respective, experimentally determined decay constants. When these are solved with the appropriate boundary conditions one obtains an expression for the time resolved emission of the lower level

$$[B] = \frac{k_A[A_0]}{k_A - k_B}[e^{-k_A t} - e^{-k_B t}] \tag{5.2}$$

If k_A and k_B are quite dissimilar, the function described by (4.2) will have a rise time determined by the larger of the two and a decay determined by the smaller rate constant. In the case of CN, for example the $v' = 3$ and $v' = 4$ $A^2\Pi$ levels are found to decay with 90 and 10-nsec lifetimes, respectively. One would therefore expect for the relaxed $v' = 3$ emission following $v' = 4$ excitation a short risetime and ≈ 90 nsec decay time. In reality, one observes an exponential decay time of 7.5 μsec, two orders of magnitude longer than predicted by the two-level model. This observation necessarily implies that relaxation proceeds via a real, long-lived intermediate level, which in this case is assigned to the $v' = 8$ level of the ground electronic state. This case can be described by a modified set of equations

$$\frac{d[A]}{dt} = -k_A[A] \tag{5.3a}$$

$$\frac{d[X]}{dt} = k_A[A] - k_X[X] \tag{5.3b}$$

$$\frac{d[B]}{dt} = k_X[X] - k_B[B] \tag{5.3c}$$

Here $[X]$ stands for the intermediate level. If again the appropriate substitutions are made and the resulting differential equation is solved, one obtains

$$[B] = \frac{k_A k_X [A_0]}{[k_X - k_A][k_A - k_B][k_B - k_X]} [(k_x - k_b)e^{-k_A t} + (k_B - k_A)e^{-k_X t} + (k_A - k_X)e^{-k_B t}] \tag{5.4}$$

for the decay of the lower level. It may be noted that if $k_A \approx k_B \gg k_X$, then (4.4) can be simplified to

$$[B] = \frac{k_X}{k_b}[A_0]e^{-k_X t} \tag{5.5}$$

That is, the lower-level emission will show a single exponential decay with the lifetime of the long-lived intermediate level. The 7.5 μsec lifetime measured in the given example is the lifetime of the $v'' = 8$ level of the ground electronic state of CN. Although we have discussed the specific example of $v' = 4$ $A^2\Pi$, the relaxation of the other $A^2\Pi$ vibrational levels can be well described using the same model by assuming that relaxation proceeds via intermediate ground-state levels.

A more direct way of confirming this relaxation mechanism would involve monitoring of the transient population of the excited vibrational levels of the $X^2\Sigma^+$ state. This was accomplished as shown in Fig. 11, using a second, probe laser to further excite this population into a higher-lying $B^2\Sigma^+$ state and observing its UV fluorescence. By varying the delay between the pump and the probe laser, the time-resolved profiles of the ground-state level populations can be obtained. The advantage of this approach is that it also permits measurement of the ground-state vibrational levels below the origin of the $A^2\Pi$ state. Results obtained in this way are shown in Fig. 12. It may be noted that

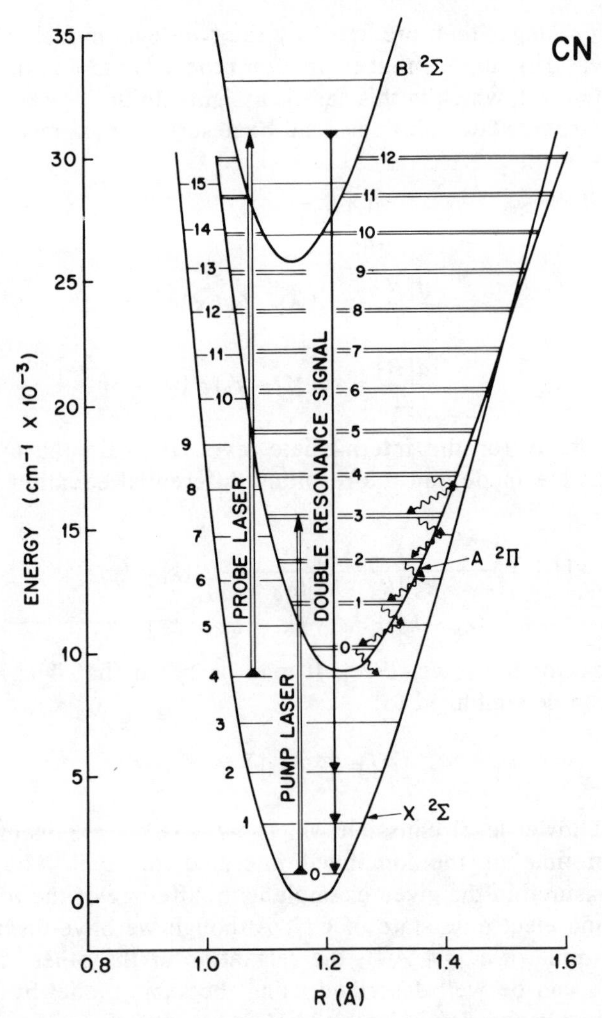

Fig. 11. CN potential curves. The optical–optical double resonance experiment is schematically illustrated.

vibrational levels below the $A^2\Pi$ origin have very long lifetimes in the millisecond range.

The general conclusions of this work were that (*a*) direct vibrational relaxation, both in the $A^2\Pi$ and $X^2\Sigma^+$ states is very inefficient and probably controlled by radiation, and (*b*) the major relaxation mechanism for the higher-lying levels above the $(v = 0)$ $A^2\Pi$ origin involves internal conversions between the $A^2\Pi$ and $X^2\Sigma^+$ manifolds. The rates of these internal conversion processes, when divided by the

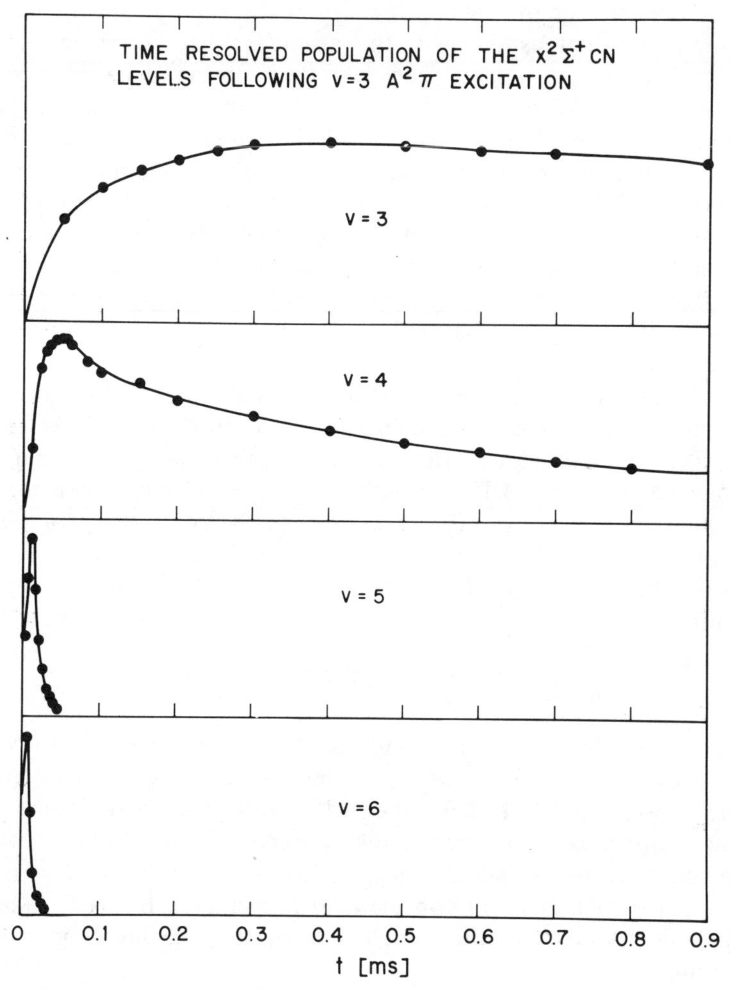

Fig. 12. Experimental time dependence for CN internal conversion cascading and vibrational relaxation. Note the considerably longer lifetimes of the $v' = 3$ and 4 levels which are located below the $A\,^2\Pi$ origin.

appropriate molecular Franck–Condon factors showed qualitatively the expected steep dependence on the size of the energy gap.

This system was also the subject of several theoretical investigations. In the earliest work the rates were calculated[130] using a simple model in which harmonic lattice phonons accept the energy and also provide the promoting modes necessary for coupling between two electronic states with different symmetries. Although the results have shown qualitatively a strong dependence of rates on the size of the energy gaps, quantiative agreement was not established. In particular, the theory predicted a considerably steeper energy gap law than observed experimentally. Later work[131] by Weisman, Nitzan, and Jortner has shown that better agreement can be obtained by considering the quadratic effects and anharmonicity. In a recent investigation, Fletcher, Fujimura, and Lin[132] have proposed that molecular rotation rather than the delocalized lattice phonons accepts the energy and also provides the "promoting mode." With this assumption a satisfactory quantitative agreement with the experiment was obtained suggesting that, at least in the simple case of diatomic molecules, calculations of absolute rates of interelectronic relaxation might be possible.

It is interesting to note that relaxation of the CN $A^2\Pi$ system by Ar atoms in the gas phase was recently studied using pulsed laser excitation.[28] Like in the matrix, the relaxation proceeds via internal conversions between the $A^2\Pi$ and $X^2\Sigma^+$ states. However, unlike the 4°K solid where the relaxation is irreversible, in the room temperature gas the reverse reactions have to be considered. The detailed balance requires, that the forward and reverse constants be related by the equation

$$\frac{k_A}{k'_A} = 2\exp\left[-\frac{\Delta E}{kT}\right] \tag{5.6}$$

where ΔE is the energy gap and the factor of 2 accounts for the degeneracy of the $A^2\Pi$ state. Experimentally, one observes strongly double exponential decays for the $A^2\Pi$ levels. The fast lifetime component corresponds to establishing a dynamic equilibrium between the given $A^2\Pi$ level and the nearest ground-state level. The longer component, which is often considerably longer than the $A^2\Pi$ radiative lifetime, corresponds to the joint decay of the two levels in thermal equilibrium.

B. Spin-Forbidden Processes

Whereas the CN case represents a particularly clear example of electronic relaxation where the spectroscopy is well understood and which was therefore best suited for theoretical studies and modeling, similar interstate relaxation processes have been observed in numerous other systems. Even more surprisingly, it was established that nonradiative electronic relaxation in matrix-isolated small molecules seems to be similarly efficient even in situations where the process is forbidden by the spin selection rules. The ability of the spin forbidden interelectronic transitions to compete favorably with direct vibrational relaxation was first demonstrated in the C_2^- molecular ion.[133] In this species, studies of the time dependence of the $B^2\Sigma_u^+$ state vibrational decay have shown that relaxation proceeds via vibrational levels of another electronic state. Using isotopic substitution it was possible to show that the $v = 0$ level of this state is located between $v' = 1$ and 0 of $B^2\Sigma_u^+$, as shown in Fig. 13. This state was then assigned as the previously unobserved $a^4\Sigma_u^+$ state predicted to be in this spectral range. In fact, the matrix data permitted the assignment of previously observed $B^2\Sigma_u^+$ rotational perturbations, and established the $a^4\Sigma_u^+$ spectroscopic constants in the gas phase.

Also the effects of the nature of the host were investigated in this system. An unexpected result was absence of any systematic variation of the efficiency of the intersystem crossing with the mass of the host. Although the rates showed some variation from matrix to matrix and from one C_2^- isotopic species to another reflecting the differential shifts of the $B^2\Sigma_u^+$ and $a^4\Sigma_u^+$ vibronic levels, overall the relaxation rates in Ne were as efficient as in the much heavier and polarizable Xe. The external

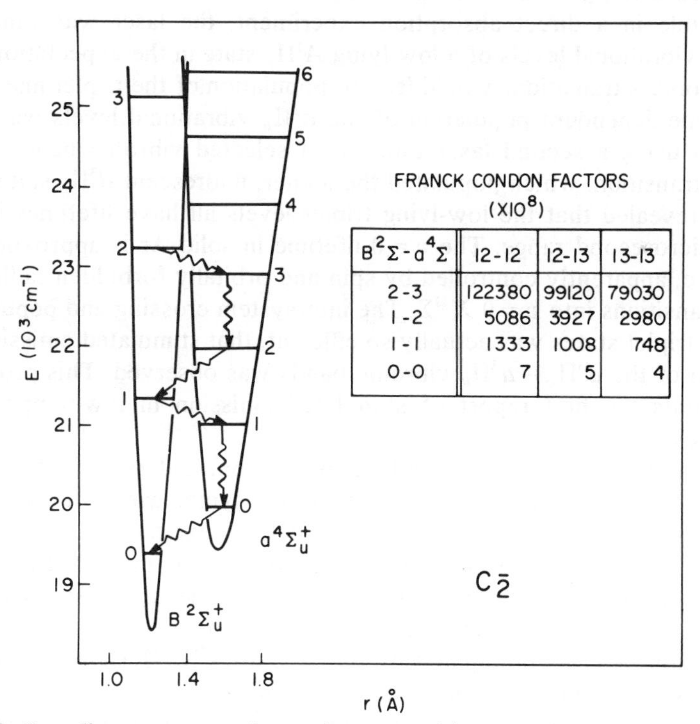

$B^2\Sigma - a^4\Sigma$	12-12	12-13	13-13
2-3	122380	99120	79030
1-2	5086	3927	2980
1-1	1333	1008	748
0-0	7	5	4

Fig. 13. Cascading sequence and gas phase potential curves for C_2^-. Broken arrows represent radiationless transitions.

heavy atom effects on the spin-orbit coupling appears to be unimportant for the relaxation processes in C_2^-.

Another interesting molecule where spin forbidden processes were found to occur in the matrix with surprising ease is the C_2 radical. This molecule has a complex spectroscopy with several low-lying electronic states. The lowest triplet state, which was for a long time assumed to be the ground state was only rather recently shown to lie some 700 cm^{-1} above the ground $X^1\Sigma_g^+$ state. The gas phase radiative lifetime of this state is undoubtedly extremely long; for some time it was, based on the electronic absorption spectra, believed that both $X^1\Sigma_g^+$ and $a^3\Pi_u$ states can be stabilized in low-temperature matrices.[134,135] Milligan and Jacox[136] have later established that the absorption bands originally assigned to the $a^3\Pi_u \rightarrow d^3\Pi_g$ Swan transition are really the Herzberg–Lagerqvist bands of C_2^-. Frosch[137] later did observe the C_2 Swan bands in emission following X-irradiation of matrices containing acetylene and he reported emission from several excited vibrational levels of the $d^3\Pi_g$ state.

The C_2 molecule was recently[38] reexamined in solid Ar and Ne matrices using pulsed laser excitation. Whereas the triplet state is not accessible in a direct absorption experiment, the laser was tuned to excite vibrational levels of a low-lying $A^1\Pi_u$ state in the expectation that radiationless transitions would lead to population of the triplet manifold. The time-dependent population of the $a^3\Pi_u$ vibrational levels was then probed using a second laser tuned to a selected vibronic band of the Swan transition, which populated the higher, fluorescent $d^3\Pi_g$ state. The study revealed that the low-lying triplet levels all have lifetimes in the low microsecond range. The $v = 0$ lifetime in solid Ar is approximately 65 μsec, apparently controlled by spin and orbitally forbidden radiationless transitions into $v = 0$ X $^1\Sigma$. The intersystem crossing and population of the triplet states was actually so efficient, that stimulated emission on several of the $d^3\Pi_g \rightarrow a^3\Pi_u$ vibronic bands was observed. This probably represents the first report of stimulated emission in low-temperature matrices.

The ability of matrices to induce interaction, and radiationless transitions, between states of different orbital symmetries was observed spectroscopically in a recent study of O_2 excited states.[138] Near 30,000 cm^{-1} O_2 has $C^3\Delta_u$ and $A^3\Sigma_u^+$ excited states with nested potential energy curves. In the nonrotating gas phase molecule, there is only a weak spin–spin interaction between them. The gas phase near UV absorption of O_2 is dominated by the $X^3\Sigma_g^- \rightarrow A^3\Sigma_u^+$ Herzberg bands. In solid N_2 host, a progression of strong perturbations between levels v of $C^3\Delta$ and $v - 2$ of $A^3\Sigma_u^+$ is observed in fluorescence excitation spectra. As shown in Fig. 14, the $X^3\Sigma_g^- \rightarrow C^3\Delta_u$ characteristic triplets actually

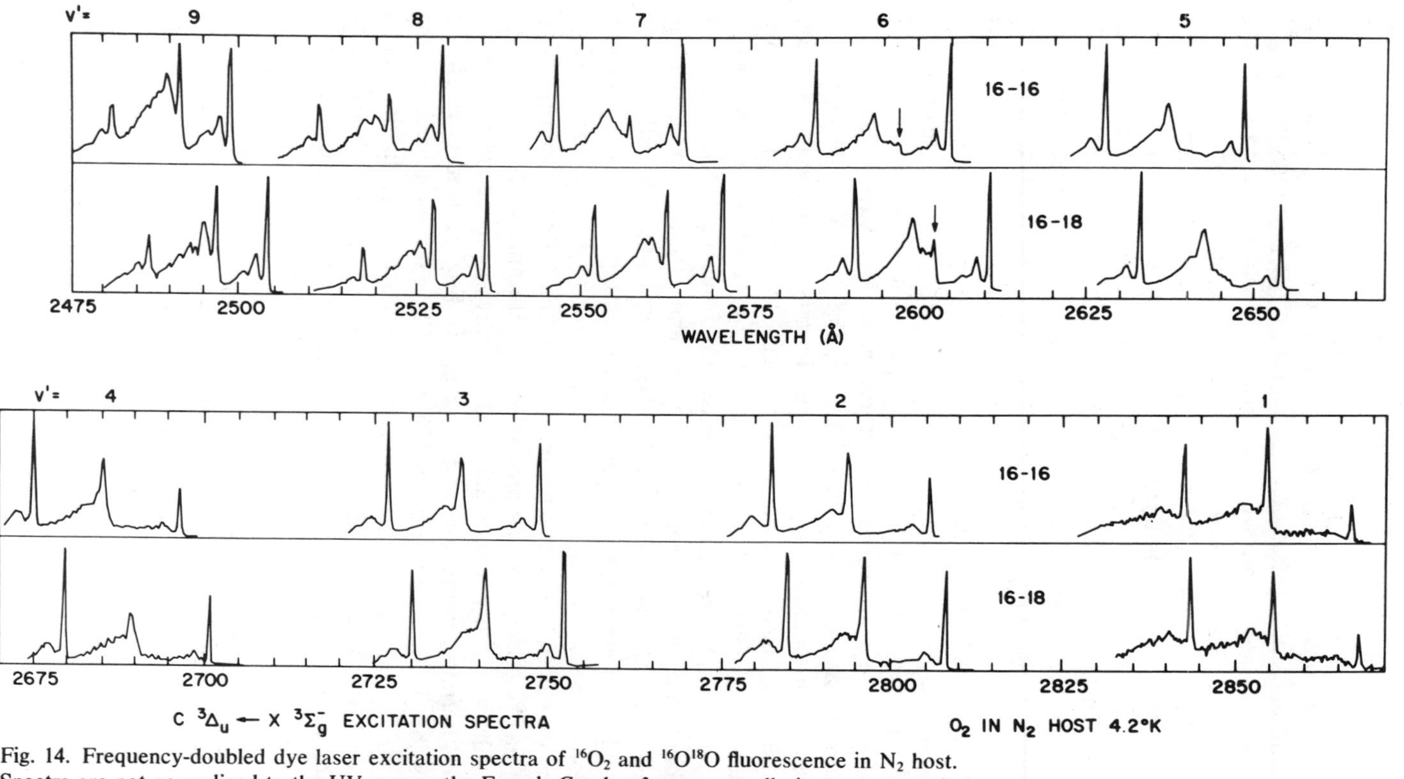

Fig. 14. Frequency-doubled dye laser excitation spectra of $^{16}O_2$ and $^{16}O^{18}O$ fluorescence in N_2 host. Spectra are not normalized to the UV power; the Franck–Condon factors actually increase strongly from low to high v. Arrows for $v = 6$ mark the location of a *second* ZPL for $^3\Delta_2$, created by the mixing with the broad underlying $A\,^3\Sigma_u^+$ absorption.

dominate the condensed phase absorption spectrum. The middle $\Omega = 2$ component of $^3\Delta$ is "pushed blue" of its zero-order position and lifetime broadened by interaction with the lower $^3\Sigma_u^+$ state. Analysis of this host-induced interaction yields a $27\ \text{cm}^{-1}$ interaction matrix element. This interaction appears to require a phonon promoting mode, as the solid-state spectroscopy is consistent with S_6 environmental symmetry in which $^3\Delta_u \rightarrow E_u$ and $^3\Sigma_u^+ \rightarrow A_u$. The fluorescence actually comes from $C^3\Delta_1$, and not from $A^3\Sigma_u^+$ as previously assigned.[139,140]

The molecules discussed previously have fairly large vibrational frequencies. It was therefore of interest to determine whether in heavier molecules with smaller vibrational spacing, vibrational relaxation would prevail over interstate electronic relaxation processes. We have therefore examined the heaviest member of the group IV diatomic molecules.[63] Pb_2 spectroscopy is very complex with a large number of low-lying and uncharacterized electronic states. We have studied emission from a state we denote $F0^+$. Excitation of individual levels of this state results in emission from the level directly excited with no vibrational relaxation in spite of the low ($\sim 150\ \text{cm}^{-1}$) vibrational spacing. Two intense infrared emission systems are seen from two low-lying electronic states populated by radiationless transitions from the $F0^+$ levels.[63,141] It was concluded that internal conversions and electronic radiationless processes prevail over vibrational relaxation. Unlike the other cases discussed previously, the Pb_2 population does not return into lower F vibrational levels. This irreversibility appears to be due to the large number of electronic states present. Once the molecule leaves the F state it has statistically a low probability of returning.

The relative facility with which internal conversions occur in the solid matrices permits population and characterization of states which are inaccessible in direct absorption. Thus in Cl_2, excitation of either the $B^3\Pi 0_u^+$ or of the $^1\Pi_1$ states produces long-lived 60 msec luminescence from the lower lying and previously unobserved $^3\Pi_2$ state.[37] The same emission is observed in thermoluminescence,[142] that is, in an experiment where a matrix containing Cl atoms is slowly warmed as to permit atom diffusion and recombination. In the heavier halogens,[70,75,143] Br_2 and ICl, the B state ($\Omega = 0$) levels fluoresce with near unity quantum yields; the relaxation into the lower $\Omega = 1$ and $\Omega = 2$ components apparently does not proceed. The larger spin–orbit coupling produces larger energy separation between the $^3\Pi\,\Omega$ components. The vanishingly small Franck–Condon factors connecting the $v = 0$ $B^3\Pi 0_u^+$ level with the high vibrational levels of the lower Ω states lead to the inefficient relaxation in this case.

A very similar situation was observed[144] in matrix-isolated S_2. The gas phase near-UV spectra of this molecule are determined by transitions into the $B^3\Sigma_u^-$ state, analogous to the Schumann–Runge bands of O_2, and the same bands are observed in matrix absorption.[145,146] A recent time-resolved study has, however, shown that the emission actually originates from a lower-lying state populated by nonradiative transition. This is probably the $B''^3\Pi_u$ state, which is believed to be responsible for the strong perturbations observed in the gas phase $B^3\Sigma_u^-$ spectrum. This new fluorescence assignment is also consistent with the fact that the same S_2 emission is also observed in matrix thermoluminescence experiments.[147,148] The $B^3\Sigma_u^-$ state correlates with an excited sulphur atom and should not be populated by recombination of ground-state S atoms.

Interesting cases of internal conversions and intersystem crossing between various electronic states have also been studied in TiO and ZrO by Broida and co-workers.[149,150] In TiO excitation of the $X^3\Delta \rightarrow B^3\Pi$ transition leads to efficient intersystem crossing and results in strong emission from the lower lying $b^1\Pi$ state. A direct observation of the weak intercombination $b^1\Pi \rightarrow X^3\Delta$ emission permitted experimental measurement of the singlet–triplet energy separation.

The dominance of interelectronic cascading over direct vibrational relaxation, and the discovery of states not accessible in direct absorption studies, are demonstrated by the unusual example of isolated $(O_2)_2$ dimers in solid neon.[108,109] These dimers in a cooperative act absorb one red photon producing an excited electronic state in which both halves of the dimer are in the metastable $a^1\Delta$ state of O_2

$$O_2(^3\Sigma_g^-) + O_2(^3\Sigma_g^-) + h\nu(\approx 6300\text{ Å}) \rightarrow O_2(^1\Delta_g) + O_2(^1\Delta_g) \qquad (5.7)$$

The relaxation mechanism of the vibrationally excited $^1\Delta + {}^1\Delta$ state actually involves intersystem crossing to a nearby vibrational level of the $^3\Sigma + {}^1\Sigma$ electronic state, direct vibrational relaxation by one quantum in this state and finally reverse crossing to the next lower $^1\Delta + {}^1\Delta$ vibrational level. The $^3\Sigma + {}^1\Sigma$ state absorbs too weakly to be seen in direct absorption. The time dependences of all levels in the cascade are directly observed via their fluorescence into the ground electronic state. The $^1\Delta + {}^1\Delta$ fluorescence terminates on the singlet component of the $^3\Sigma + {}^3\Sigma$ ground electronic state, while the $^1\Sigma + {}^3\Sigma$ emission terminates on the triplet component of $^3\Sigma + {}^3\Sigma$. In this fashion one can measure the singlet–triplet exchange splitting in the ground state of the dimer to be 55 cm^{-1}.

All examples discussed previously involve irreversible cascading of population from one level to another level lying ΔE lower in energy. Actually microscopic reversibility requires that the forward (k) and reverse (k') rates be related via

$$\frac{k'}{k} = \exp\left[-\frac{\Delta E}{kT}\right] \tag{5.8}$$

At 4.2°K k' is normally negligible. However, the $(O_2)_2$ dimers present several examples where levels from $^1\Delta + {}^1\Delta$ and $^1\Sigma + {}^3\Sigma$ states actually lie within 10 cm^{-1} of each other. Experimentally, the time-resolved emission of both components shows that the states establish a dynamic equilibrium on a subnanosecond time scale. Observation of such a dynamic equilibrium implies that k and k' are much faster than the individual decay rates of either component. As expected, this occurs whether one state is optically pumped, or the other is populated via a cascade process.

All the preceding examples indicate that electronic relaxation processes are remarkably efficient. This efficiency can be rationalized by considering the phonon line shapes and electron–phonon coupling. As discussed earlier, the guest–host interaction potential is controlled by the periphery of the electron density distribution in the guest species and is insensitive to motion of the nuclei and changes in vibrational state. Accordingly, pure vibrational transitions in matrix isolated molecules show very little intensity in the phonon wings. Since vibrational transitions generally require little rearrangement of the guest lattice atoms, they provide poor Franck–Condon factors for multiphonon relaxation. Electronic transitions, on the other hand, result in more drastic changes in electron density distribution and, accordingly, they are often accompanied by intense phonon wings and much more favorable phonon Franck–Condon factors. A similar argument has in fact been advanced[94] to suggest that delocalized lattice phonons would accept energy during electronic relaxation processes, while local modes (e.g., rotation) would accept energy during vibrational relaxation processes. Atomic relaxation studies, to be discussed in a following section, directly demonstrate the ability of lattice phonons to accept energy during electronic relaxation processes.

C. Atomic Relaxation

For some time it has been known that alkali atomic spectra in rare gas solids are broadened, and occur as triplets when initially degenerate ^2P states are involved.[9] Moskovits and Hulse[151] have recently suggested that this splitting is not a static Jahn–Teller effect due to a local environment distortion, but rather reflects a diatomic-like binding with one (perhaps fluxional) nearest-neighbor rare gas atom. In many atoms the fluorescence quantum yields appear to be low,[152–154] suggesting that very efficient nonradiative relaxation takes place. In the case of isolated Be atoms, fluorescence from metastable ^3P apparently produced by radiationless transitions from an absorbing ^1P state, has been observed.[155]

Recently tunable laser excitation has been used to prove that population of either the ^1P or ^1D states of matrix-isolated Ca atoms leads to intense ^3P emission.[156] This fast, direct multiphonon electronic relaxation, in the absence of vibrational or local mode degrees of freedom, confirms the strong coupling between electrons and lattice phonons previously postulated in the diatomic studies. In this study ZPLs were clearly resolved on the atomic bands for the first time. This proves that the line broadening mechanism is a homogeneous phonon contour. There appears to be no special source of inhomogeneous broadening for isolated atoms when compared with isolated molecules. One may, however, note in Fig. 15 that the phonon sidebands are extremely broad, in particular in the excitation spectrum. Spectra of matrix isolated atoms appear to be on the average considerably broader than spectra of molecular species. Atoms apparently interact much more strongly with the host atoms than do molecular guests. This appears to reflect the unsaturated valence electrons of reactive atoms as compared with closed shell molecules. The stronger interaction is then reflected in the extensive phonon sidebands and in the resulting apparent red shifts of the fluorescence which are often observed.[157]

The high efficiency of the intersystem crossing in Ca atom may be partially due to the presence of the nearby ^1D and ^3D states, through which the relaxation apparently proceeds. Consistent with this idea, no relaxation is observed in Mg atoms, where the D states are absent; here the ^1P state fluoresces with near unity quantum efficiency.[158,159] In view of these observations, the efficient intersystem crossing reported for matrix isolated Be atoms appears somewhat surprising.[149]

Ozin and co-workers[160,161] have discovered an intriguing "photoaggregation" phenomenon in matrices containing metal atoms. Continuous irradiation in one atomic absorption line produces specific diffusion of this atom through the matrix, and controlled formation of small metal clusters. Although this observation suggests "local melting," we have previously described how the photoselection studies of the photodissociation cage effect for diatomics show that no local melting occurs, even if 2 eV of energy is deposited in the local environment. This observation could be reconciled with the "photoaggregation" effect if the probability of atom exchange with a neighboring rare gas atom is actually very low per quantum absorbed. In this case, almost every

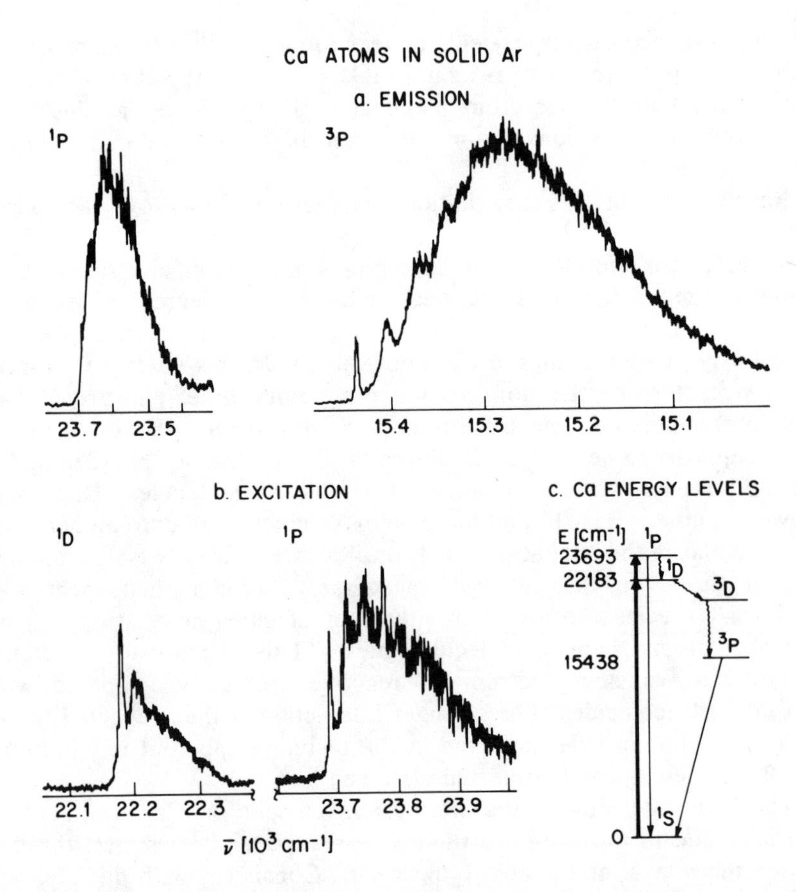

Fig. 15. Spectra of Ca atoms in solid Ar. (*a*) The emission spectrum and (*b*) the excitation spectrum obtained by monitoring the $^3 \rightarrow {}^1P$ emission intensity. The shape of the $^1S \rightarrow {}^1P$ excitation band and in particular the fast falloff at the high energy side are due to decreasing laser power. The broad phonon sideband actually extends well beyond 24,000 cm^{-1}.

relaxation event would leave the local environment unchanged in structure.

A different type of "photodiffusion" has been reported[162] for repeated pulsed infrared irradiation of SF_6 in matrices containing both $^{32}SF_6$ and $^{34}SF_6$. It is observed that both isotopic species gradually migrate to the surface and leave the matrix. Aggregates are not formed; they would form if the matrix were simply heated to the diffusion temperature. The detailed explanation of this experiment remains obscure.

VI. VIBRATIONAL ENERGY TRANSFER

A. Long-Range Multipole Interaction

We have seen that vibrational relaxation times can be very long. Even in relatively dilute samples, intermolecular vibrational energy transfer competes with relaxation and may dominate the excited state kinetics if relaxation can be neglected. In the classic studies of CO relaxation[83,105] extensive transfer of vibrational energy among dilute CO guests in solid Ar has been spectroscopically observed. Two phenomena, both related to the fact that entropy is unimportant in Boltzmann equilibria near 4.2°K, differentiate the energy flow from similar gas phase transfer and equilibration processes at 300°K. Firstly, the bulk of the energy flows into the minor isotopes $^{13}C^{16}O$ and $^{12}C^{18}O$ (present only in natural abundance) when the infrared laser initially excites $^{12}C^{16}O$. This occurs because the transfer exothermicities ΔH to $^{12}C^{18}O$ and $^{13}C^{16}O$ are large with respect to $kT \approx 3\ \text{cm}^{-1}$. In this situation, subsequent reverse transfer to $^{12}C^{16}O$ is slow. Secondly, exothermic processes of the sort $v = n + v = 1 \rightarrow v = n + 1 + v = 0$ pump considerable energy into the higher vibrational levels $v > 1$. These processes are exothermic by twice the CO vibrational anharmonicity, which is also large with respect to kT. CO ($v > 1$) excitons are effectively trapped on the particular CO in which they are formed, as the long-range dipole–dipole interaction only weakly transfers such excited states from one CO to another. Population inversions can be formed on the $v = 1 \rightarrow v = 0$ transition of the minor isotopes, and between $v > 1$ levels of $^{12}C^{16}O$.

The first time-resolved study[163] of a vibrational energy transfer rate dependence upon transfer exothermicity ΔH involved processes such as $\text{ND}^*(v = 1) + \text{CO}(v = 0) \rightarrow \text{ND}^*(v = 0) + \text{CO}(v = 1) + 78\ \text{cm}^{-1}$. The * indicates that the donor is electronically (as well as vibrationally) excited. The additional electronic excitation for the vibrational energy donor actually simplifies the energy transfer kinetics. ND* has a lower vibrational frequency than the ground electronic state of ND, so that transfer of the vibrational quantum to nearby unexcited ND molecules is endothermic and negligible at 4.2°K. Therefore the only transfer process that occurs is the desired nonresonant transfer to the added acceptor CO molecules. If the donors and acceptors have a random distribution in the solid Ar, then some ND*($v = 1$) molecules will have a nearby CO, whereas others will not. The former will have a short lifetime due to fast energy transfer, while the latter donors will have longer lifetimes. The total donor emission from the matrix will be multiexponential. Forster originally considered this situation in the context of electronic energy transfer in solution, and showed that the emission decays as

$$I(t) = \exp\left[\frac{-t}{\tau_0} - 2\gamma\sqrt{\frac{t}{\tau_0}}\right] \tag{6.1}$$

Here τ_0 is the lifetime in the absence of CO, whose normalized concentration is γ. The experimental decays actually are well described by this equation as shown in Fig. 16. By plotting the experimental γ versus [CO], the critical transfer radius R_0 in the equation

$$k_{et} = \frac{1}{\tau_0}\left(\frac{R_0}{R}\right)^6 \tag{6.2}$$

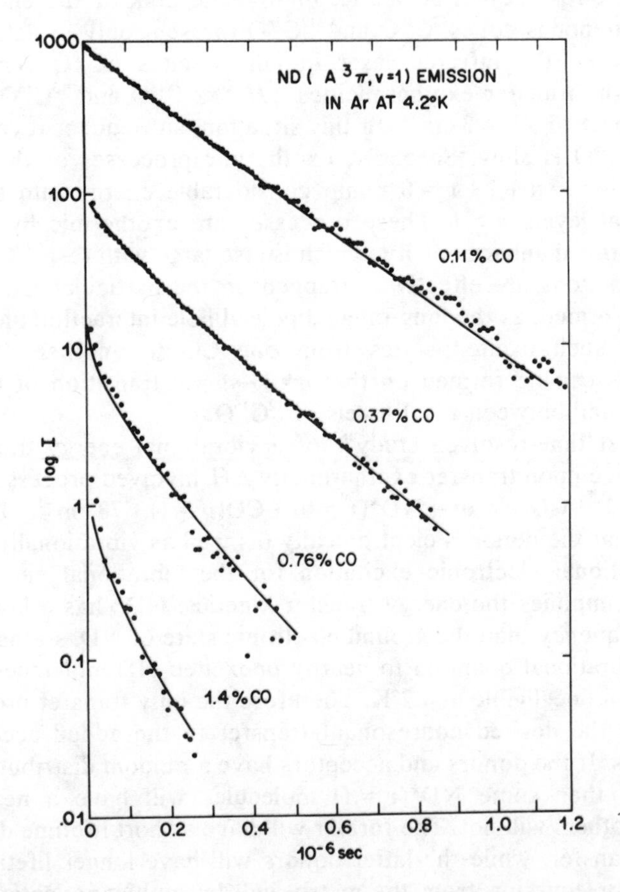

Fig. 16. Time-resolved ND* ($v = 1$) nonexponential decays as a function of CO concentration. Solid lines are theoretical fits according to (6.1). Isolated ND radicals are created by VUV photolysis of ND_3.

is obtained. Both the time-resolved data, and the quantum yield of transfer as a function of [CO], are consistent with these Forster dipole–dipole energy transfer formulae.

Figure 17 shows that by using both ^{12}CO and ^{13}CO as acceptors, and by using ND* ($v = 2$) as well as ND* ($v = 1$) as donors, the energy transfer rate dependence upon exothermicity ΔH can be explored. With ND* an exponential energy gap law of the form

$$k_{et} \,\alpha\, v_d(v_a + 1) \exp\left[\frac{-\Delta H}{28 \text{ cm}^{-1}}\right] \tag{6.3}$$

describes the three data points. However, transfer from NH* to CO does not follow this law and is anomalously fast by several orders of magnitude.

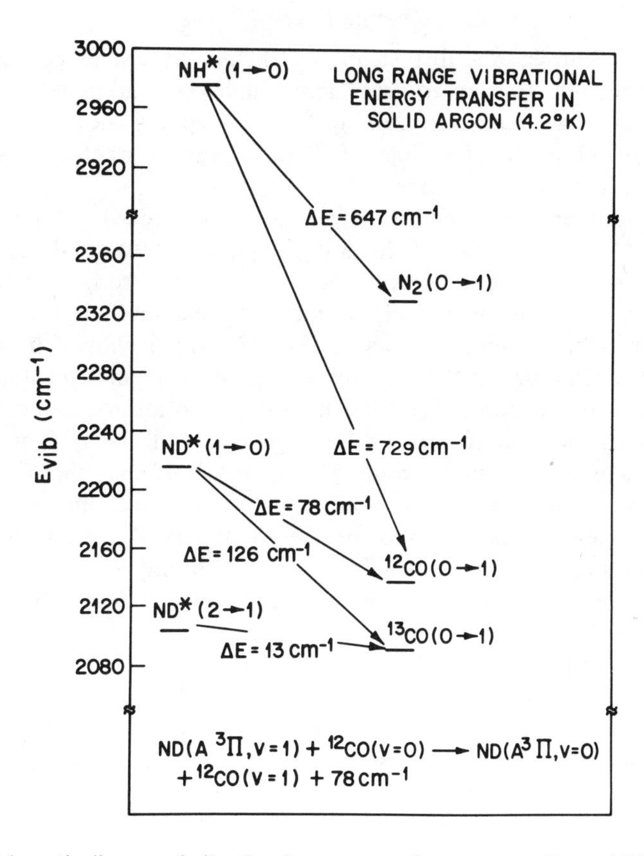

Fig. 17. Schematic diagram of vibrational energy transfer processes. Donor NH* and ND* levels are indicated on the left, and acceptor ^{12}CO, ^{13}CO, and N_2 levels on the right.

S. H. Lin and co-workers[161] have theoretically formulated the relationship between nonresonant energy transfer and vibrational relaxation. Both processes involve a guest's ability to convert internal vibration energy into lattice phonons. The discovery above that NH* is especially effective in nonresonant transfer is consistent with the previously discussed faster vibrational relaxation of NH* as compared with ND*. In this study, the rate of dissipation of $78\,cm^{-1}$ into the lattice could be compared with the vibrational relaxation data by dividing out the transition dipole factors in Lin's formulae. Very recently, Gerber and Berkowitz[162] theoretically formulated the nonresonant energy transfer process with explicit involvement of the rotational accepting modes. Excellent agreement was obtained for ratios of the rate processes experimentally measured.

B. Exchange Energy Transfer

Little is known about fast *nearest-neighbor*, exchange vibrational energy transfer. Energy transfer from highly vibrationally excited N_2 sitting next to an excited N atom in N_2 matrices has been observed.[164] A nearest-neighbor transfer time of 17 nsec was indirectly obtained for these perturbed N_2 molecules.

An experiment has been performed upon isolated $(O_2)_2$ dimers in solid neon host.[108] The strategy of this latter experiment was to detect the fast transfer as a splitting of vibronic ZPLs in the electronic spectra of the dimer. An analysis of the electronic fine structure shows the dimer geometry is rectangular D_{2h} as in Fig. 18, which shows the expected vibrational structure of both the upper and lower electronic states of the dimer. The anharmonicity of the monomer vibration produces certain splittings, such as between the levels $(2+0)$ and $(1+1)$. The notation $(n+m)$ implies that one O_2 contains n quanta, and the other O_2 contains m quanta. Levels such as $(1+0)$ are further split by vibrational energy transfer. Such splittings cannot be simply observed in absorption spectra, as the upper component is forbidden under electric dipole selection rules. However, this level is thermally populated following laser excitation of the lower component, and the splitting can be observed in the subsequent fluorescence. The exactly resonant transfer time from ${}^{16}O_2({}^1\Delta, v=1)$ to nearest neighbor ${}^{16}O_2({}^1\Delta, v=0)$ is 14 psec. An attempt to obtain the isotopic dependence via this technique failed, as isotopic substitution drastically lowers the fluorescence quantum yield due to enhanced intersystem crossing into the ${}^3\Sigma^- + {}^1\Sigma_g^+$ state. A strong energy gap law, if applicable, would yield a nonresonant transfer time from ${}^{16}O_2({}^1\Delta, v=1)$ to ${}^{18}O_2({}^1\Delta, v=0)$ of approximately 10^{-9} to 10^{-10} sec.

A quantitative analytical theory of vibrational energy transfer due to

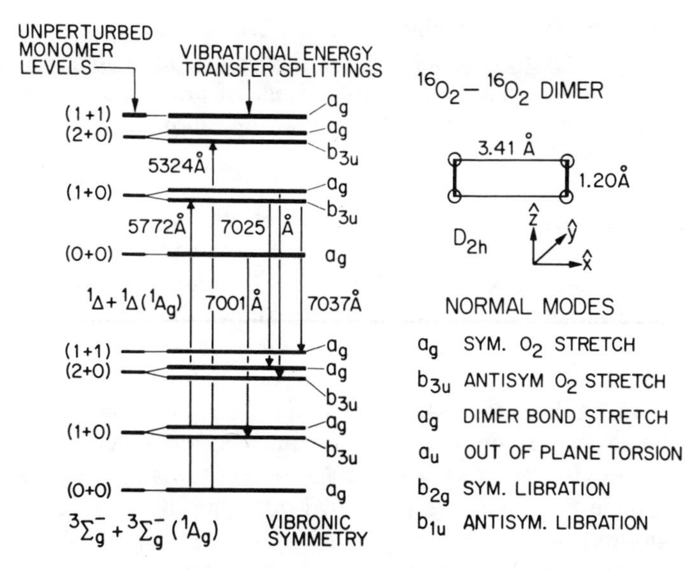

Fig. 18. Schematic diagram indicating $(O_2)_2$ dimer geometry and vibronic level structure. The correlation diagram shows how monomer states split due to vibrational energy transfer. Vertical lines indicate observed transitions. Electronic fine structure components are not shown. The bond lengths shown refer to the nearest-neighbor structure in solid α oxygen.

weak intermolecular exchange forces does not exist. In the case of a stronger interaction, intermolecular transfer begins to resemble intramolecular vibrational redistribution in a tetra-atomic molecular species. Such intramolecular redistribution is also poorly understood.

Long-range energy transfer from ground state C_2^- to C_2H_2 has been studied by Nibler and co-workers using the two-laser pump–probe technique.[165,166] They report that the transfer rate falls off much more slowly than R^{-10}, which is expected from quadrapole–quadrapole electrostatic coupling. They intriguingly suggest a "direct" through Ar interaction, with the entire system being considered as a "large" molecule. It may also be that C_2^- has a small permanent dipole in the matrix due to the influence of the nearby counterion.

VII. CONCLUSIONS.

Tunable laser sources and modern laboratory electronics have enabled a rapid advance in our understanding of small molecule radiationless transitions in rare gas matrices. This review, which is by no means complete, has described some of the considerable amount of data which is now available. These systems are attractive because of their elementary, prototypical nature which invites theoretical modeling. Nevertheless, we have shown that an enormous range of phenomena may occur, even in these simple cases. We suggest that the following generalizations seem appropriate:

1. Rare gas matrices provide extremely elastic, yet well-defined and permanent local environments. The high host compressibility, which is related to the weak host binding, allows the host to not hinder large amplitude molecular vibration. It also allows the small "bubble" to form around a $3s\sigma$ Rydberg state. Only in the case of specific nonbonded interactions, such as charge transfer or hydrogen bonding, do molecular guests show a distinct change in structure from the gas phase. Rapid deposition of several electron volts of energy in the local environment by some radiationless transition normally will not produce local melting, or anneal one type of local environment into another.
2. Direct vibrational relaxation by diatomics is characterized by an enormous range of rates ($\approx 10^{-13}$ to $\approx 10^{2}$ sec.). If excited electronic states are included, these rates are not well predicted by an energy gap law. If one were to exclude hydrides and only consider ground electronic states, then a qualitative energy gap law probably would apply. Large quanta ($\approx 2000\ cm^{-1}$) would relax in the ~1-sec range, whereas small quanta near the Debye cutoff ($\approx 65\ cm^{-1}$) would relax in the 10^{-12}-sec range. A considerable body of evidence now indicates that rotational motion accepts the energy during diatomic and polyatomic hydride relaxations, which are anomalously fast when compared with other molecules. Very limited evidence suggests lattice phonons accept energy for nonhydride molecules, except in special cases of highly asymmetric environments.
3. Interelectronic state relaxation processes (both spin forbidden and spin allowed) are normally quite efficient, especially when favorable Franck–Condon factors exist. Such processes often dominate the energy relaxation in excited electronic states. This may be related to better phonon Franck–Condon factors produced by significant resolvation around the final electronic state of the guest.
4. Mode-to-mode vibrational relaxation in small polyatomic molecules is a sensitive function of the anharmonic intermode coupling elements. No generalization seems to apply to the molecules examined to date.
5. Resonant vibrational energy transfer may dominate excited state population evolution when vibrational relaxation is slow. The small

size of kT at 4.2°K can produce unusual concentrations of energy in minor isotopic species and high vibrational levels. In nonresonant transfer, the rates depend sensitively on the exact exothermicity.

6. Significant advances have occured in theoretical modeling of diatomic vibrational relaxation and inter-electronic relaxation. These processes promise to be among the best understood molecular radiationless transitions, in the sense of quantitative agreement with experiment. The field of polyatomic molecule intermode transitions, and of exchange vibrational energy transfer ("supermolecule" effects) are poorly understood theoretically.

References

1. L. Vegard, *C. R. Acad. Sci.*, **176**, 941 (1923).
2. L. Vegard, H. Kamerlingh-Onnes, and W. H. Keesom, *C. R. Acad. Sci.*, **180**, 1084 (1925).
3. L. Vegard, *Ann. Phys.*, **6**, 487 (1930).
4. I. Norman and G. Porter, *Nature*, **174**, 508 (1954).
5. I. Norman and G. Porter, *Proc. Roy. Soc.*, **A230**, 399 (1955).
6. E. Whittle, D. A. Dows, and G. C. Pimentel, *J. Chem. Phys.*, **22**, 1943 (1954).
7. G. C. Pimentel, *Spectrochim. Acta*, **12**, 94 (1958).
8. E. D. Becker, G. C. Pimentel, and M. Van Thiel, *J. Chem. Phys.*, **26**, 195 (1957).
9. B. Meyer, *Low Temperature Spectroscopy*, Elsevier, New York, 1971.
10. H. E. Hallam (ed.), *Vibrational Spectroscopy of Transient Species*, Wiley, New York, 1973.
11. M. Moskovits and G. A. Ozin (ed.), *Cryochemistry*, Wiley, New York, 1976.
12. L. Vegard, *Z. Phys.* **75**, 30 (1932).
13. C. M. Herzfeld and H. P. Broida, *Phys. Rev.*, **101**, 606 (1956).
14. M. Peyron and H. P. Broida, *J. Chem. Phys.*, **30**, 139 (1959).
15. H. P. Broida and M. Peyron, *J. Chem. Phys.*, **32**, 1068 (1960).
16. M. McCarthy and G. W. Robinson, *J. Chim. Phys.*, **56**, 723 (1959).
17. D. S. Tinti and G. W. Robinson, *J. Chem. Phys.*, **49**, 3229 (1968).
18. D. S. Tinti, *J. Chem. Phys.*, **48**, 1459 (1968).
19. H. Y. Sun and S. A. Rice, *J. Chem. Phys.*, **42**, 3826 (1965).
20. J. S. Shirk and A. M. Bass, *J. Chem. Phys.*, **52**, 1894 (1970).
21. V. E. Bondybey and J. W. Nibler, *J. Chem. Phys.*, **56**, 4719 (1972).
22. G. W. Robinson, *J. Chem. Phys.*, **47**, 1967 (1967).
23. S. H. Lin, *J. Chem. Phys.*, **44**, 3759 (1966).
24. S. H. Lin and R. Bersohn, *J. Chem. Phys.*, **48**, 2732 (1968).
25. V. E. Bondybey and L. E. Brus, *J. Chem. Phys.*, **63**, 2223 (1975).
26. V. E. Bondybey, *J. Chem. Phys.*, **66**, 995 (1977).
27. V. E. Bondybey and T. A. Miller, *J. Chem. Phys.*, **68**, 3597 (1978).
28. D. H. Katayama, T. A. Miller and V. E. Bondybey, *J. Chem. Phys* **71**, 1662 (1979).
29. E. Weitz and G. Flynn, *Annual Rev. Phys. Chem.* Vol. 25, Annual Reviews, Palo Alto (1976).
30. V. E. Bondybey, *J. Chem. Phys.*, **66**, 4237 (1977).
31. V. E. Bondybey and J. H. English *J. Chem. Phys.*, **67**, 2868 (1977).
32. A. Nitzan, J. Jortner and P. M. Rentzepis, *Proc. Roy. Soc.*, **A328**, 367 (1972).

33. M. L. Klein and J. A. Venables, eds., *Rare Gas Solids*, Vol. 1, Academic, New York, 1976.
34. K. Rebane, *Impurity Spectra of Solids*, Plenum, New York, 1970.
35. A. Gedanken, B. Raz, and J. Jortner, *J. Chem. Phys.*, **58**, 1178 (1973).
36. B. S. Ault, W. F. Howard, and L. Andrews, *J. Mol. Spectrosc.*, **55**, 217 (1975).
37. V. E. Bondybey and C. Fletcher, *J. Chem. Phys.*, **64**, 3615 (1976).
38. V. E. Bondybey, *J. Chem. Phys.*, **65**, 2296 (1976).
39. G. W. Robinson and M. McCarthy, Jr., *J. Chem. Phys.*, **30**, 999 (1959).
40. A. J. Barnes, H. E. Hallam, and G. F. Scrimshaw, *Trans. Farad. Soc.*, **65**, 3150 (1969).
41. R. L. Redington and D. E. Milligan, *J. Chem. Phys.*, **39**, 1276 (1963).
42. J. J. Linevsky, *J. Chem. Phys.*, **38**, 658 (1963).
43. J. Y. Roncin, N. Damany, and J. Romand, *J. Mol. Spectrosc.*, **22**, 154 (1967).
44. A. Gedanken, B. Raz, and J. Jortner, *J. Chem. Phys.*, **43**, 2997 (1973).
45. E. Boursey and J. Y. Roncin, *J. Mol. Spectrosc.*, **55**, 31 (1975).
46. E. Boursey, *J. Mol. Spectrosc.*, **61**, 11 (1976).
47. M. Miladi, J. P. leFalher, J. Y. Roncin, and H. Damany, *J. Mol. Spectrosc.*, **55**, 81 (1975).
48. J. Goodman and L. E. Brus, *J. Chem. Phys.*, **67**, 933 (1977).
49. J. Goodman and L. E. Brus, *J. Chem. Phys.* **69**, 4083 (1978).
50. V. E. Bondybey, J. H. English, and T. A. Miller, *J. Am. Chem. Soc.*, **100**, 5251 (1978).
51. V. E. Bondybey, J. H. English, and T. A. Miller, *J. Chem. Phys.*, **70**, 1621 (1979).
52. V. E. Bondybey and J. H. English, *J. Chem. Phys.* **71**, 777 (1979).
53. W. H. Balfour, *Can. J. Phys.*, **54**, 1969 (1976).
54. J. H. Callomon, *Can. J. Phys.*, **34**, 1046 (1956).
55. V. E. Bondybey, J. H. English, and T. A. Miller, **70**, 1765 (1979).
56. J. Goodman and L. E. Brus, *J. Chem. Phys.*, **65**, 3808 (1976).
57. B. S. Ault and L. Andrews, *J. Chem. Phys.*, **64**, 3075 (1976).
58. J. Goodman, J. C. Tully, V. E. Bondybey, and L. E. Brus, *J. Chem. Phys.*, **66**, 4802 (1977).
59. J. C. Miller, B. S. Ault, and L. Andrews, *J. Chem. Phys.*, **67**, 2478 (1977).
60. J. C. Miller and L. Andrews, *Chem. Phys. Lett.*, **50**, 315 (1977).
61. V. E. Bondybey and C. Albiston, *J. Chem. Phys.*, **68**, 3172 (1977).
62. J. C. Miller and L. Andrews, *J. Chem. Phys.*, **69**, 2054 (1978).
63. V. E. Bondybey and J. H. English, *J. Chem. Phys.*, **67**, 3405 (1977).
64. J. Goodman and L. E. Brus, *J. Chem. Phys.*, **67**, 4858 (1977).
65. J. Goodman and L. E. Brus, *J. Chem. Phys.*, **65**, 3146 (1976).
66. L. E. Brus and V. E. Bondybey, *J. Chem. Phys.*, **65**, 71 (1976).
67. L. E. Brus and V. E. Bondybey, *Chem. Phys. Lett.*, **36**, 252 (1975).
68. V. E. Bondybey and L. E. Brus, *J. Chem. Phys.*, **62**, 620 (1975).
69. C. A. Wight, B. S. Ault, and L. Andrews, *J. Mol. Spectrosc.*, **56**, 239 (1975).
70. V. E. Bondybey and L. E. Brus, *J. Chem. Phys.*, **64**, 3724 (1976).
71. A. C. Albrecht, *J. Mol. Spectrosc.*, **6**, 684 (1961).
72. B. Dellinger and M. Kasha, *Chem. Phys. Lett.*, **38**, 9 (1976).
73. O. L. Chapman, C. L. McIntosh, and J. Pacansky, *J. Am. Chem. Soc.*, **95**, 1337 (1973).
74. O. L. Chapman, C. L. McIntosh and J. Pacansky, *J. Am. Chem. Soc.*, **95**, 614 (1973).
75. B. Davies, A. McNeish, M. Poliakoff, M. Tranquille, and J. J. Turner, *Chem. Phys., Lett.*, **52**, 477 (1977).
76. C. Y. Lin and A. Krantz, *Chem. Comm.*, **1111**, (1972).
77. B. Dellinger, D. S. King, R. M. Hochstrasser, and A. B. Smith, *J. Am. Chem. Soc.*, **99**, 7138 (1977).

78. V. E. Bondybey, S. S. Bearder, and C. Fletcher, *J. Chem. Phys.*, **64**, 5243 (1976).
79. R. I. Personov, E. I. Al'Shits, and L. A. Bykovskaya, *Opt. Commun.*, **6**, 169 (1972).
80. R. I. Personov and B. M. Kharlamov, *Opt. Commun.*, **7**, 417 (1973).
81. J. H. Eberly, W. C. McColgin, K. Kawaoka, and A. P. Marchetti, *Nature*, **251**, 215 (1974).
82. D. J. Diestler, in *Topics in Applied Physics*, Vol. XV: "Radiationless process," Springer, Berlin, 1976.
83. F. Legay, in *Chemical and Biological Applications of Lasers*, Vol. II, Academic, New York, 1977.
84. D. J. Diestler, in *Advances in Chemical Physics*,: "Potential Energy Surfaces" (Ed. K. P. Lawley), Wiley, London, 1979.
85. A. Nitzan, S. Mukamel, and J. Jortner, *J. Chem. Phys.*, **60**, 3929 (1974).
86. A. Nitzan, S. Mukamel, and J. Jortner, *J. Chem. Phys.*, **63**, 200 (1975).
87. A. Nitzan and R. S. Silbey, *J. Chem. Phys.*, **60**, 4070 (1974).
88. D. J. Diestler and R. S. Wilson, *J. Chem. Phys.*, **62**, 1572 (1975).
89. D. J. Diestler, *Chem. Phys. Lett.*, **39**, 39 (1976).
90. D. J. Diestler, *J. Chem. Phys.*, **60**, 2692 (1974).
91. S. H. Lin, *J. Chem. Phys.*, **61**, 3810 (1974).
92. S. H. Lin, *J. Chem. Phys.*, **65**, 1053 (1976).
93. L. E. Brus and V. E. Bondybey, *J. Chem. Phys.*, **63**, 786 (1975).
94. V. E. Bondybey and L. E. Brus, *J. Chem. Phys.*, **63**, 794 (1975).
95. R. B. Gerber and M. Berkowitz, *Phys. Rev. Lett.*, **39**, 1000 (1977).
96. R. B. Gerber and M. Berkowitz, *Chem. Phys. Lett.*, **49**, 260 (1977).
97. M. Berkowitz and R. B. Gerber, *Chem. Phys.*, **37**, 369 (1979).
98. K. F. Freed and H. Metiu, *Chem. Phys. Lett.*, **48**, 262 (1977).
99. K. F. Freed, D. L. Yeager, and H. Metiu, *Chem. Phys. Lett.*, **49**, 19 (1977).
100. D. J. Diestler, E. W. Knapp, and H. D. Ladouceur, *J. Chem. Phys.*, **68**, 4056 (1978).
101. M. Shugard, J. C. Tully, and A. Nitzan, *J. Chem. Phys.*, **69**, 336 (1978).
102. A. Nitzan, M. Shugard, and J. C. Tully, *J. Chem. Phys.*, **69**, 2525 (1978).
103. V. E. Bondybey, *J. Chem. Phys.*, **65**, 5138 (1976).
104. H. Dubost, L. Abouaf-Marguin, and F. Legay, *Phys. Rev. Lett.*, **22**, 603 (1973).
105. H. Dubost and R. Charneau, *Chem. Phys.*, **12**, 407 (1976).
106. L. Abouaf-Marguin, B. Gauthier-Roy, and F. Legay, *Chem. Phys.*, **23**, 443 (1977).
107. J. Wiesenfeld and C. B. Moore, *J. Chem. Phys.*, **70**, 930 (1979).
108. J. Goodman and L. E. Brus, *J. Chem. Phys.*, **67**, 4398 (1977).
109. J. Goodman and L. E. Brus, *J. Chem. Phys.*, **67**, 4408 (1977).
110. J. Goodman and L. E. Brus, *J. Chem. Phys.*, **69**, 1853 (1978).
111. L. J. Allamandola and J. W. Nibler, *Chem. Phys. Lett.*, **28**, 335 (1974).
112. L. J. Allamandola, A. M. Rohjantalab, J. W. Nibler, and T. Chappell, *J. Chem. Phys.*, **67**, 99 (1977).
113. A. Lauberau, G. Kehl, and W. Kaiser, *Opt. Comm.*, **11**, 74 (1974).
114. A. Lauberau, L. Kirschner, and W. Kaiser, *Opt. Comm.*, **9**, 182 (1973).
115. V. E. Bondybey, *J. Mol. Spectrosc.*, **63**, 164 (1976).
116. J. L. Wilkerson and W. A. Guillory, *J. Mol. Spectrosc.*, **66**, 188 (1977).
117. V. E. Bondybey and J. H. English, *J. Chem. Phys.*, **67**, 664 (1977).
118. V. E. Bondybey, *J. Chem. Phys.*, **66**, 4237 (1977).
119. D. E. Milligan and M. E. Jacox, *J. Chem. Phys.*, **47**, 5157 (1967).
120. R. N. Dixon, *Phil. Trans. Roy. Soc.*, **A252**, 165 (1960).
121. V. E. Bondybey and J. H. English, *J. Chem. Phys.*, **68**, 4641 (1977).
122. W. Weltner and D. McLeod, *J. Chem. Phys.*, **45**, 3096 (1966).

123. A. A. Gorokhovski, R. Kaarli, and L. A. Rebane, *JETP Lett.*, **20**, 216 (1974).
124. A. A. Gorokhovski, R. Kaarli, and L. A. Rebane, *Opt. Comm.*, **16**, 282 (1976).
125. S. Voelker, R. M. Macfarlane, A. Z. Genack, H. P. Trommsdorff, and J. H. van der Waals, *J. Chem. Phys.*, **67**, 1759 (1977).
126. V. E. Bondybey and J. H. English *J. Am. Chem. Soc.* **101**, 3446 (1979).
127. B. Davis, A. McNeish, M. Poliakoff, and J. J. Turner, *J. Am. Chem. Soc.*, **99**, 7513 (1977).
128. A. McNeish, M. Poliakoff, K. P. Smith, and J. J. Turner, *J. Chem. Soc. Chem. Comm.*, **859** (1976).
129. A. Baca, R. Rossetti and L. E. Brus, *J. Chem. Phys.* **70**, 5575 (1979).
130. V. E. Bondybey and A. Nitzan, *Phys. Rev. Lett.*, **38**, 889 (1977).
131. I. Weissman, J. Jortner, and A. Nitzan, *Chem. Phys.*, **26**, 413 (1977).
132. D. Fletcher, Y. Fujimura, and S. H. Lin, *Chem. Phys. Lett.*, in press.
133. V. E. Bondybey and L. E. Brus, *J. Chem. Phys.*, **63**, 2223 (1975).
134. W. Weltner, Jr., P. N. Walsh, and C. L. Angell, *J. Chem. Phys.*, **40**, 1299 (1964).
135. D. E. Milligan, M. E. Jacox, and L. Abouaf-Marquin, *J. Chem. Phys.*, **46**, 4562 (1967).
136. D. E. Milligan and M. E. Jacox, *J. Chem. Phys.*, **51** 1952 (1969).
137. R. P. Frosch, *J. Chem. Phys.*, **54**, 2660 (1971).
138. J. Goodman and L. E. Brus, *J. Chem. Phys.*, **67**, 1482 (1977).
139. L. J. Schoen and H. P. Broida, *J. Chem. Phys.*, **32**, 1184 (1960).
140. I. R. Sil'dos, L. A. Rebane, A. B. Treshchalov, and A. E. Lykhmus, *JETP Lett.*, **22**, 151 (1975).
141. R. A. Teichman and E. R. Nixon, *J. Mol. Spectrosc.*, **59**, 299 (1976).
142. R. R. Smardzewski, *Proceedings of the International Conference on Matrix Isolation Spectroscopy*, Berlin (1977).
143. B. S. Ault, W. F. Howard, and L. Andrews, *J. Mol. Spectrosc.*, **55**, 217 (1975).
144. V. E. Bondybey and J. H. English, *J. Chem. Phys.*, **69**, 1865 (1978).
145. L. E. Brewer, G. D. Brabson, and B. Meyer, *J. Chem. Phys.*, **42**, 1385 (1965).
146. L. E. Brewer, and G. D. Brabson, *J. Chem. Phys.*, **44**, 3274 (1966).
147. J. Fournier, C. Lalo, J. Deson, and C. Vermeil, *J. Chem. Phys.*, **66**, 2656 (1977).
148. S. R. Long and G. C. Pimentel, *J. Chem. Phys.*, **66**, 2119 (1977).
149. J. M. Brom and H. P. Broida, *J. Chem. Phys.*, **63**, 3178 (1975).
150. L. J. Lauchlan, J. M. Brom, and H. P. Broida, *J. Chem. Phys.*, **65**, 2672 (1976).
151. M. Moskovits and J. E. Hulse, *J. Chem. Phys.*, **67**, 4271 (1977).
152. A. A. Belyaeva, Y. B. Predtechenskii, and L. D. Shcherba, *Opt. Spektrosk.*, **34**, 21 (1973).
153. A. A. Belyaeva, R. B. Dushin, Y. B. Predtechenskii, and L. D. Shcherba, *Dokl. Akad. Nauk SSSR*, **199**, 628 (1971).
154. N. S. Mitchenko, Y. B. Predtechenskii, and L. D. Shcherba, *Opt. Spectrosc.*, **41**, 521 (1976).
155. J. M. Brown, W. D. Hewett, and W. Wetner, *J. Chem. Phys.*, **62**, 3122 (1975).
156. V. E. Bondybey, *J. Chem. Phys.*, **68**, 1308 (1978).
157. L. C. Balling, M. D. Havey, and J. F. Dawson, *J. Chem. Phys.*, **69**, 1670 (1978).
158. L. B. Knight, R. D. Brittain, M. A. Starr, and C. H. Joyner, *J. Chem. Phys.*, **61**, 5289 (1974).
159. V. E. Bondybey, to be published.
160. G. A. Ozin and H. Huber, *Inorg. Chem.*, **17**, 155 (1978).
161. W. E. Klotzbucher and G. A. Ozin, *J. Am. Chem. Soc.*, in press.
162. L. H. Jones, S. Ekberg and L. B. Asprey, *J. Chem. Phys.*, **70**, 1566 (1979).
163. J. Goodman and L. E. Brus, *J. Chem. Phys.*, **65**, 1156 (1976).
164. R. B. Gerber and M. Berkowitz, *Chem. Phys. Lett.*, **56**, 105 (1978).

DENSITY OPERATOR DESCRIPTION OF EXCITONIC ABSORPTION AND MOTION IN FINITE MOLECULAR SYSTEMS

CLAUDE ASLANGUL

Centre de Mécanique Ondulatoire Appliquée,
Laboratoire du C.N.R.S.
23 *Rue du Maroc, Paris, France*

AND

PHILEMON KOTTIS

Laboratoire d'Optique Moléculaire
(Associate Laboratory n°283 of the C.N.R.S.)
Université de Bordeaux I,
33405, *Talence, France*

CONTENTS

SYMBOLS

$\hat{\rho}$	Von Neuman's density operator
$\hat{H}_0$	Electronic Hamiltonian of an aggregate
$\hat{\bar{\rho}}$	Reduced density operator
E^0	Energy of the electronic ground state of an isolated site

E^*	Energy of the first excited electronic state of an isolated aggregate
$\hbar\omega_i^0, \hbar\omega_i^*$	Vibrational quantum of the site-lattice oscillator in the ground and first excited electronic states
Δ	Dispersion of a branch of phonons
k_B	Boltzmann's constant
T	Absolute temperature
M_p	p^{th} site of an aggregate of N sites in all
$\lvert M_0\rangle$ or $\lvert A_0\rangle$	Antisymmetrized electronic ground state of an aggregate
$\lvert M_p\rangle$	p^{th} antisymmetrized, monoexcited electronic state of an aggregate
$\hat{A}$	Antisymmetrizing operator for all the electrons in an aggregate
E_0	Matrix element $\langle M_p\lvert\hat{H}_0\rvert M_p\rangle \equiv \hbar\omega_0$
V	Matrix element (real) $\langle M_p\lvert\hat{H}_0\rvert M_{p+1}\rangle \equiv \hbar\Delta\omega$
$\hat{\epsilon}^i(t)$	Semiclassical stochastic operator of the ith aggregate
$\langle\rangle$	Statistical average of the ensemble of microsystems: $\mathcal{N}^{-1}\sum_{i=1}^{\mathcal{N}}$
$\lvert A_k\rangle$	k^{th} eigenstate of $\hat{H}_0$ $(k = 1, 2, \ldots, N)$
$\hat{H}^i(t)$	Total Hamiltonian of the ith aggregate
$\hat{\rho}(t)$	Statistical density operator, equals $\langle\hat{\rho}^i(t)\rangle$
S	Entropy function, proportional to $-\text{Tr}(\hat{\rho}\ln\hat{\rho})$
$\mathcal{A}$	The algebra of linear operators in a Hilbert space $\mathcal{H}$
$\mathcal{A}'$	Dual of $\mathcal{A}$
$\lvert X)$	Element of $\mathcal{A}$
$(X\rvert$	Element of $\mathcal{A}'$
$(X_1\lvert X_2)$	Scalar product $(\lvert X_1), \lvert X_2)$, equal to $\text{Tr}(\hat{X}_1^\dagger\hat{X}_2)$
λ^*	Complex conjugate of the scalar λ
$\sigma_{\mu\lambda}^{\tau}$	Structure constants of a Lie algebra
$\hat{\hat{\Xi}}_-$	Id-operator built from $\hat{\Xi}$
$\hat{K}$	Operator associated with the constant of movement K
$\mathcal{B}$	Eigen base of the id-operators associated with the constants of movement

$\mathscr{A}_K$	Subspace of $\mathscr{A}$ generated by the $\{\|K_i)\}_i$
$\mathscr{A}_i$	Subspace of $\mathscr{A}_K$ generated by $\|K_i)$
$\hat{P}_K$	Projector on $\mathscr{A}_K$
$\hat{\hat{P}}_K$	Supplementary projection operator of $\hat{P}_K$
$\mathscr{A}^*$	Transfer subspace
$\mathscr{A}^a$, $\mathscr{A}^e$	Absorption, emission subspaces
$\|\rho_K), \|\rho^*), \|\rho^a), \|\rho^e)$	Projections of $\|\rho)$ on $\mathscr{A}_K$, $\mathscr{A}^*$, $\mathscr{A}^a$, $\mathscr{A}^e$
$\|\tilde{\rho}(t))$	Intermediate representation of $\|\rho(t))$ with respect to $\hat{H}_0$
$\hat{U}_0(t, t_0)$	Evolution operator $\exp\{(1/i\hbar)\hat{H}_0(t-t_0)\}$
$\hat{\tilde{\epsilon}}^i_-(t)$	Intermediate representation of $\hat{\epsilon}^i_-(t)$
$\mathscr{K}_n(t_1, t_2, \ldots, t_n)$	n^{th} order cumulant
$\hat{\Gamma}$	Macroscopic relaxation operator
$\hat{H}$	Macroscopic Hamiltonian, equal to $\hat{H}_{0-} - i\hbar\hat{\Gamma}$
β	$= (k_B T)^{-1}$
$\|M^I_p\rangle$	Monoexcited, antisymmetrized electronic state built on the pth site of chain $I (I = A, B)$
$\hbar\Delta\omega$	Intrachain coupling between nearest neighbors
$\hbar\Delta\omega'$	Interchain coupling between nearest neighbors
$\|A_{kb}\rangle$	Eigenstate of $\hat{H}_0$ (double chain), of energy $\hbar\omega_{kb}$ $(1 \leqslant k \leqslant N, b = \pm 1)$
$\|A_k\rangle$	Eigenstate of $\hat{H}_0$ (simple chain), of energy $\hbar\omega_k$ $(1 \leqslant k \leqslant N)$
$\hat{P}_G$	Projection operator in $\mathscr{H}$ onto the electronic ground state
$\hat{\hat{P}}_G$	Supplementary projection operator of $\hat{P}_G$
$\hat{H}_K, \hat{H}^*, \hat{H}^a, \hat{H}^e$	Restrictions of $\hat{H}$ to $\mathscr{A}_K$, $\mathscr{A}^*$, $\mathscr{A}^a$, $\mathscr{A}^e$
$\hat{a}^+_{kb}$	"Creation" operator $\|A_{kb}\rangle\langle A_0\|$ (double chain)
$\hat{a}_{kb}$	"Annihilation" operator $\|A_0\rangle\langle A_{kb}\|$ (double chain)
$\hat{a}^+_k$	"Creation" operator $\|A_k\rangle\langle A_0\|$ (simple chain)
$\hat{a}_k$	"Annihilation" operator $\|A_0\rangle\langle A_k\|$ (simple chain)
$\hat{H}^\alpha$	$\hat{H}^a$ or $\hat{H}^e$
$\hat{Y}^{IJ}_{pq}$	Operator $\|M^I_p\rangle\langle M^J_q\|$ (double chain)
$\hat{Y}_{pq}$	Operator $\|M_p\rangle\langle M_q\|$ (simple chain)
$C_n(z)$	n^{th} order Gegenbauer polynomial
θ	angle equal to $\pi/(N+1)$ (N: number of sites)
$\mathscr{L}[f(t)]$	Laplace transform of $f(t)$, also noted $F(s)$

$\mathscr{L}^{-1}[F(s)]$	$\equiv f(t)$
$\epsilon_p^{Ii}(t)$	Local intrachain fluctuation of site M_p^I
$v_p^{Ii}(t)$	Intrachain fluctuation between M_p^I and M_{p+1}^I
$w_p^i(t)$	Interchain fluctuation between M_p^A and M_p^B
$\gamma_0, \gamma_1, \gamma_2$	Mean-square fluctuations of $\epsilon_p^{Ii}(t)$, $v_p^{Ii}(t)$, $w_p^i(t)$
$\hat{m}_p^{I+}$	Operator $\|M_p^I\rangle\langle M_0\|$
Γ_{dc}	$= \gamma_0 + 2\gamma_1 + \gamma_2$
Γ_{ch}	$= \gamma_0 + 2\gamma_1$
Γ_d	$= \gamma_0 + \gamma_1$
$\hat{W}(t)$	Semiclassical field-matter coupling operator
$\|\rho_{eq})$	Macroscopic (equilibrium) state
$\hat{G}(t)$	Causal propagator of $\hat{H}$, equal to $(i\hbar)^{-1}\theta(t)\exp\{(1/i\hbar)Ht\}$
$\|\rho^{(1)}(t))$	linear response
$\|R^{(1)}(\omega))$	Fourier transform of $\|\rho^{(1)}(t))$
$\hat{\mathscr{G}}^{(+)}(\omega)$	Fourier transform of $\hat{G}(t)$
$\hat{\mathscr{G}}^{(+)\alpha}(\omega)$	Restriction of $\hat{\mathscr{G}}^{(+)}(\omega)$ to $\mathscr{A}^\alpha$ ($\alpha = a$ or e)
$\chi(\omega)$	Complex optical susceptibility, written also as $\chi'(\omega) - i\chi''(\omega)$
$\chi^\alpha(\omega)$	$\alpha = a$ or e, absorption or emission susceptibility
$\mathbf{u}$	Electric polarization of the applied field
$\mathbf{A}(\mathbf{r}, t)$	Vector potential of the applied field
$\hat{\mathbf{d}}$	Total electron dipole moment of the aggregate
$\mathbf{m}_p$	Unit vector along the "transition" moment $\langle M_0\|\hat{\mathbf{d}}\|M_p\rangle$
d	Modulus of the "transition" moment, such that $\langle M_0\|\hat{\mathbf{d}}\|M_p\rangle = d\mathbf{m}_p$
$\mathbf{u}_k$	Unit vector along the transition moment between $\|A_0\rangle$ and $\|A_k\rangle$, equals $1/d\langle A_0\|\hat{\mathbf{d}}\|A_k\rangle$; $\mathbf{u}_{kb} = 1/d\langle A_0\|\hat{\mathbf{d}}\|A_{kb}\rangle$
ψ_{12}	$= (\mathbf{m}_1, \mathbf{m}_2)$ in a dimer
T_k^2	Weighting factors for Lorentzian resonances (single chain)
T_{kb}^2	Weighting factors for Lorentzian resonances (double chain)
$\|\psi^*\rangle$	State prepared by direct optical absorption
$S(\|\psi^*\rangle)$	Delocalization entropy of $\|\psi^*\rangle$

$\hat{\mathscr{E}}(t)$	Energy density operator
$\tilde{\mathscr{E}}(\{\mathbf{x}\}, \{\mathbf{x}'\}; t)$	Coordinate kernel of $\hat{\mathscr{E}}(t)$
$\mathscr{M}_{\lambda,uv\ldots}$	Spatial moment of order λ ($u, v, \ldots = x, y, z$)
$\rho_0(\phi, \psi; t)$	Generating function for spatial moments (double chain)
$\rho_0(\phi; t)$	Generating function for spatial moments (single chain and dimer)
$\bar{\rho}_0(\phi, \psi; s)$	Laplace transform of $\rho_0(\phi, \psi; t)$
$\bar{\rho}_0(\phi; s)$	Laplace transform of $\rho_0(\phi;t)$
$\Delta^2_{uv}(t)$	Element (u, v) of the dispersion tensor
α	angle such that $\cos\alpha = \gamma_0/2\lvert\Delta\omega\rvert$ ($\pi/2 \geqslant \alpha \geqslant 0$ or $\alpha = i\alpha''$, α'' positive real number)
τ_{coh}	"Coherence time" in a dimer
D	"Diffusion" constant
$S_{uv}(t)$	Element (u, v) of the delocalization tensor
T^*	Pseudo-period of the pure coherent movement
q	Index of the site initially excited
X	Distance between adjacent sites along Ox
Y	Distance between adjacent sites along Oy (double chain)
θ'	Fixed angle equal to π/N
$\lvert\Upsilon_j)$	Eigenstate of $\hat{\Gamma}^*$ in the subspace generated by $\lvert\{Y_{pp}\}_{1\leqslant p\leqslant N}$ ($j = 0, 1, \ldots, N-1$)
$C_{xy}(t)$	Statistical correlation coefficient between movements in the directions Ox and Oy
$\boldsymbol{\delta}_{\pm}(t)$	Eigenstates (in the plane of the double chain) of the dispersion tensor, associated with the eigenvalues $D_{\pm}$
$T^*_{x,N}$	Pseudo-period of the pure coherent movement along Ox in a double chain of $2N$ sites in all

PREFACE

Following Frenkel's[1] decisive work and especially in the fifteen or so years since Davydov's study,[2] considerable experimental and theoretical research has been done on molecular crystals.[3–6] Investigations have born on the optical response of such crystals, for their line shapes contain precious information on the nature of exciton–phonon interactions[7], and on the transfer of energy during the lifetime of the excited state, which seems to play an important part in activation, notably in some biological processes[8] such as photosynthesis.[9-11] On a more speculative note we mention work by Little et al.[12,13] on the possibility of "high" temperature superconductivity in organic salts, where excitons would fill the role of the phonons in the BCS theory.

Comparatively less attention has been paid to molecular aggregates composed of the same molecules as crystals, but containing a small number of sites.[14-27] This lack of interest, particularly in the theory, may be explained partly by the preliminary difficulty of the absence of translational symmetry in such molecular associations, quite the contrary of crystals. We shall see that this point complicates formal developments, as much for optical absorption as for the propagation of energy during the life of the excited state. Furthermore, the notion of momentum operator being inapplicable to the elementary excitations,[28-32] the very meaning of particles must be reexamined, for the quantified electronic and vibrational fields usually are interpreted in terms of particles only when their stationary states are defined by given values of the energy, without dispersion, and of the conjugate moment, the momentum.

Finite chains have several advantages: Firstly, since infinite crystals do not exist, it is tempting to try to build a formalism explicitly containing the dependence of physical properties on the number of sites. On the one hand, one could determine to what extent real crystals may be described by methods based on models of infinite crystals; and on the other, draw attention to the cases in which small crystallites must be treated exactly. Secondly, if one could construct a model varying continuously with the number of sites, one would have a tool suitable for the smallest units, dimers, or the largest, infinite crystals. Such a theory, ranging from the microscopic to the macroscopic, might be used to test quantum concepts. Finally, it is now possible to prepare very small crystallites of from 50 to 100 sites (J. Ferguson, private communication), so it is important to have a theoretical treatment specifically suited to these systems.

Despite the preceding reserves we continue to use the unfortunate but now established term "exciton" to designate elementary electronic excitations of a finite aggregate. The energy "band" is no more than a sequence of discrete levels, the density of which depends on the number of sites, the strength of their couplings, and the spectral region considered. Definition of an "effective mass" by the usual relation (reciprocal of the second derivative of energy with respect to momentum) is impossible, strictly speaking, since the dispersion "function" does not have a derivative in the usual sense. The systematic recourse to corpuscular terminology of quantum physics is acceptable provided we see in the terms used no more than a collective excitation of the aggregate, defined here by its energy since there is no degeneracy in the model retained.

We may now outline the spirit of the present work. Our analysis is founded systematically on the study of the stationary states of the system, and on states built from them. Radically different from the widespread use of explanations and mechanisms founded on bases which are not stationary for any physically significant Hamiltonian, our point of view is not mere pedantry, but simply a consequence of the rules of quantum mechanics. This attitude alone, and not assertions built on a nonstationary basis, is capable of interpreting correctly the mathematical quantities in the theory.[33–35] As a consequence of this truism, we may use only invariants in the theory of representations, or quantities invariant to an approximation less than those inherent in any model.[36] We find that the notion of collective excitation is linked to coherence as defined by Cohen–Tannoudji,[37–39] and indeed we point out two kinds of coherence: coherence between sites induced by their interaction with the same external field and coherence introduced by the interactions between parts of the same quantum system.

One might expect the optical properties of an aggregate to be similar to those of an isolated site, since interactions between sites in molecular crystals are for the same physical reasons much smaller than the energies of the first electronic transitions of the molecule (20 and 20,000 cm^{-1}, respectively). This is not so, for two equally profound reasons. The first is the aptitude of the site transition moments to react collectively to optical excitation, for example, the frequency shift due to intersite couplings; selection rules specific to the aggregate, and incomprehensible in a theory where each site constitutes an absorbing entity and is independent of its neighbors, excepting static interactions. The second is due to coupling between electronic and vibrational degrees of freedom, usually called exciton–phonon interactions. It causes characteristic broadening[7] of optical resonances and always may be represented by second quantification techniques which in addition exhibit its corpuscular aspect. This formulation, apparently devoid of empiricism, presupposes in fact a choice of the spatial functions in the coefficients of the development of the interaction operator on the creation and annihilation operators. The exact mechanism of this coupling being at present unknown, some authors take linear functions of the nuclear coordinates[30] and others quadratic functions.[33] Quite apart from any later introduction of adjustable parameters, the principles of this approach reasonably may be said to have heavy phenomenological overtones, so it cannot claim all-around superiority over other attempts at skirting the preceding difficulty.

Another approach, common in physics and other intellectual fields when the "machinery" of a phenomenon is unknown, is the statistical method. Although by its nature it has great advantages, it should not be used blindly, since it has meaning only in precise physical conditions and indeed the temperature is important in what follows. We use Haken's and Strobl's[40,41] stochastic model, based on ideas introduced by Sewell.[32]

The principal two goals of this work are a detailed study of the optical response of an ensemble of aggregates, particularly the intensity and selection rules typical of such systems, and a tentative quantum description of excitation transfer during the life of the excited state. Regarding the second point, we give the evolution of the dispersion tensor and of the center of the wave packet. The commoner treatments, for infinite chains, naturally lead to a stationary "center of mass," a considerable limit to the idea of transfer.

The general plan is as follows: In Section I we give a detailed description of the physical system, define the mathematical model used, and partly justify the stochastic model.

Section II contains the bulk of the formalism required to derive the equations for the macroscopic evolution of each class of system of interest. Fano's[43] method of super-operators is developed in as much detail as is necessary.

Section III is consecrated to the optical response of the system. After establishing a compact expression for the susceptibility in terms of super-operators, and supposing a linear response, we give numerous calculated spectra of dimers, ordinary one-branch chains, and double chains (allowing for different classes of geometrical arrangement). Finally we give spectra for a mixture of aggregates, simulating the physical situation in which a distribution of N-mers is obtained by doping the matrix.

Section IV we develop a solution of the problem of transfer during the life of the excited state. After examining the possibility of preparing a localized, nonstationary state by direct optical excitation (state for which movement will be the most obvious) we take the usual precautions to define an energy density whose role for the exciton is that of the wave function in coordinate representation for a one-body problem. We may then compute any spatial moment from this probability density of presence, particularly the mean position of the exciton and its dispersion, or extent about this point. We find that whereas initial localization of the energy may be possible, it disappears and the exciton has a complicated movement, never the same as a simple superposition of a wave and a scattering component. Globally, the momentum contained in a volume the size of a site fluctuates rapidly.

In Section V we present the principal conclusions of this work and the extensions and enlargements envisaged at present.

I. INTRODUCTION

The purpose of this introduction is to present the physical system examined and the characteristics of the model chosen to describe it. After analyzing the separation of the different degrees of freedom (electronic and vibrational) we introduce a working Hilbert space and semiclassical random interactions accounting for crystal phonons and their reaction on couplings between sites in the same aggregate. Critical analysis of the stochastic model leads to a first statement of its validity. Finally we list the specific characteristics of finite systems and give an inventory of relevant earlier works.

A. Description of the Physical System

This work concerns dilute solid solutions of molecular aggregates of identical organic molecules trapped in an ordered organic matrix. They are ideal systems[14,16–20] for studying oriented aggregates and indeed the crystal itself, in the case of an isotopic substitution leaving molecular interactions untouched.[15]

Frenkel's treatment[1] of excitons is suitable for these aggregates insofar as the electron and hole are a short distance apart,[15] the contrary of the Wannier exciton which is the analog of a charge transfer state with a hydrogenoidlike spectrum.

An important property of the matrices studied here is the difference between their electronic resonances and those of the aggregates, supposed to be very large compared to interactions between sites of the same aggregate. In the case of a hydrogen–deuterium isotopic substitution, for example, the 0–0 transition in benzene is shifted by about $200\,\mathrm{cm}^{-1}$. Since the shift decreases with increasing molecular weight, being only $60\,\mathrm{cm}^{-1}$ for anthracene, one may if necessary carry out a chemical substitution. The theoretical importance of this hypothesis is that it allows us to neglect the electronic degrees of freedom of the host crystal. Experimentally it means we may use a moderately monochromatic source to excite the electronic levels of the aggregates, with no fear of touching the exciton band of the matrix in which the molecular chains are embedded.

Essentially, the role of the crystal is to provide phonons which may interact with the electronic and vibrational degrees of freedom of the aggregates, since each aggregate site is coupled locally to the lattice.

Consequently it is unimportant to know whether the guests are aligned perfectly with the hosts, as in naphthalene in durene[44] or very imperfectly as in tetracene in anthracene,[45] though it may be important to know in a more general context, as a source of information on interactions between molecules. There is, of course, no problem in the case of isotopic substitution.[46] It suffices that the aggregate itself should be an oriented system, as is indeed the case when it is placed in a crystalline matrix. We may then calculate theoretical values for the relative absorption intensities of different electronic resonances and discuss necessary and sufficient conditions for optical preparation of a given nonstationary state.

Furthermore, we consider only aggregates with a low density of intrasite vibrational levels. In anthracene, for example, the vibrational quantum of the first electronic state is of the order of 500 cm^{-1}. Thus, it will be possible to limit discussion to only one vibrational level. This truncation may correspond to very fast or very slow local relaxation compared to other relaxation processes between sites in the aggregate and between a site and the lattice. In the first case the vibrational level is the vertical transition level, and in the second it is the lowest level of the excited electronic state.

B. Definition of the Model

1. Separation of the Degrees of Freedom

In the first place, we suppose that the Born–Oppenheimer approximation holds for each site. The degrees of freedom of the system may be divided into two classes: (*a*) electronic degrees of freedom of the aggregate; (*b*) vibrational degrees of freedom: lattice phonons, intersite vibrations of the aggregate.

Diagonalization of the electronic Hamiltonian yields a sequence of states of the rigid aggregate, a ground state, and bunches of states separated by 10,000 to 20,000 cm^{-1}, at least for those of lowest energy. The average density of each group is $(\hbar\omega/N)^{-f}$ where N is the number of sites in the aggregate and $\hbar\omega$ the magnitude of the interaction between neighboring sites. The centre of gravity of each group is shifted by a quantity traditionally called D, with respect to the electronic state of an isolated site. The lattice phonons are divided into several continuous branches with amplitudes of dispersion of the order of 100 cm^{-1}. We may add that if the cristalline molecules are distorted strongly, each branch is represented by a multiform function and gives rise to a band for each wave number. Finally, intermolecular vibrations in the aggregate, between heavy, weakly bonded bodies have very small quanta, generally of the order of 5 to 40 cm^{-1}.

2. *Mathematical Representation of the Degrees of freedom. Working Spaces*

Using the preceding considerations we may define an effective or working space for the problem of finding approximate eigenstates of the system described in Section I.A.

In the first place we neglect the electronic levels of the crystal since they are not resonant with those of the aggregate. Everything happens as if the crystal were always in its ground state.

Next we keep only 0th order states of the aggregate, antisymmetrized tensor products of singly excited sites. This is legitimate since groups of electronic states of the rigid aggregate are widely separated and if we suppose a low exciton concentration, transitions to highly excited states by fusion are very improbable.[47] This approximation is common. Moreover, as in Section I.A we keep only one intrasite vibrational level; transfer will be essentially vibronic[42], not forgetting intermolecular vibrational coupling.

C. Stochastic Representation of the Intermolecular Degrees of Freedom

1. *Justification of the Stochastic Representation*

Intuitively we see that each site is coupled locally to the lattice so that we must add to its energy that of the corresponding local harmonic oscillator, with a quantum $\hbar\omega$. The states of such an oscillator, coupled to the lattice phonons, have a finite width roughly the same as the dispersion Δ of the phonon branch with which it interacts. At sufficiently high temperatures $\hbar\omega \ll k_B T$ and $\Delta \ll k_B T$ so that the vibrational site–lattice energy may be represented by a random function of time, due to thermal fluctuations.[31,32,40,41,48,49]

This Brownian motion may be seen as a rapid oscillation (defined by fluctuating parameters) of a trapped molecule about its average, equilibrium position.[50–53] Consequently, the matrix elements representing all the interactions between two sites in the aggregate will have a fluctuating component. The intermolecular vibrations also will be defined by parameters (frequencies and equilibrium positions) modulated by the lattice–aggregate thermal agitation.

It is thus understandable to represent the low-frequency vibrations by random functions. Diagonal matrix elements on the exciton basis correspond to the energy of the local site–lattice oscillator, to which must be added the random part of D, whereas off-diagonal terms represent the fluctuation of interactions between sites of the same aggregate.

Comparison between the general characteristics of the movement of the ensemble and the movement defined by the reduced density operator[54] provides an a posteriori justification of the preceding description. If $\hat{H}_0$ is the electronic Hamiltonian, $\hat{H}_B$ the low-frequency Hamiltonian for site–site and site–lattice vibrations and $\hat{H}_{\text{int}}$ their coupling operator, Liouville's equation is:

$$i\hbar \frac{d}{dt} \hat{\rho}(t) = [\hat{H}_0 + \hat{H}_B + \hat{H}_{\text{int}}, \hat{\rho}(t)]$$

Contraction on the vibrational degrees, denoted by Tr_B, yields,

$$i\hbar \frac{d}{dt} \hat{\bar{\rho}}(t) = [\hat{H}_0, \hat{\bar{\rho}}(t)] + \text{Tr}_B([\hat{H}_B + \hat{H}_{\text{int}}, \hat{\rho}(t)])$$

where $\hat{\bar{\rho}}(t) \equiv \text{Tr}_B \hat{\rho}(t)$ is the electronic density operator obtained by reduction of the total density operator with respect to the preceding vibrational degrees of freedom.

In certain cases[54] it is possible to obtain an explicit equation for the evolution of $\hat{\rho}(t)$, containing the autocorrelation operators of the interaction representation of $\hat{H}_{\text{int}}$, the statistical mean being taken over the initial state of the entire system. Besides the assumption of separability at time $t = t_0$ (the operator $\hat{\rho}(t)$ being the tensor product of the two reduced operators), the crux of this deduction is the Markhovian hypothesis, according to which the correlation time of the reservoir, here the low-frequency degrees of freedom, is very much shorter than the relaxation time of the subsystem in which we are interested. We may then smooth out the fluctuations of the thermostat over a time long compared to its correlation time but very short compared to the characteristic times of the subsystem. The last condition obviously must hold if we are to obtain a significant differential equation. If it did not, we should obtain at best a finite difference equation. In terms of frequencies, we see the Markhovian hypothesis leads us to suppose, in particular, that the spectral width of the reservoir is very large compared to that of the subsystem. Moreover, a frequency shift, analogous to the Lamb shift, may appear. It is generally negligible[55] compared to the width, which depends hyperbolically on the relaxation time. Summing up, and provisionally calling the order of magnitude of the time constants of the sub-system, γ (defined in more detail by the parameters $\gamma_{|p-q|}$ of (1.8)), the Markhovian approximation supposes $\gamma \ll \Delta$.[56,57] With the arguments given in Section I.C.1 we have therefore:

$$\gamma \ll \Delta \ll k_B T$$

If either of the strong inequalities is violated the preceding model is entirely invalid.

Let us note one further, essential point: In the Markhovian approximation the equation finally obtained for $\hat{\rho}(t)$ is differential and not integro-differential (hereditary) as it would be without such an assumption. This forgetfulness of the subsystem is an a posteriori justification of the terminology. Thus, it makes no difference to the results whether one contracts Liouville's equation on degrees of freedom with specific spectral properties or whether one introduces the same degrees in the form of a random semiclassical interaction. In both cases we find analogous, irreversible evolutions (excepting negligible renormalizations) which are the consequence of coupling between discrete states and one or more effective continua.

Whereas the first method seems more rigorous in its principles, (though in practice its application requires some pragmatic choices), it is extremely arduous to use, even for ideally simple models, and calls on sophisticated mathematical techniques. The second approach is more phenomenological, or even empirical, but is infinitely easier to develop and may be more readily comprehensible. It has the disadvantage of eliminating a priori a form of quantification and cannot therefore explain everything, notably at very low temperatures where line shapes depend specifically on the quantum nature of exciton–phonon interactions (Lorentzians for weak coupling, Gaussians for strong coupling). From the academic standpoint it may be criticized for its hybrid quantum and classical character, like that of the magnetic theories founded on Bloch's equations, or to a lesser extent, that of the semiclassical description of the interaction between matter and radiation.

2. *Discussion of Site Energy*

We first consider coupling between a site M of an aggregate and a lattice site S. Let E^0 and E^* be the electronic energies of the first two excited states of M, in the vibrational state corresponding to a given vibronic transition with respect to the nuclear coordinates of the molecule M. With obvious notations, the energy of the transition is

$$E = E^* + \sum_i (v_i^* + \tfrac{1}{2})\hbar\omega_i^* - E^0 - \sum_i \frac{\hbar\omega_i^0}{2} \tag{1.1}$$

where the summation over i covers all the site–lattice vibrational modes. In fact, each local vibrational level is coupled step by step with those of all the sites of the lattice. In other words the oscillator M–S interacts with the lattice phonons (it being granted that molecule M may be coupled directly to several crystalline sites). Consequently, each "level" of the vibrator M–S has a width comparable with the dispersion Δ of the phonon branch interacting with it.

In the range of validity of the stochastic model, namely Δ and $\hbar\omega$

much smaller than $k_B T$, (1.1) may be rewritten

$$E = E^* - E^0 + \epsilon(t) \tag{1.2}$$

where $\epsilon(t)$ is the fluctuation of the energy of the M–S oscillator due to rapid variation of the numbers v_i^* and to modulation of the frequencies ω_i^* and ω_i^0 and of the equilibrium positions.

By analogy with thermodynamics, we may say there is an exchange of heat (changing quantum numbers) and of work (changing states).

It is important to realize that we have made an adiabatic approximation in saying that the thermal fluctuations preserve the internal vibrational state of the site M. In particular we exclude any possibility of nonradiative transitions to the ground state.

Let us now consider a dimer of two sites M_1 and M_2. Owing to the degeneracy of the monoexcited states on M_1 and M_2 there is a local deviation from the Born–Oppenheimer approximation, depending on the intersite nuclear coordinates. The Born–Oppenheimer approximation naturally remains valid when applied to the delocalized states. The total Hamiltonian may be written

$$\begin{aligned}\hat{H}(q_i, Q_i) &= \hat{H}_1(q_1, Q_1) + \hat{H}(q_2, Q_2) + \hat{V}_{12}(q_1, q_2, Q_1, Q_2) + \hat{T}_{12} \\ &\equiv \hat{H}_0(q_i, Q_i) + \hat{T}_{12}\end{aligned} \tag{1.3}$$

In these equations $\hat{V}_{12}$ is the total electrostatic interaction between M_1 and M_2 and $\hat{T}_{12}$ is the kinetic energy operator of their relative movement. Diagonalization in the absence of random forces would lead to normal modes of the dimer where both the sites would appear, with comparable weights.

On the truncated reduced basis of monoexcitations defined by

$$|M_0\rangle = \hat{A}|M_{1,0}M_{2,0}\rangle \quad |M_1\rangle = \hat{A}|M_{1,1}M_{2,0}\rangle \quad |M_2\rangle = \hat{A}|M_{1,0}M_{2,1}\rangle \tag{1.4}$$

where $\hat{A}$ is the total electronic antisymmetrizing operator, the matrix of $\hat{H}_0$ is, neglecting an additive constant:

$$\begin{bmatrix} E^0 + D_0 & 0 & 0 \\ 0 & E^* + D^* & V \\ 0 & V & E^* + D^* \end{bmatrix}$$

where

$$D_0 = \langle M_0|\hat{V}_{12}|M_0\rangle_{q_i}; \qquad D^* = \langle M_p|\hat{V}_{12}|M_p\rangle_{q_i} \quad (p = 1, 2)$$
$$V = \langle M_1|\hat{V}|M_2\rangle_{q_i}$$

We suppose the phases to be chosen such that all these matrix elements are real. V and D depend, therefore, on the intersite nuclear coordinates and in the stochastic model fluctuate about their mean values. In other words, if after diagonalizing $\hat{H}_0$, we define normal modes for the dimer by solving the nuclear equation where the eigenvalue of plays the role of a potential energy, we find the eigenfrequencies and equilibrium positions are random functions.

3. *Justification of the Stochastic Model; Domain of Validity*

Limiting ourselves to the case of the dimer, for the sake of ease of notation, the Hamiltonian of the system on the local basis is, neglecting an additive constant

$$\begin{bmatrix} 0 & 0 & 0 \\ 0 & E_0 + \epsilon_{11}(t) & V + \epsilon_{12}(t) \\ 0 & V + \epsilon_{12}(t) & E_0 + \epsilon_{22}(t) \end{bmatrix} \equiv \hat{H}_0 + \hat{\epsilon}(t) \tag{1.5}$$

The terms $\epsilon_{pq}(t)$ are electronic energy matrix elements fluctuating on account of the site–site and site–lattice vibrations.

Choice of a stochastic approach presupposes the complete definition of the process $\hat{\epsilon}(t)$, or equivalently of all its correlation operators. Firstly we suppose $\hat{\epsilon}(t)$ is centered so that,

$$\langle \hat{\epsilon}(t) \rangle = 0 \tag{1.6}$$

This is not a restriction since if it were not the case we could add a nonzero average value to the static part, $\hat{H}_0$. Secondly, we suppose $\hat{\epsilon}(t)$ to be a Gaussian process. We see in Section II.F that this assumption greatly simplifies the deduction of the explicit form of the macroscopic evolution operator. Physically we may justify this conjecture by noting that the "perturbation" is the sum of a vast number of independent fluctuations, all with the same distribution of probability, in practice the interaction of a large number of degrees of freedom (large number of molecules and or large number of modes), so that by the central limit theorem $\hat{\epsilon}(t)$ has to a very good approximation a normal distribution. Hence all the correlation functions of order higher than two may be expressed in terms of the first two.

$\hat{\epsilon}(t)$ being by convention centered, the stochastic process is completely defined by the time dependence of the second-order correlation operators. As we said, in the stochastic description the characteristic time of fluctuations is very short, and we shall idealize this situation in writing

$$\langle \hat{\epsilon}(t)\hat{\epsilon}(t') \rangle = 2\hbar^2 \hat{\gamma} \delta(t - t') \tag{1.7}$$

where $\hat{\gamma}$ is a time-independent operator. It should be noted that such a supposition gives a "hyper-Markhovian" character to the process, since true Markhovian correlation functions are exponentials.[58,59] We see in Section II.F that the macroscopic evolution equation is differential, without memory. Plainly, the Dirac function correlation is indissociable from some high temperature, difficult to estimate, and such a choice cannot be made for arbitrarily low temperatures. We have here a high-temperature approximation, of a different nature to that leading to the stochastic description of low-frequency vibrations.

The essential approximations are stated previously (recourse to a semiclassical stochastic representation, Gaussian process, and very short correlation times). The following treatment of correlations between $\epsilon_{pq}(t)$ and $\epsilon_{p'q'}(t')$ is given only to avoid unnecessary formal complications and we make a point of interpreting their physical significance. In Haken's and Strobl's model[40,41,48] one keeps only positive definite correlations and neglects the rest. Thus,

$$\langle \epsilon_{pq}(t)\epsilon_{p'q'}(t')\rangle = 2\hbar^2\gamma_{|p-q|}\delta(t-t')\{\delta_{pp'}\delta_{qq'} + \delta_{pq'}\delta_{qp'}(1-\delta_{pq})\} \qquad (1.8)$$

The two terms in curly brackets are due to the symmetry of the matrix $\hat{\epsilon}(t)$ on a real basis, so that there are obvious relations between, for example, $\langle \epsilon_{pq}(t)\epsilon_{p'q'}(t')\rangle$ and $\langle \epsilon_{pq}(t)\epsilon_{q'p'}(t')\rangle$. Hence it is necessary to interpret only the first term of (1.8), in $\delta_{pp'}.\ \delta_{qq'}$. It shows clearly that the only correlation functions declared to be nonzero are autocorrelation functions of the matrix elements. Let us examine the correlations thus neglected.

Firstly, terms such as $\langle \epsilon_{pp}(t)\epsilon_{p'p'}(t')\rangle$ where $p \neq p'$ represent correlations between fluctuations of diagonal energies on different sites. Since we have supposed the autocorrelation time for a site to be infinitely short, it seems natural to suppose there is no correlation between sites. Abandoning this condition would certainly introduce refinements out of keeping with other approximations involved.

Secondly we consider functions of the type $\langle \epsilon_{pp}(t)\epsilon_{pq}(t')\rangle$ for $p \neq q$. In practice, nearest-neighbor interactions limit q to $p \pm 1$. $\epsilon_{pp}(t)$ is the sum of energy fluctuations of the local M–S oscillator and of the fluctuations of the diagonal matrix element $\hat{V}_{12}$ defined by (1.3), and $\epsilon_{pq}(t)$ is the fluctuation of $\langle M_p|\hat{V}_{12}|M_q\rangle$. It is reasonable to neglect correlations between $\epsilon_{pq}(t')$ and the contribution of $\epsilon_{p\tilde{p}}(t)$ of the local vibration M–S. On the other hand, it is difficult to imagine no correlation between $\epsilon_{pq}(t')$ and $\langle M_p|\hat{V}_{12}|M_q\rangle$ since these quantities are of common nature and origin. Thus the overall effect of neglecting $\langle \epsilon_{pp}(t)\epsilon_{pq}(t')\rangle$ is to suppose that the preponderant contribution to $\epsilon_{pp}(t)$ is due to the local vibration. It is more a conjecture than a presumption.

As a consequence of relation (1.8), the correlation operator may be developed on the local basis as

$$\langle\langle M_p|\hat{\epsilon}(t)\hat{\epsilon}(t')|M_q\rangle\rangle = 2\hbar^2\delta(t-t')\sum_{p'}\gamma_{|p-p'|}\{\delta_{pp'}\delta_{p'q}+\delta_{pq}\delta_{p'p'}(1-\delta_{pp'})\}$$

$$= 2\hbar^2\delta(t-t')\delta_{pq}\left\{\gamma_0+\sum_{p'}(1-\delta_{pp'})\gamma_{|p-p'|}\right\} \quad (1.9)$$

Thus $\langle\hat{\epsilon}(t)\hat{\epsilon}(t')\rangle$ written on the local basis is a diagonal matrix.

It is worth remembering that the stochastic treatment of the low-frequency, high-density vibrations is justifiable only at high temperatures and we have seen several independent arguments underlining this, such as negligible dispersion and small local vibrational quanta compared to k_BT and infinitely short correlation times. Others are given later, notably in Section II.G. Hence we must not extrapolate our results to arbitrarily low temperatures.

Briefly then, the quantum system, with Hamiltonian $\hat{H}_0$ is subjected to a (semiclassical) external influence $\hat{\epsilon}(t)$. According to quantum mechanics, the physically significant quantities are eigenvalues of $\hat{H}_0$, such as the energies $E_1 = E_0 + V$ and $E_2 = E_1 - V$ associated, respectively, with the eigenstates $|A_1\rangle = 1/\sqrt{2}(|M_1\rangle + |M_2\rangle)$ and $|A_2\rangle = 1/\sqrt{2}(|M_1\rangle - |M_2\rangle$. The ground state will be written $|A_0\rangle$ or $|M_0\rangle$ depending on the notation most suitable at the time. The situation is thus comparable to the semiclassical description of matter–radiation interactions where an external field causes the system to evolve between eigenstates of the unperturbed Hamiltonian.

D. Relation to Previous Work and Peculiarities of Finite Systems

Although caution is necessary in relating earlier work on infinite crystals to the present study of finite chains, a number of concepts may be carried over from one to the other. Naturally, we later point out the characteristic features of finite chains.

Historically, the first studies bore on infinite molecular crystals, a fact related to Frenkel's[1] article on the degradation of luminous energy to heat in solids. Although he wrote of "excitation waves" and the term "exciton" does not appear in his article, the concept of the exciton is clearly present and it is natural to call the "particle" created by "local excitation" of energy with conservation of total electric charge of the altered site, a Frenkel exciton. It has been said, in pictorial terms, that the optical electron and the positive hole are on the same molecule.

Such a particle, in absence of any others, has a wave or coherent movement. In fact it was soon realized that the exciton depends on crystal phonons and that if the latter remain in thermal equilibrium, the exciton–phonon collisions result in an irreversible exciton movement, termed incoherent. The need for a theory containing coherence and incoherence on the same footing soon became clear and three approaches were developed, by Haken et al.,[40,41,48,49,60–66] by Grover and Silbey,[67–69] and by Kenkre and Knox.[70–75]

In Haken's model, the one used in the rest of this work, a microscopic equation containing random interactions leads to an equation for the macroscopic evolution enabling one to calculate absorption and transfer in infinite crystals, at the cost of some cumbersome mathematics.[48,76] Essentially, it contains two parameters γ_0 and γ_1 (see (1.8)). The relation between this equation and experimental quantities is not always easy to establish. Haken and Reineker, comparing their theoretical expressions for the linewidth and the scattering constant of anthracene crystals with experimental values,[77,78] concluded that at room temperatures $\gamma_0 = 70\,\mathrm{cm}^{-1}$ and $\gamma_1 = 0.1\,\mathrm{cm}^{-1}$. It is tempting to relate the inequality $\gamma_1 \ll \gamma_0$ to the conjecture that the principal part of $\langle M_p|\hat{\epsilon}(t)|M_q\rangle$ is due to the local oscillator M–S.

The model of Grover and Silbey[67–69] is entirely different. Taking advantage of the small energy dispersion of a free exciton, they introduce a new basis in which the exciton is "dressed" in lattice phonons. Creation and annihilation operators represent electronic excitations with vibrational distorsion of the crystal. Assimilating the phonons to a thermostat, they obtain the time dependence of the spatial dispersion of the exciton.

Kenkre,[74] who has shown that though these methods have different premises, they lead to readily comparable equations, refers to them jointly as the GSHRS (Grover–Silbey–Haker–Reineker–Strobl) theory. His joint work with Knox[71,72] is entirely different, and further, seeks only to describe energy transfer during the lifetime of the excited state. Based on a phenomenological equation derived with Zwanzig's[79–82] formalism, it leads to an integro-differential equation for diagonal matrix elements of the density operator on the local basis, $\langle M_p|\hat{\rho}(t)|M_p\rangle$. An integral in an evolution equation may indicate non-Markhovian behaviour, i.e. with memory. However, one should be cautious in affirming that Kenkre's and Knox's model introduces a memory neglected in the previous two, for the integro-differential equation is only for the *diagonal* elements of $\hat{\rho}(t)$. Now, if one takes an ordinary differential equation, without memory, for *all* the matrix elements of the density operator and if one eliminates the off-diagonal elements, one obtains a formal integro-differential equation in which the memory is only apparent, being the result of a partial integration.

As an illustration let us consider a Schrödinger equation, without memory. (We see later that Liouville's equation may be written in this form, without the commutator.)

$$i\hbar \frac{d}{dt}|\psi(t)\rangle = \hat{H}|\psi(t)\rangle$$

Let $\hat{P}$ and $\hat{Q}$ be two orthogonal, (idempotent) projection operators such that

$$\hat{P}\hat{Q} = \hat{Q}\hat{P} = \hat{0}; \qquad \hat{P} + \hat{Q} = \hat{I}$$

Suppose further that the initial state is such that

$$\hat{Q}|\psi(t=0)\rangle \equiv |\psi_Q(t=0)\rangle = 0$$

Writing $\hat{P}|\psi(t)\rangle = |\psi_P(t)\rangle$, we have by projection

$$i\hbar \frac{d}{dt}|\psi_P(t)\rangle = \hat{P}\hat{H}\hat{P}|\psi_P(t)\rangle + \hat{P}\hat{H}\hat{Q}|\psi_Q(t)\rangle$$

$$i\hbar \frac{d}{dt}|\psi_Q(t)\rangle = \hat{Q}\hat{H}\hat{P}\,|\psi_P(t)\rangle + \hat{Q}\hat{H}\hat{Q}|\psi_Q(t)\rangle$$

Formal integration of the second equation with its initial value leads to

$$|\psi_Q(t)\rangle = \int_0^{+\infty} \hat{G}_Q(t-t')\hat{Q}\hat{H}\hat{P}|\psi_P(t')\rangle dt'$$

where $\hat{G}_Q(t-t')$ is the causal Green's function, equal to $[\theta(t-t')/i\hbar]\exp[(1/i\hbar)\hat{Q}\hat{H}\hat{Q}(t-t')]$. Putting $|\psi_Q(t)\rangle$ into the first equation we have

$$i\hbar \frac{d}{dt}|\psi_P(t)\rangle = \hat{P}\hat{H}\hat{P}|\psi_P(t)\rangle + \int_0^{+\infty} \hat{P}\hat{H}\hat{Q}\hat{G}_Q(t-t')\hat{Q}\hat{H}\hat{P}|\psi_P(t')\rangle\, dt'$$

Thus, simply eliminating some of the degrees of freedom of $|\psi(t)\rangle$ leads to a "memory" in the equation of evolution of $|\psi_P(t)\rangle$, a part of $|\psi(t)\rangle$. The diagonal elements in Kenkre's and Knox's theory play the role of $|\psi_P(t)\rangle$ and the off-diagonal elements that of $|\psi_Q(t)\rangle$ and the former appear to conform to a hereditary equation. In fact the "memory" is due only so such a projection. This well-known phenomenon is found frequently in methods based on partitions. See, for example, Levine's book[83] and references cited therein.

Comparison of the memories of these two models is futile unless one has defined the nature and origin of the linear operator weighting the diagonal part of $\hat{\rho}(t)$ in the integral. Kenkre affirms[70] that it may be deduced from information in the optical spectrum, the limiting case of exponential decay being a Lorentzian line shape. He has shown[84] that the pseudo-memory of the GSHRS method contains a singularity of type

$\delta(t)$ besides the exponential term, and that it is incorrect for very short times. Of course, this may result from the Markhovian approximation in which the division of time in the evolution of the system cannot be smaller than the correlation time of the thermostat. The singularity may be the cause of Haken's and Reineker's[85] surprising affirmation that the movement is a scattering process for short times and a wave one for long ones. This is difficult to understand physically. In all events, the results of Haken et al. are partly special cases of those of Kenkre and Knox, and yet lead to the contrary property of a wave, then a scattering movement.

We note that in all these models, the role of the phonons is that of a heat bath, causing irreversibility. Whereas in Haken et al.'s theory this appears immediately in the presence of random semiclassical terms, Grover and Silbey explicitly take the statistical average over the canonical phonon states. Kenkre first carries out "course graining" to introduce irreversibility in microscopic equations symmetric with respect to the reversal of time.

The finite systems examined in the following text have several peculiarities: The familiar rule $\Delta k = 0$ for infinite crystals does not hold for their optical spectra. In Section III we discuss a series of selection rules which are often complicated and depend on the geometry of the aggregate. As a direct result of end effects in finite chains, their line shapes are not Lorentzians. Even when end effects are negligible, for $N \gtrsim 20$, in average chains there are numerous inhomogeneous broadenings leading to markedly asymmetric lines.

Regarding energy transfer in the excited state, it is no longer a trivial problem, as is the case for infinite crystals, to determine the evolution of the mean position of the center of the optically created packet of excitation. Moreover, spatial dispersion varies in such a way that keeping the usual definition for the scattering constant D, for instant just after excitation, its expression depends on which site was excited initially. We show that the scattering and wave parts of the movement are intimately related and that the movement cannot be represented as a superposition of the two.

We note that theoretically recurrences are possible in finite systems, though the characteristic time increases very rapidly with N (Section IV) and may exceed the lifetime of the excited state.

Finally, the momentum has no meaning in finite chains, which are not invariant under translation. Hence wave packets will be superpositions of discrete stationary states defined by quantum numbers (associated with the energy), each taking a finite number of integral values. For at least one reason, one cannot strictly define creation and annihilation operators: The trace of a commutator in a space of finite dimension δ is nil, whereas the trace of the identity operator is δ.

II. FORMALISM

In this section we develop the mathematical formalism, avoiding complicated methods since they tempt one to make abusive simplifications. The formalism is applied to the dimer, the single chain, and the double chain such as α-dibromonaphthalene.

A class of superoperators is presented after a brief account of quantum coherence, defined by Cohen–Tannoudji,[37] and the reasons for the unavoidable use of the density operator. Analysis of the integrals of motion leads to a mathematical space in the form of a direct sum of subspaces, each of which corresponds to a physical phenomenon: optical absorption, energy transfer during the excited state. A detailed application of the formalism to the case of the dimer is given, to allow brief treatment of more complex systems. Discussion of the general characteristics of the movement in Haken's model, leading to new conditions of validity, is facilitated by the preceding operators.

A. Necessity of Introducing the Density Operator. Quantum Coherence

Owing to the random forces in our model, the Hamiltonians of the microsystems are different. Thus, for the ith aggregate we may write, with the notation of (1.5)

$$\hat{H}^i(t) = \hat{H}_0 + \hat{\epsilon}^i(t) \tag{2.1}$$

and each microsystem at time t will be represented by a vector $|\psi^i(t)\rangle$. The result on an observable $\hat{\Omega}$, of quantum averaging and averaging over the ensemble of systems[39] is

$$\mathcal{N}^{-1}\sum_{i=1}^{\mathcal{N}}\langle\psi^i(t)|\hat{\Omega}|\psi^i(t)\rangle = \mathcal{N}^{-1}\sum_{i=1}^{\mathcal{N}}\mathrm{Tr}(\hat{\Omega}|\psi^i(t)\rangle\langle\psi^i(t)|) \equiv \mathrm{Tr}(\hat{\Omega}\hat{\rho}(t)) \tag{2.2}$$

where $\mathcal{N}$ is the size of the statistical ensemble and $\hat{\rho}(t)$ is Von Neumann's density operator defined by

$$\hat{\rho}(t) = \mathcal{N}^{-1}\sum_{i=1}^{\mathcal{N}}|\psi^i(t)\rangle\langle\psi^i(t)| \equiv \langle|\psi^i(t)\rangle\langle\psi^i(t)|\rangle \tag{2.3}$$

Plainly, the entropy S calculated by $S \propto -\mathrm{Tr}(\hat{\rho}\ln\hat{\rho})$ is not, in general, an integral of motion; a pure distribution, for example, evolving towards a mixture. Nonconservation of the initial idempotence is in itself sufficient cause to use the density operator.

We first determine the equation of motion of $\hat{\rho}(t)$, since it enables us to calculate any average quantity. Letting $\hat{\rho}^i(t) = |\psi^i(t)\rangle\langle\psi^i(t)|$ in Liouville's microscopic equation:

$$i\hbar \frac{d}{dt}\hat{\rho}^i(t) = [\hat{H}^i(t), \hat{\rho}^i(t)] \tag{2.4}$$

we have immediately

$$i\hbar \frac{d}{dt}\hat{\rho}(t) = [\hat{H}_0, \hat{\rho}(t)] + \langle[\hat{\epsilon}^i(t), \hat{\rho}^i(t)]\rangle \tag{2.5}$$

where $\hat{\rho}(t)$ is defined by (2.3). The average of the commutator on the right-hand side is found in Section II.F after some preliminary mathematics to simplify the problem as much as possible.

Beforehand, however, we must examine the concept of quantum coherence as defined by Cohen-Tannoudji[37] and determine its relation to the pure or mixed nature of a statistical ensemble.

If $|m\rangle$ and $|n\rangle$ are two stationary states of a general quantum system the coherence between them is given by $|\langle m|\hat{\rho}(t)|n\rangle|$ and plays a determinant role in the evolution of the average values of observables which are not integrals of motion. If $\hat{\Omega}$ is the operator of such an observable and if $\hat{H}$ is the Hamiltonian of an isolated system ($\partial\hat{H}/\partial t \equiv 0, \hat{H}^+ = \hat{H}$), Liouville's equation is

$$i\hbar \frac{d}{dt}\hat{\rho}(t) = [\hat{H}, \hat{\rho}(t)] \tag{2.6}$$

which may be integrated on the basis of eigenstates of $\hat{H}$ to give

$$\langle m|\hat{\rho}(t)|n\rangle = e^{i\omega_{nm}t}\langle m|\hat{\rho}(0)|n\rangle$$

where $\omega_{mn} = \hbar^{-1}(E_m - E_n)$, E_n being an eigenvalue of $\hat{H}$. The average value of the observable $\hat{\Omega}$ is given by

$$\begin{aligned}\langle\hat{\Omega}\rangle(t) \equiv \mathrm{Tr}\,\hat{\Omega}\hat{\rho}(t) &= \sum_m \sum_n \langle m|\hat{\Omega}|n\rangle\langle n|\hat{\rho}(t)|m\rangle \\ &= \sum_m \langle m|\hat{\Omega}|m\rangle\langle m|\hat{\rho}(t)|m\rangle + \sum_{m\neq n} \langle m|\hat{\Omega}|n\rangle\langle n|\hat{\rho}(t)|m\rangle \\ &\equiv C^{ste} + 2\,\mathrm{Re}\left\{\sum_{m<n} \langle m|\hat{\Omega}|n\rangle\langle n|\hat{\rho}(0)|m > e^{i\omega_{mn}t}\right\}\end{aligned} \tag{2.7}$$

Thus observation of beats in a nonconservative quantity $\hat{\Omega}(\langle m|\hat{\Omega}|n\rangle \neq 0)$ is in this case concomitant with initial coherence. Moreover, harmonic evolution at the frequency $\hbar^{-1}(E_m\text{-}E_n)$ appears if and only if the states $|m\rangle$ and $|n\rangle$ are initially coherent. In the preceeding example the coherence is a constant of motion, as is the entropy S.

The situation for a nonisolated system is different. If there is any relaxation, due say to contact with a thermostat, Liouville's equation takes the form

$$i\hbar \frac{d}{dt} \hat{\rho}(t) = [\hat{H}', \hat{\rho}(t)] - i\hbar \hat{\Gamma}_R\{\hat{\rho}(t)\} \tag{2.8}$$

where $\hat{H}'$ is a renormalized Hamiltonian and $\hat{\Gamma}_R$ is a relaxation operator acting on $\hat{\rho}(t)$. Coherence and entropy generally evolve irreversibly and the former, in particular, decays with a "lifetime" depending on the pair of states considered. Damped beats of the mean values of nonconservative quantities may be observed if the decay time is not too short compared to the reciprocal of the difference of the eigenfrequencies.

It is important to realize that coherence between states $|m\rangle$ and $|n\rangle$ requires that they should both be populated, as is shown by the relation[86]:

$$|\langle m|\hat{\rho}|n\rangle|^2 \leq \langle m|\hat{\rho}|m\rangle\langle n|\hat{\rho}|n\rangle \tag{2.9}$$

a direct consequence of the basic properties of the density operator. Thus a coherent initial state can result only from a perturbation forming linear combinations of stationary states. Hence idempotence of $\hat{\rho}(t)$, although necessary for (maximum) coherence, is not a sufficient condition.

Although the expression of coherence depends on the representation, it is in fact an invariant, since discounting degeneracy, the representation is unique, being that of the stationary states of the Hamiltonian, and it would be erroneous to believe, as some have said in the literature, that coherence may be analyzed on any representation.

Other measures of coherence, depending on the "distance" between $\hat{\rho}$ and $\hat{\rho}^2$, and invariant in the theory of representations have been more or less successful in characterizing the pure or mixed nature of the statistical ensemble. Of particular interest is the entropy $S \propto -\mathrm{Tr}(\hat{\rho} \ln \hat{\rho})$ the principal properties of which will be found in any treatise on probability theory, for example, the appendix on information theory in Renyi's book.[87] Notably, S is nil only if $\hat{\rho}$ is idempotent. This is possible only if the diagonal matrix elements between states effectively populated are nonzero; hence, these states are coherent. The preceding condition clearly is not sufficient: The matrix elements must be linked by definite equations translating the idempotence of $\hat{\rho}$. As a measure of coherence, the preceding trace is a true invariant, calculable on any basis, but it has the disadvantage of a global measure, not bringing out the precise origins of the coherence as does Cohen-Tannoudji's definition.

Two types of coherence are commonly distinguished. An example of the first is that induced in an ensemble of "independent" atoms or molecules subjected to the same electromagnetic field.[88] Without the field, there would be no coherence. The second kind, which we might call intrinsic coherence, is typified by the coherence between interacting subsystems of a quantum system, for which the total Hamiltonian is not diagonal on the basis of tensor products of states of the isolated subsystems, which have no physical meaning. The eigenbasis of the Hamiltonian is a set of collective states of the subsystems, which lose their individual characters and form a coherent "super"-system. In the present work, we are concerned only with the second kind of coherence, where the subsystems are the weakly interacting sites of an aggregate.

The preceding distinction disappears if we enlarge the definition of the system, for example, by including photons, in which case interaction between atoms occurs through the quantified field.[89,90]

We should note also that in such a field coherence is independent of the magnitude of undelayed coupling between atoms, the important point being the distance between them relative to the wavelength of the photons. In our example the sites are analogous to the atoms of Dicke's gas and we see at once that absorption by one site is an imaginary process, (see also Section IV.B), for the distance between sites, of the order of the angstrom, is much shorter than the wavelength of optical photons. This remains true even when the static interactions between sites can be neglected and the aggregate forms an oriented "gas."

B. Introduction of Superoperators

The basic idea of the method of superoperators[43] lies in the use of the vector space structure of the algebra $\mathscr{A}$ of linear operators on a Hilbert space $\mathscr{H}$, henceforth supposed to be of finite dimension δ.

Given a basis $\{\Omega_\lambda\}_\lambda$ of $\mathscr{A}$ any operator $\hat{X}$ may be decomposed as

$$\hat{X} = \sum_{\lambda=1}^{\Delta} X_\lambda \hat{\Omega}_\lambda \tag{2.10}$$

where the X_λ are the components of $\hat{X}$ on the basis and Δ is the dimension of $\mathscr{A}$. Some of the principal references for this widely used technique will be found in Refs. 43, 91–93. Quite generally, given a complete basis $\{|n\rangle\}_{1\leq n\leq\delta}$ of $\mathscr{H}$ we may write any operator $\hat{X}$ as

$$\hat{X} = \sum_m \sum_n X_{mn} \hat{S}_{mn} \tag{2.11}$$

with $\hat{S}_{mn} = |m\rangle\langle n|$, from which it is seen that the dimension Δ of $\mathscr{A}$ is δ^2.

In Lynden-Bell's notation,[94] an element X of $\mathscr{A}$ is written

$$|X) = \sum_m \sum_n X_{mn}|S_{mn}) \tag{2.12}$$

By analogy with Dirac's notation, elements of the dual of $\mathscr{A}$, $\mathscr{A}'$ will be written between (and |. The Hilbert space structure of $\mathscr{A}$ and $\mathscr{A}'$ may be defined by a definite positive Hermitian scalar product, $(|X_1), |X_2))$ or $(X_1|X_2)$ like Dirac's notation.

$$|\rangle \twoheadrightarrow |) \qquad \langle| \twoheadrightarrow (|$$

The scalar product

$$(X_1|X_2) = \mathrm{Tr}(\hat{X}_1^\dagger \hat{X}_2) \tag{2.13}$$

is particularly suitable here for if the basis is a set of operators representing observables $\hat{\Omega}_\lambda$ of a system

$$|\rho(t)) = \sum_\lambda \rho_\lambda(t)|\Omega_\lambda)$$

and if the $\hat{\Omega}_\lambda$ are an orthonormal set, $((\Omega_\lambda|\Omega_\mu) = \delta_{\lambda\mu})$

$$\rho_\lambda(t) = \mathrm{Tr}\hat{\Omega}_\lambda^\dagger \hat{\rho}(t) = \mathrm{Tr}\hat{\Omega}_\lambda \hat{\rho}(t)$$

so that the components of the density operator on this basis are the average values of the corresponding observables in the dynamic state represented by $\hat{\rho}(t)$.

Consider now two operators Ξ and $\hat{X}$ and let $\hat{Y}$ be their commutator

$$\hat{Y} = [\hat{\Xi}, \hat{X}] \tag{2.14}$$

Letting y_λ, ξ_μ and x_ν be the components of $|Y), |\Xi)$ and $|X)$ on an orthornomal basis $\{|\Omega_\lambda)\}_\lambda$ projection of (2.14) leads to

$$y_\mu = \mathrm{Tr}\{\hat{\Omega}_\mu^\dagger[\hat{\Xi}, \hat{X}]\} = \sum_\lambda x_\lambda \, \mathrm{Tr}\{\Omega_\mu^\dagger[\hat{\Xi}, \hat{\Omega}_\lambda]\} = \sum_\lambda x_\lambda \, \mathrm{Tr}\{[\hat{\Omega}_\mu, \hat{\Omega}_\lambda^\dagger]^\dagger \hat{\Xi}\}$$

in which use is made of the cyclic properties of the trace, present here since the space is of finite dimension. Let $\sigma_{\mu\lambda}^\tau$ be structure constants defined by Ref. 95:

$$[\hat{\Omega}_\mu, \hat{\Omega}_\lambda^\dagger] = \sum_\tau \sigma_{\mu\lambda}^\tau \hat{\Omega}_\tau \tag{2.15}$$

The δ^3 (complex) numbers $\sigma\mu_\lambda^\tau$ are not all independent, owing to the many relations between commutators and to the cyclic properties of the trace. Thus

$$\sigma_{\mu\lambda}^\tau = \mathrm{Tr}\{\hat{\Omega}_\tau^\dagger[\hat{\Omega}_\mu, \hat{\Omega}_\lambda^\dagger]\} = \mathrm{Tr}\{\hat{\Omega}_\lambda^\dagger[\hat{\Omega}_\tau^\dagger, \hat{\Omega}_\mu]\} \equiv -\sigma_{\mu\tau}^\lambda \tag{2.16}$$

A study of the constants of the movement shows that it is unnecessary to calculate all the Δ^2 commutators.

We may now write:

$$y_\mu = \sum_\lambda \sum_\tau x_\lambda \sigma^{\tau *}_{\mu\lambda} \operatorname{Tr}(\hat{\Omega}^\dagger_\tau \hat{\Xi}) \equiv \sum_\lambda \sum_\tau x_\lambda \sigma^{\tau *}_{\mu\lambda} (\Omega_\tau | \Xi)$$

or finally,

$$(\Omega_\mu | Y) = \sum_\lambda \sum_\tau \sigma^{\tau *}_{\mu\lambda} (\Xi | \Omega_\tau)^* (\Omega_\lambda | X)$$

Thus, defining an operator $\hat{\Xi}_-$ acting on $\mathscr{A}$ by its matrix on the $\{|\Omega_\lambda)\}_\lambda$:

$$(\Omega_\mu | \hat{\Xi}_- | \Omega_\lambda) = \sum_\tau \sigma^{\tau *}_{\mu\lambda} (\Omega_\tau | \Xi) \tag{2.17}$$

we have for all μ

$$(\Omega_\mu | Y) = \sum_\lambda (\Omega_\mu | \hat{\Xi}_- | \Omega_\lambda)(\Omega_\lambda | X) = (\Omega_\mu | \hat{\Xi}_- | X) \tag{2.18}$$

Hence

$$|Y) = \hat{\Xi}_- | X)$$

which defines a simple linear transformation between elements of $\mathscr{A}$, in place of the commutator equation (2.14). Naturally, some difficulties remain, notably the size of the matrices of operators like $\hat{\Xi}_-$ since $\mathscr{H}$ is of dimension δ and $\mathscr{A}$ of dimension δ^2. On the other hand this form is immediately amenable to the techniques of vector analysis such as variations, perturbations, and partitions. Formal expressions written in the new form are much simpler than in the initial one (2.14), since, as we see in Section III.A in an expression for the optical susceptibility, it avoids multiple commutators, for example.

Definition (2.17) of $\hat{\Xi}_-$ is characteristic of a class of superoperators, the id-operators,[96,97] "id" being derived from "interior derivation," another term for the operation of commutation.[95] In what follows, the subscript "–" denotes superoperators of this kind.

Explicit calculation of superoperators is deferred to a later section. The table of commutators $\{[\Omega_\lambda, \Omega_\mu]\}_{\lambda\mu}$ gives us a qualitative idea of the matrix of any id-operator. In particular if the commutator is zero, $\sigma^\tau_{\lambda\mu}$ is zero for all τ and we have the following proposition:

$$[\hat{\Omega}_\lambda, \hat{\Omega}^\dagger_\mu] = 0 \Rightarrow \text{for all } \hat{\Xi}\text{: } (\Omega_\lambda | \hat{\Xi}_- | \Omega_\mu) = 0 \tag{2.19}$$

Consequently, if $\hat{\Omega}_\lambda$ is Hermitian, $(\Omega_\lambda | \hat{\Xi}_- | \Omega_\lambda) = 0 \ \forall \hat{\Xi}$.

This is not a necessary condition for a given matrix element of an operator $\hat{\hat{\Xi}}_-$ to be nil, since by (2.17) it depends also on the projections $(\Omega_\tau|\Xi)$. It is easy to prove

$$\hat{\hat{\Xi}} \text{ is Hermitian} \Rightarrow \hat{\hat{\Xi}}_- \text{ is Hermitian} \tag{2.20}$$

and

$$\Xi \text{ is Hermitian} \Rightarrow (\Omega_\lambda|\hat{\hat{\Xi}}_-|\Omega_\mu) = -(\Omega_\lambda^\dagger|\hat{\hat{\Xi}}_-|\Omega_\mu^\dagger) \tag{2.21}$$

A further advantage of this technique is that the spectrum of an id-operator is formed by differences of pairs of eigenvalues of the operator from which it is derived by internal derivation. Thus, for an operator $\hat{H}$ with eigenvalues and eigenvectors E_k and $|E_k\rangle$ we have

$$[\hat{H}, |E_k\rangle\langle E_l|] = (E_k - E_l)|E_k\rangle\langle E_l| \Leftrightarrow \hat{H}_-\|E_k\rangle\langle E_l|) = (E_k - E_l)\|E_k\rangle\langle E_l|) \tag{2.22}$$

which shows the spectrum of $\hat{H}_-$ is the set of numbers $E_k - E_l$, with corresponding eigenfunctions in $\mathscr{A}$, $|E_k\rangle\langle E_l|$ acting in $\mathscr{H}$. Hence the energy of a transition is found directly, rather than by calculating the E_k separately and then subtracting them one from another and this may considerably improve the accuracy of the result.

There exists in this formalism a natural distinction between microscopic and macroscopic operators, between those describing reversible movements and those describing relaxation. The first kind are id-operators; those of the second are general superoperators.

The preceding methods are the basis of the theory of Lie algebras. References 95,98–100 contain some interesting, detailed applications to mathematical physics.

C. Choice of Basis in $\mathscr{A}$

As in any problem of representation, the basis used in is a matter of convenience, the obvious one, from the point of view of ease of calculations, being that formed by the eigenvectors of id-operators $\hat{K}_{i-}$ corresponding to constants of motion $\hat{K}_i$. This is a consequence of the proposition

$$[\hat{\hat{\Xi}}, \hat{X}] = \hat{Y} \Rightarrow [\hat{\hat{\Xi}}_-, \hat{X}_-] = \hat{Y}_- \tag{2.23}$$

easily established with Jacobi's identity:

$$[\hat{A},[\hat{B}, \hat{C}]] + [\hat{B},[\hat{C}, \hat{A}]] + [\hat{C},[\hat{A}, \hat{B}]] = 0 \tag{2.24}$$

Consider now a time-independent operator $\hat{K}$ commuting with the Hamiltonian $\hat{H}$. In Schrödinger notation

$$\frac{\partial \hat{K}}{\partial t} = 0 \qquad [\hat{H}, \hat{K}] = 0 \tag{2.25}$$

By (2.23) we have

$$[\hat{K}, \hat{H}] = 0 \Rightarrow [\hat{K}_-, \hat{H}_-] = 0 \tag{2.26}$$

Hence, if $\{\hat{K}_i\}_i$ is a complete set of commuting observables with associated id-operators $\{\hat{K}_{i-}\}_i$, and if the basis $\mathscr{B}$ is formed of their eigenvectors $|k_{1j_1}, k_{2j_2}, \ldots)$, then for all i and j_i:

$$\hat{K}_{i-}|k_{1j_1}, k_{2j_2}, \ldots) = k_{ij_i}|k_{1j_1}, k_{2j_2}, \ldots) \tag{2.27}$$

and H_- is a diagonal matrix when written on basis $\mathscr{B}$.

It may happen that we are unable to construct a complete set of integrals of motion, in which case there are generally several eigenvectors with the same sequence of quantum numbers k_{ij_i}, distinguished by a supplementary index k, and written $|kk_{1j_1}, k_{2j_2}, \ldots)$. Degeneracy of this kind occurs when operators commute with $\hat{H}$ but not among themselves.

At the very worst, the matrix of $\hat{H}$ written on $\mathscr{B}$ will be blocked out. Each block along the diagonal, corresponding to a different sequence of quantum numbers K_{ij_i}, is of dimension equal to the necessary number of values of k.

We prove next that the eigenspaces of the $\hat{K}_{i-}$ are stable under the action of $\hat{H}$. Let $\mathscr{A}_K$ be such an eigenspace. For all $|X)$ in $\mathscr{A}_K$,

$$\hat{K}_-|X) = k|X) \Leftrightarrow [\hat{K}, \hat{X}] = k\hat{X} \tag{2.28}$$

Applying Jacobi's identity to $\hat{H}$, $\hat{K}$ and $\hat{X}$, in which $\hat{H}$ and $\hat{K}$ commute by hypothesis, we have

$$[\hat{H},[\hat{K}, \hat{X}]] + [\hat{K},[\hat{X}, \hat{H}]] = 0$$

Using (2.28),

$$[\hat{K},[\hat{H}, \hat{X}]] = k[\hat{H}, \hat{X}]$$

showing that $|[\hat{H}, \hat{X}])$ is an eigenvector of $\hat{K}_-$ for the eigenvalue k and so belongs to $\mathscr{A}_K$. Hence, for all $|X)$ in $\mathscr{A}_K$, $\hat{H}_-|X)$ is in $\mathscr{A}_K$. Q.E.D.

D. Physically Invariant, Irreducible Subspaces

The basis $\mathscr{B}$ introduced in Section II.C provides a natural, physically significant partition of $\mathscr{A}$.

Consider a subspace $\mathscr{A}_K$ engendered by commuting constants of motion $\{\hat{K}_i\}_i$. For any two constants we have

$$[\hat{K}_i, \hat{K}_j] = 0 \Leftrightarrow \hat{K}_{i-}|K_j) = 0$$

so that one of the bases of $\mathscr{A}_K$ is the set of all eigenvectors $|k_{1j_1}, k_{2j_2}, \ldots)$

for which all the eigenvalues are nil. By the relation

$$[\hat{H}, \hat{K}_i] = 0 \Leftrightarrow \hat{H}_-|K_i) = 0$$

the matrix of the projection of $\hat{H}_-$ on the subspace $\mathscr{A}_K$ is the zero matrix. Taking into account Section II.B, Liouville's equation (2.4) written in terms of id-operators is

$$i\hbar \frac{d}{dt}|\rho^i(t)) = \hat{H}^{\,i}_-(t)|\rho^i(t)) \tag{2.29}$$

Let $\hat{P}_K$ be the projection operator of space $\mathscr{A}_K$ and $\hat{\bar{P}}_K$ its supplement in $\mathscr{A}$. Since, as shown previously,

$$\hat{H}^{\,i}_-(t) = \hat{\bar{P}}_K \hat{H}^{\,i}_-(t)\hat{\bar{P}}_K \tag{2.30}$$

Projecting (2.29) we find

$$\begin{aligned} i\hbar \frac{d}{dt}\hat{P}_K|\rho^i(t)) &= 0 \\ i\hbar \frac{d}{dt}\hat{\bar{P}}_K|\rho^i(t)) &= \hat{\bar{P}}_K \hat{H}_-(t)\hat{\bar{P}}_K|\rho^i(t)) \end{aligned} \tag{2.31}$$

Henceforth we refer to $\mathscr{A}_K$ as the subspace of stationary states, because it is the only one containing stationary (equilibrium) states of the statistical ensemble (system) characterized by $\langle d/dt|\rho^i(t))\rangle \equiv d/dt|\rho(t)) = 0$ and denoted by $|\rho_{eq})$. Moreover, the projection of $|\rho^i(t))$ on $\mathscr{A}_K$ is stationary.

Consider next the subspace subtended by all the eigenvectors $|k, k_{1j_1}, k_{2j_2}, \ldots)$ for a given sequence of k_{ij_i} and different k. As we saw in Section II.C, this subspace is stable under the action of $\hat{H}^{\,i}_-(t)$. Hence it contains a complete description of any physical process in the eigenvalues of the operator $\hat{K}_i$, for the associated eigenvectors and Hamiltonians have vanishing components on the generators of this subspace. We later introduce a partition:

$$\mathscr{A} = \mathscr{A}_K \oplus \mathscr{A}^* \oplus \mathscr{A}^a \oplus \mathscr{A}^e \tag{2.32}$$

where the subspaces are in order, those of the stationary states, the transfer of energy in the excited state, and optical absorption and emission.

E. Illustration: The Dimer

In this section we give a detailed treatment of the simplest problem, that of the dimer. Taking account of the restrictions made in Section I.B.2, the dimension δ of $\mathscr{H}$ is 3.

a. Hamiltonians and Integrals of Motion. With the notation introduced in Section I.C.3, we have

$$\hat{H}_0 = \hbar\omega_0(|M_1\rangle\langle M_1| + |M_2\rangle\langle M_2|) + \hbar\Delta\omega(|M_1\rangle\langle M_2| + |M_2\rangle\langle M_1|) \quad (2.33)$$

where $\hbar\omega_0$ replaces E_0 and $\hbar\Delta\omega$ replaces V. For the sake of brevity in the indices we henceforth put

$$\langle M_p|\hat{\epsilon}^i(t)|M_p\rangle = \epsilon_p^i(t) \qquad \langle M_p|\hat{\epsilon}^i(t)|M_{p+1}\rangle = v_p^i(t) \quad (2.34)$$

where it is to be understood that the phases of the basis vectors are chosen so that the Hamiltonians $\hat{H}_0$ and $\hat{\epsilon}^i(t)$ are real. The stochastic operator $\hat{\epsilon}^i(t)$ of the ith aggregate may be written

$$\hat{\epsilon}^i(t) = \epsilon_1^i(t)|M_1\rangle\langle M_1| + v_1^i(|M_1\rangle\langle M_2| + |M_2\rangle\langle M_1|) + \epsilon_2^i(t)|M_2\rangle\langle M_2| \quad (2.35)$$

We now determine the basis $\mathcal{B}$ and the commuting operators of the constants of motion.

In view of the structure of $\hat{H}^i(t) = \hat{H}_0 + \hat{\epsilon}^i(t)$ these operators must be of the form $\lambda|M_0\rangle\langle M_0| + \mu(|M_1\rangle\langle M_1| + |M_2\rangle\langle M_2|)$ where λ and μ may be chosen real so the constants of motion will be automatically Hermitian. The physical significance of this form is clear: Besides the total population, (the system is closed so Tr $\hat{\rho}(t) = \text{const.}$), the summed population of the two excited states is constant, for in setting $\langle M_0|\hat{\epsilon}^i(t)|M_p\rangle = 0$ we neglected nonradiative transitions between the excited states and the ground state.

It being convenient to use an orthonormal basis as in the derivation of (2.17), the first constant of norm one is

$$\hat{K}_1 = (3)^{-1/2}(|M_0\rangle\langle M_0| + |M_1\rangle\langle M_1| + |M_2\rangle\langle M_2|) \quad (2.36)$$

It is related to the total population, since the scalar product (2.13) yields

$$(K_1|\rho(t)) = \text{Tr}\{\hat{K}_1^\dagger\hat{\rho}(t)\} = (3)^{-1/2}\sum_{p=0}^{2}\langle M_p|\hat{\rho}(t)|M_p\rangle = (3)^{-1/2}$$

Noting that the second constant must verify

$$(K_1|K_2) = 0, \qquad (K_2|K_2) = 1$$

we have

$$\hat{K}_2 = (6)^{-1/2}(-2|M_0\rangle\langle M_0| + |M_1\rangle\langle M_1| + |M_2\rangle\langle M_2|) \quad (2.37)$$

For the reasons set out in Section I.C.3 we generally use the eigenbasis of $\hat{H}_0$:

$$\hat{K}_1 = (3)^{-1/2}(|A_0\rangle\langle A_0| + |A_1\rangle\langle A_1| + |A_2\rangle\langle A_2|) \quad (2.38)$$

$$\hat{K}_2 = (6)^{-1/2}(-2|A_0\rangle\langle A_0| + |A_1\rangle\langle A_1| + |A_2\rangle\langle A_2|) \quad (2.39)$$

b. Eigenvalue Problem of the K_{i-}. Partition of $\mathscr{A}$. Zero is a nine times ($3^2 = 9$) degenerate eigenvalue of $\hat{K}_{1-}$, the maximum possible, since $\hat{K}_{1-}$ is proportional to the identity operator of $\mathscr{H}$ and commutes with all the linear operators on $\mathscr{H}$. We saw in Section II.B that the spectrum of $\hat{K}_{2-}$ is formed of the differences of diagonal elements of $\hat{K}_1$ written on its eigenbasis. In the present case zero is five times degenerate; $\sqrt{\frac{3}{2}}$ and $-\sqrt{\frac{3}{2}}$ being twice degenerate. (Obviously, the spectrum of an id-operator is symmetric about the origin.) Since all the eigenvalues of $\hat{K}_{1-}$ are the same, the partition is determined by those of $\hat{K}_{2-}$:

$$\mathscr{A} = \mathscr{A}_0 \oplus \mathscr{A}^a \oplus \mathscr{A}^e$$

where the spaces on the right-hand side are of dimensions 5, 2, and 2 corresponding to the eigenvalues 0, $\sqrt{\frac{3}{2}}$, $-\sqrt{\frac{3}{2}}$. $\mathscr{A}_0$ contains the subspace of the integrals of motion (see Section II.D) and the stationary states. If $\mathscr{A}_i$ is engendered by $|K_i)$ the partition becomes

$$\mathscr{A} = \mathscr{A}_1 \oplus \mathscr{A}_2 \oplus \mathscr{A}^* \oplus \mathscr{A}^a \oplus \mathscr{A}^e \tag{2.40}$$

where $\mathscr{A}_1$, $\mathscr{A}_2$ and $\mathscr{A}^*$ are of dimension 1, 1, and 3. $\mathscr{A}_1 \oplus \mathscr{A}_2$ is the subspace $\mathscr{A}_K$ of (2.32). Each subspace is degenerate with respect to $\hat{K}_{1-}$ and $\hat{K}_{2-}$. To determine a basis of $\mathscr{A}^*$ we must find three operators Ω^* such that

$$\hat{K}_{2-}|\Omega^*) = 0 \Leftrightarrow [\hat{K}_2, \hat{\Omega}^*] = 0$$

The operator corresponding to the difference of the populations of the excited states $|A_1\rangle$ and $|A_2\rangle$ is a solution,

$$\hat{\Omega}_1^* = (2)^{-1/2}(|A_1\rangle\langle A_1| - |A_2\rangle\langle A_2|) \tag{2.41}$$

orthogonal to $\hat{K}_1$ and $\hat{K}_2$, of unit length and commutes with $\hat{H}_0$. $\hat{K}_1, \hat{K}_2$ and $\hat{\Omega}_1^*$ are a complete set of commuting operators for H_0; that is, $|K_1)$, $|K_2)$ and $|\Omega_1^*)$ are a complete basis for $|H_0)$. Hence $\hat{H}_{0-}$ is a diagonal matrix on the basis of common eigenvectors of $\hat{K}_{1-}$, $\hat{K}_{2-}$ and $\hat{\Omega}_{1-}^*$. $\hat{\Omega}_2^*$ and $\hat{\Omega}_3^*$ defined by

$$\hat{\Omega}_2^* = |A_1\rangle\langle A_2|; \qquad \hat{\Omega}_3^* = |A_2\rangle\langle A_1| \tag{2.42}$$

are orthonormal, linearly independent and commute with $\hat{K}_2$.

The operators $|\Omega^a)$ required for the subspace $\mathscr{A}^a$ corresponding to the eigenvalue $\sqrt{\frac{3}{2}}$ of $\hat{K}_{2-}$ verify

$$\hat{K}_{2-}|\Omega^a) = \frac{\sqrt{3}}{\sqrt{2}}|\Omega^a) \Leftrightarrow [\hat{K}_2, \hat{\Omega}^a] = \frac{\sqrt{3}}{\sqrt{2}}\hat{\Omega}^a$$

Putting $\hat{\Omega}^a = \Sigma_{k,l}\Omega_{kl}|A_k\rangle\langle A_l|$, expanding, with form (2.39) of $\hat{K}_2$ and carrying out the nine projections one finds that the only nonzero Ω_{kl} are

Ω_{10} and Ω_{20}, which are undetermined. Quite generally, therefore, where λ and μ are scalars, and we may take

$$\hat{\Omega}_1^a = |A_1\rangle\langle A_0|; \qquad \hat{\Omega}_2^a = |A_2\rangle\langle A_0| \tag{2.43}$$

so that the $|\Omega_k^a)$, $k = 1, 2$, are eigenvectors of $\hat{H}_{0-}$.

Similarly, in subspace $\mathscr{A}^e$ the operators

$$\hat{\Omega}_1^e = |A_0\rangle\langle A_1|; \qquad \hat{\Omega}_2^e = |A_0\rangle\langle A_2| \tag{2.44}$$

are eigenvectors of $\hat{H}_{0-}$ but with the opposite eigenvalues to those of the $|\Omega_k^a)$ since $\hat{\Omega}_k^e = \hat{\Omega}_k^{a\dagger}$; see (2.21).

c. Physical Meaning of the Subspaces of $\mathscr{A}$. We come now to identifying the physical phenomena associated with each subspace, though first the equation of motion must be integrated in each subspace.

Suppose that the system evolves under a general Hamiltonian $\hat{H}''$ of the form

$$\hat{H}''(t) = \hat{H}^i(t) + \hat{W}(t)$$

where $\hat{H}^i(t)$ is defined by (2.1). The derivative of $(K_2|\rho^i(t)) = \mathrm{Tr}\{\hat{K}_2\hat{\rho}^i(t)\}$, or the average value of K_2 results from Liouville's equation

$$i\hbar \frac{d}{dt}|\rho^i(t)) = \hat{H}_-''(t)|\rho^i(t))$$

so that

$$\frac{d}{dt}(K_2|\rho^i(t)) = (i\hbar)^{-1}(K_2|\hat{H}_-^i(t) + \hat{W}_-(t)|\rho^i(t))$$

$\hat{K}_2$ is a constant for the movement governed by $\hat{H}^i(t)$, so $\hat{H}_-^i(t)|K_2) = 0$ and the preceding equation may be simplified to

$$\frac{d}{dt}(K_2|\rho^i(t)) = (i\hbar)^{-1}(K_2|\hat{W}_-(t)|\rho^i(t)) \tag{2.45}$$

Straightforward manipulation yields

$$(K_2|\hat{W}_-(t)|\rho^i(t)) = -(W(t)|\hat{K}_{2-}|\rho^i(t)) \tag{2.46}$$

Projecting $|\rho^i(t))$ on each eigenspace of $\hat{K}_{2-}$,

$$|\rho^i) = |\rho_1) + |\rho_2) + |\rho^*) + |\rho^a) + |\rho^e)$$

where the superscript i has been dropped on the right-hand side. Carrying the eigenvalues of each subspace into (2.46) and (2.45) leads to

$$\frac{d}{dt}(K_2|\rho^i(t)) = (i\hbar)^{-1}\frac{\sqrt{3}}{\sqrt{2}}\{(W(t)|\rho^e) - (W(t)|\rho^a)\}$$

It follows from the definitions of $\hat{\Omega}_k^a$ and $\hat{\Omega}_k^e$ ((2.43), (2.44)), that $\hat{\rho}^{e\dagger} = \hat{\rho}^a$ so that, $\hat{W}$ being Hermitian, $(W|\rho^e) = (W|\rho^a)^*$. Finally,

$$\frac{d}{dt}(K_2|\rho^i(t)) = \hbar^{-1}\sqrt{6}\,\mathrm{Im}(W(t)|\rho^a(t)) = -\hbar^{-1}\sqrt{6}\,\mathrm{Im}(W(t)|\rho^e(t))$$

Hence $(K_2|\rho^i(t))$ varies only under the influence of operators W such that $(W(t)|\rho^\alpha) \neq 0$, $\alpha = a$ or e; $(K_2|\rho^i(t))$ is of course constant if $\hat{W}(t)$ is absent. $|\rho^a)$ may be written

$$|\rho^a) = c_1|\Omega_1^a) + c_2|\Omega_2^a)$$

whence

$$\begin{aligned}(W(t)|\rho^a(t)) &= c_1(t)(W(t)|\Omega_1^a) + c_2(t)(W(t)|\Omega_2^a)\\ &= c_1(t)\langle A_0|\hat{W}(t)|A_1\rangle + c_2(t)\langle A_0|\hat{W}(t)|A_2\rangle\end{aligned}$$

where the coefficients $c_k(t)$ are arbitrary, since they depend on the initial state which is completely unrestricted. Consequently the necessary and sufficient condition for time dependence of $(K_2|\rho^i(t))$ is that one at least of $\langle A_0|\hat{W}(t)|A_1\rangle$ and $\langle A_0|\hat{W}(t)|A_2\rangle$ should differ from zero, in which case the time dependence of this average value is a function only of $|\rho^a)$ or $|\rho^e)$. An operator of this kind must represent the semiclassical coupling between field and matter. As for given k, $\hat{\Omega}_k^a$ and $\hat{\Omega}_k^e$ are Hermitian conjugates, the matrices in $\mathscr{A}^a$ and $\mathscr{A}^e$ of an id-operator are opposite. See (2.21). The poles of the corresponding resolvents are opposite, one space corresponding to positive resonance frequencies, or absorption, and the other to negative ones, or emission. $\mathscr{A}^a$ and $\mathscr{A}^e$ completely define the optical susceptibility.

$\mathscr{A}^*$ is easily interpreted. When $\hat{W}(t) = 0$, $\hat{K}_2$ is a constant and the quantities $(\Omega^*|\rho)$ depend only on the $\langle A_k|\hat{\rho}|A_l\rangle$ $k, l = 1, 2$.

$\mathscr{A}^*$ describes transfer of energy during the life of the excited state and will be called the transfer space.

d. $\hat{H}_{0-}$ and $\hat{\epsilon}_-^i(t)$. $\hat{H}_{0-}$ and $\hat{\epsilon}_-^i(t)$ may be found from $\hat{H}_0$ and $\hat{\epsilon}^i(t)$ by the methods of Section II.B. From (2.33), (2.35), (2.38), (2.39), and (2.41) to (2.43) we have

$$\begin{aligned}\hat{H}^i(t) = &\frac{1}{\sqrt{3}}\{2\hbar\omega_0 + \epsilon_1^i(t) + \epsilon_2^i(t)\}\Big(\hat{K}_1 + \frac{1}{\sqrt{2}}\hat{K}_2\Big) + \sqrt{2}\{\hbar\Delta\omega + v_1^i(t)\}\hat{\Omega}_1^*\\ &+ \tfrac{1}{2}\{\epsilon_1^i(t) - \epsilon_2^i(t)\}(\hat{\Omega}_2^* + \hat{\Omega}_3^*)\end{aligned} \tag{2.47}$$

We already know it suffices to calculate the diagonal blocks in the table of commutators corresponding to the partition of $\mathscr{A}$. The results displayed in Tables I to III are derived from the properties described in Section II.B, especially (2.19) to (2.21).

TABLE I
Commutators in the Transfer Subspace; for instance: $[\hat{\Omega}_2^*, \hat{\Omega}_1^{*\dagger}] = -\sqrt{2}\,\hat{\Omega}_2^*$

	$\hat{\Omega}_1^{*\dagger}$	$\hat{\Omega}_2^{*\dagger}$	$\hat{\Omega}_3^{*\dagger}$
$\hat{\Omega}_1^*$	0	$-\sqrt{2}\hat{\Omega}_3^*$	$\sqrt{2}\hat{\Omega}_2^*$
$\hat{\Omega}_2^*$	$-\sqrt{2}\hat{\Omega}_2^*$	$\sqrt{2}\hat{\Omega}_1^*$	0
$\hat{\Omega}_3^*$	$\sqrt{2}\hat{\Omega}_3^*$	0	$-\sqrt{2}\hat{\Omega}_1^*$

TABLE II
Commutators in the Absorption Subspace

	$\hat{\Omega}_1^{a\dagger}$	$\hat{\Omega}_2^{a\dagger}$
$\hat{\Omega}_2^a$	$\frac{\sqrt{3}}{\sqrt{2}}\hat{K}_2 + \frac{1}{\sqrt{2}}\hat{\Omega}_1^*$	0
$\hat{\Omega}_2^a$	0	$\frac{\Sigma 3}{\Sigma 2}\hat{K}_2 - \frac{1}{\sqrt{2}}\hat{\Omega}_1^*$

TABLE III
Commutators in the Emission Subspace

	$\hat{\Omega}_1^{e\dagger}$	$\hat{\Omega}_2^{e\dagger}$
$\hat{\Omega}_1^e$	$-\frac{\sqrt{3}}{\sqrt{2}}\hat{K}_2 - \frac{1}{\sqrt{2}}\hat{\Omega}_1^*$	0
$\hat{\Omega}_1^e$	0	$-\frac{\sqrt{3}}{\sqrt{2}}\hat{K}_2 + \frac{1}{\sqrt{2}}\hat{\Omega}_1^*$

Using definition (2.17) we obtain $\hat{H}_-^i(t)$ in each subspace by replacing an operator $\hat{\Omega}_\lambda$ in the corresponding commutation table by the projection of $\hat{H}^i(t)$ on this operator. By (2.47),

$$\hat{H}_-^{i*}(t) = \begin{bmatrix} 0 & \frac{-1}{\sqrt{2}}\Delta\epsilon^i & \frac{+1}{\sqrt{2}}\Delta\epsilon^i \\ \frac{-1}{\sqrt{2}}\Delta\epsilon^i & +2(v_1^i + V) & 0 \\ \frac{+1}{\sqrt{2}}\Delta\epsilon^i & 0 & -2(v_1^i + V) \end{bmatrix} \qquad (\Delta\epsilon^i = \epsilon_1^i - \epsilon_2^i) \tag{2.48}$$

$$\hat{H}^{ia}_{-}(t) = \begin{bmatrix} \frac{1}{2}\sum \epsilon^i + v_1^i + \hbar\omega_1 & 0 \\ 0 & \frac{1}{2}\sum \epsilon^i - v_1^i + \hbar\omega_2 \end{bmatrix} \qquad (\sum \epsilon^i = \epsilon_1^i + \epsilon_2^i) \tag{2.49}$$

$$\hat{H}^{ie}_{-}(t) = \begin{bmatrix} -\frac{1}{2}\sum \epsilon^i - v_1^i - \hbar\omega_1 & 0 \\ 0 & -\frac{1}{2}\sum \epsilon^i + v_1^i - \hbar\omega_2 \end{bmatrix} \tag{2.50}$$

where $V = \hbar\omega$, and $\omega_1 = \omega_0 + \Delta\omega$, $\omega_2 = \omega_0 - \Delta\omega$ are the eigenfrequencies of $\hbar^{-1}\hat{H}_0$. The matrix of $\hat{H}^i_{-}(t)$ projected on $\mathscr{A}^e$ follows from $\hat{H}^{ia}_{-}(t)$ by (2.21), changing the signs throughout.

The preceding equations reveal one or two points. Thus, from (2.48) we see it is because of the difference of local fluctuations, $\epsilon_1^i(t) - \epsilon_2^i(t)$ (lack of resonance) that the $|A_k\rangle$ are not instantaneous eigenvectors of $\hat{H}^i_{-}(t)$. If the $\epsilon_p^i(t)$ were nil all the states $|A_k\rangle$, $k = 0, 1, 2$ would have constant macroscopic populations. We shall see that the difference of the populations of $|A_1\rangle$ and $|A_2\rangle$ has a life of γ_0^{-1} a measure of the mean square fluctuation of the $\epsilon_p^i(t)$. The matrices (2.49), (2.50) are diagonal and we should expect some very simple line shapes. In Section III.B we see that the line shapes of the dimer are Lorentzian depending on the parameters γ_0 and γ_1.

F. Derivation of the Macroscopic Equation[97]

Given Liouville's equation for the ith microsystem,

$$i\hbar \frac{d}{dt} |\rho^i(t)) = \hat{H}^i_{-}(t) |\rho^i(t)) \tag{2.51}$$

we are to derive the evolution of the density operator $|\rho(t))$ of the statistical ensemble, defined by (2.3). Extensive use of superoperators greatly simplifies the problem.

In interaction representation relative to $\hat{H}_0$,

$$|\tilde{\rho}(t)) = \exp\left[\frac{1}{i\hbar}\hat{H}_0(t - t_0)\right] |\rho^i(t)) \equiv U_0^\dagger(t, t_0) |\rho^i(t)) \tag{2.52}$$

whence

$$i\hbar \frac{d}{dt} |\tilde{\rho}^i(t)) = \hat{\tilde{\epsilon}}^i_{-}(t) |\tilde{\rho}^i(t)) \tag{2.53}$$

with

$$\hat{\hat{\epsilon}}_-^i(t) = \hat{U}_0^\dagger(t, t_0)\hat{\epsilon}_-^i(t)\hat{U}_0(t, t_0) \tag{2.54}$$

Formal integration of (2.53) by Feynman's method,[101] in which the ordering parameter is the time, leads to

$$|\tilde{\rho}^i(t)) = \{\exp\left(\frac{1}{i\hbar}\int_{t_0}^t \hat{\hat{\epsilon}}_-^i(t')dt'\right)\}|\rho^i(t_0)) \tag{2.55}$$

Supposing that the preparation is pure, ($\hat{\rho}^i(t_0)$ independent of i and $\hat{\rho}^2(t_0) = \hat{\rho}(t_0)$), because all the microsystems have the same initial state, excepting phase, the statistical average of (2.55) is

$$|\langle\tilde{\rho}^i(t)\rangle) = \langle\exp\left(\frac{1}{i\hbar}\int_{t_0}^t \hat{\hat{\epsilon}}_-^i(t')dt'\right)\rangle|\rho(t_0)) \tag{2.56}$$

The left-hand side is the same as

$$\mathcal{N}^{-1}\sum_{i=1}^{\mathcal{N}} \hat{U}_0^\dagger(t, t_0)|\rho^i(t)) = \hat{U}_0^\dagger|\langle\rho^i(t)\rangle) = \hat{U}_0^\dagger|\rho(t)) \equiv |\tilde{\rho}(t)) \tag{2.57}$$

The right-hand side may be evaluated using the cumulants[102–104] of the probability laws of $\hat{\hat{\epsilon}}_-^i(t)$. By definition of the cumulant $\hat{\mathcal{K}}_n$

$$\langle\exp\left(\frac{1}{i\hbar}\int_{t_0}^t \hat{\hat{\epsilon}}_-^i(t')dt'\right)\rangle = \exp\left\{\sum_{n=1}^{\infty}(i\hbar)^{-n}\int_{t_0}^t dt_1\int_{t_0}^{t_1} dt_2\cdots\int_{t_0}^{t_{n-1}} dt_n\,\hat{\mathcal{K}}_n(t_1, t_2, \ldots, t_n)\right\} \tag{2.58}$$

For the normally distributed process, explained in Section I.C.3, cumulants beyond the second are all nil.[105] In fact, $\hat{\epsilon}^i(t)$ was taken, without loss of generality, to be centered, so $\langle\hat{\epsilon}^i(t)\rangle = 0$ implies $\langle\hat{\hat{\epsilon}}^i(t)\rangle = 0$. The right-hand side of (2.58) becomes, therefore,

$$\langle\exp\left(\frac{1}{i\hbar}\int_{t_0}^t \hat{\hat{\epsilon}}_-^i(t')dt'\right)\rangle = \exp\left\{-\hbar^{-2}\int_{t_0}^t dt_1\int_{t_0}^{t_1} dt_2\hat{\mathcal{K}}_2(t_1, t_2)\right\} \tag{2.59}$$

in which

$$\begin{aligned}\hat{\mathcal{K}}_2(t_1, t_2) &= \langle\hat{\hat{\epsilon}}_-^i(t_1)\hat{\hat{\epsilon}}_-^i(t_2)\rangle\\ &= \hat{U}_0^\dagger(t_1, t_0)\langle\hat{\epsilon}_-^i(t_1)\hat{U}_0(t_1, t_0)\hat{U}_0^\dagger(t_2, t_0)\hat{\epsilon}_-^i(t_2)\rangle\hat{U}_0(t_2, t_0)\end{aligned}$$

But by (1.7) all the matrix elements of the operator between $\langle$ and $\rangle$ are nil for $t_1 \neq t_2$. Moreover, for $t_1 = t_2$, $\hat{U}_0(t_1, t_0)U_0^\dagger(t_2, t_0)$ is the identity operator so

$$\begin{aligned}\hat{\mathcal{K}}_2(t_1, t_2) &= \hat{U}_0^\dagger(t_1, t_0)\langle\hat{\epsilon}_-^i(t_1)\hat{\epsilon}_-^i(t_2)\rangle\hat{U}_0(t_1, t_0)\\ &\equiv 2\hbar^2\hat{\hat{\Gamma}}(t_1)\delta(t_1 - t_2)\end{aligned} \tag{2.60}$$

where $\hat{\tilde{\Gamma}}(t_1)$ is the interaction representation of some time-independent operator, $\hat{\Gamma}$, in Schrödinger representation. Notice that $\hat{\mathscr{K}}_2(t_1, t_2) = \hat{\mathscr{K}}_2(t_2, t_1)$. Hence, putting (2.60) in the right-hand side of (2.59) and using (2.57) we have,

$$|\tilde{\rho}(t)) = \left\{\exp\left(-\int_{t_0}^{t} \tilde{\Gamma}(t')dt'\right\}|\rho(t_0)\right)$$

By definition of the interaction representation this means

$$\begin{aligned}|\rho(t)) &= \left\{\exp\left[\frac{1}{i\hbar}\hat{H}_{0-}(t-t_0)\right]\right\}\left\{\exp\left[-\int_{t_0}^{t}\hat{\tilde{\Gamma}}(t')dt'\right]\right\}|\rho(t_0)) \\ &\equiv \left\{\exp\left[\frac{1}{i\hbar}(\hat{H}_{0-} - i\hbar\hat{\Gamma})(t-t_0)\right]\right\}|\rho(t_0))\end{aligned}$$

In other words $|\rho(t))$ is a solution of the differential equation:

$$i\hbar\frac{d}{dt}|\rho(t)) = (\hat{H}_{0-} - i\hbar\hat{\Gamma})|\rho(t)) \equiv \hat{H}|\rho(t)) \tag{2.61}$$

in which $\hat{\Gamma}$ is a time-independent operator defined by

$$2\hbar^2\hat{\Gamma}\delta(t-t') = \langle\hat{\epsilon}^{i}_{-}(t)\hat{\epsilon}^{i_s}_{-}(t')\rangle \tag{2.62}$$

The preceding derivation is simpler than Reineker's,[48] which involves the functional derivative of a generalized density operator, based on $\hat{\epsilon}^{i}(t)$.

The operators $\hat{H}$ and $\hat{\Gamma}$ do not warrant the subscript $_-$ since they are superoperators, but not id-operators, there being no operator $\hat{\Gamma}^f$ on $\mathscr{H}$ generating an operator $\hat{\Gamma}^f_-$ in $\mathscr{A}$, according to (2.17), such that $\hat{\Gamma}^f_- \equiv \hat{\Gamma}$. This follows from the fact that the Lie algebra of id-operators is not stable with respect to multiplication of two operators on $\mathscr{H}$, followed by statistical averaging. Indeed we shall see $\hat{\Gamma}$ may have nonzero matrix elements corresponding to elements which are always nil for an id-operator, by the terms of (2.19) (see (2.121) and Table I).

Physically this means $\hat{\Gamma}$ must be a macroscopic operator. Symmetry with respect to the reversal of time and a Ritz principle of combination of relaxation constants are lost when we consider a statistical system containing random interactions ($\hat{\Gamma}$ is by definition, positive). Fano[106] has made a similar point in discussing pressure broadening. Thus the essential distinction introduced by superoperators is that generators of reversible movements are id-operators whereas generators of irreversible movements are general superoperators. The algebraic structure of the former and their links with commutators lead to a Ritz combination principle, whereas for the latter no such rule of association exists.

The relation (2.63), valid for all $|X)$ in $\mathscr{A}$, is useful in what follows:

$$
\begin{aligned}
2\hbar^2\hat{\Gamma}\delta(t-t')|X) &= \langle \hat{\epsilon}^i_-(t)\hat{\epsilon}^i_-(t')\rangle|X) \\
&= |\{\langle \hat{\epsilon}^i(t)\hat{\epsilon}^i(t')\rangle, \hat{X}\}_+ - 2|\langle \hat{\epsilon}^i(t)\hat{X}\hat{\epsilon}^i(t')\rangle) \qquad (2.63)
\end{aligned}
$$

where $\{\hat{A}, \hat{B}\}_+ = \hat{A}\hat{B} + \hat{B}\hat{A}$. We have used the parity property of Dirac's distribution. The preceding relation is useful in determining an expression for the relaxation operator, $\hat{\Gamma}$, especially in exhibiting its scalar component. Equation 2.63 shows also that any attempt at writing relaxation in $\mathscr{H}$ as a commutator is bound to fail.

G. General Discussion of the Movement. More Conditions of Validity

We are here concerned with some aspects of the movement determined by the macroscopic equation (2.62) and with some consequent conditions of validity of the stochastic model.

By definition, the projection of $\hat{\Gamma}$ on the orthogonal subspace of $\mathscr{A}_K$ (space of equilibrium states), is positive definite, and further, the eigenvalues of the projection of $\hat{H}$ are all in the lower half of the complex plane. In general $\hat{H}_{0-}$ and $\hat{\Gamma}$ do not commute so that

$$
e^{1/i\hbar(\hat{H}_{0-} - i\hbar\hat{\Gamma})(t-t_0)} \neq e^{1/i\hbar\hat{H}_{0-}(t-t_0)}e^{-\hat{\Gamma}(t-t_0)}
$$

Consequently the movement is not the superposition of a reversible (coherent) evolution and an irreversible (incoherent) one. Hence the matrix of $\hat{\Gamma}$ on the eigenbasis of $\hbar^{-1}\hat{H}_{0-}$ generally is not diagonal. The diagonal elements of $\hat{\Gamma}$ are analagous to the reciprocals of relaxation times like T_1 in Bloch's equations and the off-diagonal terms are analagous to T_2^{-1}. Plainly we cannot interpret these matrix elements directly; it being necessary to find the irreversible part of the evolution of an observable to define the lifetimes, for example. In the dimer γ_0^{-1} is the life of the difference of population of the excited states $|A_1\rangle$ and $|A_2\rangle$, as may be surmised from the matrix in (2.121).

Another consequence of the fact that $\hat{\Gamma}$ is positive is that $\hat{H}_{0-}$ is not renormed, there being no shifts in the levels. This is an approximation known to be reasonable by comparison with more elaborate models.

Generalizing the discussion of the dimer in Section II.E, there are always two integrals of motion, associated with the conservation of the total population and the summed populations of the excited states. The equilibrium states are

$$
|\rho(t=+\infty)) \equiv |\rho_{\text{eq}}) = \lambda|I) + \mu|\bar{I}) \qquad (2.64)
$$

in which $\hat{I}$ is the identity equal to $\Sigma_{k\geqslant 0}|A_k\rangle\langle A_k|$ and $\hat{\bar{I}} = \hat{I} - |A_0\rangle\langle A_0|$. Thus in any equilibrium state the $|A_k\rangle$, $k\rangle 0$ are all equally populated. Since the dispersion in energy of these states is roughly $|\Delta\omega|$ the model, compared to a Boltzmann distribution, makes sense only of $|\Delta\omega| \ll k_B T$. This is an extra condition, independent of those in Sections I.C.1 and I.C.3. We notice that the equal populations would be a serious problem if we were interested in low-frequency transitions between these states, for the response of the system would vanish to all orders of perturbation, as all operators commute with the identity. The model would not work, or the theory of the response of the system, generally developed for an equilibrium initial state before perturbation, would have to be reformulated. Indeed, if the probability of absorption between $t = 0$ and $t = \gamma_0^{-1}$ is large, the sample draws energy from the field.

Not surprisingly the total energy of the ensemble $\mathrm{Tr}\{\hat{H}_0\hat{\rho}(t)\}$ is not constant, since the system is not isolated, being in contact with a heat bath. The energy is

$$\langle E^i \rangle(t) \equiv \mathrm{Tr}\{H_0\hat{\rho}(t)\} = (H_0|\rho(t)) \tag{2.65}$$

and its derivative follows from (2.61):

$$\frac{d}{dt}\langle E^i \rangle(t) = \frac{1}{i\hbar}(H_0|\hat{H}_{0-} - i\hbar\hat{\Gamma}|\rho(t)) = -(H_0|\hat{\Gamma}|\rho(t)) \tag{2.66}$$

The right-hand matrix element generally differs from zero for $t < +\infty$, except if $|\rho(t))$ lies entirely in $\mathscr{A}_K$ in which case $|\rho(t)) = |\rho_{\mathrm{eq}})$ and all average values are stationary. Thus the system, when out of equilibrium, does indeed exchange energy with the thermostat. If we suppose the statistical and quantum average of fluctuations to be identically zero, then the equation of conservation of energy is

$$\mathrm{Tr}\left\langle \hat{\rho}^i(t)\frac{d}{dt}\hat{\epsilon}^i(t)\right\rangle = -(H_0|\hat{\Gamma}|\rho(t)) \tag{2.67}$$

The trace of $\hat{\rho}^2(t)$ is nonincreasing with time: Using the scalar product (2.13) and the Hermitian character of $\hat{\rho}(t)$ and $\hat{H}_0$, the derivative becomes

$$\frac{d}{dt}\mathrm{Tr}\{\hat{\rho}^2(t)\} \equiv \frac{d}{dt}(\rho(t)|\rho(t)) = -2(\rho(t)|\hat{\Gamma}|\rho(t)) \tag{2.68}$$

$\hat{\Gamma}$ being by definition positive, the right-hand side is less than or equal to zero, and zero only if $|\rho(t))$ is a macroscopic equilibrium state lying entirely in $\mathscr{A}_K$.

If this is not so, $\mathrm{Tr}\{\hat{\rho}^2(t)\}$ decreases with time. Consequently,

$$\mathrm{Tr}\{\hat{\rho}^2(t)\} < \mathrm{Tr}\{\hat{\rho}^2(0)\} = \mathrm{Tr}\{\hat{\rho}(0)\} = 1 \qquad \forall t > 0 \tag{2.69}$$

Now $\mathrm{Tr}\{\hat{\rho}^2(t)\} < 1 \Leftrightarrow \hat{\rho}^2(t) < \hat{\rho}(t)$ which shows that the relaxation operator $\hat{\Gamma}$ causes an initially pure state to evolve towards a mixed state, destroying any initial coherence. If $\hat{H}_{0-}$ and $\hat{\Gamma}$ commuted we should have

$$\mathrm{Tr}\{\hat{\rho}^2(t)\} = (\rho(0)|e^{-2\hat{\Gamma}t}|\rho(0))$$

An inherent defficiency of this type of stochastic model is that it does not predict Boltzmann equilibria: Let

$$|\rho(t)) = |\rho(\beta)) + |\rho'(t))$$

where $\beta = (k_B T)^{-1}$ and $\hat{\rho}(\beta) = \exp(-\beta\hat{H}_0)/\mathrm{Tr}\{\exp(-\beta\hat{H}_0)\}$. Putting $|\rho(t))$ in (2.61) we obtain

$$i\hbar \frac{d}{dt}|\rho'(t)) = -i\hbar\hat{\Gamma}|\rho(\beta)) + (\hat{H}_{0-} - i\hbar\hat{\Gamma})|\rho'(t))$$

In a Boltzmann equilibrium, $|\rho'(t))$ vanishes identically, so $\hat{\Gamma}$ must verify:

$$0 = -i\hbar\hat{\Gamma}|\rho(\beta)) \tag{2.70}$$

Moreover, if $\hat{P}_K$ is the projection operator on $\mathscr{A}_K$ (see Section II.D), $\hat{\epsilon}^i_-(t) = \hat{\hat{P}}_K\hat{\epsilon}^i_-(t)\hat{\hat{P}}_K$ (see (2.30)). Hence, by definition (2.62), $\hat{\Gamma} = \hat{\hat{P}}_K\hat{\Gamma}\hat{\hat{P}}_K$ and (2.70) becomes ($\hat{\hat{P}}^2_K = \hat{\hat{P}}_K$):

$$0 = -i\hbar(\hat{\hat{P}}_K\hat{\Gamma}\hat{\hat{P}}_K)\hat{\hat{P}}_K|\rho(\beta)) \tag{2.71}$$

Now $\hat{\Gamma}$ is by definition strictly positive in the supplementary space of $\mathscr{A}_K$ so $\hat{\hat{P}}_K\hat{\Gamma}\hat{\hat{P}}_K$ is nonsingular and (2.71) implies

$$0 = \hat{\hat{P}}_K|\rho(\beta))$$

or ($\hat{P}_K + \hat{\hat{P}}_K = \hat{I}$):

$$|\rho(\beta)) \equiv (\hat{P}_K + \hat{\hat{P}}_K)|\rho(\beta)) = \hat{P}_K|\rho(\beta))$$

$|\rho(\beta))$ is unique, there being no degeneracy in the energy, so the last equation implies

$$\hat{P}_K = \frac{|\rho(\beta))(\rho(\beta)|}{(\rho(\beta)|\rho(\beta))}$$

The preceding form of $\hat{P}_K$ is clearly incompatible with the definition already given:

$$\hat{P}_K = |K_1)(K_1| + |K_2)(K_2|$$

This is hardly surprising since thermodynamic principles and the temperature did not enter into the derivation of the macroscopic equation (2.61). Indeed it would be curious if our model were compatible with thermodynamics. It would be futile to improve the stochastic model in trying to overcome this difficulty.

It is essential to bear in mind the following point. We have seen that the restriction of $\hat{\rho}(+\infty)$ to the excitation subspace of $\mathcal{H}$ is proportional to the corresponding identity operator, or in other words, that in an equilibrium state all the diagonal terms of the restriction are equal, the off-diagonal terms being nil of course. This is true on any basis, localized, delocalized, or derived from them by a unitary transformation; see Appendix B in Avakian et als.' article.[107] From our point of view, this property, being the result of unfortunate mathematical characteristics of the model, is insufficient grounds for concluding that any basis, such as the localized basis, can be given a simple physical interpretation. We continue to regard stationary states of $\hat{\mathcal{H}}_0$ as the only basis on which matrices have any meaning and as the only one on which to discuss populations and coherence. The "degeneracy" of bases results solely from a quantitative simplification, justifiable at high temperatures, but which does not support any qualitative conclusions. In the limiting case $\Delta\omega = 0$, all the bases are equivalent with regard to average values of the energy. Yet one still cannot favor the localized basis. The difficulty is more than academic when one considers quantities other than the energy. The two attitudes lead to radically different conclusions about the spatial distribution of the energy:

1. If we for a moment suppose that in a mixed equilibrium state of the ensemble all the aggregates are in state $|M_p)$ (which is not a stationary state of $\hat{H}_0$) with a constant probability independent of ρ, the spatial distribution is as follows: The exciton has a constant probability of being on any given site of the aggregate, but is always localized on one of them. Such are the consequences of favoring the basis $\{|M_p\rangle\}_p$. Conversely, if we suppose that the energy is localized, then the aggregate may be in an equilibrium state which is not stationary of its own Hamiltonian.

2. If, on the other hand, as we continue to do, we suppose that in the equilibrium state of the statistical ensemble there is a constant probability, independent of k, that an aggregate is in a state $|A_k\rangle$, then the energy is in general very delocalized. The delocalization varies slightly with k but is always close to its maximum value.

These two points are discussed in more detail in Section IV.B.

One can hardly imagine more contrary points of view. In the absence of any proof, especially when $\gamma_0 \ll |\Delta\omega|$, we avoid statements like: "The phonons localise the exciton",[108] based on intuitive manipulation of Heisenberg's uncertainty relations, which cannot really be called proofs but are at best pictorial descriptions of the as yet unproved. Classical analogies, like the one of two weakly coupled, randomly perturbed pendula, are even less convincing. Their imposition of macroscopic concepts on quantum systems is perhaps a denial of quantum characteristics.[109] Indeed it is hard to conceive how quantum mechanical principles could agree with pictorial descriptions of the spacio-temporal process or relocalization. A consequence of these principles[110–112] is that two systems in interaction, or which once interacted, are not separable and that an analysis in terms of sites is at odds with the premises. (To refute this is to refute all quantum mechanics, for inseparability is an immediate consequence of Schrödinger's equation. Such a study could be carried out only with the aid of a reduced density operator for each site, for which one would have to write the equation of motion, integrate it, and discuss its solution, no mean task,[113] since the electrons of all the sites are identical particles. Moreover, even if there were no direct interactions, coupling via the field would persist as the distance between sites is small compared to the wavelengths of the characteristic radiation of the system. The preceding arguments are independent of any strong inequality between $|\Delta\omega|$ and the relaxation constants expressed in terms of γ_0 and γ_1). We see in Section IV that this problem is not inevitable. It then seems something of a false problem. No decisive experimental test has been made, since it would require a very fine probe in both space and time. The essential theoretical interpretation of its interaction with a crystal or an aggregate, would be but a facet of the general theory of measurement in quantum mechanics, a domain which is as yet unsatisfactorily developed.[114–115] Indeed, one could well ask "To what degree may our macroscopic ideas of time and space be extended to quantum levels?"[109] Is not the very essence of quantum phenomena in their inadequacy? What do electrons in atoms or molecules "do", if indeed they "do" anything at all?

H. Mathematical Details

In this section we extend the ideas of Sections II.C and II.D to ordinary and double chains. The results obtained are immediate generalizations of those proved in detail in Section II.E.

1. Constants of Motion. Partition of $\mathscr{A}$

The system being closed, with only radiative transitions between the ground state and the excited states, there must be two constants, the total population and the population of all the excited states. In turn the population of the ground state is constant. The effective dimension of the state space of a double chain composed of two chains A and B, of N sites each, is $2N+1$. The constant $\hat{K}_1$ corresponding to the conservation of the total population is

$$\hat{K}_1 = (2N+1)^{-1/2}\left(|M_0\rangle\langle M_0| + \sum_{I=A,B}\sum_{p=1}^{N} |M_p^I\rangle\langle M_p^I|\right) \tag{2.72}$$

where the $\{|M_p^I\rangle\}_{1\leq p\leq N}^{I=A,B}$ are a straightforward generalization of the singly excited antisymmetrized states introduced in (1.4). The extra index I indicates to which chain the singly excited site p belongs. An extra index is needed also to specify the stationary states $|A_{kb}\rangle$ of the static Hamiltonian $\hat{H}_0$, (defined in Section II.I.1). k takes integral values between 1 and N and b has two values, conventionally taken to be $+1$ and -1. Plainly, therefore,

$$\hat{K}_1 = (2N+1)^{-1/2}\left(|A_0\rangle\langle A_0| + \sum_{b=\pm 1}\sum_{k=1}^{N} |A_{kb}\rangle\langle A_{kb}|\right) \tag{2.73}$$

The second constant is of the form

$$\hat{K}_2 = \lambda |A_0\rangle\langle A_0| + \mu \sum_{b=\pm 1}\sum_{k=1}^{N} |A_{kb}\rangle\langle A_{kb}|$$

in which λ and μ are scalars determined by

$$(K_2|K_1) = 0; \qquad (K_2|K_2) = 1$$

Hence

$$\hat{K}_2 = \{2N(2N+1)\}^{-1/2}\left(-2N|A_0\rangle\langle A_0| + \sum_{b=\pm 1}\sum_{k=1}^{N} |A_{kb}\rangle\langle A_{kb}|\right) \tag{2.74}$$

Furthermore $\hat{K}_2$ may be expressed as

$$\hat{K}_2 = \{2N(2N+1)\}^{-1/2} \sum_{b=\pm 1}\sum_{k=1}^{N} \left(|A_{kb}\rangle\langle A_{kb}| - |A_0\rangle\langle A_0|\right) \tag{2.75}$$

from which it is clear that $(K_2|\rho(t))$ is proportional to the statistical average of the mean difference of population between the ground state and each of the $2N$ excited states.

The partition of $\mathscr{A}$ is based on the classification of eigenvectors of $\hat{K}_{1-}$ and $\hat{K}_{2-}$. Letting $|k_1, k_2)$ be an eigenvector, we have by Section II.C:

$$(k_1, k_2|\hat{H}_-^i(t)|k_1'k_2') = 0 \tag{2.76}$$

if k_1' and, or k_2' differ from k_1 and, or k_2. This property remains true of the macroscopic operator $\hat{H}$ defined by (2.61). We must first find the eigenvalues of $\hat{K}_{1-}$ and $\hat{K}_{2-}$. The former operator being proportional to the identity, we have

$$\hat{K}_{1-}|X) = 0; \ |X) \text{ in } \mathscr{A}$$

which shows that $\hat{K}_{1-}$ is completely degenerate and that it is $\hat{K}_{2-}$ which partitions $\mathscr{A}$. The discussion of the dimer given in Section II.E is easily generalized to the present case.

Let $\hat{P}_G$ be the projection operator on $\mathscr{H}$ of the ground state and $\hat{\hat{P}}_G$ its supplement.

$$\hat{P}_G = |A_0\rangle\langle A_0|; \qquad \hat{\hat{P}}_G = \hat{I} - \hat{P}_G$$

There are three classes of operator acting on H: $\hat{X}_G$, $\hat{\hat{X}}_G$ and $\hat{X}^\alpha$ defined by the following relations;

$$\hat{X}_G = \hat{P}_G \hat{X}_G \hat{P}_G; \qquad \hat{\hat{X}}_G = \hat{\hat{P}}_G \hat{\hat{X}}_G \hat{\hat{P}}_G$$

$$\hat{X}^\alpha = \hat{P}_G \hat{X}^\alpha \hat{\hat{P}}_G + \hat{\hat{P}}_G \hat{X}^\alpha \hat{P}_G$$

$\hat{K}_2$ satisfies

$$\hat{K}_2 = \hat{P}_G \hat{K}_2 \hat{P}_G + \hat{\hat{P}}_G \hat{K}_2 \hat{\hat{P}}_G$$

(see (2.75)). Thus

$$[\hat{K}_2, \hat{X}_G] = 0 \Leftrightarrow \hat{K}_{2-}|X_G) = 0$$

$$[\hat{K}_2, \hat{\hat{X}}_G] = 0 \Leftrightarrow \hat{K}_{2-}|\bar{X}_G) = 0$$

There are $(2N)^2$ operators of type $\hat{\hat{X}}_G$ and one of type $\hat{X}_G$. Hence the spectrum of $\hat{K}_{2-}$ has zero as an eigenvalue at least $4N^2 + 1$ times. The eigenvectors are arbitrary combinations of $|X_G)$ and the $|\bar{X}_G)$.

By definition (2.75) of $\hat{K}_2$,

$$\hat{P}_G \hat{K}_2 \hat{P}_G = -(2N)^{1/2}(2N+1)^{-1/2}\hat{P}_G; \qquad \hat{\hat{P}}_G \hat{K}_2 \hat{\hat{P}}_G = (2N)^{-1/2}(2N+1)^{-1/2}\hat{\hat{P}}_G$$

There are, therefore, two types of operator $\hat{X}^\alpha$:

$$\hat{X}_a \text{ such that: } \hat{X}^a = \hat{\hat{P}}_G \hat{X}^a \hat{P}_G,\ \hat{K}_{2-}|X^a) = \left(\frac{2N+1}{2N}\right)^{1/2}|X^a)$$

$$\hat{X}^e \text{ such that: } \hat{X}^e = \hat{P}_G \hat{X}^e \hat{\hat{P}}_G,\ \hat{K}_{2-}|X^e) = -\left(\frac{2N+1}{2N}\right)^{1/2}|X^e)$$

Clearly there are $4N$ operators $\hat{X}^\alpha$; $2N$ of kind $\hat{X}^a$ and $2N$ of kind $\hat{X}^e$.

The following resumes their properties. The figures in parentheses are the dimensions of the corresponding subspace.

$$\hat{K}_{2-}|X_G) = 0 \qquad (1) \tag{2.77a}$$

$$\hat{K}_{2-}|\bar{X}_G) = 0 \qquad (4N^2) \tag{2.77b}$$

$$\hat{K}_{2-}|X^a) = \left(\frac{2N+1}{2N}\right)^{1/2}|X^a) \qquad (2N) \tag{2.77c}$$

$$\hat{K}_{2-}|X^e) = -\left(\frac{2N+1}{2N}\right)^{1/2}|X) \qquad (2N) \tag{2.77d}$$

In the partition of $\mathscr{A}$

$$\mathscr{A} = \mathscr{A}_0 \oplus \mathscr{A}^a \oplus \mathscr{A}^e \tag{2.78}$$

the dimensions of the subspaces on the right are $4N^2 + 1, 2N$, and $2N$, totaling $(2N+1)^2$, the dimension of $\mathscr{A}$. In fact two directions in $\mathscr{A}_0$ are generated by the constants of movement $|K_1)$ and $|K_2)$, so $\mathscr{A}$ may be further decomposed into spaces of dimension $1, 1, 4N^2 - 1, 2N$, and $2N$.

$$\mathscr{A} = \mathscr{A}_1 \oplus \mathscr{A}_2 \oplus \mathscr{A}^* \oplus \mathscr{A}^a \oplus \mathscr{A}^e \tag{2.79}$$

(2.79) is entirely analogous to (2.40) for the dimer, $\mathscr{A}_i$ being generated by $|K_i)$, $\mathscr{A}^*$ by the vectors $|\bar{X}_G)$ with the additional condition $\mathrm{Tr}\,\hat{\bar{X}}_G = 0$ by the orthogonality of $|K_2)$. $\mathscr{A}^\alpha (\alpha = a$ or $e)$ is generated by the $|X^\alpha)$, of which there are $2N$ for each value of α.

The reasoning by which each subspace is identified is identical to that for the dimer in Section II.E and is omitted here. $\mathscr{A}^*$ is the transfer space, $\mathscr{A}^a$ the absorption space, and $\mathscr{A}^e$ the emission space. The direct sum of $\mathscr{A}_1$ and $\mathscr{A}_2$, $\mathscr{A}_K$, is the space of stationary states. In what follows $\hat{H}^{\,i}_{-}(t)$ is decomposed into the direct sum:

$$\hat{H}^{\,i}_{-}(t) = \hat{H}^{\,i}_{-K}(t) \oplus \hat{H}^{\,i}_{-}{}^{*}(t) \oplus \hat{H}^{\,ia}_{-}(t) \oplus \hat{H}^{\,ie}_{-}(t) \tag{2.80a}$$

and similarly $\hat{H}$ defined by (2.51) is

$$\hat{H} = \hat{H}_K \oplus \hat{H}^* \oplus \hat{H}^a \oplus \hat{H}^e \tag{2.80b}$$

$$\hat{H}_{0-} = \hat{H}_{0-K} \oplus \hat{H}^*_{0-} \oplus \hat{H}^a_{0-} \oplus \hat{H}^e_{0-} \tag{2.80c}$$

$$\hat{\Gamma} = \hat{\Gamma}_K \oplus \hat{\Gamma}^* \oplus \hat{\Gamma}^a \oplus \hat{\Gamma}^e \tag{2.80d}$$

Operators with the subscript K are null operators in $\mathscr{A}_K$; see (2.30). If we take conjugate bases of operators in $\mathscr{A}^a$ and $\mathscr{A}^e$, $(\hat{X}^{a\dagger} = \hat{X}^e)$, then by (2.21) the matrices of $\hat{H}^{\,ia}_{-}(t)$ and $\hat{H}^{\,ie}_{-}(t)$ are of opposite sign.

We give without proof expressions for operators appropriate to an ordinary chain of N sites. N replaces $2N$ and the index b becomes superfluous.

$$\hat{K}_1 = (N+1)^{-1/2}\left(|A_0\rangle\langle A_0| + \sum_{k=1}^{N} |A_k\rangle\langle A_k|\right) \tag{2.81}$$

$$\hat{K}_2 = \{N(N+1)\}^{-1/2}\left(-N|A_0\rangle\langle A_0| + \sum_{k=1}^{N} |A_k\rangle\langle A_k|\right) \tag{2.82}$$

$$\hat{K}_{2-}|X_G) = 0 \qquad (1) \tag{2.83a}$$

$$\hat{K}_{2-}|X_G) = 0 \qquad (N^2) \tag{2.83b}$$

$$\hat{K}_{2-}|X^a) = \left(\frac{N+1}{N}\right)^{1/2}|X^a) \qquad (N) \tag{2.83c}$$

$$\hat{K}_{2-}|X^e) = -\left(\frac{N+1}{N}\right)^{1/2}|X^e) \qquad (N) \tag{2.83d}$$

As before, the figures in parentheses are the degeneracies of the eigenvalue in each equation.

2. Choice of Bases

Before writing the equation of evolution of each physical system, we must define the bases used in each of the subspaces on the right-hand side of (2.79).

Since the partition is determined by eigenvalues of $\hat{K}_{2-}$ the same kind of reasoning cannot be used to choose one basis in favor of another in each degenerate subspace. We must invoke general principles or intuitively pick the most convenient basis.

In the case of $\mathscr{A}^a$ or $\mathscr{A}^e$, which describe the response of the system to an exterior field, it is important to introduce at the outset eigenvectors of $\hat{H}^{\alpha}_{0-}(\alpha = a \text{ or } e)$ since the eigenvalues of this operator are the resonance frequencies of the aggregate in the absence of any relaxation. Generalizing Section II.E on the dimer, ((2.43) and (2.44)), we define bases of operators $\hat{a}^{+}_{kb}$ and a_{kb} in $\mathscr{A}^a$ and $\mathscr{A}^e$ by

$$\hat{a}^{+}_{kb} = |A_{kb})(A_0|; \qquad \hat{a}_{kb} = |A_0)(A_{kb}|; \qquad \hat{a}^{\dagger}_{kb} = \hat{a}^{\dagger}_{kb} \tag{2.84}$$
$$(k = 1, 2, \ldots, N; \qquad b = \pm 1)$$

Relation (2.84) generalizes definitions (2.43) and (2.44) for the dimer. Our choice of basis reflects the idea that the sites of the aggregate absorb or emit collectively, so that eigenstates of $\hat{H}^{\alpha}_{0-}$ must be brought to the fore.

$\mathscr{A}_0$ in (2.78) contains the transfer subspace. The restrictions on $\hat{\epsilon}^i(t)$ were founded on arguments written in terms of the local basis $\{|M^I_p)\}_{1\leq p\leq N; I=A,B}$. Moreover, since the concept of transfer is based on this localization we shall use the natural basis $\{|Y^{IJ}_{pq})\}_{1\leq p,q\leq N; I,J=A,B}$ defined by

$$\hat{Y}^{IJ}_{pq} = |M^I_p)(M^J_q| \qquad (p, q = 1, 2, \ldots, N; I,J = A,B) \tag{2.85}$$

On occasion it will be convenient to introduce operators whose components on the $\{|Y^{IJ}_{pq})\}_{p,q;\,I,J}$ are continuous functions, such that the vectors thus defined in $\mathscr{A}_0$ are everywhere dense. Instead of solving differential matrix equations of high order ($\sim N^2$) we shall be able to limit the problem to a smaller order ($\sim N$) integro-differential system. Both methods are used in Section IV, when we deal with transfer of energy during the life of the excited state.

I. Macroscopic Equations

In this section expression (2.61) is developed for the double chain so that equations for the ordinary chain (Section II.I.5) and the dimer (Section II.I.6) may be deduced as special cases.

Static interactions in $\hat{H}_0$ will be limited to nearest-neighbor coupling in the same chain and between sites opposite each other on different chains. These interactions suffice to ensure that a given site depends on all the others, which is the main point. For two chains A and B of N sites each,

$$\hat{H}_0 = \hbar \sum_{I=A,B} \left\{ \omega_0 \sum_{p=1}^{N} |M_p^I\rangle\langle M_p^I| + \Delta\omega \sum_{p=1}^{N-1} (|M_{p+1}^I\rangle\langle M_p^I| + |M_p^I\rangle\langle M_{p+1}^I|) \right\}$$
$$+ \hbar\Delta\omega' \sum_{p=1}^{N} (|M_p^A\rangle\langle M_p^B| + |M_p^B\rangle\langle M_p^A|) \tag{2.86}$$

which amounts to supposing that intrachain ($\hbar\Delta\omega$) and interchain ($\hbar\Delta\omega'$) interactions are homogeneous throughout the double chain.

An obvious basis is $\{|M_{pb}\rangle\}_{pb}$ defined by

$$|M_{pb}\rangle = \frac{1}{\sqrt{2}}(|M_p^A\rangle + b|M_p^B\rangle) \qquad (b = \pm 1) \tag{2.87}$$

On this basis,

$$\hat{H}_0 = \hbar \sum_{b=\pm 1} \left\{ \sum_{p=1}^{N} (\omega_0 + b\Delta\omega')|M_{pb}\rangle\langle M_{pb}| \right.$$
$$\left. + \Delta\omega \sum_{p=1}^{N-1} (|M_{p+1b}\rangle\langle M_{pb}| + |M_{pb}\rangle\langle M_{p+1b}|) \right\} \tag{2.88}$$

The matrix for $\hat{H}_0$ is thus the direct sum of two matrices corresponding to each of the two values of b. Hence,

$$\hat{H}_0 = \hat{H}_{0,+1} \oplus \hat{H}_{0,-1} \equiv \bigoplus_{b=\pm 1} \hat{H}_{0,b} \tag{2.89}$$

where $\hat{H}_{0,b}$ is a Hamiltonian for an ordinary chain with diagonal elements on the localized basis equal to $\omega_0 + b\Delta\omega' = \omega_b$.

1. Eigenstates of $\hat{H}_0$

The two restrictions of $\hat{H}_0$ differing only in their scalar parts, there is an operator $\hat{h}_0$ independent of b such that

$$\hat{H}_{0,b} = \hbar(\omega_b \hat{I} + \Delta\omega \hat{h}_0) \tag{2.90}$$

in which $\hat{I}$ is the identity and $\hat{h}_0$ is a matrix whose elements on the basis $\{|M_{pb}\rangle\}_{pb}$ are all nil, excepting those just above and just below the diagonal which are equal to 1. The eigenvalue problem of $\hat{H}_{0,+1}$ and $\hat{H}_{0,-1}$ is thus simplified to that of $\hat{h}_0$. The secular equation of this operator is

$$\hbar^N \det(\Delta\omega\hat{h}_0 - \omega\hat{I}) = 0 \Leftrightarrow D_N(\omega) = 0$$

The following inductive relation holds for the determinant

$$D_N(\omega) + \hbar\omega D_{N-1}(\omega) - \hbar^2\Delta\omega^2 D_{N-2}(\omega) = 0$$

This is the same as the relation between successive Gegenbauer polynomials, $C_N(z)$,[116,117]

$$C_N(z) - 2zC_{N-1}(z) - C_{N-2}(z) = 0 \tag{2.91}$$

The formulae are identical if

$$z = -\frac{\omega}{2\Delta\omega}; \qquad D_N(\omega) = \hbar^N \Delta\omega^N C_N\left(-\frac{\omega}{2\Delta\omega}\right) \tag{2.92}$$

Letting $z = \cos x$ (x is real as may be shown by induction), $C_N(z)$ may be written[116]

$$C_N(\cos x) = \frac{\sin(N+1)x}{\sin x} \tag{2.93}$$

The secular equation reduces to

$$\sin(N+1)x = 0 \Leftrightarrow x = k\frac{\pi}{N+1} \equiv k\theta \quad (2\pi) \tag{2.94}$$

k taking integral values which may be chosen between 1 and N. By (2.93) the eigenvalues of $\hat{H}_{0,b}$ are:

$$\hbar\omega_{kb} = \hbar(\omega_0 + b\Delta\omega') + 2\hbar\Delta\omega\cos k\theta \qquad (k = 1, 2, \ldots, N) \tag{2.95}$$

The difference system for the eigenvectors,

$$\{c_p + \lambda c_{p+1} + c_{p+2} = 0\}_{0\leq p\leq N-1}$$

where $\lambda = -2\cos k\theta$, $c_0 = 0$ and c_1 is arbitrary, may be solved in several ways, for example, using the Laplace transform.[118] The eigenvectors of $\hat{H}_{0,b}$ are

$$|A_{kb}\rangle = \mathcal{N}_k \sum_{p=1}^{N} \sin kp\theta |M_{pb}\rangle \tag{2.96}$$

$\mathcal{N}_K$ is determined by the following relation, in which L is an integer,

$$\sigma_L \equiv \sum_{p=1}^{N} e^{iL\theta} = \frac{1+(-1)^L}{2}\left\{-1 + (N+1)\sum_{K=-\infty}^{K=+\infty} \delta_{L,2K(N+1)}\right\} + i\frac{1-(-1)^L}{2}\cot k\frac{\theta}{2} \tag{2.97}$$

Since k lies between 1 and N, $\mathcal{N}_k = (2/N+1)^{1/2}$; finally:

$$|A_{kb}\rangle = \left(\frac{1}{N+1}\right)^{1/2} \sum_{p=1}^{N} \sin kp\theta(|M_p^A\rangle + b|M_p^B\rangle) \tag{2.98}$$

It will be noticed that there are no Bloch-type translational factors such as are found in the usual treatment of infinite systems. However, the dispersion in energy is the same as that of an unbounded crystal with one molecule per unit cell, in which there are only nearest-neighbor interactions. The difference is that here k takes discrete values, whereas in an infinite crystal it would vary continuously.

2. *Projection of $\hat{H}_{0-}$ on Each Invariant Subspace of $\mathscr{A}$*

First we determine the matrix of $\hat{H}_{0-}^{a}$, (see (2.80c)) on the basis generated by the $\{|a_{kb}^{+})\}_{kb}$ associated with the operators defined in the first of relations (2.84). That of $\hat{H}_{0-}^{e}$ on the $\{|a_{kb})\}_{kb}$ has the opposite sign. Now,

$$[\hat{H}_0, \hat{a}_{kb}^{+}] = [\hat{H}_0, |A_{kb}\rangle\langle A_0|] = \hbar\omega_{kb}|A_{kb}\rangle\langle A_0|$$

where ω_{kb} is defined in (2.98). Hence,

$$(a_{kb}^{+}|\hat{H}_{0-}|a_{k'b'}^{+}) \equiv \mathrm{Tr}\{\hat{a}_{kb}[\hat{H}_0, \hat{a}_{k'b'}^{+}]\} = \hbar\omega_{kb}\,\mathrm{Tr}(\hat{a}_{kb}\hat{a}_{k'b'}^{+}) = \hbar\omega_{kb}\delta_{kk'}\delta_{bb'} \tag{2.99}$$

$\hat{a}_{kb}^{+}$ and a_{kb} being Hermitian conjugates, we have (see (2.21))

$$(a_{kb}|\hat{H}_{0-}|a_{k'b'}) = -(a_{kb}^{+}|\hat{H}_{0-}|a_{k'b'}^{+}) = -\hbar\omega_{kb}\delta_{kk'}\delta_{bb'}$$

where $\bar{I}$ and $\bar{J}$ are opposite to I and J ; see Table 2.4.
Thus,

$$\begin{aligned}(Y_{p'q'}^{I'J'}|\hat{H}_{0-}|Y_{pq}^{IJ}) &\equiv \mathrm{Tr}\{Y_{p'q'}^{I'J'\dagger}[\hat{H}_0, \hat{Y}_{pq}^{IJ}]\} \\ &= \hbar\Delta\omega\delta_{II'}\delta_{JJ'}\{\delta_{qq'}(\delta_{pp'+1}+\delta_{pp'-1}) - \delta_{pp'}(\delta_{qq'+1}+\delta_{qq'-1})\} \\ &\quad + \hbar\Delta\omega'\delta_{pp'}\delta_{qq'}(\delta_{I'\bar{I}}\delta_{JJ'} - \delta_{J'\bar{J}}\delta_{II'})\end{aligned} \tag{2.100}$$

3. *The Relaxation Operator $\hat{\Gamma}$ on Each Invariant Subspace of $\mathscr{A}$*

We use (2.63) to calculate the matrices of $\hat{\Gamma}$, defined by (2.62). Generalizing the dimer case (see (1.5) and notations in (2.34)), $\hat{\epsilon}^{i}(t)$ may

TABLE IV

I	A	B
$\bar{I}$	B	A

be written,

$$\hat{\epsilon}^i(t) = \sum_{I=A,B} \left\{ \sum_{p=1}^{N} \epsilon_p^{Ii}(t)|M_p^I\rangle\langle M_p^I| + \sum_{p=1}^{N-1} v_p^{Ii}(t)(|M_{p+1}^I\rangle\langle M_p^I| + |M_p^I\rangle\langle M_{p+1}^I|) \right\}$$
$$+ \sum_{p=1}^{N} w_p^i(t)(|M_p^A\rangle\langle M_p^B| + |M_p^B\rangle\langle M_p^A|) \tag{2.101}$$

where $\epsilon_p^{Ii}(t)$ is the local fluctuation, and $v_p^{Ii}(t)$ and $w_p^{Ii}(t)$ are nonlocal intrachain and interchain fluctuations. Generalizing the arguments set forth for the dimer in Section I.C.3 the correlation functions are:

$$\langle \epsilon_p^{Ii}(t)\epsilon_{p'}^{I'i}(t')\rangle = 2\hbar^2\gamma_0\delta_{pp'}\delta_{II'}\delta(t-t') \tag{2.102a}$$
$$\langle v_p^{Ii}(t)v_{p'}^{I'i}(t')\rangle = 2\hbar^2\gamma_1\delta_{pp'}\delta_{II'}\delta(t-t') \tag{2.102b}$$
$$\langle w_p^{i}(t)w_{p'}^{i}(t')\rangle = 2\hbar^2\gamma_2\delta_{pp'}\delta(t-t') \tag{2.102c}$$

It follows that

$$\langle \hat{\epsilon}^i(t)\hat{\epsilon}^i(t')\rangle = 2\hbar^2\delta(t-t')\left\{(\gamma_0 + 2\gamma_1 + \gamma_2)\sum_{I=A,B}\sum_{p=1}^{N}|M_p^I\rangle\langle M_p^I| - \gamma_1\hat{P}_L\right\} \tag{2.103}$$

where $\hat{P}_L$ is the projection operator introduced in $\mathscr{H}$ by the boundary conditions of the finite chain:

$$\hat{P}_L = \sum_{I=A,B} |M_1^I\rangle\langle M_1^I| + |M_N^I\rangle\langle M_N^I| \tag{2.104}$$

The term in braces on the right of (2.103) may just as well be written:

$$(\gamma_0 + \gamma_1 + \gamma_2)\sum_{I=A,B}\sum_{p=1}^{N}|M_p^I\rangle\langle M_p^I| + \gamma_1\sum_{I=A,B}\sum_{p=2}^{N-1}|M_p^I\rangle\langle M_p^I|$$

No more physical meaning should be attached to the extra term in both these equations than that it is due to the finite length of the chain. Henceforth we use the first form, in which the extra term is negligible for large N. The matrix of the restriction $\hat{\Gamma}^a$, of $\hat{\Gamma}$ to $\mathscr{A}^a$ may be found as follows. The first term on the right of (2.63) is ($|X) = |a_{kb}^+)$):

$$|\langle \hat{\epsilon}^i(t)\hat{a}_{kb}^+\hat{\epsilon}^i(t')\rangle) = |\langle \hat{\epsilon}^i(t)|A_{kb}\rangle\langle A_0|\hat{\epsilon}^i(t')\rangle) = 0 \tag{2.105}$$

since $\hat{\epsilon}^i(t)$ is supposed not to couple the ground state to the excited states. Indeed,

$$\{\hat{P}_L, \hat{a}_{kb}^+\}_+ = \hat{P}_L\hat{a}_{kb}^+ + \hat{a}_{kb}^+\hat{P}_L = \hat{P}_L\hat{a}_{kb}^+$$
$$= \sum_{I=A,B} (\langle M_1^I|A_{kb}\rangle|M_1^I\rangle\langle M_0| + \langle M_N^I|A_{kb}\rangle|M_N^I\rangle\langle M_0|)$$

Letting $\hat{m}_p^{I+} = |M_p^I\rangle\langle M_0|$,

$$|\{\hat{P}_L, \hat{a}_{kb}^+\}_+) = \sum_{I=A,B} (m_1^{I+}|a_{kb}^+)|m_1^{I+}) + (m_N^{I+}|a_{kb}^+)|m_N^{I+})$$

which means

$$|\{\hat{P}_L, \hat{a}_{kb}^+\}_+) = \hat{P}|a_{kb}^+)$$

where $\hat{P}$ is the projection operator on the subspace $\mathscr{A}^a$, defined by

$$\hat{P} = \sum_{I=A,B} |m_1^{I+})(m_1^{I+}| + |m_N^{I+})(m_N^{I+}| \tag{2.106}$$

$\hat{P}$ may be expressed as the sum of four orthogonal (idempotent) projectors $\hat{P}_{cb}$

$$\hat{P} = \sum_{c=0,1}\sum_{b=\pm1} \hat{P}_{cb}; \qquad \hat{P}_{cb}\hat{P}_{c'b'} = \delta_{bb'}\delta_{cc'}\hat{P}_{cb}; \qquad \mathrm{Tr}\,\hat{P}_{cb} = 1 \tag{2.107}$$

$$\hat{P}_{cb} = |\pi_{cb})(\pi_{cb}|; \qquad |\pi_{cb}) = \tfrac{1}{2}\{|m_1^{A+}) + b|m_1^{B+}) + (2c-1)[|m_N^{A+}) + b|m_N^{B+})]\}$$

Finally, by (2.63),

$$(a_{kb}^+|\hat{\Gamma}^a|a_{k'b'}^+) = (\gamma_0 + 2\gamma_1 + \gamma_2)\delta_{kk'}\delta_{bb'} - \gamma_1(a_{kb}^+|\hat{P}|a_{k'b'}^+)$$

One of the consequences of the finite length of the chain is that the relaxation matrix written on the eigenbasis of $\hat{H}_{0-}$ is not diagonal. The extra terms for such a system may indeed be called end effects.

We turn now to the restriction $\hat{\Gamma}_0$, of $\hat{\Gamma}$ to subspace $\mathscr{A}_0$ which contains the transfer space, $\mathscr{A}^*$. By (2.103),

$$\begin{aligned}(Y_{p'q'}^{I'J'}|\{\langle\hat{\epsilon}^i(t)\hat{\epsilon}^i(t')\rangle, \hat{Y}_{pq}^{IJ}\}_+) &= \mathrm{Tr}(|M_{q'}^{J'}\rangle\langle M_{p'}^{I'}|\{\langle\hat{\epsilon}^i(t)\hat{\epsilon}^i(t')\rangle, |M_p^I\rangle\langle M_q^J|\}_+)\\ &= \langle M_{p'}^{I'}|\{\langle\hat{\epsilon}^i(t)\hat{\epsilon}^i(t')\rangle, |M_p^I\rangle\langle M_q^J|\}_+|M_{q'}^{J'}\rangle\end{aligned}$$

The analog of (1.9) for a double chain is

$$\langle\langle M_p^I|\hat{\epsilon}^i(t)\hat{\epsilon}^i(t')|M_{p'}^{I'}\rangle\rangle = 2\hbar^2\delta(t-t')\delta_{II'}\delta_{pp'}\left\{\gamma_0 + \sum_q (1-\delta_{pq})\gamma_{|p-q|} + \gamma_2\right\}$$

If we introduce only γ_0 and γ_1 for each chain, which is logical if the interactions go no further than nearest neighbors, the braces equal $\gamma_0 + 2\gamma_1 + \gamma_2$. γ_2 should not be confused with $\gamma_{|p-q|}, |p-q| = 2$, since in view of what has been said above, the latter parameter cannot occur. Projecting the first important quantity on the right of (2.63) onto $|Y_{p'q'}^{I'J'})$ ($|X) = |Y_{pq}^{IJ})$)

$$\begin{aligned}&(Y_{p'q'}^{I'J'}|\{\langle\hat{\epsilon}^i(t)\hat{\epsilon}^i(t')\rangle, \hat{Y}_{pq}^{IJ}\}_+)\\ &\quad = 2\hbar^2\delta(t-t')\delta_{II'}\delta_{JJ'}\delta_{pp'}\delta_{qq'}\{2(\gamma_0 + 2\gamma_1 + \gamma_2) - \gamma_1(\delta_{p1} + \delta_{q1} + \delta_{pN} + \delta_{qN})\}\end{aligned} \tag{2.108}$$

The second term in braces is an end effect. Now,

$$
\begin{aligned}
(Y^{I'J'}_{p'q'}|\langle\hat{\epsilon}^i(t)Y^{IJ}_{pq}\hat{\epsilon}^i(t')\rangle) &= \mathrm{Tr}(|M^{J'}_{q'}\rangle\langle M^{I'}_{p'}|\langle\hat{\epsilon}^i(t)|M^I_p\rangle\langle M^J_q|\hat{\epsilon}^i(t')\rangle) \\
&= \langle M^{I'}_{p'}|(\langle\hat{\epsilon}^i(t)|M^I_p\rangle\langle M^J_q|\hat{\epsilon}^i(t')\rangle)|M^{J'}_{q'}\rangle \\
&= \langle(\epsilon^{Ii}_p\delta_{pp'}\delta_{II'} + v^{Ii}_p\delta_{p'p+1}\delta_{II'} + v^{Ii}_{p-1}\delta_{p'p-1}\delta_{II'} + w^i_p\delta_{pp'}\delta_{I\bar{I}'}) \\
&\quad \times(\epsilon^{J'i}_{q'}\delta_{qq'}\delta_{JJ'} + v^{J'i}_{q'}\delta_{qq'+1}\delta_{JJ'} + v^{J'i}_{q'-1}\delta_{qq'-1}\delta_{JJ'} \\
&\quad + w^i_{q'}\delta_{qq'}\delta_{J'\bar{J}})\rangle \\
&= 2\hbar^2\delta(t-t')\{(\gamma_0\delta_{II'}\delta_{JJ'}\delta_{IJ} + \gamma_2\delta_{I\bar{I}'}\delta_{J\bar{J}'})\delta_{pq}\delta_{pp'}\delta_{qq'}\} \\
&\quad + 2\hbar^2\delta(t-t')\gamma_1\delta_{II'}\delta_{JJ'}\delta_{IJ}\delta_{pq}\delta_{p'q'}(\delta_{p'p+1} + \delta_{p'p-1}) \\
&\quad + 2\hbar^2\delta(t-t')\gamma_1\delta_{II'}\delta_{JJ'}\delta_{IJ}(\delta_{pq-1}\delta_{p'q'+1}\delta_{p'p+1} \\
&\quad + \delta_{pq+1}\delta_{p'q'-1}\delta_{p'p-1})
\end{aligned} \tag{2.109}
$$

4. *Equation of Evolution of the Double Chain*

Assembling the preceding results, we obtain an expression for the Hamiltonian of the macroscopic equation (2.61) of a double chain. Collecting terms in each subspace, according to (2.61) for $\hat{H}$ and to (2.63) for $\hat{\Gamma}$, we have, in $\mathscr{A}^a$ (by (2.99), (2.103), (2.105), and (2.106)):

$$
\begin{aligned}
(a^+_{kb}|\hat{H}^a|a^+_{k'b'}) &\equiv (a^+_{kb}|\hat{H}|a^+_{k'b'}) \\
&= \hbar\omega_{kb}\delta_{kk'}\delta_{bb'} - i\hbar(\gamma_0 + 2\gamma_1 + \gamma_2)\delta_{kk'}\delta_{bb'} + i\hbar\gamma_1(a^+_{kb}|\hat{P}|a^+_{k'b'})
\end{aligned} \tag{2.110}
$$

$\hat{P}$ is defined by (2.106).

Besides the diagonal, scalar relaxation there is a "cross" relaxation of four blocks, one for each pair of values of b and of the parity of k. This is another end effect of the finite aggregate.

Collecting (2.100), (2.108), and (2.109), the corresponding formula in $\mathscr{A}_0$ is:

$$
\begin{aligned}
&(Y^{I'J'}_{p'q'}|\hat{H}|Y^{IJ}_{pq}) \\
&= \hbar\Delta\omega\delta_{II'}\delta_{JJ'}\{\delta_{qq'}(\delta_{pp'+1} + \delta_{pp'-1}) - \delta_{pp'}(\delta_{qq'+1} + \delta_{qq'-1})\} \\
&\quad + \hbar\Delta\omega'\delta_{pp'}\delta_{qq'}(\delta_{I'\bar{I}}\delta_{JJ'} - \delta_{J'\bar{J}}\delta_{II'}) \\
&\quad - 2i\hbar\delta_{pp'}\delta_{qq'}\{\gamma_0(1 - \delta_{pq}\delta_{IJ})\delta_{II}\delta_{JJ'} + \gamma_2(\delta_{II'}\delta_{JJ'} - \delta_{pq}\delta_{J\bar{J}'}\delta_{J\bar{J}'})\} \\
&\quad - i\hbar\gamma_1\delta_{II'}\delta_{JJ'}\delta_{pp'}\delta_{qq'}(4 - \delta_{p1} - \delta_{q1} - \delta_{pN} - \delta_{qN}) \\
&\quad + 2i\hbar\gamma_1\delta_{IJ}\delta_{II'}\delta_{JJ'}\{\delta_{pq}\delta_{p'q'}(\delta_{p'p+1} + \delta_{p'p-1}) + \delta_{pq-1}\delta_{p'p+1}\delta_{q'q-1} + \delta_{pq+1}\delta_{p'p-1}\delta_{q'q+1}\}
\end{aligned} \tag{2.111}
$$

Hereafter, we write

$$
\Gamma_{dc} = \gamma_0 + 2\gamma_1 + \gamma_2 \tag{2.112}
$$

5. Equation of the One-branch Chain

The equation of evolution of a one branch chain follows from (2.110) and (2.111) in which $\Delta\omega' = 0$, $\gamma_2 = 0$, and I and J are now superfluous:

$$(a_k^+|\hat{H}|a_{k'}^+) = \hbar\omega_k\delta_{kk'} - i\hbar(\gamma_0 + 2\gamma_1)\delta_{kk'} + i\hbar\gamma_1(a_k^+|\hat{P}|a_{k'}^+) \quad (2.113)$$

$\hat{P}$ becomes $\hat{P}_0 + \hat{P}_1$ where $\hat{P}_c (c = 0,1)$ is a projector verifying

$$(a_\kappa^+|P_c|a_{\kappa'}^+) = \frac{4}{N+1}\sin\kappa\theta \sin\kappa'\theta \quad (2.114)$$

where κ and κ' are both even or both odd for $c = 0$ or $c = 1$, respectively. Similarly (2.111) becomes

$$\begin{aligned}(Y_{p'q'}|\hat{H}|Y_{pq}) &= \hbar\Delta\omega\{\delta_{qq'}(\delta_{pp'+1} + \delta_{pp'-1}) - \delta_{pp'}(\delta_{qq'+1} + \delta_{qq'-1})\} \\ &\quad - i\hbar\delta_{pp'}\delta_{qq'}\{2\gamma_0(1 - \delta_{pq}) + \gamma_1(4 - \delta_{p1} - \delta_{q1} - \delta_{pN} - \delta_{qN})\} \\ &\quad + 2i\hbar\gamma_1\{\delta_{pq}\delta_{p'q'}(\delta_{p'p+1} + \delta_{p'p-1}) + \delta_{pq-1}\delta_{p'p+1}\delta_{q'q-1} + \delta_{pq+1}\delta_{p'p-1}\delta_{q'q+1}\}\end{aligned} \quad (2.115)$$

Henceforth we write

$$\Gamma_{ch} = \gamma_0 + 2\gamma_1 \quad (2.116)$$

6. Equation of the Dimer

The Hamiltonian $\hat{H}$ follows from Section II.I.5, with $N = 2$. The projectors are particularly simple:

$$(a_1^+|\hat{P}_1|a_1^\dagger) = \tfrac{4}{3}\sin^2\frac{\pi}{3} = 1$$

$$(a_2^+|\hat{P}_1|a_2^+) = \tfrac{4}{3}\sin^2 2\frac{\pi}{3} = 1$$

so that

$$\hat{P}_1 = |a_1^+)(a_1^+| \qquad \hat{P}_0 = |a_2^+)(a_2^+|$$

Hence,

$$(a_k^+|\hat{H}|a_{k'}^+) = \hbar\omega_k\delta_{kk'} - i\hbar(\gamma_0 + \gamma_1)\delta_{kk'} \quad (2.117)$$

Compared to the ordinary chain we see that $\hat{\Gamma}$ is now diagonal and that the scalar component of $\hat{\Gamma}^a$ is $\gamma_0 + \gamma_1$, instead of $\gamma_0 + 2\gamma_1$, a point we take up later. We show that the extra term, proportional to $\hat{P}_c$, alters the resonances of the ordinary chain causing stronger absorption in the center and less in the wings than for a Lorentzian, so that the line is narrowed. When $N = 2$, the resonances are Lorentzians of width $2\gamma_0 + 2\gamma_1$.

Setting $N = 2$ in (2.115) leads to

$$\begin{aligned}(Y_{p'q'}|\hat{H}|Y_{pq}) &= \hbar\Delta\omega\{\delta_{qq'}(\delta_{pp'+1} + \delta_{pp'-1}) - \delta_{pp'}(\delta_{qq'+1} + \delta_{qq'-1})\} - i\hbar\delta_{pp'}\delta_{qq'}\{2\gamma_0(1 - \delta_{pq}) + 2\gamma_1\} \\ &\quad + 2i\hbar\gamma_1\{\delta_{pq}(\delta_{p'p+1} + \delta_{p'p-1}) + \delta_{pq-1}\delta_{p'p+1}\delta_{q'q-1} + \delta_{pq+1}\delta_{p'p-1}\delta_{q'q+1}\}\end{aligned} \quad (2.118)$$

Henceforth we write

$$\Gamma_d = \gamma_0 + \gamma_1 \quad (2.119)$$

The dimension of $\mathscr{A}_0 = (N+1)^2 - 2N = 5$ being small, the matrix of $\hat{H}$ may be written in full on the basis $|K_1), |Y_{11}), |Y_{22}), |Y_{12}), |Y_{21})$; we find:

$$\hbar\begin{bmatrix} 0 & 0 & 0 & 0 & 0 \\ 0 & -2i\gamma_1 & +2i\gamma_1 & -\Delta\omega & +\Delta\omega \\ 0 & +2i\gamma_1 & -2i\gamma_1 & +\Delta\omega & -\Delta\omega \\ 0 & -\Delta\omega & +\Delta\omega & -2i(\gamma_0 + \gamma_1) & 2i\gamma_1 \\ 0 & +\Delta\omega & -\Delta\omega & 2i\gamma_1 & -2i(\gamma_0 + \gamma_1) \end{bmatrix} \quad (2.120)$$

Similarly, we may write the matrix of $\hat{H}$ in subspace $\mathscr{A}^*$, which is three dimensional; on the basis $\{|\Omega_\lambda)\}_{\lambda=1,2,3}$, it results from (2.41) and (2.42):

$$\hbar\begin{bmatrix} -2i\gamma_0 & 0 & 0 \\ 0 & -i(\gamma_0 + 4\gamma_1) + 2\Delta\omega & +i\gamma_0 \\ 0 & +i\gamma_0 & -i(\gamma_0 + 4\gamma_1) - 2\Delta\omega \end{bmatrix} \quad (2.121)$$

Comparing this matrix with Table I, we note that the relaxation operator $\hat{\Gamma}$ does indeed have nonzero elements where *every* id-operator has nil ones. This confirms that $\hat{\Gamma}$ is not an id-operator.

III. RESPONSE TO A HOMOGENEOUS, CLASSICAL, OPTICAL EXCITATION

In this section we apply the methods described previously to find the optical susceptibility of two classes of aggregates. The first is that of the single chain of weakly interacting molecules, for example, tetrachlorobenzene, the smallest such chain being the dimer. The second is the double chain composed of two finite coupled lattices, for example, α-dibromonaphthalene.

We first apply the method of superoperators to determine the optical susceptibility. The system may be supposed to be linear since we envisage classical sources of excitation. Besides being esthetically pleasing, this compact expression of $\chi(\omega)$ avoids complicated mathematics like functional derivatives. Straightforward algebra leads to an exact expression of the susceptibility. It is shown that the non-Lorentzian line shape is a direct result of the finite length of the chain.

Another consequence of the finite length is a set of special selection rules. These are determined for three kinds of configuration: parallel chains whose limiting case $N \to \infty$ is a one-dimensional crystal with one molecule per unit cell; alternating chains, analagous to a crystal with two molecules per unit cell; and helical chains, common in biological systems. Whereas deviation from the Lorentzian form of each homogeneous line is in fact negligible for $N \gtrsim 20$, considerable inhomogeneous, asymmetric broadening appears in the strongest resonance of medium-sized chains, $N \sim 50$.

The real situation, obtained by doping a crystal, is a mixture of dimers, trimers, and so on, so we present a spectrum calculated for such a mixture. At high temperatures, static interactions are small compared to fluctuations, leading to a line shape that is a combination of two Lorentzians with the same center but different widths. This section closes with a simulated spectrum for one of the preceding configurations of double chain, the selection rules being the "product" of those of a dimer and of a double chain.

A. Optical Susceptibility

We derive in this section an expression for the susceptibility of a linear system in the formation of Section II.

The evolution of the ith microsystem, perturbed by a semiclassical, homogeneous field $\hat{W}(t)$ is given by

$$i\hbar \frac{d}{dt} |\rho^i(t)) = [\hat{H}_{0-} + \hat{\epsilon}_-^i(t) + \hat{W}_-(t)] |\rho^i(t)) \equiv \hat{H}_-'^i(t) |\rho^i(t)) \tag{3.1}$$

The quantum average of the energy $E^i(t)$ is

$$E^i(t) = \mathrm{Tr}\{\hat{H}''^i(t)\hat{\rho}^i(t)\} \equiv (H''^i(t)|\rho^i(t))$$

whence

$$\frac{d}{dt} E^i(t) = \left(\frac{d}{dt} H''^i(t)|\rho^i(t)\right) + \left(H''^i(t)|\frac{d}{dt}\rho^i(t)\right) = \left(\frac{d}{dt} H''^i(t)|\rho^i(t)\right)$$

since for an id-operator Ξ_- we have $\Xi_-|\Xi) \equiv 0$. Putting $E(t) = \langle E^i(t) \rangle$,

$$\frac{d}{dt}E(t)=\frac{d}{dt}\langle E^i(t)\rangle=\left\langle\frac{d}{dt}E^i(t)\right\rangle=\left\langle\left(\frac{d}{dt}W(t)+\frac{d}{dt}\epsilon^i(t)\right)|\rho^i(t)\right\rangle$$
$$=\left(\frac{d}{dt}W(t)|\rho(t)\right)+\left\langle\left(\frac{d}{dt}\epsilon^i(t)\right)|\rho^i(t)\right\rangle$$

The first term is the variation of energy due to exchanges with the field. The second describes interactions with the heat bath. The energy, ΔE, drawn from or given to the field is

$$\Delta E=\int_{-\infty}^{+\infty}\left(\frac{d}{dt}W(t)|\rho(t)\right)dt \tag{3.2}$$

where $-\infty$ and $+\infty$ stand, respectively, for times long before and long after the perturbation was switched on and off.

We assume the usual conditions of linear response: Small perturbation, short duration compared to times in which the populations may alter appreciably. Then only off-diagonal terms vary much. Let

$$\hat{\rho}(t)\simeq\hat{\rho}^{(0)}(t)+\hat{\rho}^{(1)}(t)$$

Suppose that at time $t=t_0$, before the perturbation, the system is in an equilibrium state:

$$|\rho(t_0))=|\rho_{eq})$$

Hence

$$|\rho^{(0)}(t_0))=|\rho_{eq});\qquad |\rho^{(1)}(t_0))=0$$

Introducing the causal propagator of $\hat{H}$,

$$\hat{G}(t)=\frac{1}{i\hbar}\theta(t)e^{(1/i\hbar)\hat{H}t}=\frac{1}{i\hbar}\theta(t)e^{(1/i\hbar)(\hat{H}_0-i\hbar\hat{\Gamma})t} \tag{3.3}$$

where $\theta(t)$ is Heaviside's function, the (first-order) solution of the problem with the given initial values is:

$$|\rho^{(0)}(t))=|\rho_{eq})$$

$$|\rho^{(1)}(t))=\lim_{t_0\to-\infty}i\hbar\int_{t_0}^{+\infty}\hat{G}(t-t')\hat{W}_-(t')\hat{G}(t'-t_0)|\rho_{eq})dt'$$

Since $|\rho_{eq})$ is an equilibrium state it follows that

$$|\rho^{(1)}(t))=\int_{-\infty}^{+\infty}\hat{G}(t-t')\hat{W}_-(t')|\rho_{eq})dt' \tag{3.4}$$

Integration by parts shows that $|\rho^{(0)}(t))=|\rho_{eq})$ contributes nothing to integral (3.2), for $\hat{W}(t)$ vanishes at the bounds of integration. Putting

(3.4) in (3.2) we have

$$\Delta E = \int_{-\infty}^{+\infty} dt \int_{-\infty}^{+\infty} dt' \left(\frac{d}{dt} W(t)|\hat{G}(t-t')\hat{W}_{-}(t')|\rho_{\text{eq}})\right) \tag{3.5}$$

Now $\hat{W}(t)$ may be expressed as $f(t)\hat{W}$ where $\hat{W}$ is time independent and $f(t)$ is a real function. f and its Fourier transform are related by

$$F(\omega) = \int_{-\infty}^{+\infty} dt\, e^{i\omega t} f(t) \Leftrightarrow f(t) = \frac{1}{2\pi}\int_{-\infty}^{+\infty} d\omega\, e^{-i\omega t} F(\omega) \tag{3.6}$$

Letting $\hat{\mathscr{G}}^{(+)}(\omega)$ denote the frequency transform of the propagator $\hat{G}(t)$,

$$\hat{\mathscr{G}}^{(+)}(\omega) = \int_{-\infty}^{+\infty} \hat{G}(t) e^{i\omega t}\, dt \Leftrightarrow \hat{G}(t) = \frac{1}{2\pi}\int_{-\infty}^{+\infty} \hat{\mathscr{G}}^{(+)}(\omega) e^{-i\omega t}\, d\omega$$

we have[119]

$$\hat{\mathscr{G}}^{(+)}(\omega) = \lim_{\eta\to 0_+}\{(\hbar\omega + i\eta)\hat{I} - (\hat{H}_{0-} - i\hbar\hat{\Gamma})\}^{-1} \tag{3.7}$$

Replacing time-dependent quantities in (3.5) by their frequency transforms and applying the convolution theorem (see Levine[83])

$$\Delta E = \int_{-\infty}^{+\infty} dt \int_{-\infty}^{+\infty} d\omega\, \frac{i\omega}{2\pi} e^{i\omega t} F^*(\omega)(W| \frac{1}{2\pi}\int_{-\infty}^{+\infty} d\omega' \hat{\mathscr{G}}^{(+)}(\omega') F(\omega')\hat{W}_{-}|\rho_{\text{eq}}) e^{-i\omega' t}$$

For real $f(t)$, $F(\omega) = F^*(-\omega)$. Carrying out a substitution gives

$$\Delta E = \frac{i}{2\pi}\int_{0}^{+\infty} d\omega\, \omega |F(\omega)|^2 (W|\{\hat{\mathscr{G}}^{(+)}(\omega) - \hat{\mathscr{G}}^{(+)}(-\omega)\}\hat{W}_{-}|\rho_{\text{eq}}) \tag{3.8}$$

The energy exchanged with the field is an integral of a function characteristic of the system, weighted by the spectral density of the field. Let us introduce the response function to a white excitation:

$$\chi(\omega) = (W|\hat{\mathscr{G}}^{(+)}(\omega)\hat{W}_{-}|\rho_{\text{eq}}) \equiv \chi'(\omega) - i\chi''(\omega) \tag{3.9}$$

Moreover, $\chi(\omega)$ is the Fourier transform of a real function so, $\chi(\omega) = \chi^*(-\omega)$. It follows that

$$\chi(\omega) = \chi^*(-\omega) \Leftrightarrow \chi'(\omega) = \chi'(-\omega) \text{ and } \chi''(\omega) = -\chi''(-\omega) \tag{3.10}$$

The spectral response to the perturbation $\hat{W}(t)$ with spectrum $|F(\omega)|^2$ is $I(\omega) = -\omega|F(\omega)|^2\,\text{Im}(\chi(\omega))$ an example of the well-known result that the response depends principally on the imaginary part of the generalized susceptibility.[120–123]

It will be noted that the id-operator formalism leads quickly and easily to a simple expression for the susceptibility, (3.9): This derivation compares favorably with the ponderous one in Reineker's thesis[48] which involves calculation of the autocorrelation functions of the electric dipoles, necessitating a generalized density operator and functional derivatives to define them.

B. Spectrum of the Dimer[124,97]

We derive in this section the susceptibility (3.9) of a dimer. The semiclassical interaction with a field of vector potential $\mathbf{A}(\mathbf{r}, t)$, with Coulomb's gauge, is

$$\hat{W}(t) = -\frac{e}{mc}\sum_j \mathbf{A}(\mathbf{r}_j, t)\cdot\hat{\mathbf{p}}_j \tag{3.11}$$

where summation is over all electrons, e and m are the electronic charge, and mass and $\mathbf{p}_j$ is the jth conjugate moment. Now,

$$i\hbar\frac{\hat{\mathbf{p}}_j}{m} = [\hat{\mathbf{r}}_j, \hat{H}_0]$$

so

$$\hat{W}(t) = \frac{i}{\hbar c}\sum_j \mathbf{A}(\mathbf{r}_j, t)\cdot[e\hat{\mathbf{r}}_j, \hat{H}_0]$$

For large wavelengths this becomes

$$\hat{W}(t) = \frac{i}{\hbar c}\mathbf{A}(\mathbf{r}_0, t)\cdot[\hat{\mathbf{d}}, \hat{H}_0] \tag{3.12}$$

where $\mathbf{r}_0$ is a position in the middle of the system and $\mathbf{d}$ is the total electronic dipole moment. This approximation limits the size of the aggregates, which must be much smaller than the wavelengths, of the order of 5000 Å. Let

$$\frac{1}{c}\mathbf{A}(\mathbf{r}_0, t) = f(t)\mathbf{u}$$

in which $\mathbf{u}$ is the unit vector in the direction of polarization of the electric field. If $|A_k\rangle$, $k = 1,2$, is an eigenstate of $\hat{H}_0$, then

$$\langle A_0|\hat{W}(t)|A_k\rangle = i\omega_k f(t)\langle A_0|\mathbf{u}\cdot\hat{\mathbf{d}}|A_k\rangle \equiv i\omega_k d f(t)\mathbf{u}\cdot\boldsymbol{\mu}_k \tag{3.13}$$

Moment $\mathbf{d}_{\mu k}$ is between the ground state and excited state $|A_k\rangle$. d is the length of the "transition" moment to a localized state:

$$\langle A_0|\hat{\mathbf{d}}|M_p\rangle \equiv \langle M_0|\hat{\mathbf{d}}|M_p\rangle = d\mathbf{m}_p \qquad \|\mathbf{m}_p\| = 1 \tag{3.14a}$$

$$\boldsymbol{\mu}_1 = \frac{1}{\sqrt{2}}(\mathbf{m}_1 + \mathbf{m}_2) \qquad \boldsymbol{\mu}_2 = \frac{1}{\sqrt{2}}(\mathbf{m}_1 - \mathbf{m}_2) \qquad \boldsymbol{\mu}_1\cdot\boldsymbol{\mu}_2 = 0 \tag{3.14b}$$

Using the scalar product (2.13), we have

$$|W(t)) = f(t) \sum_{k=1}^{2} i\omega_k d\mathbf{u} \cdot \boldsymbol{\mu}_k\{|a_k) - |a_k^{+})\} \equiv f(t)|W) \tag{3.15}$$

where $\hat{a}_k$ and $\hat{a}_k^{+}$ are defined in (2.84) to the dimer (b superfluous, $k = 1,2$).

The expansion of the equilibrium vector on the vectors defined in (2.38) and (2.39) (see Section II.G and (2.64)), is

$$|\rho_{\text{eq}}) = \frac{1}{\sqrt{3}}|K_1) - \frac{r}{\sqrt{6}}|K_2) \tag{3.16}$$

r is given by

$$-\frac{r}{\sqrt{6}} = \text{Tr}(\hat{K}_2\hat{\rho}_{\text{eq}}) \Leftrightarrow r = 2\langle A_0|\hat{\rho}_{\text{eq}}|A_0\rangle - \langle A_1|\hat{\rho}_{\text{eq}}|A_1\rangle - \langle A_2|\hat{\rho}_{\text{eq}}|A_2\rangle$$

Conversely,

$$\langle A_0|\hat{\rho}_{\text{eq}}|A_0\rangle = \tfrac{1}{3} + \frac{r}{3}; \qquad \langle A_k|\hat{\rho}_{\text{eq}}|A_k\rangle = \tfrac{1}{3} - \frac{r}{6}$$

Since the diagonal elements are real numbers between 0 and 1 it follows that $-1 \leqslant r \leqslant +2$. If $r = -1$ only the excited states are populated. These populations are equal in the equilibrium states of this model. On the other hand, if $r = +2$, only the ground state is occupied, corresponding to absorption.

Negative values of r correspond to an inversion of population in which each excited state is more populated than the ground state.

Now $|\rho_{\text{eq}})$ has nonzero components only on $|K_1)$ and $|K_2)$ so by (3.9) we need not bother to calculate all the matrix. Its action on these vectors is in the first two columns. The first column is nil since $\hat{K}_1$ is proportional to the identity. As for $\hat{K}_2$,

$$\hat{W}_{-}|K_2) = |[\hat{W}, \hat{K}_2]) = -\left|\left[\hat{K}_2, \sum_{k=1}^{2} i\omega_k d\mathbf{u}\cdot\boldsymbol{\mu}_k(\hat{a}_k - \hat{a}_k^{+})\right]\right)$$

$\hat{a}_k$ and $\hat{a}_k^{+}$ are by definition eigenvectors of $\hat{K}_{2-}$ for the eigenvalues $-\sqrt{\tfrac{3}{2}}$ and $+\sqrt{\tfrac{3}{2}}$; see (2.83c) and (2.83d). The preceding expression may be put in the form

$$\hat{W}_{-}|K_2) = \frac{\sqrt{3}}{\sqrt{2}} \sum_{k=1}^{2} i\omega_k d\mathbf{u}\cdot\boldsymbol{\mu}_k\{|a_k) + |a_k^{+})\} \tag{3.17}$$

Using the partition of $\mathscr{A}$ (2.32), the total susceptibility (3.9) is the sum of two contributions: $\chi(\omega) = \chi^a(\omega) + \chi^e(\omega)$, where

$$\chi^{\alpha}(\omega) = (W^{\alpha}|\hat{\mathcal{G}}^{(+)\alpha}(\omega)\hat{W}_{-}^{\alpha}|\rho_{eq}) \qquad (\alpha = a \text{ or } e)$$

$\hat{\mathcal{G}}^{(+)\alpha}(\omega)$ is the restriction of $\hat{\mathcal{G}}^{(+)}(\omega)$ to $\mathcal{A}^{\alpha}$ and $\hat{W}_{-}^{\alpha}|\rho_{eq})$ and $|W_{\alpha})$ are the projections of $\hat{W}_{-}|\rho_{eq})$ and $|W)$. By (2.117), (2.120), definition (3.7) of $\hat{\mathcal{G}}^{(+)}(\omega)$ and (3.15) to (3.17), we have[119]

$$\begin{aligned}\chi^{a}(\omega) &= \sum_{k=1}^{2} (i\omega_k d\mathbf{u}\cdot\boldsymbol{\mu}_k)\{\hbar\omega - (\hbar\omega_k - i\hbar\Gamma_d)\}^{-1}\left(\frac{-r}{\sqrt{6}}\frac{\sqrt{3}}{\sqrt{2}} i\omega_k d\mathbf{u}\cdot\boldsymbol{\mu}_k\right) \\ &= \frac{rd^2}{2\hbar}\sum_{k=1}^{2}(\mathbf{u}\cdot\boldsymbol{\mu}_k)^2\frac{\omega_k^2}{\omega - \omega_k + i\Gamma_d}\end{aligned}$$

Now $|1 - \omega_k/\omega_0|$ is small for all k (of the order of 10^{-3} to 10^{-4}) so with a very good approximation,

$$\chi^{a}(\omega) = \frac{rd^2\omega_0^2}{2\hbar}\sum_{k=1}^{2}(\mathbf{u}\cdot\boldsymbol{\mu}_k)^2\frac{1}{\omega - \omega_k + i\Gamma_d} \tag{3.18}$$

Similarly,

$$\chi^{e}(\omega) = -\frac{rd^2\omega_0^2}{2\hbar}\sum_{k=1}^{2}(\mathbf{u}\cdot\boldsymbol{\mu}_k)^2\frac{1}{\omega + \omega_k + i\Gamma_d} \tag{3.19}$$

Plainly $\chi^{a}(\omega)$ represents absorption. Its sign is that of r and is positive when the ground state is more populated than each excited state. It varies rapidly for $\omega \sim \omega_k$. $\chi^{e}(\omega)$ corresponds to emission, being positive if r is negative, that is, ground state less populated than each excited state. It varies sharply at $\omega \sim -\omega_k < 0$. Thus the absorption of a white source is

$$\chi^{a\prime\prime}(\omega) = \frac{r\omega_0^2 d^2}{2\hbar}\sum_{k=1}^{2}(\mathbf{u}\cdot\boldsymbol{\mu}_k)^2\frac{\Gamma_d}{(\omega - \omega_k)^2 + \Gamma_d^2} \tag{3.20}$$

There are two resonant frequencies at ω_1 and ω_2, with Lorentzian line shapes of half-peak width $2\Gamma_d$. Their weights, determine their relative maximum intensities.

Introducing the vectors $\mathbf{m}_p (p = 1,2)$ defined by (3.14a) and an angle $\theta = \pi/(N+1)$ for an aggregate of N sites, here $\pi/3$, (3.20) becomes

$$d\mathbf{u}\cdot\boldsymbol{\mu}_k = \frac{\sqrt{2}}{\sqrt{3}}\sum_{p=1}^{2}\sin kp\theta\, d\mathbf{u}\cdot\mathbf{m}_p \equiv d\frac{\sqrt{2}}{\sqrt{3}}T_k$$

$$\chi^{a\prime\prime}(\omega) = \frac{r\omega_0^2 d^2}{3\hbar}\sum_{k=1}^{2}T_k^2\frac{\Gamma_d}{(\omega - \omega_k)^2 + \Gamma_d^2} \tag{3.21}$$

where the sum T_k depends on the geometry of each aggregate.

Let

$$(\mathbf{m}_1, \mathbf{m}_2) = \psi_{12}; \qquad \boldsymbol{\mu}_k = \mu_k\mathbf{e}_k; \qquad \|\mathbf{e}_k\| = 1$$

Then $\chi^{a''}(\omega)$ becomes

$$\chi^{a''}(\omega) = \frac{r\omega_0^2 d^2}{\hbar}\Gamma_d \left\{ \frac{\cos^2 \frac{\psi_{12}}{2}}{(\omega-\omega_1)^2+\Gamma_d^2}(\mathbf{u}\cdot\mathbf{e}_1)^2 + \frac{\sin^2 \frac{\psi_{12}}{2}}{(\omega-\omega_2)^2+\Gamma_d^2}(\mathbf{u}\cdot\mathbf{e}_2)^2 \right\} \tag{3.22}$$

At the risk of triviality it is worth pointing out precautions concerning the complex phases of the matrix elements of $\hat{H}_0$ and the dipole moment $\mathbf{d}$ in a given representation, to be taken in using the different expressions of the susceptibility. The phase of off-diagonal elements is arbitrary, so that the signs of $\hbar\omega$ and vectors like $\mathbf{m}_p$ is of no direct physical significance. One should avoid reasoning too "classically" on these quantities[125,126] whose phases disappear in the final result. Conversely if we fix the sign of $\hbar\Delta\omega$, that of $\mathbf{m}_p$ is no longer arbitrary. The phases are completely related and depend solely on the representation.

Some apparent paradoxes disappear when we consider the strict relation between the phases of the $\mathbf{m}_p$ and $\hbar\Delta\omega$, which depend only on the initial choice of the vectors $\{|M_p\rangle\}_{p=1,2}$. In a real representation, $\hbar\Delta\omega$ and $\mathbf{m}_p$ are real. For a given sign of the off-diagonal element of $\hat{H}$, the angle between $\mathbf{m}_1$ and $\mathbf{m}_2$ can vary in an interval of length π compatible with the sign of $\hbar\Delta\omega$. It is of course possible to adopt the reverse procedure, common in experimental literature[107,127] of choosing first the $\mathbf{m}_p$ and then determining the sign of $\hbar\Delta\omega$. In keeping with our first results, we adopt the former procedure, taking $\hbar\Delta\omega$ to be negative. The eigenvalues of $\hat{H}_0$ increase with k: $\omega_k < \omega_{k+1}$. Henceforth angles like ψ_{12} are restricted to an interval of length π compatible with the negative sign of $\hbar\Delta\omega$.

There can be no doubt in practice, if the wave functions are good. Proper wave functions define the size and sign of $\Delta\omega$ unambiguously and the direction cosines of $\mathbf{m}_p$ and hence the susceptibility. If an exact calculation gives a positive value for $\hbar\Delta\omega$, opposite to our convention, the spectra may be found from those following by reversing the direction of the frequency axis.

The relative intensities of the two resonances depend principally on two parameters: the angle ψ_{12}, which depends on the geometrical configuration of the dimer; and the scalar products $\mathbf{u}\cdot\mathbf{e}_k$ which may be varied by turning the electric polarization of the incident field.

Fixing $\mathbf{u}$ at $\mathbf{u} = \mathbf{m}_1$,

$$\mathbf{u}\cdot\boldsymbol{\mu}_1 = \frac{1}{\sqrt{2}}(1+\mathbf{m}_1\cdot\mathbf{m}_2) = \sqrt{2}\cos^2\frac{\psi_{12}}{2}$$

$$\mathbf{u}\cdot\boldsymbol{\mu}_2 = \frac{1}{\sqrt{2}}(1-\mathbf{m}_1\cdot\mathbf{m}_2) = \sqrt{2}\sin^2\frac{\psi_{12}}{2}$$

whence (3.20) becomes

$$\chi^{a''}(\omega) = \frac{r\omega_0^2 d^2}{2\hbar}\Gamma_d\left\{\frac{\cos^4\frac{\psi_{12}}{2}}{(\omega-\omega_1)^2+\Gamma_d^2} + \frac{\sin^4\frac{\psi_{12}}{2}}{(\omega-\omega_2)^2+\Gamma_d^2}\right\}$$

so that for $\Gamma_d \ll |\Delta\omega|$,

$$\frac{\chi^{a''}(\omega_2)}{\chi^{a''}(\omega_1)} \simeq tg^4 \frac{\psi_{12}}{2} \tag{3.23}$$

The dependence of absorption on the polarization $\mathbf{u}$ for given ψ_{12} follows from (3.22), with $\Gamma_d \ll |\Delta\omega|$,

$$\frac{\chi^{a''}(\omega_2)}{\chi^{a''}(\omega_1)} \simeq tg^2 \frac{\psi_{12}}{2} \frac{(\mathbf{u}\cdot\mathbf{e}_2)^2}{(\mathbf{u}\cdot\mathbf{e}_1)^2} \tag{3.24}$$

Equation 3.24 shows that, except for parallel sites, $\psi_{12} = 0$, π, any ratio may be obtained by turning the polarization $\mathbf{u}$. The special polarizations $\mathbf{u}\|\mathbf{e}_1$ and $\mathbf{u}\|\mathbf{e}_2$ give

$$\frac{\chi_{\mathbf{e}_2}^{a''}(\omega_2)}{\chi_{\mathbf{e}_1}^{a''}(\omega_1)} \simeq tg^2 \frac{\psi_{12}}{2} \tag{3.25}$$

Thus nonparallel sites may be optically excited to any linear combination of $|A_1\rangle$ and $|A_2\rangle$ by varying the polarization $\mathbf{u}$ of a field centered close to ω_0 and of large spectral width compared to $|\Delta\omega|$. Conversely parallel sites may be excited only to one of the states $|A_k\rangle$. Thus a localized excitation cannot be created by absorption on a parallel site dimer. The development of such a state contains *both* $|A_1\rangle$ and $|A_2\rangle$ with the same weight. If one of them is a forbidden transition one cannot reconstruct a localized state by interference. Generalizations to larger chains are given in Section IV on transfer. Section IV.B is devoted to the necessary and sufficient conditions in which an initially localized state can be created by optical excitation.

C. Spectrum of an Ordinary Chain

Since the derivation of the susceptibility of a chain is naturally more complicated than for the dimer, it is convenient to divide this section into three parts. We find the line shape is not Lorentzian but that the deviation disappears as we consider longer and longer chains.

1. Calculation of the Susceptibility

Starting from expression (3.11) and employing methods like those in the beginning of the last section, we obtain the generalized semiclassical matter–field interaction $\hat{W}(t)$ (3.15) in the form

$$W(t)) = f(t) \sum_{k=1}^{N} i\omega_k d\mathbf{u}\cdot\boldsymbol{\mu}_k\{|a_k) - |a_k^+)\} \equiv f(t)|W) \tag{3.26}$$

The equilibrium states are

$$|\rho_{\text{eq}}) = (N+1)^{-1/2}|K_1) - r\{N(N+1)\}^{-1/2}|K_2) \tag{3.27}$$

r is a real number given by

$$r = -\{N(N+1)\}^{1/2}\,\text{Tr}(\hat{K}_2\hat{\rho}_{\text{eq}}) = N < A_0|\hat{\rho}_{\text{eq}}|A_0\rangle - \sum_{k=1}^{N}\langle A_k|\hat{\rho}_{\text{eq}}|A_k\rangle$$

Also,

$$\langle A_0|\hat{\rho}_{\text{eq}}|A_0\rangle = \frac{1}{N+1} + \frac{r}{N+1}$$

$$\langle A_k|\hat{\rho}_{\text{eq}}|A_k\rangle = \frac{1}{N+1} - \frac{r}{N(N+1)} \qquad (k = 1, 2, \ldots, N)$$

Hence $-1 \leq r \leq N$. $r = -1$ corresponds to an empty ground state and each excited state with a population $1/N$. Only the ground state is occupied when $r = N$. For negative r the ground state is less populated than each excited state. These remarks generalize those made about the dimer.

Generalizing (3.17) leads to

$$\hat{W}_-|K_2) = \frac{(N+1)^{1/2}}{N^{1/2}} \sum_{k=1}^{N} i\omega_k d\mathbf{u}\cdot\boldsymbol{\mu}_k\{|a_k) + |a_k^+)\} \tag{3.28}$$

so that for $\alpha = a$ or e

$$\chi^\alpha(\omega) = (W^\alpha|\hat{\mathscr{G}}^{(+)}(\omega)\hat{W}_-^\alpha|\rho_{\text{eq}}) = \sum_{k=1}^{N}\sum_{k'=1}^{N}(W^\alpha|\bar{a}_k^\alpha)(\bar{a}_k^\alpha|\hat{\mathscr{G}}^{(+)\alpha}(\omega)|\bar{a}_{k'}^\alpha)(\bar{a}_{k'}^\alpha|\hat{W}_-^\alpha|\rho_{\text{eq}})$$

In this expression $|\bar{a}_k^\alpha)$ is $|a_k^+)$ (respectively $|a_k)$) for $\alpha = a$ (respectively $\alpha = e$). As for the dimer, ((3.18) and (3.19)), χ^e follows from χ^a by changing the signs of r and the ω_k, so we consider $\chi^a(\omega)$. Putting expressions (3.26) to (3.28) in the last form of $\chi^a(\omega)$ gives

$$\chi^a(\omega) = \frac{rd^2}{N}\sum_{k=1}^{N}\sum_{k'=1}^{N}\omega_k\omega_{k'}(\mathbf{u}\cdot\boldsymbol{\mu}_k)(\mathbf{u}\cdot\boldsymbol{\mu}_{k'})(a_k^+|\hat{\mathscr{G}}^{(+)a}(\omega)|a_{k'}^+) \tag{3.29}$$

$\chi^a(\omega)$ depends on the resolvent $\hat{\mathscr{G}}^{(+)a}(\omega)$ whose matrix on the $\{|a_k^+)\}_{1\leq k\leq N}$ is not diagonal owing to the term representing end effects in $\hat{H}^a$; see (2.113). However, the remarkable properties of $\hat{P}$ enable us to calculate exactly quantities of the form $(a_k^+|\hat{\mathscr{G}}^{(+)a}(\omega)|a_{k'}^+)$

Let

$$\hat{H}^{a'}_{0-} = \sum_{k=1}^{N} \hbar(\omega_k - \omega_0)|a_k^+)(a_k^+|; \qquad \hat{I}^a = \sum_{k=1}^{N} |a_k^+)(a_k^+|$$

By (2.113) (with the notation of (2.116)),

$$\hat{H}^a = \hbar(\omega_0 - i\Gamma_{ch})\hat{I}^a + \hat{H}^{a'}_{0-} + i\hbar\gamma_1(\hat{P}_0 + \hat{P}_1) \tag{3.30}$$

$\hat{P}_0$ and $\hat{P}_1$ are defined in (2.114). Thus,

$$\mathscr{G}^{(+)a}(\omega) = \{z\hat{I}^a - \hat{H}^{a'}_{0-} - \lambda(\hat{P}_0 + \hat{P}_1)\}^{-1} \tag{3.31}$$

in which z stands for $\hbar(\omega - \omega_0 + i\Gamma_{ch})$ and λ for $i\hbar\gamma_1$. Putting the resolvent,

$$\hat{\mathscr{G}}^{(+)a}_0(\omega) = \{(z\hat{I}^a - \hat{H}^{a'}_{0-}\}^{-1}$$

into the preceding expression yields a power series in λ for $\mathscr{G}^{(+)a}(\omega)$:

$$\begin{aligned}\hat{\mathscr{G}}^{(+)a}(\omega) &= \hat{\mathscr{G}}^{(+)a}_0(\omega) + \lambda\hat{\mathscr{G}}^{(+)a}_0(\omega)(\hat{P}_0 + \hat{P}_1)\hat{\mathscr{G}}^{(+)a}_0(\omega)\\ &\quad + \lambda^2\hat{\mathscr{G}}^{(+)a}_0(\omega)(\hat{P}_0 + \hat{P}_1)\hat{\mathscr{G}}^{(+)a}_0(\omega)(\hat{P}_0 + \hat{P}_1)\hat{\mathscr{G}}^{(+)a}_0(\omega) + \cdots\end{aligned}$$

Expression (2.107) for $\hat{P}_c$, $(c = 0, 1)$, gives

$$\hat{P}_c\hat{\mathscr{G}}^{(+)a}_0(\omega)\hat{P}_c = |\pi_c)(\pi_c|\hat{\mathscr{G}}^{(+)a}_0(\omega)|\pi_c)(\pi_c| = g_c(\omega)|\pi_c)(\pi_c| \equiv g_c(\omega)\hat{P}_c$$

$g_c(\omega)$ is the diagonal element of $\hat{\mathscr{G}}^{(+)}_0(\omega)$ on $|\pi_c)$ (defined by (2.107)), to be calculated later. All the operators $\hat{P}_c\mathscr{G}^{(+)a}(\omega)\hat{P}_{c'}$ are nil for $c \neq c'$, since each $\hat{P}_0$ projects into a subspace of $\mathscr{A}^a$ for which k has a definite parity. Hence $\hat{H}^{a|}_{0-}$ and thus $\hat{\mathscr{G}}^{(+)a}_0(\omega)$ are diagonal on the basis $\{|a_k^+)\}_k$. The perturbation series is easily re-summed to:

$$\hat{\mathscr{G}}^{(+)a}(\omega) = \hat{\mathscr{G}}^{(+)a}_0(\omega) + \sum_{c=0,1} \frac{\lambda}{1 - \lambda g_c(\omega)} \hat{\mathscr{G}}^{(+)a}_0(\omega)\hat{P}_c\hat{\mathscr{G}}^{(+)a}_0(\omega) \tag{3.32}$$

All that remains to determine $\hat{\mathscr{G}}^{(+)a}(\omega)$ is to calculate the $g_c(\omega)$. On the basis $\{|m_p^+)\}_{1 \leq p \leq N}$ $\hat{H}^{a'}_{0-}$ is tridiagonal with a nil principal diagonal, because by definition,

$$\begin{aligned}\hat{H}^{a'}_{0-} &= \hbar \sum_{k=1}^{N}\sum_{p=1}^{N}\sum_{p'=1}^{N} 2\Delta\omega \cos k\theta \left(\frac{2}{N+1}\right) \sin kp\theta \sin kp'\theta |m_p^+)m_{p'}^+|\\ &= \hbar \sum_{p=1}^{N}\sum_{p'=1}^{N} \frac{4\Delta\omega}{N+1} \sum_{k=1}^{N} \cos k\theta \sin kp\theta \sin kp'\theta |m_p^+)(m_{p'}^+|\end{aligned}$$

Now, by form (2.86) of $\hat{H}_0$, relations (2.98) and (2.95) adapted to the ordinary chain (I, J and b superfluous), or by an easy calculation using the sum σ_L of (2.97), it may be seen that the sum over k multiplied by $4/(N + 1)$ is $\delta_{p'p+1} + \delta_{p'p-1}$. Therefore,

$$\hat{H}^{a'}_{0-} = \hbar\Delta\omega \sum_{p=1}^{N-1} \{|m_p^+)(m_{p+1}^+| + |m_{p+1}^+)(m_p^+|\}$$

By definition $g_c(\omega) = (\pi_c|\mathscr{G}_0^{(+)a}(\omega)|\pi_c)$; the vector $|\pi_c)$ for the ordinary chain is

$$|\pi_c) = \frac{1}{\sqrt{2}}\{|m_1^+) + (2c-1)|m_N^+)\}$$

See (2.106), in which b, A, and B are here superfluous, and (2.107) renormalized to ensure Tr $\hat{P}_c = 1$.

Now, $(2c-1)^2 = 1$ for $c = 0, 1$, so

$$g_c(\omega) = \tfrac{1}{2}\{(m_1^+|\hat{\mathscr{G}}_0^{(+)a}(\omega)|m_1^+) + (m_N^+|\mathscr{G}_0^{(+)a}(\omega)|m_N^+)\}$$
$$+\tfrac{1}{2}(2c-1)\{(m_1^+|\mathscr{G}_0^{(+)a}(\omega)|m_N^+) + (m_N^+|\hat{\mathscr{G}}_0^{(+)a}(\omega)|m_1^+)\}$$

Owing to the tri-diagonal form of $\hat{H}_{0-}^{a'}$ the preceding matrix elements of the resolvent can be expressed simply, in terms of the Gegenbauer polynomials, C_n.[116,117] Collecting terms, we have

$$g_c(\omega) = \frac{1}{\hbar\Delta\omega}\frac{C_{N-1}(\zeta) + (-1)^{1+c}}{C_N(\zeta)}$$

In this formula ζ denotes $z/2\Delta\omega$. The nth degree, dimensionless polynomial $C_n(\zeta)$ is given by

$$C_n(\omega) = \frac{y_1^{n+1}(\omega) - y_2^{n+1}(\omega)}{y_1(\omega) - y_2(\omega)}$$

where $y_i (i = 1, 2)$ are the solutions of

$$Y^2 - 2\zeta Y + 1 = 0$$

In terms of unreduced variables, $g_c(\omega)$ is

$$g_c(\omega) = \frac{1}{\hbar}\frac{\lambda_1^N(\omega) - \lambda_2^N(\omega) + (-1)^{1+c}\Delta\omega^{N-1}\{\lambda_1(\omega) - \lambda_2(\omega)\}}{\lambda_1^{N+1}(\omega) - \lambda_2^{N+1}(\omega)} \tag{3.33}$$

in which

$$\lambda_{1,2} = \tfrac{1}{2}(\omega - \omega_0 + i\Gamma_{ch}) \pm \tfrac{1}{2}\{(\omega - \omega_0 + i\Gamma_{ch})^2 - 4\Delta\omega^2\}^{1/2}$$

The real and imaginary parts of $g_c(\omega)$ have been traced in the region of ω_0, for $N = 10$, in Figs. 1 and 2. These diagonal elements of $\mathscr{G}_0^{(+)a}(\omega)$ are analytic in the upper half of the complex plane. They therefore verify the Kramers–Kronig dispersion relation,[122,132,133]

The obvious functional correspondance between the real and imaginary parts, see Figs. 1 and 2, of $g_c(\omega)$ lies in this relation. Moreover, twice applying the closure relation and noting that $\mathscr{G}_0^{(+)a}(\omega)$ is diagonal on the $\{|a_k^+)\}_k$, we arrive at

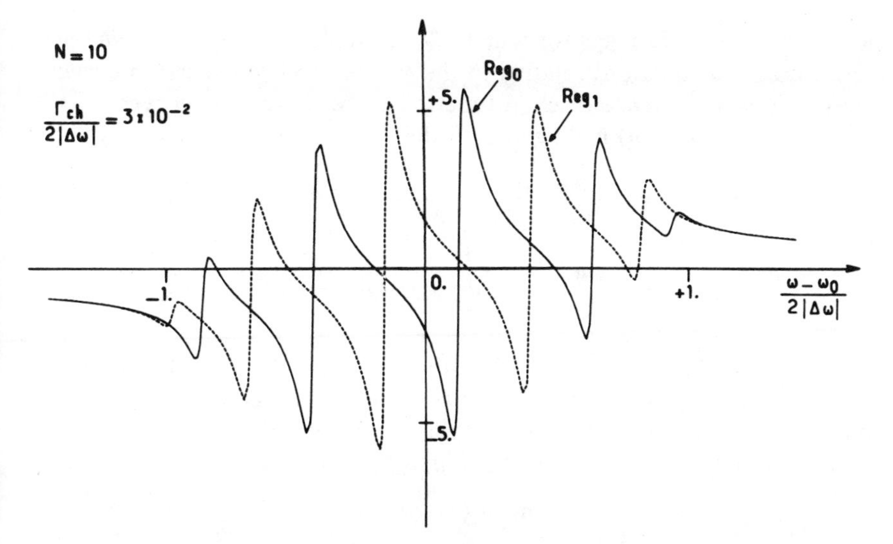

Fig. 1. Real part of the functions $g_c(\omega)(c=0,1)$ defined by (3.33); $N=10$.

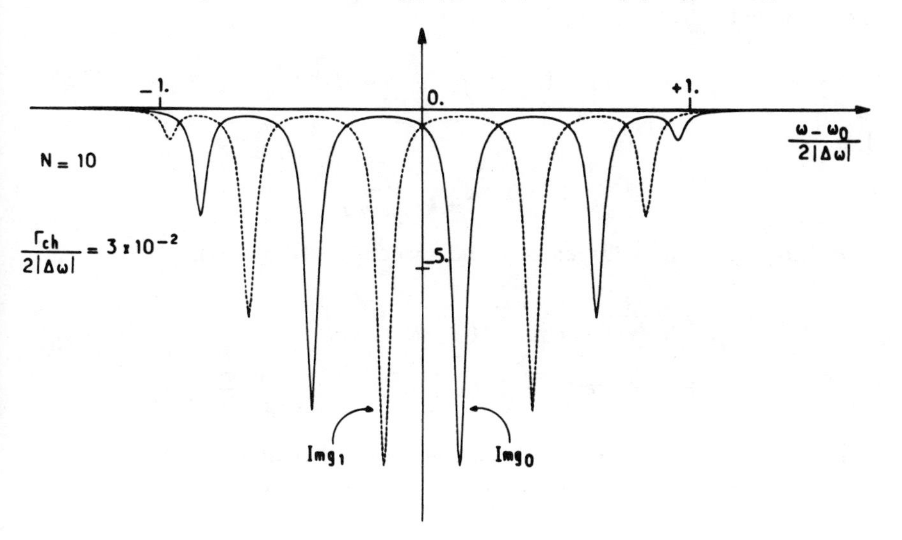

Fig. 2. Imaginary part of $g_c(\omega)$.

$$
\begin{aligned}
g_c(\omega) &= \sum_{k=1}^{N} (\pi_c|a_k^+)(a_k^+|\hat{\mathscr{G}}_0^{(+)a}(\omega)|a_k^+)(a_k^+|\pi_c) \\
&= \frac{1}{\hbar}\sum_{k=1}^{N} (\pi_c|a_k^+)\frac{1}{\omega-(\omega_0+2\Delta\omega\cos k\theta)+i\Gamma_{ch}}(a_k^+|\pi_c)
\end{aligned}
$$

Thus $g_c(\omega)$ is a sum of functions like $(\omega-\omega_k+i\Gamma_{ch})^{-1}$ weighted by

$|(\pi_c|a_k^+)|^2$. In the drawings for which $\Gamma_{ch} \ll 2|\Delta\omega|$, $\text{Re}(g_c(\omega))$ is a sequence of several odd functions relative to the ω_k resembling dispersion curves. $\text{Im}(g_c(\omega))$ is a sequence of even functions, like absorption curves.

Now, since $\hat{\mathscr{G}}_0^{(+)a}(\omega)$ is diagonal on the $\{|a_k^+)\}_k$ (3.32) and (2.114) yield

$$(a_k^+|\hat{\mathscr{G}}^{(+)a}(\omega)|a_{k'}^+) = \frac{\hbar^{-1}\delta_{kk'}}{\omega - \omega_k + i\Gamma_{ch}} + \frac{4}{N+1}\sin k\theta \times \sin k'\theta \frac{i\hbar\gamma_1}{1 - i\hbar\gamma_1 g_c(\omega)} \frac{\hbar^{-1}}{\omega - \omega_k + i\Gamma_{ch}} \frac{\hbar^{-1}}{\omega - \omega_{k'} + i\Gamma_{ch}} \tag{3.34}$$

c is 1 for k and k' both even, and 0 otherwise; (it will be recalled that if k and k' are of opposite parity the corresponding matrix element of $\hat{H}^a$ is nil, by (2.114) and so is that of $\hat{\mathscr{G}}^{(+)a}(\omega)$).

Let $\mathbf{m}_p$ be a unit vector along $\langle M_0|\mathbf{d}|M_p\rangle$; cf. dimer case:

$$d\mathbf{m}_p = \langle M_0|\mathbf{d}|M_p\rangle$$

with $\mathbf{d}$ the length of the electronic transition moment to the vector $|M_p\rangle$. The moment of stationary state $|A_k\rangle$ is then

$$\boldsymbol{\mu}_k = \left(\frac{2}{N+1}\right)^{1/2} \sum_{p=1}^{N} \sin kp\theta\, \mathbf{m}_p \tag{3.35}$$

Let

$$F_c(\omega) = \frac{1}{i + \hbar\gamma_1 g_c(\omega)}$$

Putting (3.34) in (3.29) leads to the susceptibility, $\chi^{a\prime\prime}(\omega)$:

$$\chi^{a\prime\prime}(\omega) = \frac{2r\omega_0^2 d^2}{\hbar N(N+1)} \left\{ \sum_{k=1}^{N} T_k^2 \frac{\Gamma_{ch}}{(\omega - \omega_k)^2 + \Gamma_{ch}^2} + \frac{4\gamma_1}{N+1} \text{Im}\left(\sum_{c=0,1} F_c(\omega) \left(\sum_{\kappa \geq 1}^{[N+c/2]} T_\kappa \frac{\sin \kappa\theta}{(\omega - \omega_\kappa) + i\Gamma_{ch}} \right)^2 \right) \right\} \tag{3.36}$$

$[x]$ is the integral part of x, κ is an integer between 1 and N, even for $c = 0$, odd for $c = 1$. T_k is a geometrical factor given by

$$T_k = \sum_{p=1}^{N} \mathbf{u}\cdot\mathbf{m}_p \sin kp\theta \tag{3.37}$$

Expression (3.36) reveals two contributions to $\chi^{a\prime\prime}(\omega)$. The first is a sum of Lorentzian centered on the eigenfrequencies ω_k of $\hat{H}_0$, of half-widths Γ_{ch} and weights T_k^2. We call this part $\chi_L^{a\prime\prime}(\omega)$. The second part, say $\chi_{nL}^{a\prime\prime}(\omega)$, depending on the finite length of the chains, contains functions fluctuating sharply at the ω_k. The coefficient $4/(N+1)$ suggests that $\chi_{nL}^{a\prime\prime}(\omega)$ is an end effect which decreases with increasing chain length. This conjecture is confirmed: See Section III.C.2 for a detailed account of $\chi^{a''}_{nL}(\omega)$. We note in passing that the Lorentzian and non-Lorentzian parts in the dimer combine to give a Lorentzian line shape. When $N=2$ the two parts are of the same order of magnitude. If Γ_d, $\Gamma_{0h} \ll |\Delta\omega|$ then ω for close to ω_k we may say:

$$\text{Lorentzian of half-width } \gamma_0+\gamma_1 = \text{Lorentzian of half-width } \gamma_0+2\gamma_1 + \text{extra term}$$

2. *Behavior of the Non-Lorentzian Term* $\chi^{a''}_{nL}(\omega)$

Although $\chi^{a''}_{nL}(\omega)$ is characteristic of finite systems, the division of $\chi^{a''}(\omega)$ into $\chi^{a''}_{L}(\omega)$ and $\chi^{a''}_{nL}(\omega)$ is arbitrary, as we pointed out after (2.104). In this section we seek its behavior, reserving physical interpretation for the total susceptibility $\chi^{a''}(\omega)$.

$\chi^{a''}_{nL}(\omega)$ is proportional to γ_1, so the value of γ_1 controls the size of the non-Lorentzian contribution. We show that $\chi^{a''}_{nL}(\omega k)/\chi^{a''}_{L}(\omega_k)$ for given $N>2$, k and $\Gamma_{\text{ch}} \ll |\Delta\omega|$ fixed, depends on γ_1/γ_0. Generally $\gamma_0 \ll \gamma_1$ so $\chi^{a''}_{nL}(\omega)$ contributes relatively little, but in a system for which the inequality does not hold, $\chi^{a''}_{nL}(\omega)$ will be very important.

$\chi^{a''}_{nL}(\omega)$ is also proportional to the $F_c(\omega)$. In view of what has been said previously about the $g_c(\omega)$, $F_0(\omega)$ ($F_1(\omega)$ resp.) will vary rapidly close to ω_{2k}(ω_{2k-1} resp.) since the same is true of the $g_c(\omega)$. The amplitude of these variations will, however, be less because of the factor $\hbar\gamma_1$ in the denominator of $F_c(\omega)$ (see Fig. 3) where $\gamma_1/2|\Delta\omega| = 10^{-2}$, $F_c(\omega)$ fluctuates

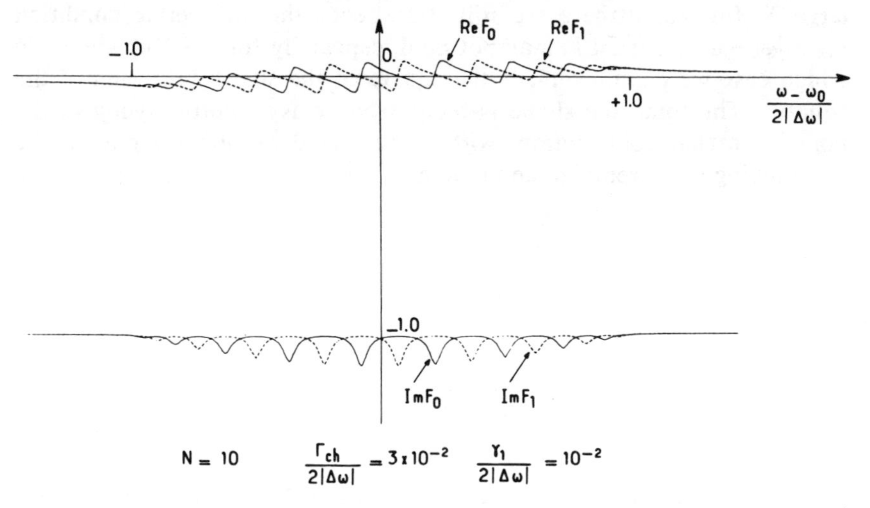

Fig. 3. Variations of the functions $F_c(\omega)(c=0, 1)$ appearing in (3.36) which gives the total optical susceptibility. The number of sites is equal to 10 and $\gamma_1/2|\Delta\omega| = 10^{-2}$.

about $-i$. For a given value of the parameter the minima and maxima are less pronounced for large N. When $N \geqslant 20$, $F_c(\omega)$, $c = 0$ and $c = 1$, stays very close to $-i$ over all the interval of interest.

In order to discuss qualitatively the non-Lorentzian contribution we suppose $\gamma_1 \ll |\Delta\omega|$. Replacing $F_c(\omega)$ by its approximate value $-i$ (3.36) yields

$$\chi_{nL}^{a''}(\omega) = \frac{2r\omega_0^2 d^2}{\hbar N(N+1)} \times \frac{-4\gamma_1}{N+1} \operatorname{Re}\left\{ \sum_{\kappa \geqslant 1}^{[N+c/2]} T_\kappa \frac{\sin \kappa\theta}{(\omega - \omega_\kappa) + i\Gamma_{ch}} \right\}^2 \qquad (3.38)$$

If N is not too large, indeed if it were there would be no end effects, Γ_{ch} may be chosen much smaller than the smallest difference $|\omega_k - \omega_{k+1}|$. An approximate form of (3.38) close to each resonance ω_k is

$$\chi_{nL}^{a''}(\omega \simeq \omega_\kappa) \simeq \frac{2r\omega_0^2 d^2}{\hbar N(N+1)} \times \frac{-4\gamma_1}{N+1} \operatorname{Re}\left\{ T_\kappa^2 \frac{\sin^2 \kappa\theta}{(\omega - \omega_\kappa + i\Gamma_{ch})^2} \right\}$$

$$= \frac{2r\omega_0^2 d^2}{\hbar N(N+1)} \times \frac{4\gamma_1}{N+1} \times \frac{\Gamma_{ch}^2 - (\omega - \omega_\kappa)^2}{\{(\omega - \omega_\kappa)^2 + \Gamma_{ch}^2\}^2} \times T_\kappa^2 \sin^2 \kappa\theta \qquad (3.39)$$

Figure 4 shows the variations of the third factor as a function of ω. It is positive if and only if $|\omega - \omega_k| < \Gamma_{ch}$ and reaches a maximum height of Γ_{ch}^{-2}, much higher than that of a Lorentzian of the same width, which would be Γ_{ch}^{-1}. Thus resonance is stronger than for a Lorentzian in the region $|\omega - \omega_k| < \Gamma_{ch}$, and weaker outside it, and the line is narrower. As long as $\Gamma_{ch} \ll |\Delta\omega|$, it is symmetric. On the other hand, for reasonably large N, for which there are still end effects, the no overlap condition $\Gamma_{ch} \ll |\omega_k - \omega_{k+1}|$ for all k, may not hold, especially for $k \sim 1$ or $k \sim N$, in which case we cannot replace the square of the sum by the sum of the squares. The total line shape is then strongly asymmetric owing to the non-Lorentzian contribution, with some added asymmetry due to the overlapping of Lorentzian components.

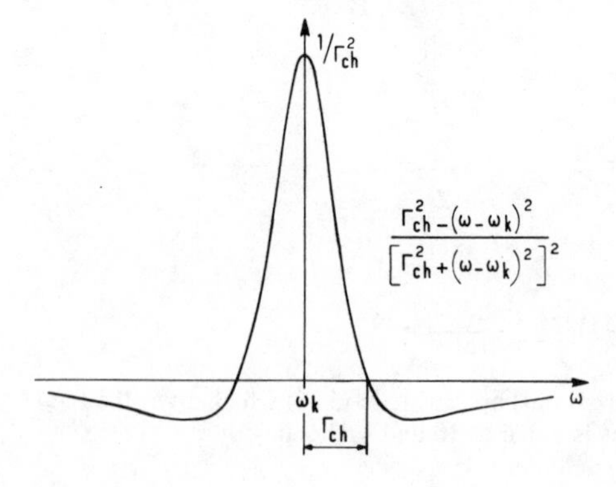

Fig. 4. Typical approximate shape of the non-Lorentzian contribution to the susceptibility (see (3.39)).

Supposing perfect resolution, the relative importance of $\chi_{nL}^{a''}(\omega)$ at ω_k follows from

$$\chi^{a''}(\omega_k) \simeq \frac{2r\omega_0^2 d^2}{\hbar N(N+1)} \left(T_k^2 \Gamma_{ch}^{-1} + \frac{4\gamma_1}{N+1} T_k^2 \Gamma_{ch}^{-2} \sin^2 k\theta \right)$$
$$= \frac{2r\omega_0^2 d^2}{\hbar N(N+1)} \frac{T_k^2}{\Gamma_{ch}} \left(1 + \frac{4}{N+1} \frac{\gamma_1}{\gamma_0 + 2\gamma_1} \sin^2 k\theta \right)$$

The ratio of the Lorentzian and non-Lorentzian parts at the center of the line, $4/(N+1)\ (\gamma_1/\Gamma_{ch}) \sin^2 k\theta$, decreases with increasing N. As pointed out previously, it is proportional to γ_1/Γ_{ch}. Hence $\chi_{nL}^{a''}(\omega)$ will be small in most systems. Finally, the relative importance of the end effects depends on the resonance k. It is larger near ω_0 than at the ends of the spectrum. The overall deviation from the Lorentzian spectrum are largest for strong central resonances.

Figures 5 and 6, derived from the exact formula (3.36), illustrate the preceding effects. The first represents the theoretical absorption spectrum of a "parallel" chain in which all the $\mathbf{m}_p$ are parallel. Figure 6 shows the effect of an orientational trap on one end of the chain. All but

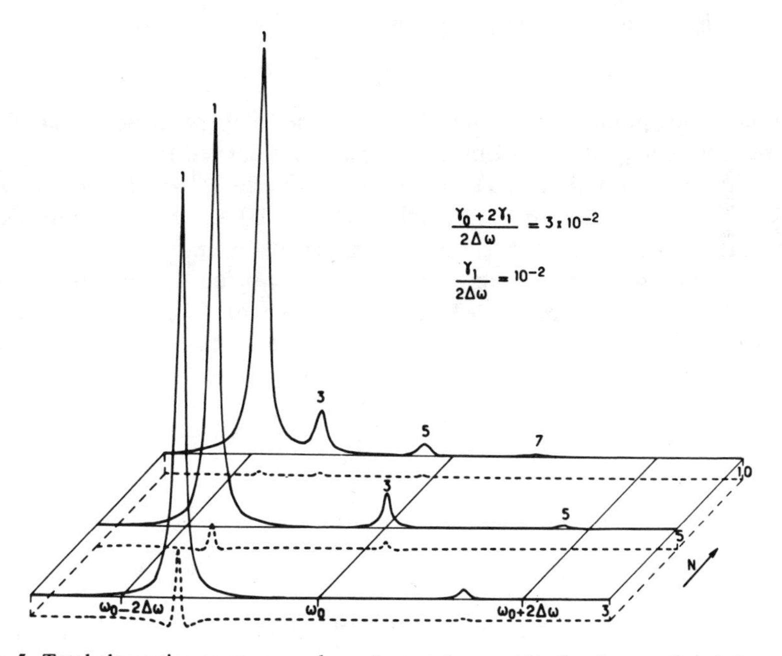

Fig. 5. Total absorption spectrum and non-Lorentzian contribution for parallel chains with $N = 3, 5, 10$. Here $\Gamma_{ch}/2|\Delta\omega| = 3 \times 10^{-2}$, $\gamma_1/2|\Delta\omega| = 10^{-2}$.

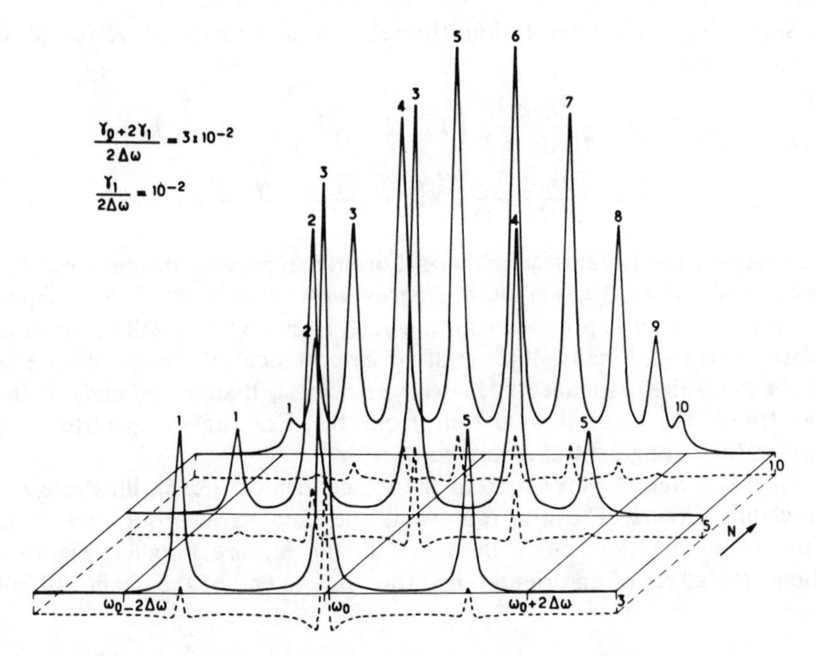

Fig. 6. Same as Fig. 5 but for chains with an oriented trap at one end.

one of the $\mathbf{m}_p$ point in the same direction. The total spectrum is traced in a continuous line, the non-Lorentzian part in a dashed line.

It will be noted that $\chi_{nL}^{a''}(\omega)$ decreases relative to $\chi^{a''}(\omega)$ as the number of sites increases, becoming negligible for $N \geqslant 20$. Case $N = 10$ in Fig. 6 shows that its contribution to the resonances at the ends of the spectrum is not symmetric; (the Lorentzian components being widely separated).

It is noteworthy that while $\chi_{nL}^{a''}(\omega)$ is a sum of squared amplitudes, $\chi_{nL}^{a''}(\omega)$ is the square of a sum of amplitudes, like an interference phenomenon. The effect would be strongest for a large number of intense lines. $\chi_{nL}^{a''}(\omega)$ vanishes in infinite chains, by definition, and because only the first resonance is allowed. The next section deals quantitatively with the strength of each line.

3. *Selection Rules Characteristic of Finite Systems*

The selection rules and the strengths of the different resonances depend on the factor T_k^2 in (3.37). T_k^2 depends on the geometry of the chain through the scalar products $\mathbf{u}\cdot\mathbf{m}_p$, the number of sites and angle θ. The lines are spread over an interval of $4|\Delta\omega|$ centered on ω_0. The selection rules following are specific to an aggregate. They are incompatible with the naive model of photon absorption by each site, independently of the rest, eventually followed by redistribution through the residual interactions. In fact, the distance between sites being minute compared to the significant wavelengths, (here of the order of several thousand angstroms), all the sites react together to the excitation, just as would all the electrons of an atom or of a molecule. The selection rules reflect the aggregate's ability to resound as a whole. Each site is coupled to all the others, so the exterior field interacts with the resulting collective modes of the aggregate.

The selection rules for the line strength depend on the arrangement of the $\mathbf{m}_p$. We examine three classes of aggregate, for which $\langle M_p|\hat{H}_0|M_{p+1}\rangle$ is independent of p:

1. Parallel chains, with all the $\mathbf{m}_p$ parallel.
2. Alternating chains, with all the odd order $\mathbf{m}_p$ parallel to one another and all the even order ones parallel in another direction. The aggregate is then composed of two parallel chains.
3. Helical chains in which there is a fixed spatial rotation between $\mathbf{m}_p$ and $\mathbf{m}_{p+1}$.

We calculate T_k $(k = 1, 2, \ldots, N)$ for each of these configurations, making liberal use of relation (2.97). If non-Lorentzian terms are small, the strength of each maximum is proportional to T_k^2. The approximation, which is better for large N, is completely acceptable for $N \geqslant 20$.

a. Parallel Chains. In a parallel chain,

$$(\mathbf{u}, \mathbf{m}_p) = \psi = C^{ste}$$

so that

$$T_k = \sum_{p=1}^{N} \mathbf{u} \cdot \mathbf{m}_p \sin kp\theta = \cos\psi \operatorname{Im} \sigma_k = \cos\psi \frac{1-(-1)^k}{2} \cot k\frac{\theta}{2} \tag{3.40}$$

where σ_k is defined by (2.97). Use has been made of the fact that k lies between 1 and N.

Even-order resonances are forbidden, as may be verified in the original form of T_k, by collecting the two terms $\sin kp\theta$ and $\sin(N+1-p)k\theta$ whose sum, $2\sin k(\pi/2)\cdot\cos(k(\pi/2) - kp\theta)$ is indeed zero for even k. The strength of odd-order resonances falls off quickly as $\cot^2(k\theta/2)$, $(k = 1 \Rightarrow k\theta/2 = \pi/2(N+1), k = N \Rightarrow k\theta = (N/(N+1)\pi/2)$. As an example Table V gives values of $\cot^2 k\theta/2$ for $N = 3$ and $N = 10$.

As N increases, the number of strong lines decreases, only one line $k = 1$ persisting for very large N. The limiting case of the infinite crystal cannot be studied in this way because we supposed the wavelength

TABLE V
$\mathrm{Cot}^2(k\theta/2)$ for Given N

k	1	2	3	4	5	6	7	8	9	10
$N = 3$	5.82	1.00	0.17	—	—	—	—	—	—	—
$N = 10$	48.4	11.6	4.79	2.42	1.33	0.75	0.41	0.21	0.09	0.03

absorbed to be large compared to the distance between sites. N should not exceed a few thousand sites. The density of states on the high energy side of the $k = 1$ resonance grows with N, so the spectrum becomes difficult to resolve. The resonance near $\omega_0 - 2|\Delta\omega|$ has marked inhomogeneous asymmetry, with the intensity falling off more slowly on the high energy side than on the low energy one.

Figure 7 shows the total spectrum for $\Gamma_{ch}/2|\Delta\omega| = 10^{-2}$ and $\gamma_1/2|\Delta\omega| = 10^{-3}$. Each line is indexed by k. For $N \leqslant 10$, all the odd-order lines are present, whereas for $N = 50$, those beyond $k \simeq 17$ are negligible. Larger widths give strong inhomogeneous broadening for $N \geqslant 10$.

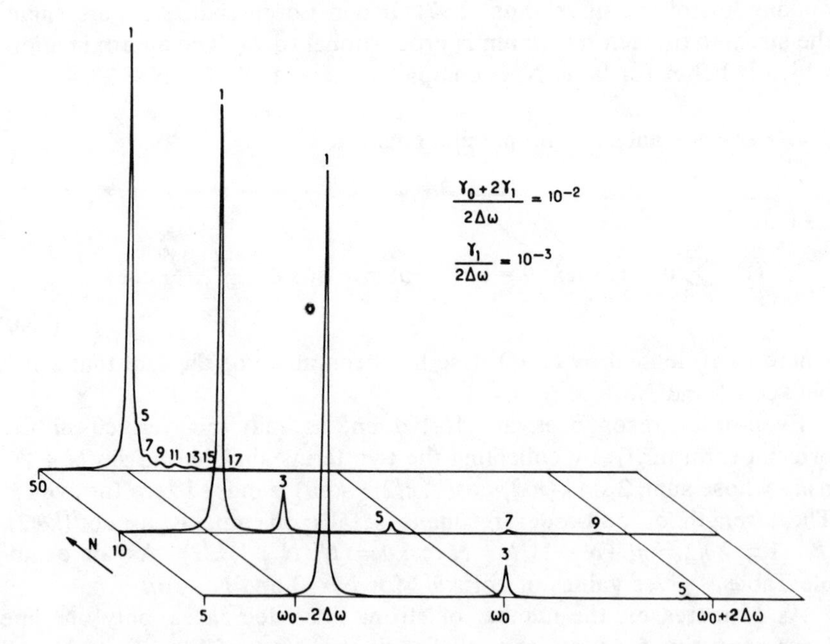

Fig. 7. Total absorption spectrum for parallel chains with $N = 5, 10, 50$. The fluctuation parameters are such that $\Gamma_{ch}/2|\Delta\omega| = 10^{-2}$ and $\gamma_1/2|\Delta\omega| = 10^{-3}$.

It will be appreciated that the preceding selection rules can be ignored only in crude interpretations of experimental spectra. While the non-Lorentzian terms are negligible for N greater than 20, the spectrum still exhibits peculiarities of a finite system for larger values of N, for example, the asymmetric, inhomogeneous broadening of the strongest line.

b. Alternating Chains. In an alternating chain,

$$(\mathbf{u}, \mathbf{m}_{2p-1}) = \psi' \qquad (\mathbf{u}, \mathbf{m}_{2p}) = \psi'' \qquad (p \geqslant 1)$$

By (2.97),

$$T_k = \frac{1-(-1)^N}{2}\,\frac{1-(-1)^k}{2}\,\frac{\cos\psi' + \cos k\theta\cos\psi''}{\sin k\theta} + \frac{1+(-1)^N}{2}\,\frac{1-(-1)^k}{2}\,\frac{\cos k\theta\cos\psi' - (-1)^k\cos\psi''}{\sin k\theta} \tag{3.41}$$

Equation 3.41 gives the same formula as (3.40) if we put $\psi' = \psi'' = 0$. Generally, even-order lines are forbidden for odd N. The even and odd lines are, respectively, proportional to $(\cos\psi' - \cos\psi'')^2\tan^2(k\theta/2)$ and $(\cos\psi' + \cos\psi'')^2\cot^2(k\theta/2)$, for even N. Large values of N lead to an accumulation of lines around $\omega_0 \pm 2|\Delta\omega|$ with strong inhomogeneous asymmetry reminiscent of an infinite crystal with two nonequivalent molecules per unit cell.

Considering two resonances symmetrically placed with respect to ω_0, say the kth and the $(N+1-k)^{\text{th}}$, we have

$$\frac{T_{N+1-k}}{T_k} = \frac{1-(-1)^N}{2}\,\frac{\cos\psi' - \cos k\theta\cos\psi''}{\cos\psi' + \cos k\theta\cos\psi''} + \frac{1+(-1)^N}{2}\,\frac{\cos\psi' + (-1)^k\cos\psi''}{\cos\psi' - (-1)^k\cos\psi''} \tag{3.42}$$

If the field is perpendicular to the even-order sites ($\cos\psi'' = 0$) $T_{N+1-k}/T_k = 1$ for all N; and if it is perpendicular to the odd sites $T_{N+1-k}/T_k = -1$ for all N. Both cases have spectra symmetric with respect to the central frequency ω_0.

Figure 8 shows a spectrum for $\psi' = \pi/6$ and $\psi'' = 2\pi/3$; then:

$$T_k = \frac{1-(-1)^N}{2}\,\frac{1-(-1)^k}{2}\,\frac{\cos\psi' - \cos k\theta\sin\psi'}{\sin k\theta} + \frac{1+(-1)^N}{2}\,\frac{1-(-1)^k}{2}\,\frac{\cos k\theta\cos\psi' + (-1)^k\cos\psi''}{\sin k\theta}$$

As mentioned previously, the central resonances disappear as N increases. The lines are noticeably asymmetric for large N. When $\mathbf{m}_p\cdot\mathbf{m}_{p+1} = 0$ and $\mathbf{u}\|\mathbf{m}_1$, the spectrum is indeed symmetric about ω_0.

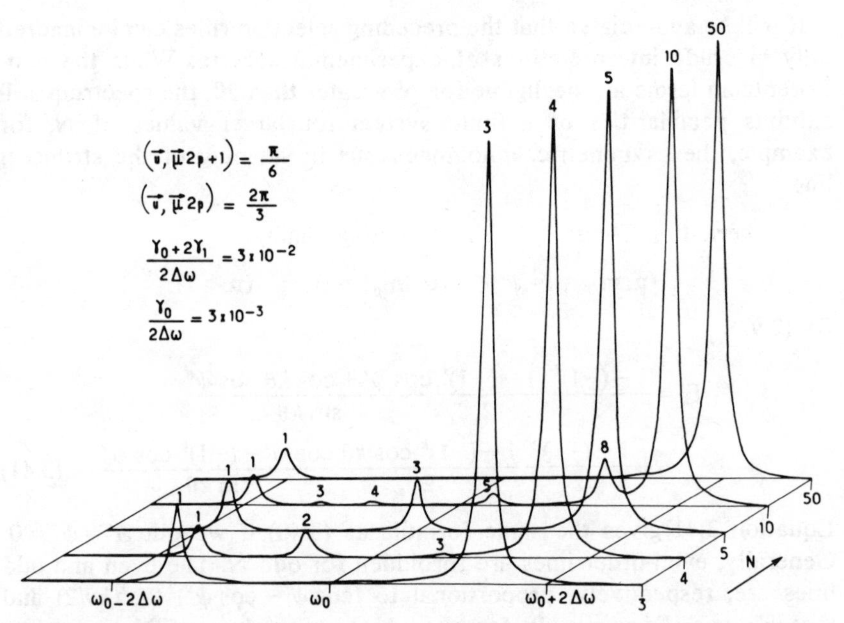

Fig. 8. Total absorption spectrum for alternating chains with $N = 3, 4, 5, 10, 50$. Here $\psi' = \pi/6$, $\psi'' = 2\pi/3$, $\Gamma_{ch}/2|\Delta\omega| = 3 \times 10^{-2}$, and $\gamma_1/2|\Delta\omega| = 3 \times 10^{-3}$.

c. Helical Chains. Consider a helical chain along Oz and such that $\mathbf{m}_1$ is parallel to Ox, so that $\mathbf{m}_1$ lies in the xOz plane. Let ψ be the angle between $\mathbf{m}_p$ and Oz, independent of p by definition; and let $\Delta\psi$ be the angle through which the projection of $\mathbf{m}_p$ on xOy must be turned to bring it onto that of $\mathbf{m}_{p+1}$. Then,

$$\mathbf{u}\cdot\mathbf{m}_p = u_x \sin\psi \cos(p-1)\Delta\psi + u_y \sin\psi \sin(p-1)\Delta\psi + u_z \cos\psi$$

Using (2.97) we have the following relations for the three Cartesian polarizations:

$$\mathbf{u}\|\mathrm{O}x \qquad T_{k,x} = \sin k\theta \, \frac{\cos\Delta\psi - (-1)^k \cos N\Delta\psi}{2(\cos\Delta\psi - \cos k\theta)} \underset{\Delta\psi\to k\theta}{\to} \frac{N+1}{2} \sin k\theta$$

$$\mathbf{u}\|\mathrm{O}y \qquad T_{k,y} = -\sin k\theta \, \frac{\sin\Delta\psi + (-1)^k \sin N\Delta\psi}{2(\cos\Delta\psi - \cos k\theta)} \underset{\Delta\psi\to k\theta}{\to} \frac{N+1}{2} \cos k\theta$$

$$\mathbf{u}\|\mathrm{O}z \qquad T_{k,z} = \frac{1-(-1)^k}{2} \cot k\frac{\theta}{2} \tag{3.43}$$

$T_{k,z}$ has of course the same form as a parallel chain. If $\Delta\psi = 0$, T_{kx} has the same expression and T_{ky} is identically zero, ($\mathbf{u}$ being perpendicular to all the $\mathbf{m}_p$).

Very special selection rules apply when the elementary rotation, $\Delta\psi$ is a multiple of θ, say $k_0\theta$. $T_{k,x}$ and $T_{k,y}$ take on the form

$$T_{k,x} = (1-\delta_{kk_0})\frac{1-(-1)^{k+k_0}}{2}\frac{\sin k\theta \cos k_0\theta}{\cos k_0\theta - \cos k\theta} + \delta_{kk_0}\frac{N+1}{2}\sin k_0\theta$$

$$T_{k,y} = (1-\delta_{kk_0})\frac{1-(-1)^{k+k_0}}{2}\frac{\sin k\theta \sin k_0\theta}{\cos k_0\theta - \cos k\theta} + \delta_{kk_0}\frac{N+1}{2}\cos k_0\theta$$

Note the Kronecker deltas. All lines of the same parity as k_0 are forbidden, except line k_0 itself, whose intensity relative to the strongest line may be large or small depending on N; see Figs. 9 and 10 which were drawn with the electric field along Ox, and $\Delta\psi = \theta$ and $\Delta\psi = 2\theta$, respectively. The strength of line k_0 depends heavily on N. The strongest line is very asymmetric for $N = 50$.

Experimental measurements could be used to accurately determine the geometry of the chain, for the relative intensities of the lines depend very much on the ratio $\Delta\psi/\theta$. Figure 11 shows the value of T_k^2 in function of $\Delta\psi/\theta$, for $N = 3, 5, 10$.

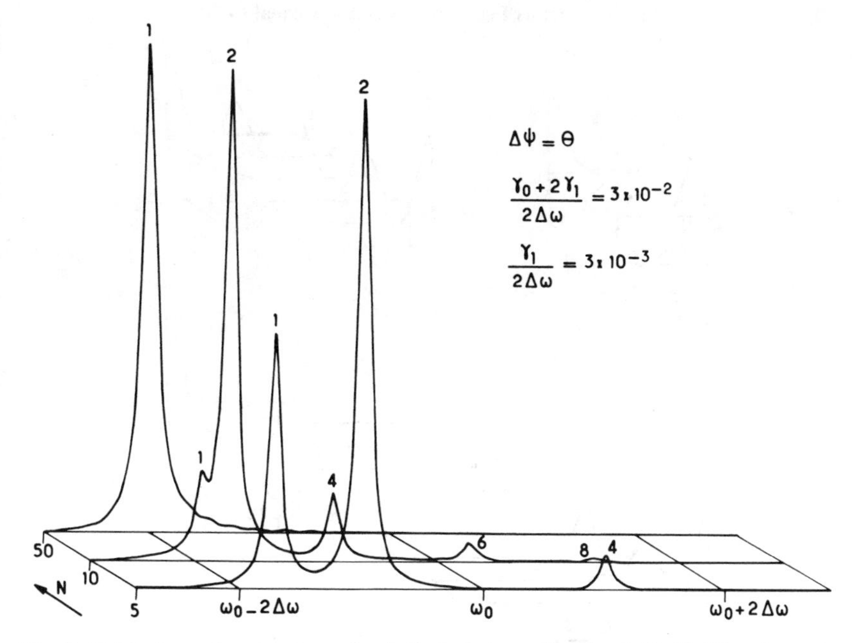

Fig. 9. Total absorption spectrum for helical chains with $N = 5, 10, 50$. The elementary rotation $\Delta\psi$ is precisely equal to $\pi/(N+1) = \theta$; the fluctuation parameters γ_0 and γ_1 are the same as for Fig. 8.

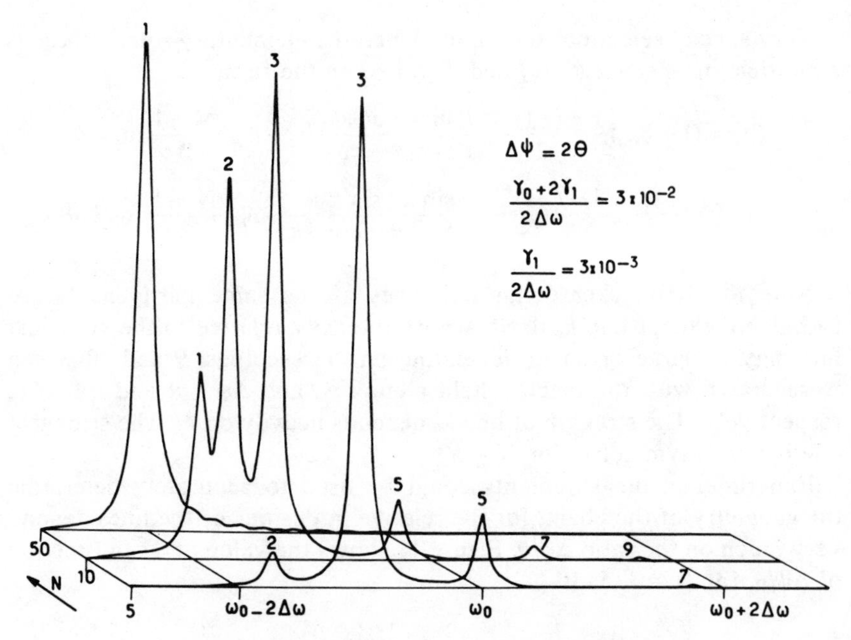

Fig. 10. Same as Fig. 9, but now $\Delta\psi$ is equal to 2θ.

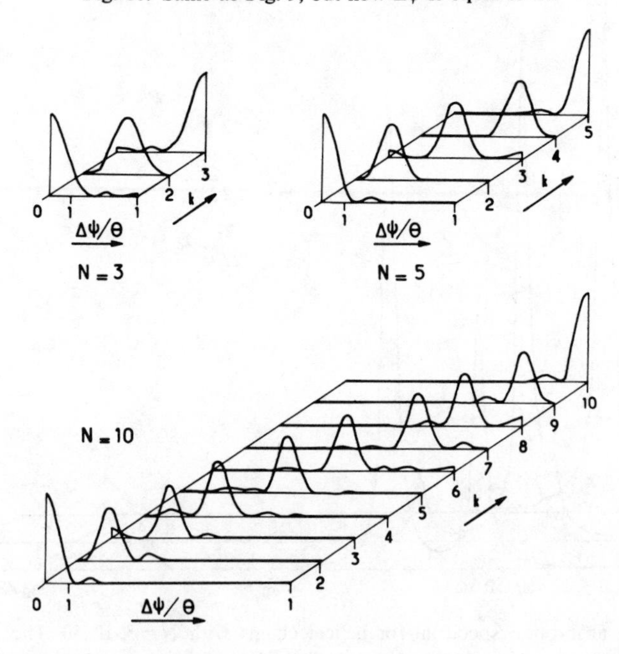

Fig. 11. Variations of the geometrical factors T_k^2 for helical chains as a function of the ratio $\Delta\psi/\theta$ for $N = 3, 5, 10$.

When N is large, though not infinite as we have pointed out, because of the assumption of large wavelengths relative to the aggregate, the asymptotic selection rule of a parallel chain,

$$T_k \propto \delta_{k1}$$

resembles that of an infinite crystal: $\Delta k = 0$.

We have already seen that the central resonances of a large, alternating chain vanish, so we may restrict our attention to T_1 and T_N. Equation 3.41 yields.

$$\frac{T_N}{T_1} = \frac{1-(-1)^N}{2} \frac{\cos \psi' - \cos \theta \cos \psi''}{\cos \psi' + \cos \theta \cos \psi''} + \frac{1+(-1)^N}{2} \frac{\cos \psi' - \cos \psi''}{\cos \psi' + \cos \psi''}$$

When N is large, there are two "Davydov components."

$$\frac{T_N}{T_1} \simeq \frac{\cos \psi' - \cos \psi''}{\cos \psi' + \cos \psi''}$$

The formulae for a helical chain will be discussed for three cases. If $\Delta\psi \ll \pi/2$, then by (3.43)

$$T_{k,x} \simeq \sin k\theta \frac{1-(-1)^k \cos N\Delta\psi}{2 \cos \Delta\psi - 2 \cos k\theta}$$

$$T_{k,y} \simeq \sin k\theta \frac{(-1)^{k+1} \sin N\Delta\psi}{2 \cos \Delta\psi - 2 \cos k\theta}$$

The asymptotic forms for very large N are shown in Table VI.

It should not be forgotten that special rules apply to the case $\Delta\psi = k_0\theta$.

When $\Delta\psi = \pi/2$, (3.43) becomes

$$T_{k,x} = \frac{(-1)^k \cos(N\pi/2)}{2} \tan k\theta; \quad T_{k,y} = \frac{1+(-1)^k \sin(N\pi/2)}{2} \tan k\theta$$

$$|T_{k,x}| = \frac{1+(-1)^N}{4} |\tan k\theta|; \quad |T_{k,y}| = \frac{2+(-1)^k i^{N-1}\{1-(-1)^N\}}{4} |\tan k\theta|$$

TABLE VI
Asymptotic Behavior of $|T_{k,x}|$ and $|T_{k,y}|$ for $\Delta\psi \ll \frac{\pi}{2}$

$\Delta\psi$	$\ll \frac{\pi}{2N}$	$\simeq \frac{\pi}{2N}\left(\simeq \frac{\theta}{2}\right)$	$\simeq \frac{\pi}{N} (\simeq \theta)$
$\lvert T_{k,x}\rvert$	$\frac{1-(-1)^k}{2} \cot k \frac{\theta}{2}$	$\frac{1}{2} \cot k \frac{\theta}{2}$	$\frac{1+(-1)^k}{2} \cot k \frac{\theta}{2}$
$\lvert T_{k,y}\rvert$	$\simeq 0$	$\frac{1}{2} \cot k \frac{\theta}{2}$	$\simeq 0$

These formulae, which are exact for all N, show that its parity is an important factor in absorption along Ox, even for large N. For odd N all the $T_{k,x}$ are nil. For even N the spectrum is symmetric with respect to ω_0. The spectrum for large even N is essentially a strong central resonance with symmetric, inhomogeneous broadening. Line strength, for a field polarized along Oy is proportional to $\{1+(-1)^{N'+k}\}\tan^2 k\theta$ if N is of the form $2N'+1$, N' being an integer; and to $\{1-(-1)^{N'+k}\}\tan^2 k\theta$ if N is of the form $2N'+3$.

The last case, $\Delta\psi = 2\pi/N$ corresponds to a configuration in which the $(N+1)^{\text{th}}$ site, if it existed, would have its $\mathbf{m}_{N+1}$ identical to $\mathbf{m}_1$. Then

$$T_{k,x} = \sin k\theta \frac{\cos(2\pi/N) - (-1)^k}{2\cos(2\pi/N) - 2\cos k\theta}$$

$$T_{k,y} = -\sin k\theta \frac{\sin(2\pi/N)}{2\cos(2\pi/N) - 2\cos k\theta}$$

When N is very large,

$$|T_{1,x}| \simeq \frac{4N}{3\pi}\frac{1-(-1)^k}{2}; \quad |T_{k\sim N/2,x}| \simeq \frac{1-(-1)^k}{2}; \quad |T_{k\sim N,x}| \simeq \frac{\pi}{N}\frac{1-(-1)^k}{2}$$

$$|T_{1,y}| \simeq \tfrac{2}{3}; \qquad |T_{k\sim N/2,y}| \simeq \frac{2\pi}{N}; \quad |T_{k\sim N,y}| \simeq \frac{\pi^2}{N^2}$$

$$|T_{1,u}| \ll |T_{k\sim N/2,u}| \ll |T_{k\sim N,u}| \qquad (u = x, y)$$

4. *Absorption Spectrum of a Mixture of Aggregates*

The results preceding may be used to simulate the absorption spectrum of a mixture of aggregates. Let c_N be the concentration of N-mers. Supposing that all the chains are oriented in the same way with respect to the incident field and are in their ground states before it is switched on ($r = N$), then the susceptibility of the mixture is

$$\chi_M^{a''}(\omega) = \sum_{N\geq 2} c_N \chi^{a''}(\omega)$$

where $\chi^{a''}(\omega)$ is given by (3.36), which applies to the dimer also with $r = N$.

Obviously there will be many resonances, since there are several for each N; not just one line as one might assume by "extrapolating" the rules for an infinite crystal. $\chi^{a''}(\omega)$ depends on N in several ways, such as angle θ, the factor $1/(N+1)$, the T_k, and so on, so that the line strengths are not simply proportional to the concentrations. Conversely, deducing the concentrations from the spectrum is much more difficult than it at first seems. General arguments cannot be given, since there is

TABLE VII
$T_k^2/(N+1)$ for Parallel Chains

N	1	2	3	4	5
2	2.000	0	—	—	—
3	2.914	0	0.086	—	—
4	3.789	0	0.211	0	—
5	4.609	0	0.333	0	0.024

such a variety of forms of T_k. As an illustration we examine the case of the parallel chains.

The T_k for a parallel chain of length N are given by expression (3.40). That the intensity is not proportional to the concentration follows from Table VII in which $T_k^2/(N+1)$ is tabulated for different values of k and N.

Thus, even if the pentamers were 2 times less numerous than the dimers their $k = 1$ lines would be of approximately equal strength. The link between intensity and concentration is, therefore, complex.

Figure 12 shows the spectrum of a mixture containing species up to

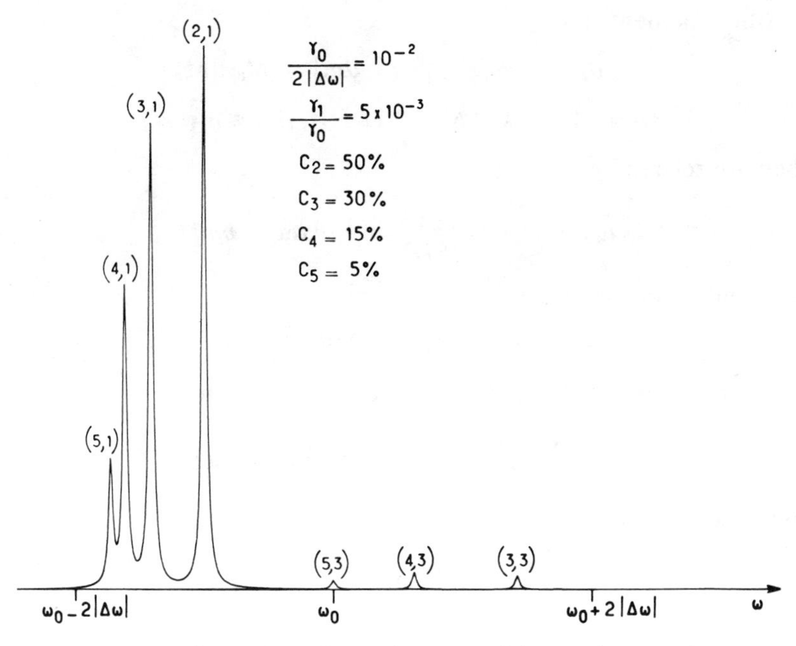

Fig. 12. Total absorption spectrum for a mixture containing species up to the pentamer.

the pentamer. The distribution of concentrations is realistic insofar as it decreases as N increases. The lines are indexed by (N, k). The lines $k = 1$ decrease in intensity as N increases, but not as fast as the concentrations, for example, line (5, 1) is not 10 times weaker than line (2, 1), although there are 10 times fewer pentamers than dimers. Another illustration is the behavior of a given line (k fixed), such as the first. For large N its strength is proportional to $(4N/\pi^2)c_N$. Its importance depends on the evolution of c_N with N as N becomes very large.

D. Spectrum of the Double Chain

The calculations for the double chain are so long that we divide this section into two parts.

1. Calculation of Susceptibility

Starting from (3.11) and generalizing (3.26) we have the following, semiclassical, matter–field interaction operator:

$$|W(t)) = f(t) \sum_{b=\pm 1} \sum_{k=1}^{N} i\omega_{kb} d\mathbf{u}\cdot\boldsymbol{\mu}_{kb}(|a_{kb}) - |a_{kb}^{+})) \equiv f(t)|W) \tag{3.44}$$

ω_{kb} and the $\hat{a}_{kb}$, $\hat{a}_{kb}^{+}$ are defined, respectively, in (2.95) and (2.84). The transition moments are

$$d\boldsymbol{\mu}_{kb} = \langle A_0|\hat{\mathbf{d}}|A_{kb}\rangle \qquad d\mathbf{m}_p^I = \langle M_0|\hat{\mathbf{d}}|M_p^I\rangle$$

$$(b = \pm 1, 1 \leqslant k \leqslant N) \qquad (I = A, B, 1 \leqslant p \leqslant N)$$

They are related by

$$\boldsymbol{\mu}_{kb} = (N+1)^{-1/2} \sum_{p=1}^{N} \sin kp\theta\, (\mathbf{m}_p^A + b m_p^{\mathbf{B}})$$

The equilibrium states are

$$|\rho_{eq}) = (2N+1)^{-1/2}|K_1) - r\{2N(2N+1)\}^{-1/2}|K_2) \tag{3.45}$$

where $|K_i)$ is explained in (2.73) and (2.75). r is given by

$$r = \sum_{b=\pm 1} \sum_{k=1}^{N} (\langle A_0|\hat{\rho}_{eq}|A_0\rangle - \langle A_{kb}|\hat{\rho}_{eq}|A_{kb}\rangle)$$

Conversely,

$$\langle A_0|\hat{\rho}_{eq}|A_0\rangle = \frac{1}{2N+1} + \frac{r}{2N+1}$$

$$\langle A_{kb}|\hat{\rho}_{eq}|A_{kb}\rangle = \frac{1}{2N+1} - \frac{r}{2N(2N+1)}$$

r is a real number between -1 and N. $r = -1$ corresponds to an empty ground state and equally populated excited states, with populations $1/2N$. If $r = 2N$, only the ground state is occupied.

$\hat{W}_-$, corresponding to $\hat{W}$ $(|W))$ in (3.44), acts on $|K_2)$, (see (2.77c) and (2.77d)), to give

$$\hat{W}_-|K_2) = \left(\frac{2N+1}{2N}\right)^{1/2} \sum_{b=\pm 1} \sum_{k=1}^{N} i\omega_{kb} d\mathbf{u}\cdot\boldsymbol{\mu}_{kb}\{|a_{kb}) + |a^+_{kb})\} \tag{3.46}$$

The general expression for $\chi^a(\omega)$, (see (3.9)), written in the notation of Section III.C.1 is, after applying (3.45),

$$\chi^a(\omega) = \frac{rd^2}{2N} \sum_{k=1}^{N} \sum_{k'=1}^{N} \sum_{b=\pm 1} \sum_{b'=\pm 1} \omega_{kb}\omega_{k'b'}(\mathbf{u}\cdot\boldsymbol{\mu}_{kb})(\mathbf{u}\cdot\boldsymbol{\mu}_{k'b'})(a^+_{kb}|\hat{\mathscr{G}}^{(+)a}(\omega)|a^+_{k'b'}) \tag{3.47}$$

Just as in Section III.C.1, the matrix elements of the resolvent are known exactly. Letting

$$\hat{H}^{a'}_{b0-} = \sum_{k=1}^{N} \hbar(\omega_{kb} - \omega_0)|a^+_{kb})(a^+_{kb}|; \qquad \hat{I}^a_b = \sum_{k=1}^{N} |a^+_{kb})(a^+_{kb}|$$

$\hat{H}^a$ may be derived from (2.110):

$$\hat{H}^a = \sum_{b=\pm 1} \hbar(\omega_0 - i\Gamma_{dc})\hat{I}^a_b + \hat{H}^{a'}_{b0-} + i\hbar\gamma_1 \sum_{c=0,1} \hat{P}_{cb} \tag{3.48}$$

See (2.112) for notations. The $\hat{P}_{cb}$ $(c = 0, 1;\ b = \pm 1)$ are projectors defined in (2.106) and (2.107). The sum over b is a direct sum, so

$$\hat{H}^a = \bigoplus_{b=\pm 1} \hat{H}^a_b$$

There is an expression (3.32) for each term of the direct sum, so we straightway use (3.32) to arrive at

$$\hat{\mathscr{G}}^{(+)a}(\omega) = \sum_{b=\pm 1} \left\{ \hat{\mathscr{G}}^{(+)}_{b0}(\omega) + \sum_{c=0,1} \frac{\lambda}{1 - \lambda g_{cb}(\omega)} \hat{\mathscr{G}}^{(+)a}_{b0}(\omega)\hat{P}_{cb}\hat{\mathscr{G}}^{(+)a}_{b0}(\omega) \right\}$$

with the notation:

$$\hat{\mathscr{G}}^{(+)a}_{b0}(\omega) = (z\hat{I}^a_b - \hat{H}^{a'}_{b0-})^{-1}; \qquad z = \hbar(\omega - \omega_0 + i\Gamma_{dc})$$

$$g_{cb}(\omega) = (\pi_{cb}|\hat{\mathscr{G}}^{(+)a}_{b0}(\omega)|\pi_{cb})$$

Vector $|\pi_{cb})$ is defined by (2.107). Calculation of $g_{cb}(\omega)$ is the same as that given in Section III.C.1. Introducing the Gegenbauer polynomials once more,

$$g_{cb}(\omega) = \frac{1}{\hbar\Delta\omega} \frac{C_{N-1}(\zeta_b) + (-1)^{1+c}}{C_N(\zeta_b)}$$

C_N is the Nth degree Gegenbauer polynomial and $\zeta b = (z + b\Delta\omega')/(2\Delta\omega)$. That is,

$$C_n(\zeta_b) = \frac{y_1^{n+1}(\zeta_b) - y_2^{n+1}(\zeta_b)}{y_1(\zeta_b) - y_2(\zeta_b)}$$

where y_i $(i = 1, 2)$ are the solutions of

$$Y^2 - 2\zeta_b Y + 1 = 0$$

The $g_{cb}(\omega)$ exhibit the same variations as the $g_c(\omega)$ in Section III.C.1. The relevant intervals are centered on $\omega_b = \omega_0 + b\Delta\omega'$, $b = \pm 1$.

Arguments similar to those for the ordinary chain (Section III.C.1), applied to (3.47) result in the following form of the susceptibility

$$\chi^{a''}(\omega) = \chi^{a''}_{+1}(\omega) + \chi^{a''}_{-1}(\omega) \tag{3.49}$$

where $\chi^{a''}_b(\omega)$ $(b = \pm 1)$ is the analog of $\chi^{a''}(\omega)$ in (3.36). Defining T_{kb} by

$$T_{kb} = \sum_{p=1}^{N} \mathbf{u}\cdot(\mathbf{m}_p^A + b\mathbf{m}_p^B) \sin kp\theta \tag{3.50}$$

the susceptibility is

$$\chi^{a''}_b(\omega) = \frac{2r\omega_0^2 d^2}{\hbar N(N+1)} \left\{ \sum_{k=1}^{N} \frac{T_{kb}^2}{4} \frac{\Gamma_{dc}}{(\omega - \omega_{kb})^2 + \Gamma_{dc}^2} \right.$$

$$\left. + \frac{4\gamma_1}{N+1} \operatorname{Im}\left[\sum_{c=0,1} F_{cb}(\omega) \left(\sum_{\kappa \geq 1}^{[(N+c)/2]} \frac{T_{\kappa b}}{2} \frac{\sin \kappa\theta}{(\omega - \omega_{\kappa b}) + i\Gamma_{dc}} \right)^2 \right] \right\} \tag{3.51}$$

$$F_{cb}(\omega) = \{i + \hbar\gamma_1 g_{cb}(\omega)\}^{-1}$$

Other notations used are found in (3.36).

2. *Discussion of the Spectrum.*

Two terms, one for each value of b, appear in expression (3.49) of $\chi^{a''}(\omega)$. The spectrum is thus the superposition of the corresponding spectra, for $b = -1$ and $b = \pm 1$, centered, respectively, on $\omega_{-1} = \omega_0 - \Delta\omega'$ and $\omega_{+1} = \omega_0 + \Delta\omega'$. Neglecting selection rules, the spectrum of each sub-chain of N sites has N lines spread over an interval of roughly $4|\Delta\omega|$.

We omit discussion of the Lorentzian and non-Lorentzian contributions since they are the same as those for the ordinary chain, (see Section III.C.2).

The selection rules follow from the values of the T_{kb} defined in (3.50). The T_{kb} may be simply expressed in terms of the corresponding quantities in Section III.C.3. Let

$$T_k^I = \sum_{p=1}^{N} \mathbf{u} \cdot \mathbf{m}_p^I \sin kp\theta \qquad (I = A, B) \tag{3.52}$$

Then

$$T_{kb} = T_k^A + bT_k^B$$

The geometrical factor T_k^I of subchain I is exactly the same as the one defined without superscript in (3.37). The "longitudinal" selection rules associated with T_k^I are determined by the calculation in Section III.C.3. The "transverse" selection rules associated with $\mathbf{u} \cdot (\mathbf{m}_p^A + \mathbf{m}_p^B)$ recall to memory those of the dimer, Section III.B. The global selection rules are the "product" of the two.

The general behavior of the total absorption is as follows: (*1*) If $|\Delta\omega'| \ll |\Delta\omega|$, the spectrum is essentially that of the ordinary chain, with a (not always resolved) sequence of doublets with components in the same ratio as that of the dimer. (*2*) If $|\Delta\omega| = |\Delta\omega'|$ the two spectra are intermingled, that corresponding to $b = +1$ stretching from $\omega_0 - 3|\Delta\omega|$ to $\omega_0 + |\Delta\omega|$, that corresponding to $b = -1$ from $\omega_0 - |\Delta\omega|$ to $\omega_0 + 3|\Delta\omega|$ ($\Delta\omega'$ negative by convention). (*3*) If $|\Delta\omega| \ll |\Delta\omega'|$ the dimer spectrum predominates but each resonance has a complicated structure, which in the neighborhood of $\omega_0 + b\Delta\omega'$ resembles that of the ordinary chain.

Figures 13 to 15 show the double chain spectra for each case. By

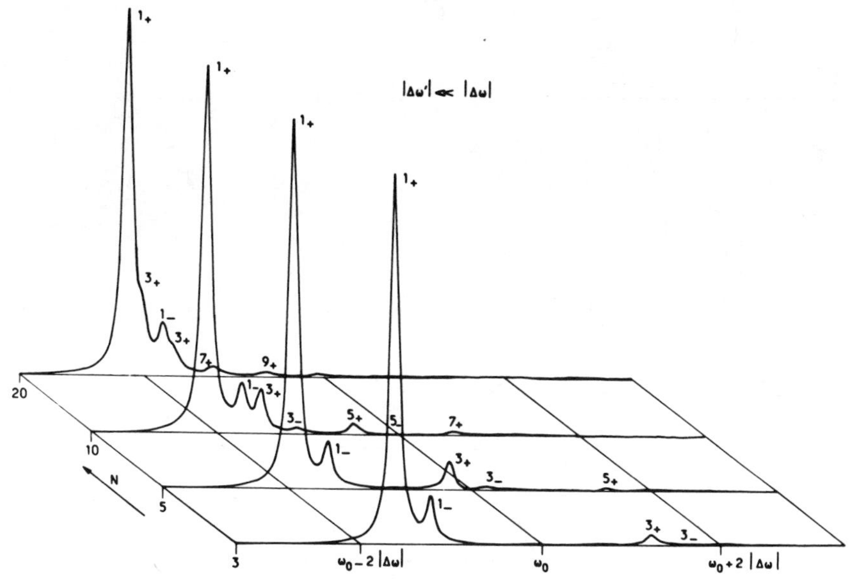

Fig. 13. Total absorption spectrum for a double chain made up with two parallel subchains; here $\psi^A = 0$, $\psi^B = \pi/3$, and $|\Delta\omega'| \ll |\Delta\omega|$.

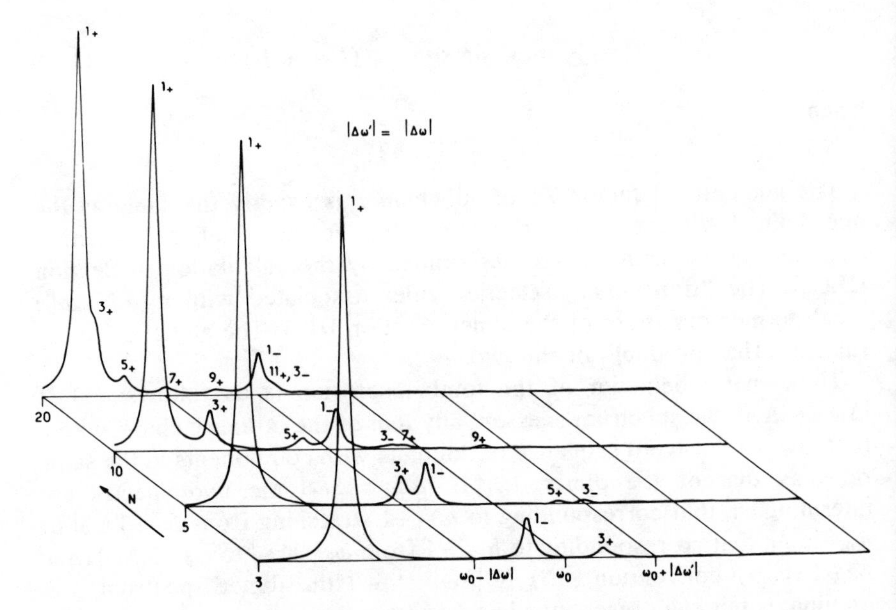

Fig. 14. Same as Fig. 13, but now $|\Delta\omega'| = |\Delta\omega|$.

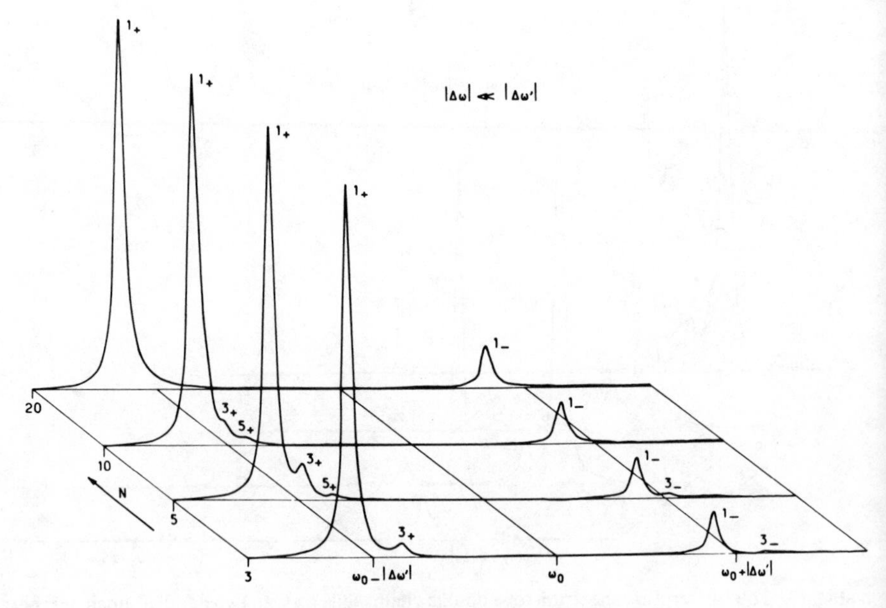

Fig. 15. Same as Fig. 13, but now $|\Delta\omega'| \gg |\Delta\omega|$.

definition,

$$(\mathbf{u}, \mathbf{m}_p^I) = \psi^I \qquad (I = A, B)$$

with $\psi^A = 0$ and $\psi^B = \pi/3$. In Fig. 13 we see that the bulk of the resonances in case (*1*) are around $\omega_0 - 2|\Delta\omega| - |\Delta\omega'|$, especially for large N. The general impression is of an ordinary chain spectrum, (cf. Fig. 7), with a doublet of separation $2|\Delta\omega'|$ at each resonance. For given b, the intensity decreases with k according to $(1-(-1)^k)/2 \cot^2(k\theta/2)$. For given k, the ratio between components $b = +1$ and $b = -1$ is

$$\left(\frac{\cos \psi^A + \cos \psi^B}{\cos \psi^A - \cos \psi^B}\right)^2$$

where we have neglected non-Lorentzian terms. They are not shown on this figure or the following one, but were taken into account when calculating the susceptibility.

In Fig. 14, (case (*2*)), $|\Delta\omega| = |\Delta\omega'|$. The bulk of the lines lie between $\omega_0 - 3|\Delta\omega|$ and ω_0, especially for large N. The doublets are less clearly differentiated and there is sometimes considerable overlapping between line (k, b) and line (k', b').

Figure 15 shows case (*3*). There are two resonances near $\omega_0 \pm |\Delta\omega'|$ with a lot of fine structure. The fine structure which is most obvious in small chains, gives rise to considerable inhomogeneous broadening when N is large.

IV. ENERGY TRANSFER DURING THE LIFE OF THE EXCITED STATE

This section is an attempt to describe quantitatively energy transfer "from site to site" during times very small compared to the life of the excited state.

The picture of site-to-site transfer in a solid or an aggregate is a direct application of concepts generally associated with dilute gases. Each atom or molecule in a tenuous gas may absorb a photon and subsequently pass on the energy to another molecule during a collision. A little reflection, however, suffices to show that extension of this concept to solids is not justifiable. It seems to have benefited from a situation in which experimental results were not sensitive enough to compel recognition of certain simple theoretical considerations. Nowadays, it is more accepted that all the sites of an aggregate react collectively to certain excitations just as would all the electrons of an atom. At the same time certain intuitive ideas, which used to form the premises of naive models, are no longer considered to be self-evident, but are seen as consequences of better-founded theories.

The following results obtained should be interpreted with due respect to their double statistical nature. Since they depend on the essentially statistical nature of quantum mechanics and on the stochastic interactions in the model, the results are characteristic of quantities calculated by laws of probability.

The spatial moments (position, dispersion) derived in this section are the average situation in all the microsystems, not what is "really" happening in a particular one. Even if we could do the impossible of "working backwards" from the statistical averages, we should need an ergodic assumption to determine how the quantum averages are realized in a microsystem. At present this assumption has not been firmly established.

Our purpose is not to give a real description of energy transfer in the sense of propagation in a crystal, for example; rather to extract as much as possible from quantum mechanics, keeping in mind the gap between it and the macroscopic level. We do not seek out any systematic relation with the existing phenomenological theories[47,107,134–136] which describe macroscopic transfer in molecular associations, aggregates, and crystals. The experimental evidence of microscopic transfer, such as emission by the crystal from points distant from the point of absorption, is indirect. The preceding use of "point" is mischievous since the "point" could in fact be a huge volume compared to a site, unless it is a permanent deep trap. It would be better to talk of domains of coherence, containing many sites. Quantities apparently as simple as the exciton density may be far from clear and even may be imprecisely defined, on the scale of the angstrom.

Intuitively, it seems movement will be most obvious for initially localised excitations. A quantitative measure of the localization is presented in Section IV.A. Ways of optically preparing a localised state are discussed in Section IV.B. In Section IV.C we introduce the equivalent of a wave function with a view to answering questions like: "Where is the exciton?" or "What is its spatial extent?" General expressions for the spatial moments of the exciton are derived in Section IV.D. The last three sections deal with transfer in dimers, ordinary chains, and double chains.

A. A Measure of the Localization of a State

One of the most sensible ways of defining localization is with respect to scalar products of the kind $\langle M_p|\psi\rangle$ between $|\psi\rangle$ and localised excitations $|M_p\rangle$.

Any optically excited state $|\psi^*\rangle$ may be expanded on the singly excited site states $\{|M_p\rangle\}_{p\geq 1}$ as:

$$|\psi^*\rangle = \sum_{p=1}^{N} c_p|M_p\rangle; \qquad \sum_{p=1}^{N} |c_p|^2 = 1 \tag{4.1}$$

with $c_p = \langle M_p|\psi^*\rangle$, by the orthogonality relations assumed throughout this work. We must now find a function of the c_p which quantitatively describes the idea of localization. It should be a minimum when all but one of the c_p are nil, the other being one, and a maximum when they are all of equal modulus. Many such functions exist. One of them, that measuring the missing information in information theory, is well known in statistical mechanics because it coincides with the entropy (to within a multiplicative factor depending on the unit of information). This kind of function appeared in the discussion of coherence in Section II.A.

It should be understood that this function is no more than a measure of the uniformity of the coefficients c_p. As we stated earlier, the $|c_p|^2$ are not probabilities, but numbers whose physical meaning does not follow directly from the quantum postulates $|c_p|^2$. They are mathematical, not physical probabilities.

The measure of delocalization, S, adopted here, is defined for a state $|\psi^*\rangle$ (see (4.1)) by

$$S(|\psi^*\rangle) = -\frac{1}{\log N}\sum_{p=1}^{N} |c_p|^2 \log |c_p|^2 \tag{4.2}$$

The factor $(\log N)^{-1}$ is to norm $S(|\psi^*\rangle)$. $S(|\psi^*\rangle)$ is greatest for delocalized states. S has the following properties:

$$0 \leq S(|\psi^*\rangle) \leq 1$$

$$S(|\psi^*\rangle) = 0 \Leftrightarrow \exists p_0, 1 \leq p_0 \leq N, |c_p| = \delta_{pp_0} \tag{4.3}$$

$$S(|\psi^*\rangle) = 1 \qquad \forall p, 1 \leq p \leq N, |c_p| = N^{-1/2} \tag{4.4}$$

It follows that there are N states of maximum localization ($S = 0$) and an infinity of states of minimum localization ($S = 1$).

Besides having the necessary and sufficient conditions of significance, S also takes into account all the moments of the distribution of the $|c_p|^2$, not just the first two, as would the usual definition of the dispersion.

B. One Photon Preparation of a Localized State[131,137]

We discuss here optical preparation of a localized state with a classical, optical source, though there are of course other methods, like causing relaxation into a deep trap. The discussion is based on the aptitude of the chain to resound at the eigenfrequencies ω_k of the

Hamiltonian $\hat{H}_0$ with eigenvectors $\{|A_k\rangle\}_{1\leq k\leq N}$:

$$\hat{H}_0|A_k\rangle = \hbar\omega_k|A_k\rangle \Leftrightarrow \left[\begin{array}{l} \omega_k = \omega_0 + 2\Delta\omega\cos k\theta \\ |A_k\rangle = \left(\dfrac{2}{N+1}\right)^{1/2} \sum\limits_{p=1}^{N} \sin kp\theta |M_p\rangle \end{array}\right. \tag{4.5}$$

The initial state may be expanded on the $|A_k\rangle$:

$$|\psi^*\rangle = \sum_{k=1}^{N} d_k|A_k\rangle \tag{4.6}$$

The transformation between bases is defined by

$$d_k = \left(\frac{2}{N+1}\right)^{1/2} \sum_{p=1}^{N} \sin kp\theta\, c_p; \qquad c_p = \left(\frac{2}{N+1}\right)^{1/2} \sum_{k=1}^{N} \sin kp\theta\, d_k \tag{4.7}$$

Let us examine two kinds of excitation:

1. Excitation with a very narrow line of width $\Delta\omega_0$ much smaller than the smallest difference between neighboring eigenvalues. If the duration of the interaction ($\sim \Delta\omega_0^{-1}$) is much longer than ω_k^{-1}, then

$$|\psi^*\rangle \simeq |A_k\rangle \tag{4.8}$$

where $|A_k\rangle$ is the vector corresponding to the frequency of the field.

$|\psi^*\rangle$ is quite a delocalized state for which $S(|\psi^*\rangle)$ is around 1 because nearly all the localized states appear in its expansion, see Table VIII. $S(|A_k\rangle)$ is never less than 0.80 for values of $N > 10$.

2. White light centered on ω_0, with width $4|\Delta\omega|$ and a very short excitation time compared to $|\Delta\omega|^{-1}$. The first-order perturbation development of $|\psi^*\rangle$ is

TABLE VIII
$S(|A_k\rangle)$ for N between 2 and 10

k \ N	1	2	3	4	5	6	7	8	9	10
1	1.00	1.00	0.95	0.93	0.92	0.91	0.91	0.91	0.91	0.91
2		1.00	0.63	0.93	0.86	0.91	0.89	0.91	0.90	0.91
3			0.95	0.93	0.68	0.91	0.90	0.86	0.91	0.91
4				0.93	0.86	0.91	0.71	0.91	0.90	0.91
5					0.92	0.91	0.90	0.91	0.73	0.91
6						0.91	0.89	0.86	0.90	0.91
7							0.91	0.91	0.91	0.91
8								0.91	0.90	0.91
9									0.91	0.91
10										0.91

$$|\psi^*\rangle \propto \sum_{k=1}^{N} |A_k\rangle\langle A_k|\hat{W}|A_0\rangle$$

where $\langle A_k|\hat{W}|A_0\rangle$ is the average value of $\langle A_k|\hat{W}(t)|A_0\rangle$ over the (short) interval of excitation. Introducing the localized states:

$$|\psi^*\rangle \propto \sum_{k=1}^{N} |A_k\rangle \sum_{p=1}^{N} \langle A_k|M_p\rangle\langle M_p|\hat{W}|M_0\rangle \tag{4.9}$$

The sum over p depends on the geometry of the chain. $\langle M_p|\hat{W}|M_0\rangle$ is independent of p in a parallel chain like the system analysed by Whiteman.[19] Hence

$$|\psi^*\rangle \propto \sum_{k=1}^{N} |A_k\rangle \sum_{p=1}^{N} \langle A_k|M_p\rangle = \sum_{p=1}^{N} |M_p\rangle$$

Norming $|\psi^*\rangle$,

$$|\psi^*\rangle = N^{-1/2} \sum_{p=1}^{N} |M_p\rangle$$

It follows from (4.4) that $|\psi^*\rangle$ is one of the most delocalized states ($S(|\psi^*\rangle) = 1$). Thus localized states in a parallel chain are inaccessible to any kind of optical excitation.

We now look at things in a different order: What are the necessary and sufficient conditions on the aggregate and the field in order to obtain a localized state $|M_p\rangle$ by interference of the $|A_k\rangle$? Mathematically, this amounts to conditions on the d_k in (4.6) so that they are all proportional to sin $kp\theta$, the constant of proportionality being independent of k.

Nearly all the stationary states of $\hat{H}_0$ appear in the expansion of $|M_p\rangle$ on the $\{|A_k\rangle\}_{1\leq k\leq N}$. The first condition must be, therefore, that the field be roughly uniform in an interval of length $4|\Delta\omega|$ centered on ω_0. Its coherence time, τ_0, should be shorter such that

$$\tau_0 \ll \frac{2\pi}{4|\Delta\omega|}$$

Further the system will resound only if it has time to "count" the periods. All the ω_k are of the same order as ω_0, so

$$\tau_0 \gg \frac{2\pi}{\omega_0}$$

Putting the inequalities together,

$$\frac{2\pi}{\omega_0} \ll \tau_0 \ll \frac{2\pi}{4|\Delta\omega|} \tag{4.10}$$

Typical values are $4|\Delta\omega| \sim 20\ \text{cm}^{-1}$, $\omega_0 \sim 20000\ \text{cm}^{-1}$, for which

$$10^{-15}\ \text{sec} \ll \tau_0 \ll 10^{-12}\ \text{sec}$$

The preceding conditions on the photon are necessary, but not sufficient, for the system must be such that all, or nearly all, the $|A_k\rangle$ are accessible. The selection rules must not be too severe. We have shown already that such is not the case for a parallel chain. It cannot be optically excited to a stationary state. Intuitively,[138] this is understandable as the following rough arguments show: Let $\Delta p \cdot \Delta q \sim \hbar$ and $E = pc$ for the photon. We must have, from (4.10),

$$4\hbar|\Delta\omega| \ll \Delta E \ll \hbar\omega_0$$

The numerical values preceding lead to

$$10^{+3}\ \text{Å} \ll \Delta q \ll 10^{+6}\ \text{Å}$$

The "spatial extension" of the photon is thus enormous, so one can hardly see why it should always (statistically) relax on *one particular* site though it is identical to all the others. One could use Dicke's reasoning[88] since all the sites are very close together compared to the significant wavelengths.

The answer to the preceding question depends on the arrangement of the $\mathbf{m}_p$ (defined in Section III.C.1). A simple calculation shows that if

$$\mathbf{u}\cdot\mathbf{m}_p \propto \delta_{pp_0} \qquad p_0 \text{ fixed, } p = 1, 2, \ldots, N$$

a field of spectral width greater than $4|\Delta\omega|$ will be capable of creating localized states $|M_p\rangle$ by interference of the $|A_k\rangle$. This configuration is that of a chain with a trap oriented differently from the rest of the sites: All the $\mathbf{m}_p$ perpendicular to the polarization $\mathbf{u}$ of the field, except $\mathbf{m}_{p_0}$ which we may assume without loss of generality to be parallel to $\mathbf{u}$.

Direct optical preparation of a localized state depends, therefore, on two kinds of conditions on the field and the aggregate: The spectrum of the field must be wide compared to $|\Delta\omega|$ and it must be polarized in a particular way relative to the chain.

As illustrative examples we consider the spectra of chains with one oriented trap.[139] Figure 16 shows the case of a trimer with such a trap at one end of the chain. Figure 17 shows the corresponding spectrum of a chain of 50 sites. $\Gamma_{ch}/2|\Delta\omega| = 10^{-2}$ and $\gamma_1/2|\Delta\omega| = 10^{-3}$ in both figures. The continuous curve is $\chi^{a''}(\omega)$, the total response, and the dashed one is $\chi^{a''}_{nL}(\omega)$. The latter curve has been translated for clarity. Nearly all the intensities are of the same order of magnitude so that provided the conditions on polarization and frequency ((4.10)) are met, a localized state can be created in this kind of configuration. The same is true of a

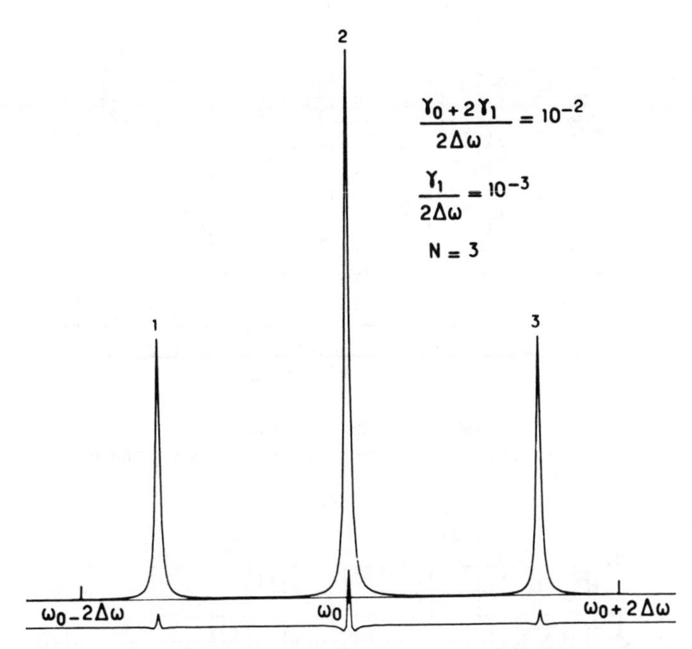

Fig. 16. Total absorption spectrum and non-Lorentzian contribution of a trimer with oriented trap at one end.

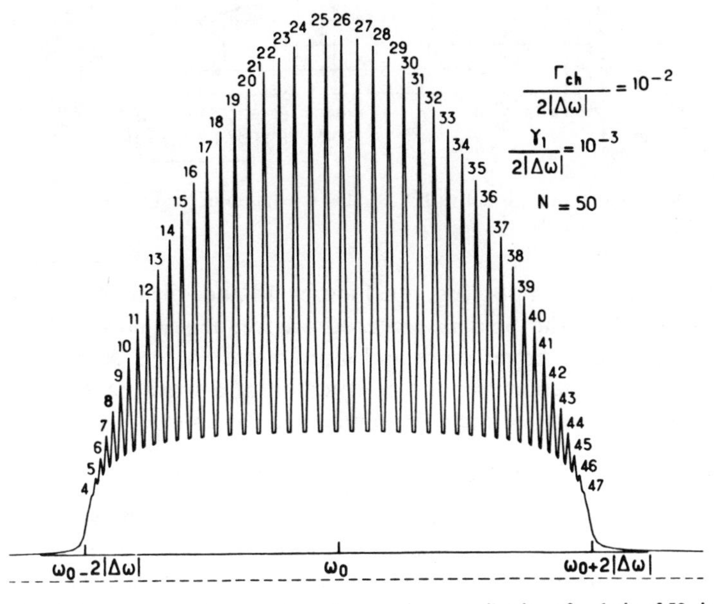

Fig. 17. Total absorption spectrum and non-Lorentzian contribution of a chain of 50 sites with an oriented trap at one end.

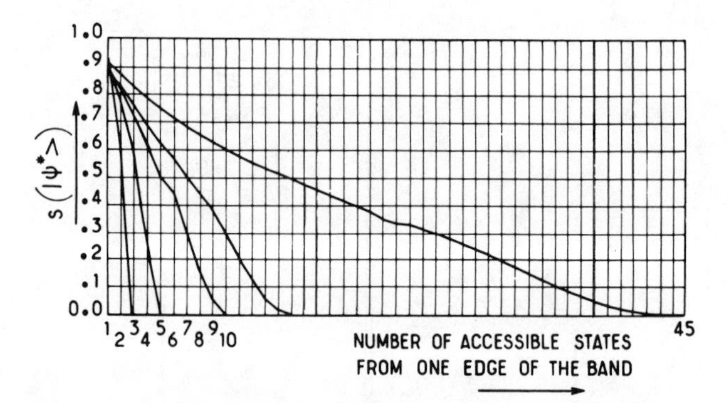

Fig. 18. Variation with the spectral width of the source (number of states covered from $|A_1>$) of the delocalization entropy of an optically prepared state. The excitation is prepared on one end of the chain.

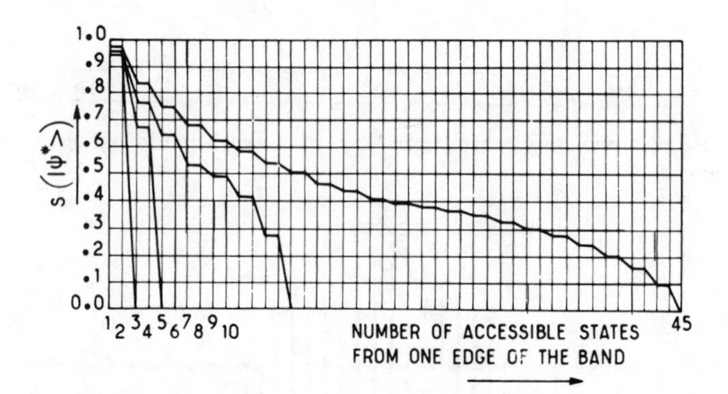

Fig. 19. Same as Fig. 18, but with the initial excitation in the middle of the chain.

double chain with faults, Fig. 20. Figures 18 and 19 show the evolution of $S(|\psi^*\rangle)$ when the field is centered on frequency ω_1 of $|A_1\rangle$ and its width increased to cover more and more states $|A_k\rangle$. S decreases as the width increases, rather like the contraction of a wave packet when one increases the spread of the complementary variable.

C. Definition of an Energy Density[140]

Before describing the movement of the exciton we must first be in a position to calculate its mean place, spatial extension, and higher moments, by means of a probability density of presence filling the role of a wave function. As the exciton is a "particle" of energy we must define an energy density.

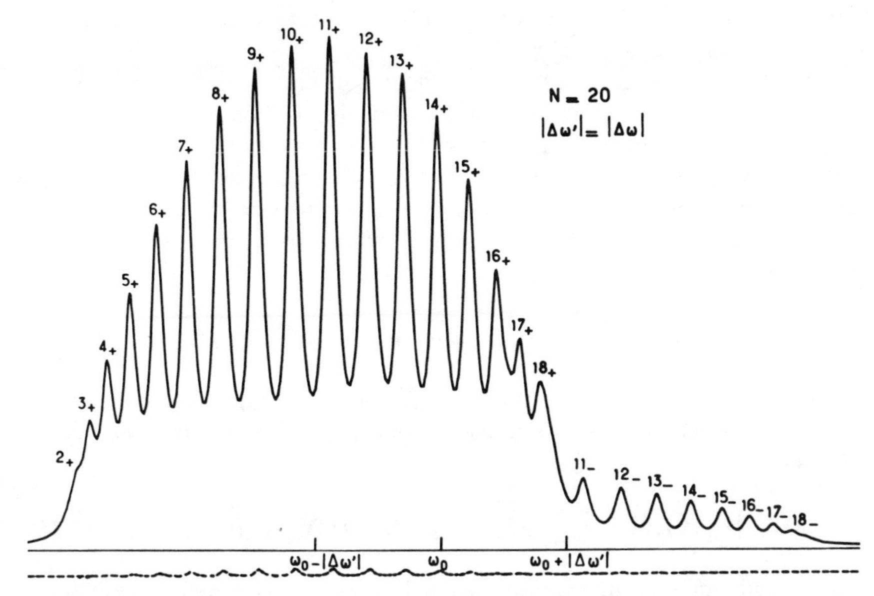

Fig. 20. Total absorption spectrum and non-Lorentzian contribution of a double chain of length 20 sites, with an oriented trap on one end of both chains.

Such a definition raises some of the great problems of quantum mechanics. The existence and uniqueness of such a function have yet to be shown. We are incompetent to discuss such matters and adopt a density having the minimal set of properties set out as follows.

Jauch[141] has defined a density for any kind of observable when the system is in a pure dynamic state. We propose to prolong his definition to a general mixed state represented by $\hat{\rho}$. Let $\hat{\mathscr{E}}$ be the energy density:

$$\hat{\mathscr{E}} = \tfrac{1}{2}(\hat{H}\hat{\rho} + \hat{\rho}\hat{H}) \tag{4.11}$$

where $\hat{H}$ is the Hamiltonian of an isolated system represented by $\hat{\rho}$. $\hat{\mathscr{E}}$, which is in the usual symmetrical form adopted for noncommuting variables,[142] has certain properties ensuring it may be interpreted as an energy density: $\hat{\mathscr{E}}$ is Hermitian. Its trace verifies

$$\operatorname{Tr} \hat{\mathscr{E}} = \operatorname{Tr}(\hat{\rho}\hat{H}) = \text{average energy} \tag{4.12}$$

The average quantum energy may be calculated by (4.12). It is easily shown that

$$i\hbar \frac{d\hat{\mathscr{E}}}{dt} = [\hat{H}, \hat{\mathscr{E}}] \tag{4.13}$$

This equation expresses the conservation of energy in an isolated system with instantaneous generator $\hat{H}$.

The preceding properties are insufficient to ensure the uniqueness of $\hat{\mathscr{E}}$, whose existence we have not proved either. We avoid these fundamental points, being satisfied with an a posteriori justification by quantitatively reasonable results. Let us also recall that it is usual to define angular momentum[146,147] or lagrangian densities.[148]

$\hat{\mathscr{E}}$ may be developed on the discrete, stationary representation $\{|A_k\rangle\}$ as

$$\hat{\mathscr{E}}(t) = \sum_k \sum_{k'} \tfrac{1}{2}(E_k + E_{k'})\langle A_k|\hat{\rho}(t)|A_{k'}\rangle |A_k\rangle\langle A_{k'}| \tag{4.14}$$

Given a set of wave functions $\{A_k(\mathbf{x})\}_k$ representing the $\{|A_k\rangle\}_k$ where $\mathbf{x} = \{\mathbf{r}, \mathbf{s}\}$, spatial and spin coordinates of all the electrons, then the kernel of $\tilde{\mathscr{E}}$ is

$$\tilde{\mathscr{E}}(\{\mathbf{x}\}, \{\mathbf{x}'\}; t) = \sum_k \sum_{k'} \tfrac{1}{2}(E_k + E_{k'})\langle A_k|\hat{\rho}(t)|A_{k'}\rangle A_k(\{\mathbf{x}\}) A_{k'}^*(\{\mathbf{x}'\}) \tag{4.15}$$

$\hat{\mathscr{E}}$ is defined in configuration space. Just as one can define a reduced electronic density in atomic or molecular physics[93,143–145] one may define a reduced density for a body in $\mathbb{R}^3$:

$$\mathscr{E}(\mathbf{r}, \mathbf{r}'; t) = \sum_k \sum_{k'} \tfrac{1}{2}(E_k + E_{k'})\langle A_k|\hat{\rho}(t)|A_{k'}\rangle \delta_{kk'}(\mathbf{r}, \mathbf{r}') \tag{4.16}$$

where the kernel $\delta_{kk'}(\mathbf{r}, \mathbf{r}')$ is defined by ($\mathbf{S}_{s,\mathbf{r}} \equiv \Sigma_s \int d\mathbf{r}$):

$$\delta_{kk'}(\mathbf{r}, \mathbf{r}') = \sum_s \sum_{s'} \underset{s_2,\mathbf{r}_2}{\mathbf{S}} \underset{s_3,\mathbf{r}_3}{\mathbf{S}} \cdots \underset{s_n,\mathbf{r}_n}{\mathbf{S}} A_k(\mathbf{x}, \mathbf{x}_2, \mathbf{x}_3, \ldots, \mathbf{x}_n) A_{k'}^*(\mathbf{x}', \mathbf{x}_2, \mathbf{x}_3, \ldots, \mathbf{x}_n)$$

for a total of n electrons. This kernel cannot in general be reduced to a product of functions of separated variables as is the case for a one body system in a stationary state.

D. Exciton Spatial Moments

We may now apply the "wave function" of the exciton to the calculation of its spatial moments. The λth moments (λ a positive whole number or zero) are

$$\mathscr{M}_{\lambda,uv\ldots}(t) = \frac{\displaystyle\int d\mathbf{r}\{uv\ldots\mathscr{E}(\mathbf{r}, \mathbf{r}'; t)\}_{\mathbf{r}'=\mathbf{r}}}{\displaystyle\int d\mathbf{r}\{\mathscr{E}(\mathbf{r}, \mathbf{r}'; t)\}_{\mathbf{r}'=\mathbf{r}}} \tag{4.17}$$

$u, v, \ldots = x, y, z$. The notation used is the traditional one.[149] In what follows we are interested principally in the second and higher order moments. The mean position of the exciton is

$$\mathcal{M}_{1,u}(t) = \frac{\int d\mathbf{r}\{\hat{u}\mathscr{E}(\mathbf{r}, \mathbf{r}'; t)\}_{\mathbf{r}'=\mathbf{r}}}{\int d\mathbf{r}\{\mathscr{E}(\mathbf{r}, \mathbf{r}'; t)\}_{\mathbf{r}'=\mathbf{r}}} \qquad (u = x, y, z) \tag{4.18}$$

The dispersion tensor, $\Delta^2_{uv}(t)$, is given by

$$\Delta^2_{uv}(t) = \mathcal{M}_{2,uv}(t) - \mathcal{M}_{1,u}(t)\mathcal{M}_{1,v}(t) \qquad (u, v = x, y, z) \tag{4.19}$$

The eigenvectors of $\Delta^2_{uv}(t)$ are the principle directions of spread at time t, the eigenvalues the dispersion along these directions and their square roots the spread of the exciton about its mean position.

The energy density operator of a simple chain with Hamiltonian $\hat{H}_0$ and semiclassical stochastic perturbation $\hat{\epsilon}^i(t)$ may be expanded on the local basis as

$$\hat{\mathscr{E}}(t) = \sum_{p=1}^{N} \sum_{p'=1}^{N} |M_p\rangle\langle M_p| \tfrac{1}{2}\{\hat{H}_0\hat{\rho}(t) + \hat{\rho}(t)\hat{H}_0\} |M_{p'}\rangle\langle M_{p'}|$$

It follows from form (2.86) of the model Hamiltonian in which $\Delta\omega' = 0$ and I is superfluous for an ordinary chain, that $\hat{\mathscr{E}}(t)$ may be expressed in terms of the kernel ($\rho_{pq}(t) = \langle M_p|\hat{\rho}(t)|M_q\rangle$),

$$\begin{aligned}\tilde{\mathscr{E}}(\mathbf{r}, \mathbf{r}'; t) &= \sum_{p=1}^{N} \sum_{p'=1}^{N} \hbar\omega_0 \rho_{pp'}(t)\delta_{pp'}(\mathbf{r}, \mathbf{r}') \\ &+ \frac{\hbar\Delta\omega}{2} \sum_{p=1}^{N} \sum_{p'=1}^{N} \{\rho_{p+1p'}(t)(1-\delta_{pN}) + \rho_{p-1p'}(1-\delta_{p1})\}\delta_{pp'}(\mathbf{r}, \mathbf{r}') \\ &+ \frac{\hbar\Delta\omega}{2} \sum_{p=1}^{N} \sum_{p'=1}^{N} \{\rho_{pp'+1}(t)(1-\delta_{p'N}) + \rho_{pp'-1}(1-\delta_{p'1})\}\delta_{pp'}(\mathbf{r}, \mathbf{r}')\end{aligned}$$

$\delta_{pp'}(\mathbf{r}, \mathbf{r}')$ is defined by

$$\delta_{pp'}(\mathbf{r}, \mathbf{r}') = \sum_{s} \sum_{s'} \mathop{\mathbf{S}}_{s_2,\mathbf{r}_2} \mathop{\mathbf{S}}_{s_3,\mathbf{r}_3} \cdots \mathop{\mathbf{S}}_{s_n,\mathbf{r}_n} M_p(\mathbf{x}, \mathbf{x}_2, \mathbf{x}_3, \ldots, \mathbf{x}_n) M^*_{p'}(\mathbf{x}', \mathbf{x}_2, \mathbf{x}_3, \ldots, \mathbf{x}_n)$$

with $M_p(\mathbf{x}_1, \mathbf{x}_2, \ldots, \mathbf{x}_n)$ the wave function of $|M_p\rangle$.

The energy of the denominator of (4.17) is the quantum average energy per microsystem of the ensemble. It was shown in Section II.G to be a nonconservative quantity. However, the $|A_k\rangle$ are equally populated at equilibrium, so its variation is of the order of the dispersion, $4|\Delta\omega|$, of the stationary states about the central frequency ω. Usually,

$|\Delta\omega| \ll \omega_0$, so

$$\begin{aligned}\mathcal{M}_1(t) &= \sum_{p=1}^{N}\sum_{p'=1}^{N} \rho_{pp'}(t) \int d\mathbf{r}\{\hat{\mathbf{r}}\delta_{pp'}(\mathbf{r},\mathbf{r}')\}_{\mathbf{r}'=\mathbf{r}} \\ &+ \sum_{p=1}^{N}\sum_{p'=1}^{N} \{\rho_{p+1p'}(t)(1-\delta_{pN}) + \rho_{p-1p'}(t)(1-\delta_{p1}) \\ &+ \rho_{pp'+1}(t)(1-\delta_{p'N}) + \rho_{pp'-1}(t)(1-\delta_{p'1})\} \int d\mathbf{r}\{\hat{\mathbf{r}}\delta_{pp'}(\mathbf{r},\mathbf{r}')\}_{\mathbf{r}'=\mathbf{r}}\end{aligned}$$

is a very good approximation of $\mathcal{M}_1(t)$.

Further simplification results from neglecting differential overlap between sites whether they are close together or far apart. It is generally very small if they are well localized.[150] Were the overlap exactly nil (molecules in distinct domains), then the $\delta_{pp'}(\mathbf{r},\mathbf{r}')$ would vanish identically. As $\mathbf{r}$ is a local operator, so would $\int d\mathbf{r}\{\mathbf{r}\delta_{pp'}(\mathbf{r},\mathbf{r}')\}_{\mathbf{r}=\mathbf{r}'}$, $\forall p' \neq p$. The preceding approximation is not essential. It leads to a first moment of the form:

$$\mathcal{M}_1(t) = \sum_{p=1}^{N} \rho_{pp}(t) \int d\mathbf{r}\{\hat{\mathbf{r}}\delta_{pp}(\mathbf{r},\mathbf{r}')\}_{\mathbf{r}'=\mathbf{r}} \tag{4.20}$$

Thus, in the case of a narrow exciton band ($|\Delta\omega| \ll \omega_0$) and sites well localized relative to one another (these conditions may be interdependent), the first moment depends explicitly only on the diagonal elements of the density operator on the local basis, in agreement with earlier definitions.[61–64,15] Naturally this is no justification of generally privileging the local basis, which is used merely as a matter of convenience. We could just as well have used the stationary states $\{|A_k\rangle\}_k$ in which case the first moment would have contained their coherences $|\langle A_k|\hat{\rho}(t)|A_{k'}\rangle|$. The second moments are

$$\mathcal{M}_{2,uv}(t) = \sum_{p=1}^{N} \rho_{pp}(t) \int d\mathbf{r}\{\hat{u}\hat{v}\delta_{pp}(\mathbf{r},\mathbf{r}')\}_{\mathbf{r}'=\mathbf{r}} \tag{4.21}$$

The spatial integrals in (4.20) and (4.21) depend on the configuration of the chain and on the internal (electronic) structure of each site. We discuss next an ordinary, linear chain simulating tetrachlorobenzene and a double, linear chain simulating α-dibromonaphthalene. For a single chain

$$\int d\mathbf{r}\{\hat{\mathbf{r}}\delta_{pp}(\mathbf{r},\mathbf{r}')\}_{\mathbf{r}'=\mathbf{r}} = \mathbf{R}_0 + p\mathbf{R} + \delta\mathbf{R}_p \tag{4.22}$$

$\mathbf{R}_0$ depends on the arbitrary origin of the coordinate system, $\mathbf{R}$ is the displacement vector between neighboring sites and $\delta\mathbf{R}_p$ is the electronic center of mass of the pth molecule relative to its origin. Well-localized sites obey $\|\delta \mathbf{R}_p\| \ll \|\mathbf{R}\|$, although this condition could be easily discarded. Similar remarks hold for the squared coordinates in the second moments.

Let the chains be oriented along Ox, successive sites being separated by an interval X. Let Y be the distance along Oy between opposite sites on double chains so that the coordinates of the pth site are

$$\text{Chain } A\text{: } x = \left(p - \frac{N+1}{2}\right)X; \quad y = \tfrac{1}{2}Y$$

$$\text{Chain } B\text{: } x = \left(p - \frac{N+1}{2}\right)X; \quad y = -\tfrac{1}{2}Y$$

Transfer along Ox will be called longitudinal, that along Oy transversal.

The abscissa of the pth site of an ordinary chain is $(p - (N+1)/2)X$.

It follows from (4.17) to (4.22) that the moments of a double chain are $(\rho_{pq}^{IJ}(t) = \langle M_p^I | \hat{\rho}(t) | M_q^J \rangle)$,

$$\mathcal{M}_{1,x}(t) = X \sum_{p=1}^{N} \left(p - \frac{N+1}{2}\right)\{\rho_{pp}^{AA}(t) + \rho_{pp}^{BB}(t)\}$$

$$\mathcal{M}_{1,y}(t) = \frac{Y}{2} \sum_{p=1}^{N} \{\rho_{pp}^{AA}(t) - \rho_{pp}^{BB}(t)\}$$

$$\mathcal{M}_{2,xx}(t) = X^2 \sum_{p=1}^{N} \left(p - \frac{N+1}{2}\right)^2 \{\rho_{pp}^{AA}(t) + \rho_{pp}^{BB}(t)\}$$

$$\mathcal{M}_{2,xy}(t) = X \frac{Y}{2} \sum_{p=1}^{N} \left(p - \frac{N+1}{2}\right)\{\rho_{pp}^{AA}(t) - \rho_{pp}^{BB}(t)\}$$

$$\mathcal{M}_{2,yy}(t) = \frac{Y^2}{4} \sum_{p=1}^{N} \{\rho_{pp}^{AA}(t) + \rho_{pp}^{BB}(t)\} = \frac{Y^2}{4}$$

The last equation follows from the conservation of the total excited population, conventionally fixed at 1.

These expressions may be derived by differentiation of the generating function, $\rho_0(\phi, \psi; t)$:

$$\rho_0(\phi, \psi; t) = \sum_{b=\pm 1} \sum_{p=1}^{N} e^{i\{b(Y/2)\psi + (p-(N+1)/2)X\phi\}} \langle M_p^I | \hat{\rho}(t) | M_p^I \rangle \tag{4.23}$$

$$\equiv \sum_{b=\pm 1} \sum_{p=1}^{N} \Phi_{bp}(\phi, \psi) \rho_{pp}^{II}(t) \tag{4.24}$$

where $I(b = +1) = A$ and $I(b = -1) = B$. We have, for example,

$$\mathcal{M}_{1,x}(t) = i^{-1} \left\{ \frac{\partial}{\partial \phi} \rho_0(\phi, \psi; t) \right\}_{\phi=\psi=0} \tag{4.25}$$

The essential quantity to be calculated is thus the generating function. We find that it is easier to tackle the problem this way than to find the matrix elements $\langle M_p^I|\hat{\rho}(t)|M_q^J\rangle$ and to sum over p and I.

The fact that the approximations made lead to a distribution of energy depending explicitly only on the $\rho_{pp}^{II}(t)$, in a certain way renews their meaning as probabilities. Moreover, they appear only in discrete sums. The entropy tensor S_{uv} whose elements are entropy functions similar to those previously introduced (Section IV.A, (4.22)) can be used as a measure of delocalization. For the same reasons as were given then, it is a better measure of dispersion than the tensor Δ_{uv}^2 because it takes account of all the moments, (therefore, all the higher correlations), whereas Δ_{uv}^2 takes into account only the first two.

E. Transfer in the Dimer[97,140]

The generating function of a dimer along Ox is (see (4.23))

$$\begin{aligned}\rho_0(\phi; t) &= \sum_{p=1}^{2} e^{i(p-3/2)X\phi}\langle M_p|\hat{\rho}(t)|M_p\rangle \\ &= e^{-i(\phi/2)X}\langle M_1|\hat{\rho}(t)|M_1\rangle + e^{i(\phi/2)X}\langle M_2|\hat{\rho}(t)|M_2\rangle \\ &\equiv e^{-i(\phi/2)X}\,\rho_{11}(t) + e^{i(\phi/2)X}\,\rho_{22}(t)\end{aligned} \tag{4.26}$$

We integrate the evolution equation of this function in Section IV.E.1 and discuss the consequences in Section IV.E.2.

1. The Generating Function

The equation for $\rho_0(\phi; t)$ follows from the projection $\hat{H}^*$ of the macroscopic Hamiltonian $\hat{H}$ on the transfer space $\mathscr{A}^*$:

$$\begin{aligned}i\hbar\frac{\partial}{\partial t}\rho_0(\phi; t) &= -4\hbar\gamma_1 \sin\frac{\phi}{2}X(\langle M_1|\hat{\rho}(t)|M_1\rangle - \langle M_2|\hat{\rho}(t)|M_2\rangle) \\ &\quad + 4i\hbar\Delta\omega\,\frac{\langle M_1|\hat{\rho}(t)|M_2\rangle - \langle M_2|\hat{\rho}(t)|M_1\rangle}{2}\sin\frac{\phi}{2}X\end{aligned}$$

It follows from (4.26) that

$$\langle M_p|\hat{\rho}(t)|M_p\rangle = \frac{X}{2\pi}\int_0^{2\pi/X} e^{-i(p-3/2)\phi X}\rho_0(\phi; t)\,d\phi$$

Writing $\rho_1(t) = \frac{1}{2}(-\langle M_2|\hat{\rho}(t)|M_1\rangle + \langle M_1|\hat{\rho}(t)|M_2\rangle)$, we have

$$\begin{aligned}\frac{\partial}{\partial t}\rho_0(\phi; t) &= -\frac{4\gamma_1}{\pi}X\int_0^{2\pi/X}\sin\frac{\phi}{2}X\,\sin\frac{\phi'}{2}X\rho_0(\phi'; t)d\phi' \\ &\quad + 4\Delta\omega\,\sin\frac{\phi}{2}X\rho_1(t)\end{aligned} \tag{4.27}$$

$\rho_1(t)$ obeys the differential equation

$$\frac{d}{dt}\rho_1(t) = -2(\gamma_0+2\gamma_1)\rho_1(t) - \frac{\Delta\omega}{\pi}X\int_0^{2\pi/X}\sin\frac{\phi'}{2}X\rho_0(\phi';t)\,d\phi' \tag{4.28}$$

Equations 4.27 and 4.28 can be solved analytically. Let $\bar{\rho}_0(\phi;s)$ and $\bar{\rho}_1(s)$ be the Laplace transforms of $\rho_0(\phi;t)$ and $\rho_1(t)$. As before, we illustrate the movement as clearly as possible, by supposing $\rho_1(t=0)=0$. Then,

$$\bar{\rho}_0(\phi;s) = \frac{\rho_0(\phi;0)}{s} + \frac{4\Delta\omega}{s}\sin\frac{\phi}{2}X\bar{\rho}_1(s) - \frac{4\gamma_1}{\pi s}X\int_0^{2\pi/X}\sin\frac{\phi}{2}X\sin\frac{\phi'}{2}X\bar{\rho}_0(\phi';s)\,d\phi'$$

$$\bar{\rho}_1(s) = -\frac{2}{s}(\gamma_0+2\gamma_1)\bar{\rho}_1(s) - \frac{\Delta\omega}{\pi s}X\int_0^{2\pi/X}\sin\frac{\phi'}{2}X\bar{\rho}_0(\phi';s)\,d\phi'$$

Eliminating $\bar{\rho}_1(s)$ in one of these equations, we have

$$\bar{\rho}_0(\phi;s) = \frac{\rho_0(\phi;0)}{s} - \frac{4}{\pi s}X\sin\frac{\phi}{2}X\left\{\frac{\Delta\omega^2}{s+2(\gamma_0+2\gamma_1)} + \gamma_1\right\}\int_0^{2\pi/X}\sin\frac{\phi'}{2}X\bar{\rho}_0(\phi';s)\,d\phi' \tag{4.29}$$

This second kind of Fredholm equation may be solved by the usual methods.[152] Let

$$I(s) = \int_0^{2\pi/X}\sin\frac{\phi'}{2}X\bar{\rho}_0(\phi';s)\,d\phi'$$

$$\bar{\rho}_0(\phi;s) = \frac{\rho_0(\phi;0)}{s} - \frac{4}{\pi s}X\sin\frac{\phi}{2}X\left\{\frac{\Delta\omega^2}{s+2(\gamma_0+2\gamma_1)} + \gamma_1\right\}I(s) \tag{4.30}$$

By (4.29),

$$I(s) = \frac{\displaystyle\int_0^{2\pi/X}\sin\frac{\phi'}{2}X\frac{\rho_0(\phi';0)}{s}\,d\phi'}{\displaystyle 1+\frac{4}{\pi s}X\left\{\frac{\Delta\omega^2}{s+2(\gamma_0+2\gamma_1)}+\gamma_1\right\}\int_0^{2\pi/X}\sin^2\left(\frac{\phi'}{2}X\right)d\phi'}$$

Simple manipulations lead to

$$I(s) = -i\pi\{\langle M_1|\hat{\rho}(0)|M_1\rangle - \langle M_2|\hat{\rho}(0)|M_2\rangle\} \times \frac{s+2(\gamma_0+2\gamma_1)}{s^2+2s(\gamma_0+4\gamma_1)+4\Delta\omega^2+8\gamma_1(\gamma_0+2\gamma_1)}$$

If the initial state is $\hat{\rho}(0) = |M_1\rangle\langle M_1|$, then $\langle M_p|\hat{\rho}(0)|M_p\rangle = \delta_{p1}$ and

$\bar{\rho}_0(\phi; s)$ is of the form

$$\bar{\rho}_0(\phi; s) = \frac{e^{-i(\phi/2)X}}{s} + \frac{4i}{s} \sin \frac{\phi}{2} X \frac{\Delta\omega^2 + \gamma_1\{s + 2(\gamma_0 + 2\gamma_1)\}}{s^2 + 2s(\gamma_0 + 4\gamma_1) + 4\Delta\omega^2 + 8\gamma_1(\gamma_0 + 2\gamma_1)}$$

Taking the inverse Laplace transform,

$$\rho_0(\phi; t) = \rho_0(\phi; 0) + i \sin \frac{\phi}{2} X \left\{1 - e^{-(\gamma_0+4\gamma_1)t} \frac{\sin(2\Delta\omega t \sin \alpha + \alpha)}{\sin \alpha}\right\} \tag{4.31}$$

α lies between 0 and $\pi/2$ for $\cos \alpha = \gamma_0/2|\Delta\omega| < 1$. If $\cos \alpha > 1$, α is of the form $i\alpha''$ where α'' is a positive real number. Since $2|\Delta\omega| \sin \alpha = \gamma_0 \tan \alpha$, the real number expression of (4.31) when $\cos \alpha > 1$ is ($\alpha = i\alpha''$)

$$\rho_0(\phi; t) = \rho_0(\phi; 0) + i \sin\frac{\phi}{2} X \left\{1 - e^{-(\gamma_0+4\gamma_1)t} \frac{\sinh(2\Delta\omega t \sinh \alpha'' + \alpha'' + \alpha'')}{\sinh \alpha''}\right\} \tag{4.32}$$

A further transformation, using γ_0 then $\alpha'' = 2|\Delta\omega| \sinh \alpha''$, yields an expression containing only γ_0 and γ_1.

Section IV.E.2 is based on the preceding expressions. We examine also the entropy function $S_{xx}(t)$:

$$S_{xx}(t) = -\frac{1}{\log 2} \sum_{p=1}^{2} \rho_{pp}(t) \log \rho_{pp}(t) \tag{4.33}$$

The $\rho_{pp}(t)$ in terms of the generating function are:

$$\rho_{pp}(t) = \frac{X}{2\pi} \int_0^{2\pi/x} e^{-i(p-3/2)\phi X} \rho_0(\phi; t)\, d\phi \tag{4.34}$$

2. Movement of the Exciton.

Repeated derivation of the generating function at $\phi = 0$ yields

$$\mathcal{M}_{1,x}(t) = -\frac{X}{2} e^{-(\gamma_0+4\gamma_1)t} \frac{\sin(2\Delta\omega t \sin \alpha + \alpha)}{\sin \alpha} \tag{4.35}$$

$$\mathcal{M}_{2,xx}(t) = \frac{X^2}{4}$$

whence

$$\Delta^2_{xx}(t) = \frac{X^2}{4}\left\{1 - e^{-2(\gamma_0+4\gamma_1)t} \frac{\sin^2(2\Delta\omega t \sin \alpha + \alpha)}{\sin^2 \alpha}\right\} \tag{4.36}$$

The expression of $\mathcal{M}_{1,x}(t)$ when $\rho_{pp}(t = 0) = \delta_{p2}$ follows from (4.35) by changing the sign. $\Delta^2_{xx}(t)$ is, of course, unchanged. Just after $t = 0$,

$$\mathcal{M}_{1,x}(t=0_+) \simeq 2\gamma_1 Xt$$
$$\Delta^2_{xx}(t=0_+) \simeq 2\gamma_1 X^2 t \equiv 2Dt \tag{4.37}$$

The "diffusion" constant $D = \gamma_1 X^2$ is comparable to the corresponding result in an infinite crystal.[63,64] This kind of comparison is valid only close to $t = 0$, since by definition the moments in an infinite crystal are unbounded, whereas they are perforce bounded in a finite chain.

Various cases may occur depending on the value of $\cos\alpha$ relative to 1.

a. $\alpha = \pi/2$ ($\gamma_0 = \gamma_1 = 0$). This case is generally called the completely coherent case, for there are no stochastic interactions to destroy the initial coherence. During a time short compared to the life of the excited state,

$$\mathcal{M}_{1,x}(t) = -\frac{X}{2}\cos 2\Delta\omega t$$
$$\Delta^2_{xx}(t) = \frac{X^2}{4}\sin^2 2\Delta\omega t \tag{4.38}$$

The movement in this extreme case is exactly periodic, of period $\pi/|\Delta\omega|$.

$$\mathcal{M}_{1,x}(t_n) = (-1)^{n+1}\frac{X}{2} \qquad \Delta^2_{xx}(t_n) = 0 \qquad t_n = n\frac{\pi}{2|\Delta\omega|}$$

An image of this is a "particle" in simple harmonic motion at the frequency $2|\Delta\omega|$ between the two sites, shrinking each time it visits a site. We find later that the shrinking is associated not with the sites generally, but with the ends of the chain. Shrinking on *every* site is specific to the dimer.

b. $\pi/2 > \alpha > 0 \Leftrightarrow 0 < \gamma_0 < 2|\Delta\omega|$ (Figs. 21 and 22). This case is called coherent. After $t = 0$, $\mathcal{M}_{1,x}(t)$ never attains the values $\pm X/2$, but lies between the exponentials $\pm X/2e^{-(\gamma_0+4\gamma_1)t}$. The exciton is never on average entirely on one site. Its distances of closest approach, which increase with time, occur at times t_k:

$$\mathrm{tg}(\gamma_0 t_k tg\alpha + \alpha) = \frac{\gamma_0}{\gamma_0 + 4\gamma_1}\,\mathrm{tg}\alpha$$

The first t_k is approximately $\pi/\gamma_0 \tan\alpha$ if $\gamma_1 \ll \gamma_0$.

$\Delta^2_{xx}(t)$ and $S_{xx}(t)$ pass through minima at the times t_k. The minima increase with time. When $t \gg (\gamma_0 + 4\gamma_1)^{-1}$ the exciton lies half way

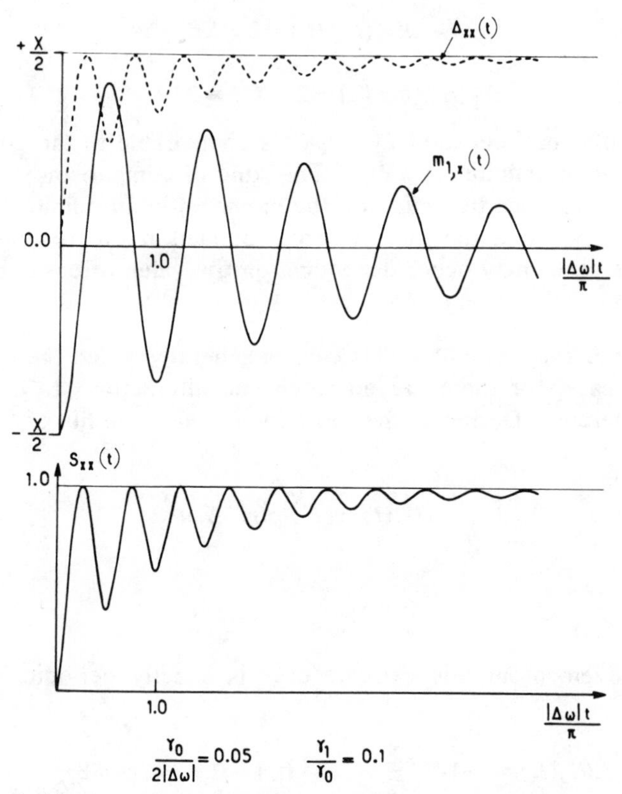

Fig. 21. Position, spread, and delocalization of a very coherent exciton in a dimer.

between the sites ($\mathcal{M}_{1,x} = 0$), its size is of the order of the distance between them ($\Delta_{xx} = X/2$) and the delocalization entropy is 1. At equilibrium the exciton is completely delocalized and the sample is in a uniform statistical mixture of states $|A_k\rangle$; see the discussion at the end of Section II.G.

It is apparent from Fig. 22 that the oscillations disappear very quickly when $\cos\alpha$ is increased. If $\cos\alpha = 0.5$ there are hardly any in $\Delta_{xx}^2(t)$ and $S_{xx}(t)$. Moreover, as expected, $S_{xx}(t)$ is a better measure of localization than is $\Delta_{xx}(t)$, because although they behave similarly its extrema are more pronounced. The overall movement is a damped oscillation of pseudo-period $\pi/(|\Delta\omega|\sin\alpha)$ or $2\pi/\gamma_0\tan\alpha$, increasing as fluctuations become larger.

c. $\alpha = 0 \Leftrightarrow \gamma_0 = 2|\Delta\omega|$ (Fig. 23). Letting α tend to zero in (4.35) and

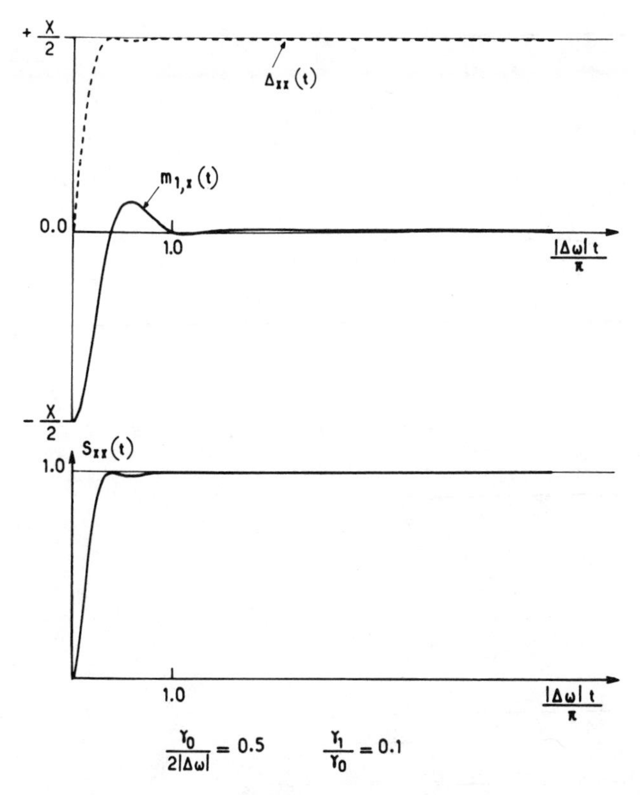

Fig. 22. Same as Fig. 21, coherent case.

(4.36) we find

$$\mathcal{M}_{1,x}(t) = -\frac{X}{2}\, e^{-(\gamma_0+4\gamma_1)t}(1 + 2\Delta\omega t) \equiv -\frac{X}{2}\, e^{-(\gamma_0+4\gamma_1)t}(1 + \gamma_0 t)$$

$$\Delta_{xx}(t) = \frac{X^2}{4}\{1 - e^{-2(\gamma_0+4\gamma_1)t}(1 + 2\Delta\omega t)^2\}$$

$$\equiv \frac{X^2}{4}\{1 - e^{-2(\gamma_0+4\gamma_1)t}(1 + \gamma_0 t)^2\} \tag{4.39}$$

These are monotonic increasing functions of time, with the instant $\pi/\gamma_0 \tan\alpha \to \infty$ and an infinite pseudo-period. $\mathcal{M}_{1,x}(t)$ has a point of inflexion at $t = (\gamma_0 - 4\gamma_1)/\gamma_0(\gamma_0 + 4\gamma_1)$ or roughly $\gamma_0^{-1}(1 - 8\gamma_1/\gamma_0)$ when $\gamma_0 \ll \gamma_1$.

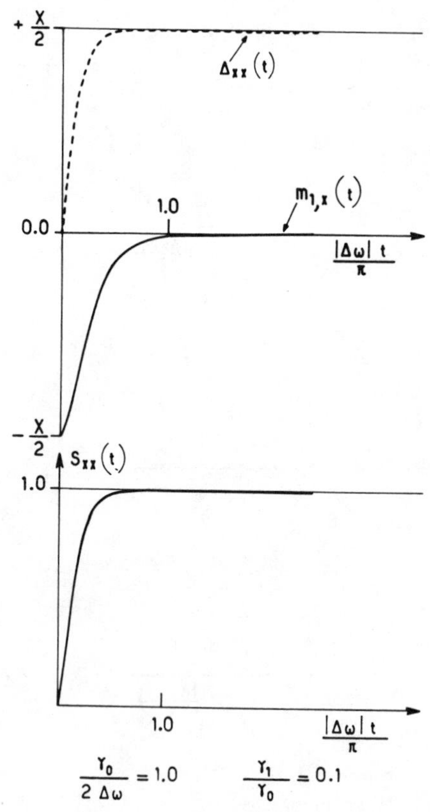

Fig. 23. Same as Fig. 21, intermediate case.

d. $\alpha = i\alpha''$, α'' a positive real number $\Leftrightarrow \gamma_0 \rangle 2\Delta\omega|$ (Figs. 24 and 25). This is the incoherent case, in which none of the quantities have the slightest oscillations or inflections. $\mathcal{M}_{1,x}(t)$ and $\Delta_{xx}(t)$ tend smoothly towards their limits, which correspond to a completely delocalized state. Figure 25 shows the variations of $\mathcal{M}_{1,x}(t)$ and $\Delta_{xx}(t)$ for fixed γ_0 and several values of γ_1. In the incoherent case, the movement is extremely sensitive to small changes in γ_1. Delocalization sets in faster for large values of γ_1.

e. $\alpha \rightarrow i\infty \Leftrightarrow \gamma_0 > 2|\Delta\omega| \rightarrow 0$. This case corresponds to an oriented gas in which the average interaction between sites is nil but the mean square fluctuation is not. It is called the completely incoherent case. The limiting forms of (4.35) and (4.36) are:

$$\mathcal{M}_{1,x}(t) = -\frac{X}{2}\, e^{-4\gamma_1 t}; \qquad \Delta^2_{xx}(t) = \frac{X^2}{4}(1 - e^{-8\gamma_1 t}) \tag{4.40}$$

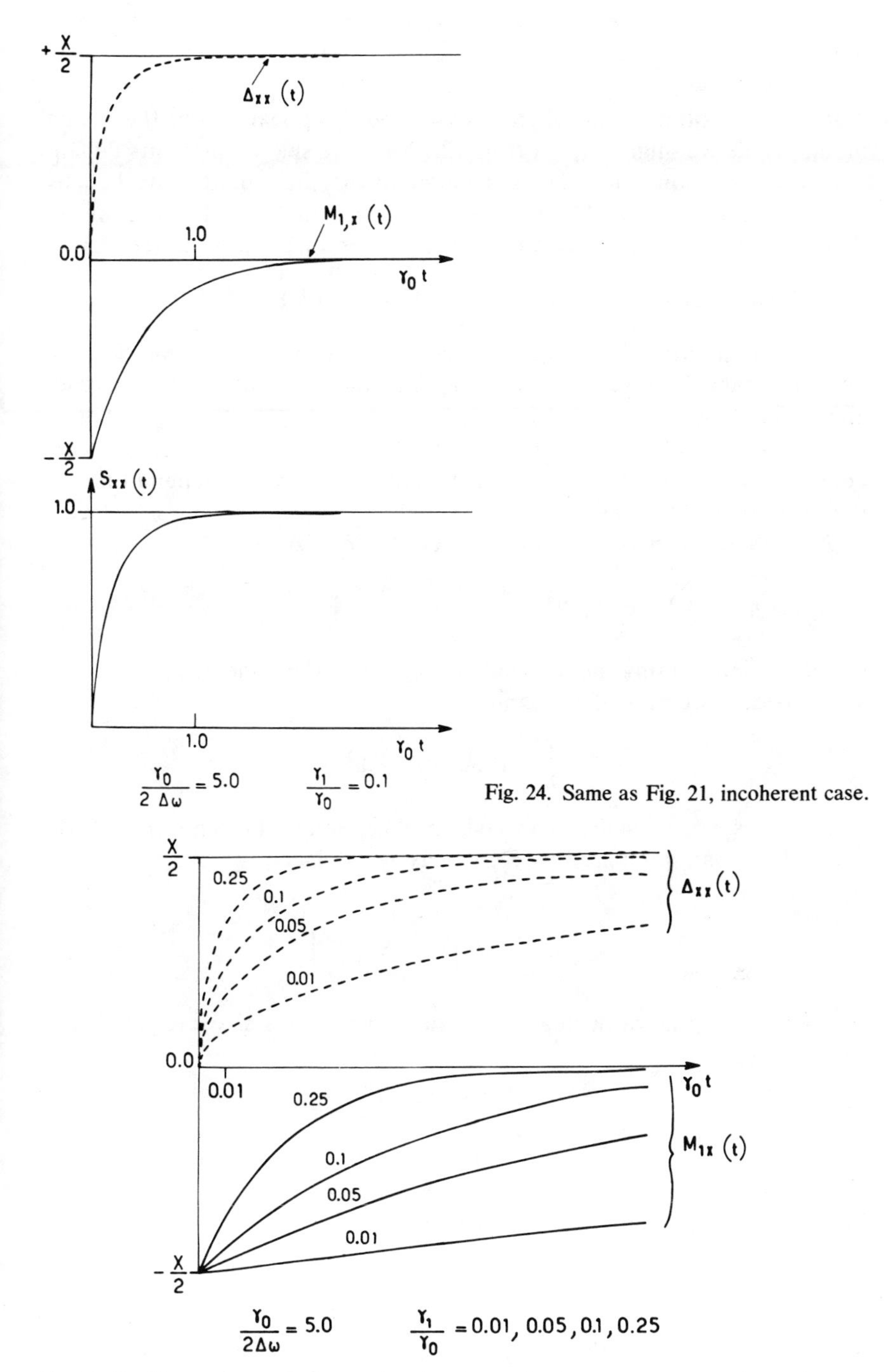

Fig. 24. Same as Fig. 21, incoherent case.

Fig. 25. Position and spread of an exciton in a dimer; incoherent case; for various values of the mean square fluctuation of the interaction between the two sites.

It will be noticed that when $|\Delta\omega| = 0$, γ_0 disappears from the spatial moments, as we might expect from (2.120) if the spatial moments depend explicitly only one the diagonal elements of $\hat{\rho}(t)$ on the local basis. When $\Delta\omega \neq 0$, γ_0 is still important, as it measures the lifetime of the difference of the populations of states $|A_1\rangle$ and $|A_2\rangle$; see matrix (2.121).

$$\langle A_1|\hat{\rho}(t)|A_1\rangle - \langle A_2|\hat{\rho}(t)|A_2\rangle = e^{-2\gamma_0 t}(\langle A_1|\hat{\rho}(0)|A_1\rangle - \langle A_2|\hat{\rho}(0)|A_2\rangle)$$

The initial condition $\hat{\rho}(0) = |M_1\rangle\langle M_1|$ is partly responsible for the "disappearance" of γ_0. The populations of the $|A_k\rangle$'s at $t = 0$ are already equilibrium values.

Generally speaking, γ_0 is the essential factor in damping in the very coherent case whereas, on the other hand, the moments depend on γ_1 in the very incoherent case.

The coherence between $|A_1\rangle$ and $|A_2\rangle$, $\langle A_1|\hat{\rho}(t)|A_2\rangle$ is

$$\langle A_1|\hat{\rho}(t)|A_2\rangle = e^{-(\gamma_0+4\gamma_1)t}\,\frac{\sin(2\Delta\omega t \sin\alpha + \alpha) - i\sin(2\Delta\omega t \sin\alpha)}{\sin\alpha}$$

Before determining the interval during which the coherence is notable, we consider integrals of the kind:

$$\int_0^{t_1} dt|\langle A_1|\hat{\rho}(t)|A_2\rangle|^2$$

where t_1 is very small compared to the excited lifetime, τ^*. If the integral is small compared to τ^*, t_1 may be replaced by $+\infty$. τ_{coh} defined by

$$\tau_{\text{coh}} = \int_0^{+\infty} dt \left|\frac{\langle A_1|\hat{\rho}(t)|A_2\rangle}{\langle A_1|\hat{\rho}(0)|A_2\rangle}\right|^2$$

measures the interval during which the coherence is sizeable relative to its initial value:

$$\tau_{\text{coh}} = \frac{(\gamma_0 + 4\gamma_1)(\gamma_0 + 2\gamma_1) + 2\Delta\omega^2}{4(\gamma_0 + 4\gamma_1)\{2\gamma_1(\gamma_0 + 2\gamma_1) + \Delta\omega^2\}}$$

Let us consider some particular cases, starting with $\gamma_0, \gamma_1 \ll 2|\Delta\omega|$. Developing τ_{coh} locally as a power series,

$$\tau_{\text{coh}} \simeq \frac{1}{2(\gamma_0 + 4\gamma_1)} \gg \frac{1}{|\Delta\omega|}$$

Thus coherence in the completely coherent case is very long compared to $\hbar/V$ where V is the magnitude of interaction between neighboring sites. The intermediate case, $\gamma_0 = 2|\Delta\omega|$, yields

$$\tau_{coh} \simeq \frac{3}{2\gamma_0} = \frac{3}{4|\Delta\omega|}$$

if $\gamma_0 \ll \gamma_1$ as well. The coherence lasts roughly $\hbar/V$. τ_{coh} in the very incoherent case is

$$\tau_{coh} \simeq \frac{1}{8\gamma_1}$$

Once more γ_0 has disappeared. τ_{coh} is inversely proportional to the square fluctuation of the interaction between two sites, which vanishes in a completely coherent movement. τ_{coh} increases with decreasing γ_1 becoming infinite, when $\gamma_1 = 0$, which is hardly surprising since $\langle A_1|\hat{\rho}(t)|A_2\rangle$ is then constant as are the populations of all the $|A_k\rangle$ and the $\rho_{pp}(t)$. The initial state is an equilibrium state for which the averages of all observables are independent of time.

Summing up this section, the general impression is that the image of a particle jumping from site to site is neither indispensable nor ineluctable. We have already pointed out that the statistical model, precisely because it is statistical, is not a model of what is happening in *one* dimer. The "hopping model" is not incompatible with the results of the very incoherent case, for the distribution of the macroscopic equilibrium state could be calculated in two ways:

1. Complete delocalization in each aggregate.
2. Equiprobable localization on one of two possible sites in every aggregate.

Methods *1* and *2* clearly lead to the same spatial moments. However, the hopping model, based on affirmations going beyond the equations, cannot be justified. It was shown at the end of Section II.G to be incompatible with the principles of quantum mechanics. Although it may give a useful average (over the sample) description in certain conditions, its classical characteristics do not apply to specifically quantum phenomena.

F. Transfer in an Ordinary Chain[158]

This section contains analytical solutions of the extreme cases of very coherent and very incoherent conditions; and a numerical solution of the intermediate case.

1. Pure Coherent Transfer[137]

In this case all the squared fluctuations are nil. Using the general expression of $\hat{H}^*$ (projection of the macroscopic "Hamiltonian" defined in (2.51) on the transfer space $\mathscr{A}^*$), we find the following (see (2.115)) differential equation for the matrix elements of the statistical density operator on the local basis, in which $\rho_{pq} = \langle M_p|\hat{\rho}(t)|M_q\rangle$.

$$\frac{d}{dt}|\rho^*(t)) = -2i\Delta\omega \sum_{l=1}^{N}\sum_{m=l}^{N} |X_{lm})(\cos l\theta - \cos m\theta)(X_{lm}|\rho^*(t)) \tag{4.41}$$

In terms of $\hat{X}_{lm} = |A_l\rangle\langle A_m|$, where $|A_k\rangle$ is an eigenvector of $\hat{H}_0$,

$$|\rho^*(t)) = \sum_{l=1}^{N}\sum_{m=1}^{N} |X_{lm})e^{-2i\Delta\omega t(\cos l\theta - \cos m\theta)}(X_{lm}|\rho^*(0))$$

Suppose the excitation is initially on the qth site. Then

$$\hat{\rho}(0) = |M_q\rangle\langle M_q| \Leftrightarrow (Y_{rr'}|\rho^*(0)) = \delta_{rq}\delta_{r'q}$$

Applying the unitary transformation (4.7),

$$\begin{aligned}\rho_{pp}(t) &= \left(\frac{2}{N+1}\right)^2 \sum_{l,l'} \sin lp\theta \sin lq\theta \sin l'p\theta \sin l'q\theta\, e^{-2i\Delta\omega t(\cos l\theta - \cos l'\theta)} \\ &\equiv \left|\frac{2}{N+1}\sum_{l=1}^{N} \sin lp\theta \sin lq\theta\, e^{-2i\Delta\omega t \cos l\theta}\right|^2\end{aligned} \tag{4.42}$$

The generating function is

$$\rho_0(\phi; t) = \sum_{p=1}^{N} e^{i(p-(N+1)/2)\phi X}\, \rho_{pp}(t) \tag{4.43}$$

Combining it with formula (4.42) we may find any moment by differentiation at $\phi = 0$. The entropy function is

$$S_{xx}(t) = \frac{-1}{\log N}\sum_{p=1}^{N} \rho_{pp}(t) \log \rho_{pp}(t) \tag{4.44}$$

It follows from (4.42) that the movement is not generally periodic since $\rho_{pp}(t)$ is a linear combination of harmonic functions of noncommensurable periods. Thus, although there can be no ineversibility (by definition) in this extreme case, exact recurrence cannot occur, except in a chain of $N = 3$ sites, when if $q = 1$ (excitation on one end),

$$\mathscr{M}_{1,x}(t) = -X \cos\sqrt{2}\Delta\omega t; \qquad \Delta_{xx}^2(t) = \frac{X^2}{2}\sin^2\sqrt{2}\Delta\omega t \tag{4.45a}$$

and if $q = 2$ (excitation prepared on the middle site),

$$\mathscr{M}_{1,x}(t) = 0 \forall t; \qquad \Delta_{xx}^2(t) = X^2 \sin^2\sqrt{2}\Delta\omega t \tag{4.45b}$$

The movement in a trimer is thus periodic of period $\pi\sqrt{2}/|\Delta\omega|$. The amplitude of the oscillation $\Delta_{xx}(t)$ is $\sqrt{2}$ times greater for an excitation in the middle of the chain, which "may move off in both directions" than for one prepared on one end of the chain, $\hat{\rho}(0) = |M_1\rangle\langle M_1|$, which cannot. This shows the exciton tends to fill the available space as soon as possible.

Although strict recurrence cannot occur in chains of $N > 3$, we may define a pseudo-period by

$$\mathcal{M}_{1,x}(T^*) \simeq \mathcal{M}_{1,x}(0) \qquad \Delta_{xx}(T^*) \ll X \qquad S_{xx}(T^*) \simeq 0$$

T^* may be determined numerically as follows. Consider times t_n defined by

$$t_n = n \frac{2\pi}{2|\Delta\omega| \cos[N/2]\theta}$$

where n is an integer. Then

$$2|\Delta\omega| t_n \cos k\theta = 2\pi n \frac{\cos k\theta}{\cos[N/2]\theta} \equiv 2\pi x_{kn}$$

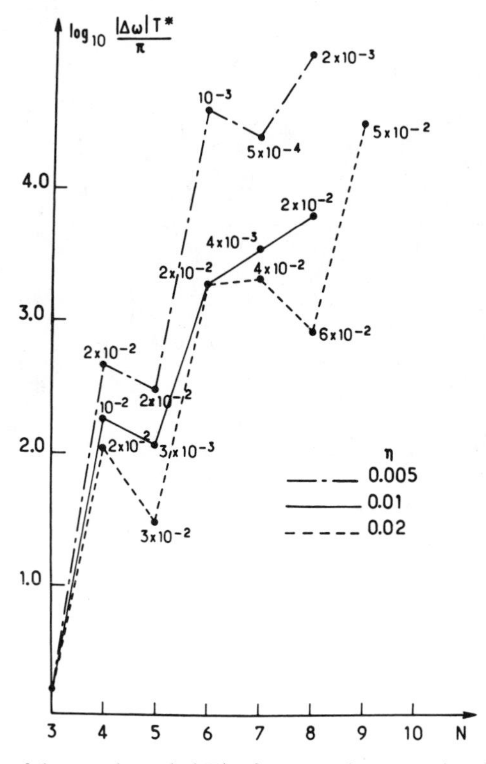

Fig. 26. Logarithm of the pseudo-period T^*, of a pure coherent exciton in an ordinary chain, as a function of the length of the chain. Values by each point are $S_{xx}(T^*)$.

Putting $x_{kn} = [x_{kn}] + n_{kn}$,

$$2|\Delta\omega|t_n \cos k\theta = 2\pi\eta_{kn}$$

to within 2π.

If $\eta_{kn} < \eta$, for all k, for some given η, t_n is the pseudo-period T^*. T^* depends on η. η is to a certain extent arbitrary because it is hard to foresee the error in T^* due to an error of at most η in the phase factors of each term of (4.42). One way around this is to choose T^* such that $S(T^*)$ is very small. Figure 26 shows $\log_{10} T^*$ (T^* in units of $2\pi/2|\Delta\omega|$) in function of N for different values of η. The initial excitation is taken to be on the end of the chain. T^* increases very rapidly with N, for all η such that $S(T^*)$ is small. When $N \sim 10$, $T^* \gg 10^4\ \pi/|\Delta\omega|$ and may be comparable with the excited lifetime, in which case the recurrence may not have time to occur. T^* may be shown to depend on, the site initially excited, q. This is clearly true in (4.45), of the true period of the movement in a trimer.

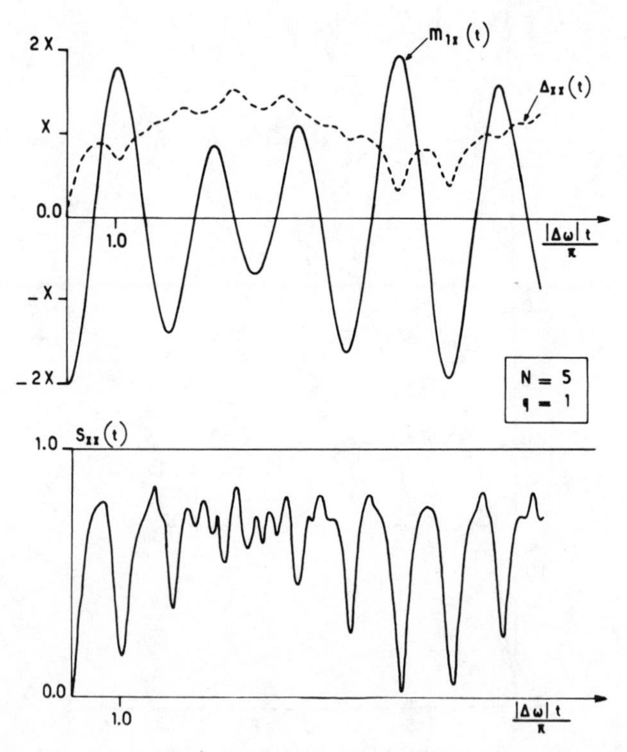

Fig. 27. Position, spread, and delocalization of a pure coherent exciton prepared on one end of a pentamer.

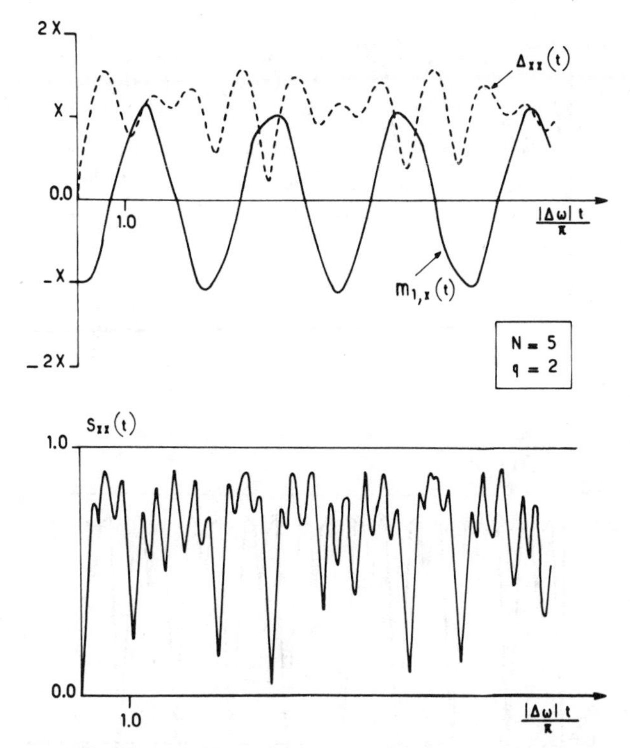

Fig. 28. Same as Fig. 27, but with the initially excited site next to the end of the chain.

Figures 27 to 29 show $\mathcal{M}_{1,x}(t)$, $\Delta_{xx}(t)$ and $S_{xx}(t)$ for $N = 5$ and for each value of q.

We note that contraction occurs only at the ends of the chain and not on every site and that even then it is not complete. This is a very different picture from that of a particle hopping from site to site. In fact, most of the time, the excitation is on average delocalized, and localized (sometimes poorly at that) only when it can go no further and must turn back. The reversing may occur elsewhere besides the end of the chain, see Fig. 27 for example: Even for small N it is wrong to say the exciton by and large goes back and forth between the ends of the chain. It may hesitate in the middle and Δ_{xx} and S_{xx} may be large at such times. In Fig. 28, $N = 5$, $q = 2$, the center of the excitation, $\mathcal{M}_{1,x}(t)$, stays between $\pm X$, so it can reach the ends only when its dispersion Δ_{xx} is X at the same time. $\mathcal{M}_{1,x}(t)$ is constant if the excitation is prepared in the middle of the chain. The exciton "breathes," like a balloon fixed in the middle of the chain, inflated and deflated regularly, so that all the sites are affected.

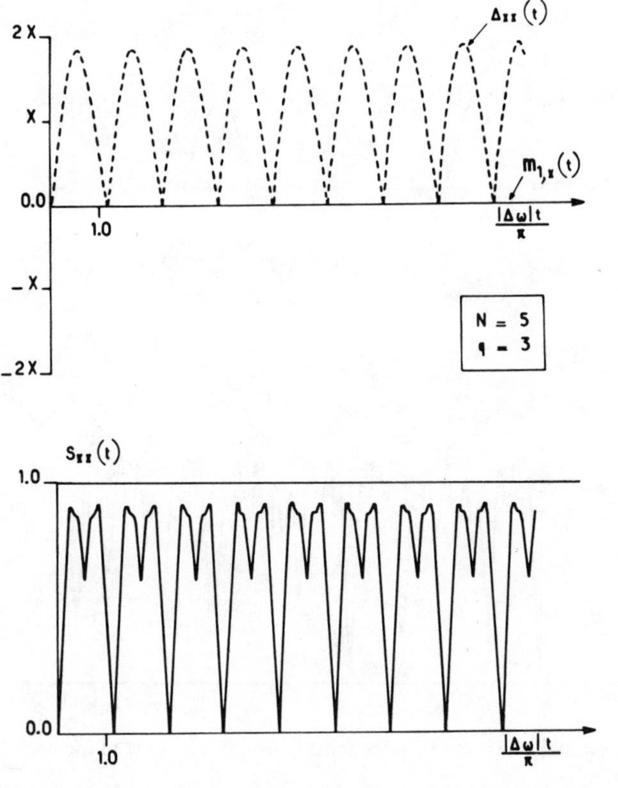

Fig. 29. Same as Fig. 27, but excitation prepared in the middle of the chain.

The second moments tend to be larger when the amplitude of the oscillations of the center of the packet is small so that overall the exciton is more often than not delocalized.

We may better trace the spatial distribution of the energy, defined by

$$\text{Prob}\left\{\left(p-\frac{N+1}{2}-\frac{1}{2}\right)X \leqslant x \leqslant \left(p-\frac{N+1}{2}+\frac{1}{2}\right)X\right\} = \rho_{pp}(t)$$

Figures 30 and 31 show the quantitative evolution of the extent of the exciton as a function of time. The variation is simple in this case, that of a trimer, but should not be extrapolated to larger aggregates.

Figures 32 and 33 illustrate that for large N the oscillations of $\mathcal{M}_{1,x}(t)$ and $\Delta_{xx}(t)$ are larger. The time scale is the same as before. Apart from that, the important points mentioned for $N = 5$ are still true. The initial

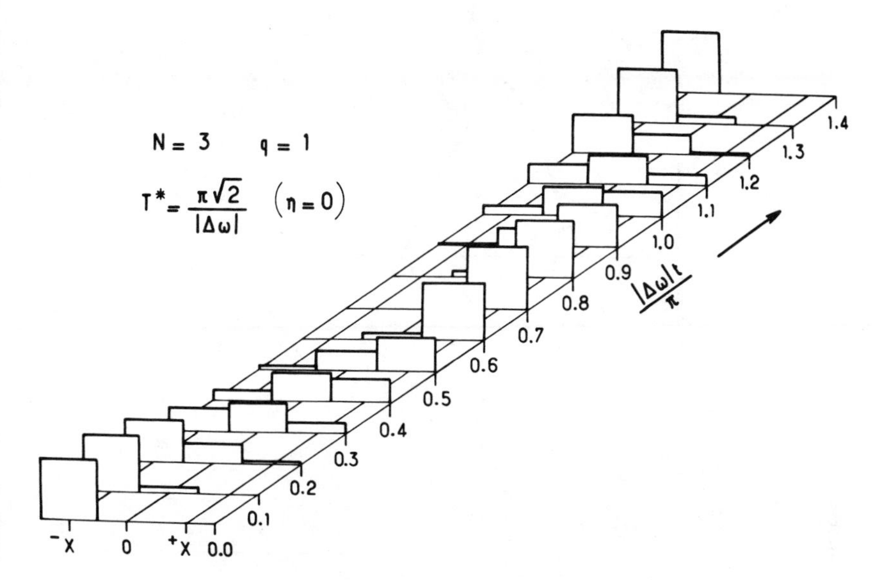

Fig. 30. Spatial distribution of a completely coherent trimer exciton prepared on one end of the chain.

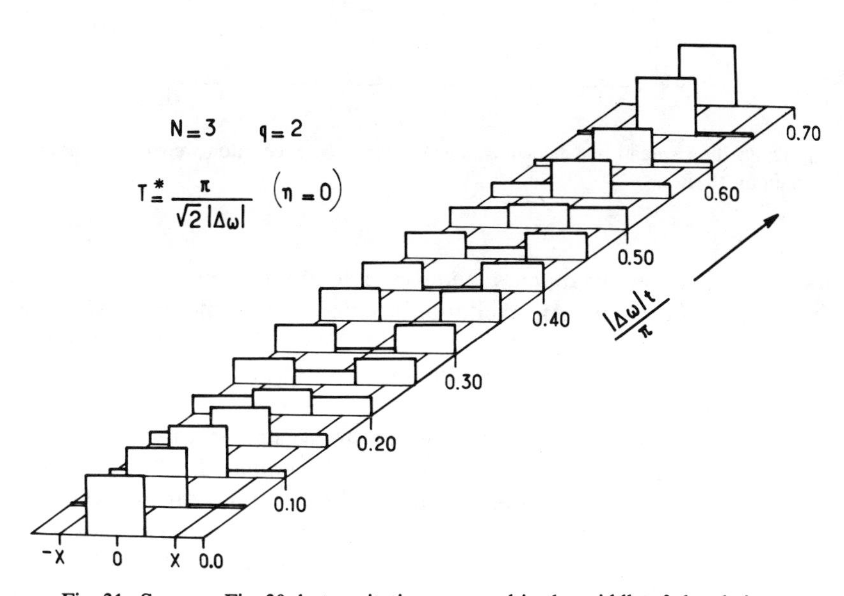

Fig. 31. Same as Fig. 30, but excitation prepared in the middle of the chain.

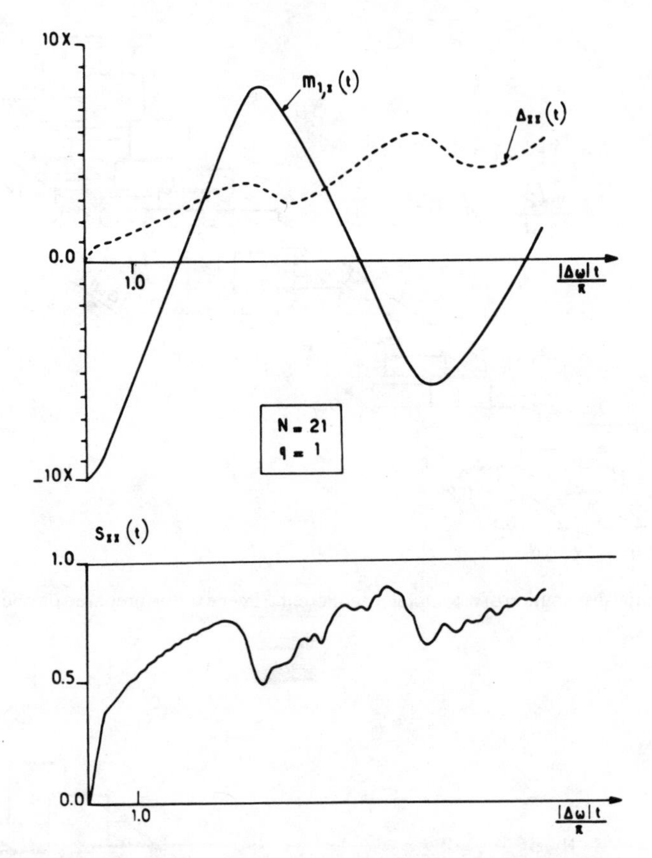

Fig. 32. Position, spread, and delocalization of a pure coherent exciton prepared on one end of a chain of 21 sites.

localization soon disappears and the exciton is spread over a large fraction of the chain. An excitation created in the middle of the chain expands and contracts regularly (see Fig. 33) $q = 11$, though its minimum volume is relatively much larger than it would be in a small chain; see Fig. 27.

Comparison with earlier work on infinite crystals is valid only around $t = 0$. There all moments in our model are bounded, contrary to the infinite chain. The first moment in an infinite chain is constant whereas it is not constant in a finite aggregate except if the initial excitation is in the middle. Everywhere in an infinite chain is "in the middle" and the ends do not exist. This comparison is easier to make as a special case of the general case transfer in Section II.F. 3. However, it is easy to show

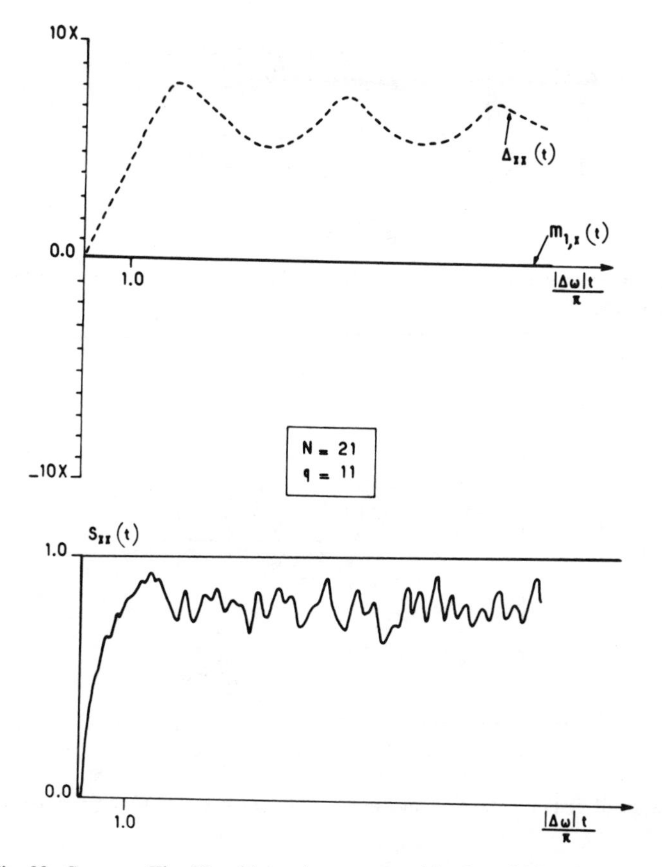

Fig. 33. Same as Fig. 32, with exciton prepared in the middle of the chain.

here that the initial values of the derivatives of all the moments are nil, for all q:

$$\left\{\frac{d}{dt}\rho_{pp}(t)\right\}_{t=0} = -2i\Delta\omega\left(\frac{2}{N+1}\right)^2 \sum_{l,l'} \sin lp\theta \sin l'p\theta \sin lq \sin l'q\theta(\cos l\theta - \cos l'\theta)$$

Arranging the indices l and l' one finds the sum is nil. It follows that the time derivative of the generating function at $t=0$ is nil. The derivative of $\Delta_{xx}(t)$ is not always finite, being discontinuous whenever $\Delta_{xx}(t)=0$. Δ_{xx}^2 increases close to the origin as t^2 at the most, a characteristic of the wavelike[68] movement in the pure coherent case. Figure 34 shows the variation of $\mathcal{M}_{1,x}(\pi/|\Delta\omega|) - \mathcal{M}_{1,x}(0)$ and $\Delta_{xx}(\pi/|\Delta\omega|)$ as functions of N.

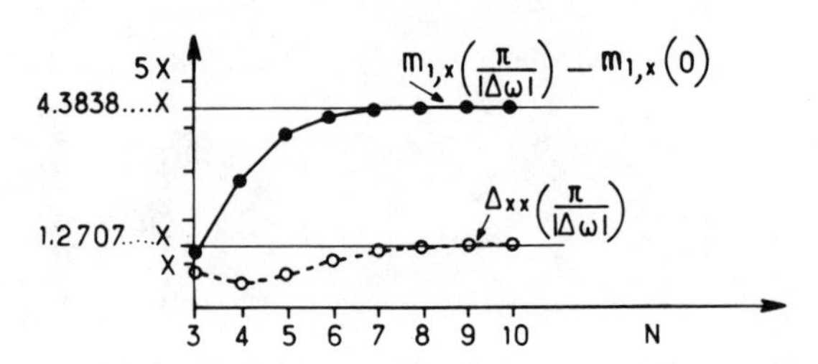

Fig. 34. Distance traveled and spread at time $t = \pi/|\Delta\omega|$ for a very coherent exciton prepared on one end of the chain of length N.

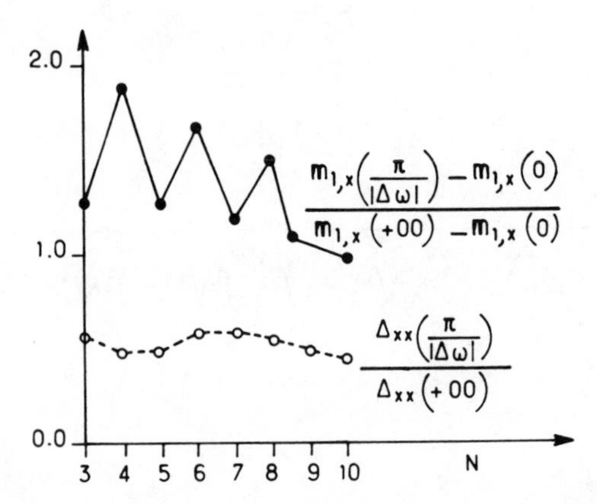

Fig. 35. Ratios of distance traveled and spread at time $t = \pi/|\Delta\omega|$ to their values at $t = +\infty$.

They are independent of N for $N \geqslant 10$. Their relative variations are roughly decreasing functions of the number of sites (Fig. 35).

2. *Completely Incoherent Transfer*

This case is that of a vanishing static interaction, but nonzero quadratic fluctuations measured by γ_1, so that $\hat{H}^* = -i\hbar\hat{\Gamma}^*$. Expanding this operator on the $|Y_{pq})$'s (see (2.115) and (2.85), in which I and J are superfluous for an ordinary chain), we have

$$\hat{H}^* = -i\hbar \sum_{p,q} |Y_{pq})\{2\gamma_0(1-\delta_{pq}) + \gamma_1(4 - \delta_{p1} - \delta_{pN} - \delta_{q1} - \delta_{qN})\}(Y_{pq}|$$

$$+ 2i\hbar\gamma_1 \sum_{p} \sum_{p'} |Y_{pp})(\delta_{p'p+1} + \delta_{p'p-1})(Y_{p'p'}|$$

$$+ 2i\hbar\gamma_1 \sum_{p,q} \sum_{p',q'} |Y_{pq})(\delta_{pq-1}\delta_{p'p+1}\delta_{q'q-1} + \delta_{pq+1}\delta_{p'p-1}\delta_{q'q+1})(Y_{p'q'}|$$

It follows that the diagonal elements $\rho_{pp}(t) = (Y_{pp}|\rho^*(t))$ are coupled only among themselves and not to other $\rho_{pq}(t)$ for $p \neq q$. Supposing the moments depend only on the diagonal elements of $\hat{\rho}(t)$ on the local basis, and that the initial state is localized, the effective space is the subspace of $\mathscr{A}^*$, of dimension N generated by the $\{|Y_{pp})\}_{1 \leq p \leq N}$. Let $\hat{H}^{*\prime}$ be the projection of $\hat{H}^*$ on this space. Then

$$\hat{H}^{*\prime} = -2i\hbar\gamma_1 \sum_p \sum_{p'} |Y_{pp})\{(2 - \delta_{p1} - \delta_{pN})\delta_{pp'} - (\delta_{p'p+1} + \delta_{p'p-1})\}(Y_{p'p'}| \tag{4.46}$$

As for the dimer (Section IV.E.2) in these conditions, γ_0 does not enter into the spatial moments. The effective Hamiltonian is proportional to γ_1, a far cry from the approach to infinite chains in which γ_1 is sometimes taken to be zero.[64]

If $\hat{\rho}(0) = |M_q\rangle\langle M_q|$ the off-diagonal terms of $\hat{\rho}(t)$ are nil for all t, but the same is not true of the $\langle A_k|\hat{\rho}(t)|A_{k'}\rangle$ so that there is some (oriented gas type) coherence for $t < +\infty$. In the present case (see Section II.G)

$$\mathrm{Tr}\{\hat{\rho}^2(t)\} = (\rho(0)|e^{-2\hat{\Gamma}t}|\rho(0)) = (Y_{qq}|e^{-2\hat{\Gamma}t}|Y_{qq})$$

This shows the measure of incoherence $1 - \mathrm{Tr}\{\hat{\rho}^2(t)\}$ depends on the initial state and increases with time.

Introducing an operator $\hat{h}^*$ whose matrix on the basis $|Y_{11}), |Y_{22}), \ldots, |Y_{NN})$ is

$$\begin{matrix}
1 & 1 & 0 & 0 & 0 \cdots 0 & 0 & 0 \\
1 & 0 & 1 & 0 & 0 \cdots 0 & 0 & 0 \\
0 & 1 & 0 & 1 & 0 \cdots 0 & 0 & 0 \\
0 & 0 & 1 & 0 & 1 \cdots 0 & 0 & 0 \\
0 & 0 & 0 & 1 & 0 \cdots 0 & 0 & 0 \\
\vdots & \vdots & \vdots & \vdots & \vdots & \vdots & \vdots \\
0 & 0 & 0 & 0 & 0 \cdots 0 & 1 & 0 \\
0 & 0 & 0 & 0 & 0 \cdots 1 & 0 & 1 \\
0 & 0 & 0 & 0 & 0 \cdots 0 & 1 & 1
\end{matrix}$$

we have

$$\hat{H}^{*\prime} = -4i\hbar\gamma_1 \sum_{p=1}^{N} |Y_{pp})(Y_{pp}| + 2i\hbar\gamma_1\hat{h}^* \tag{4.47}$$

The first operator in this equation is a scalar, so we need only solve the eigenvalue problem of $\hat{h}^*$. The secular equation is

$$|\hat{h}^* - \lambda I^*| \equiv \Delta_N(\lambda) = 0$$

Expanding the determinant along its first line,

$$\Delta_N(\lambda) = (1-\lambda)\left\{C_{N-1}\left(-\frac{\lambda}{2}\right) + C_{N-2}\left(-\frac{\lambda}{2}\right)\right\} - \left\{C_{N-2}\left(-\frac{\lambda}{2}\right) + C_{N-3}\left(-\frac{\lambda}{2}\right)\right\}$$

where the $C_n(z)$ are the Gegenbauer polynomials.[116,117] Using the inductive relation between them,

$$C_n(z) = 2zC_{n-1}(z) + C_{n-2}(z)$$

Then,

$$\Delta_N(\lambda) = 2\left(1 - \frac{\lambda}{2}\right)C_{N-1}\left(-\frac{\lambda}{2}\right)$$

from which the eigenvalues are[116]:

$$\lambda_j = 2\cos j\theta'; \qquad \theta' = \frac{\pi}{N}; \qquad j = 0, 1, 2, \ldots, N-1$$

The eigenvalues of $\hat{H}^*$ are, therefore,

$$-4i\hbar\gamma_1 + 4i\hbar\gamma_1 \cos j\theta' = -8i\hbar\gamma_1 \sin^2 j\frac{\theta'}{2}; \quad 0 \leqslant j \leqslant N-1 \qquad (4.48)$$

The eigenvectors are, (see Section II.I.1):

$$|\Upsilon_j) = \left[\frac{2-\delta_{j0}}{N}\right]^{1/2} \sum_{p=1}^{N} \cos(p - \tfrac{1}{2})j\theta' |Y_{pp}) \qquad (0 \leqslant j \leqslant N-1) \quad (4.49)$$

The inverse transformation is

$$|Y_{pp}) = \sum_{j=0}^{N-1} \left[\frac{2-\delta_{j0}}{N}\right]^{1/2} \cos(p - \tfrac{1}{2})j\theta' |\Upsilon_j) \qquad (1 \leqslant p \leqslant N) \quad (4.50)$$

The eigenvalue 0 corresponds to the eigenvector $\Sigma_{p=1}^{N} |Y_{pp})$ corresponding to the conservation of the total excited population.

Integration of the equation for the projections of $|\rho^*(t))$ on the $\{|\Upsilon_j)\}_{0 \leqslant j \leqslant N-1}$, is now immediate:

$$(\Upsilon_j|\rho^*(t)) = e^{-8\gamma_1 t \sin^2 j(\theta'/2)}(\Upsilon_j|\rho^*(0))$$

Putting $\hat{\rho}(0) = |M_q\rangle\langle M_q|$ and using (4.49) and (4.50) we find the following expression of the generating function (defined by (4.43)):

$$\rho_0(\phi; t) = \frac{\sin N\dfrac{\phi}{2}X}{N\sin\dfrac{\phi}{2}X} - \frac{2}{N} i \sin\frac{\phi}{2} X \sum_{j=1}^{N-1} \frac{e^{-iN(\phi/2)X} - (-1)^j e^{+iN(\phi/2)X}}{\cos\phi X - \cos j\theta'} E_{jq}(t)$$
$$(4.51)$$

TABLE IX
The First Two Moments of the Pure Incoherent Movement ($|\Delta\omega| = 0$) of an Exciton in a Trimer and a Tetramer.

		$q = 1$	$q = 2$
$N = 3$	$\mathcal{M}_{1,x}(t)$	$-Xe^{-2\gamma_1 t}$	0
	$\mathcal{M}_{2,xx}(t)$	$\frac{2}{3}X^2(1 + \frac{1}{2}e^{-6\gamma_1 t})$	$\frac{2}{3}X^2(1 - e^{-6\gamma_1 t})$
$N = 4$	$\mathcal{M}_{1,x}(t)$	$-\frac{X}{2}(2ch2\sqrt{2}\gamma_1 t + 2\sqrt{2}sh2\sqrt{2}\gamma_1 t)e^{-4\gamma_1 t}$	$-\frac{X}{2}(ch2\sqrt{2}\gamma_1 t - \sqrt{2}sh2\sqrt{2}\gamma_1 t)e^{-4\gamma_1 t}$
	$\mathcal{M}_{2,xx}(t)$	$\frac{5}{4}X^2(1 + \frac{4}{5}e^{-4\gamma_1 t})$	$\frac{5}{4}X^2(1 - \frac{4}{5}e^{-4\gamma_1 t})$

where $E_{jq}(t)$ stands for $\cos(j(\theta'/2))\cos[j(q - \frac{1}{2})\theta']\exp[-8\gamma_1 t \sin^2(j(\theta'/2))]$. Repeated derivation at $\phi = 0$ yields

$$\mathcal{M}_{1,x}(t) = -\frac{X}{N}\sum_{j=1}^{N-1}\frac{1-(-1)^j}{2}\frac{E_{jq}(t)}{\sin^2 j\dfrac{\theta'}{2}} \tag{4.52}$$

$$\mathcal{M}_{2,xx}(t) = \frac{N^2-1}{12}X^2 + X^2\sum_{j=1}^{N-1}\frac{1+(-1)^j}{2}\frac{E_{jq}(t)}{\sin^2 j\dfrac{\theta'}{2}} \tag{4.53}$$

Table IX shows the first two moments for $N = 3, 4$ and $q = 1, 2$. The general formulae show that the moments are linear combinations of exponentials whose time constants $[8\gamma_1 \sin^2(j(\theta'/2))^{-1} \equiv \tau_j$, decrease as j increases. The coefficients of the exponentials are very complicated functions of j and depend too on the site initially excited, q. If $q = 1$, they are $\cot^2(j(\theta'/2))$, a rapidly decreasing function of j. For $N = 5$, the ratio of the two first nonzero terms of $\mathcal{M}_{1,x}(t)$ is $[\cot(\pi/10)/\cot(3\pi/10)]^2 \simeq 18$.

$\mathcal{M}_{1,x}(t)$, $\Delta_{xx}(t)$, and $S_{xx}(t)$ for $N = 5$ and $N = 10$ are shown in Figs. 36 and 37. The curves are indexed by q the site initially excited. The preceding quantities tend smoothly towards their limiting values, the convergence being faster when the site initially excited is near the center of the chain. $(d/dt)\mathcal{M}_{1,x}(t = 0) = 0$ except for $q = 1$ or N. This is proved as a special case of the general transfer problem, Section IV.F.3. The distance traveled between $t = 0$ and $t = \gamma_1^{-1}$ is nearly independent of N.

$$\mathcal{M}_{1,x}(\gamma_1^{-1}) - \mathcal{M}_{1,x}(0) = \frac{X}{N}\sum_{j=1}^{N-1}\frac{1-(-1)^j}{2}\frac{E_{jq}(0) - E_{jq}(\gamma_1^{-1})}{\sin^2 j\dfrac{\theta'}{2}} \tag{4.54}$$

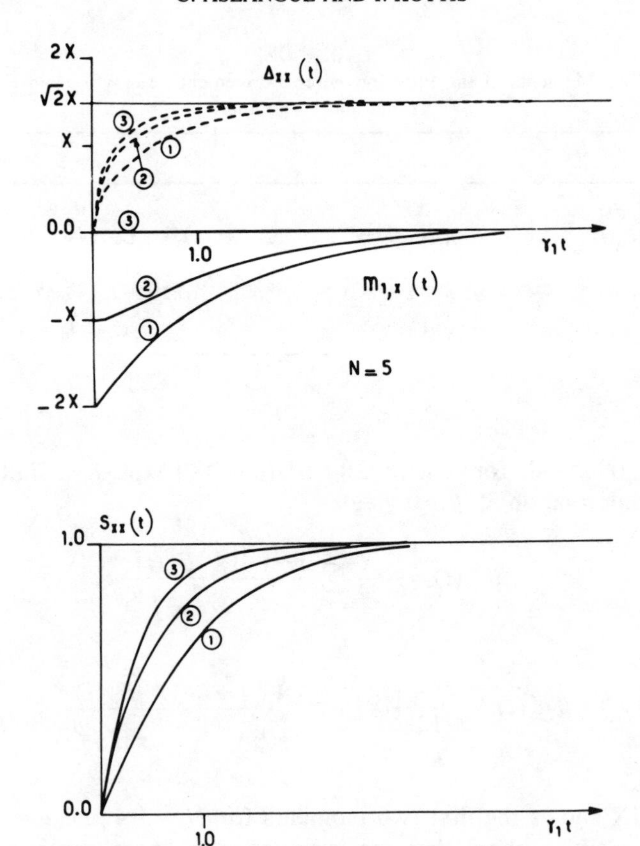

Fig. 36. Position, spread, and delocalization of a very incoherent exciton prepared in a pentamer. The various circled figures refer to the site initially excited.

The asymptotic value 1.1465 is attained for $N \sim 10$; see Fig. 38. The average speed, $\gamma_1 X$, between these times, is roughly independent of N, and independent when $N > 10$. Correspondingly, the portion of chain covered in the same time decreases rapidly with N (Fig. 39).

The derivatives at the origin follow easily from (4.46) ($\hat{\rho}(0) = |M_q\rangle\langle M_q|$):

$$\left\{\frac{\partial}{\partial t}\rho_0(\phi; t)\right\}_{t=0} = -8\gamma_1 \sin^2\left(\frac{\phi}{2}X\right) e^{i(q-(N+1)/2)\phi X}$$
$$+ 4i\gamma_1 \sin\frac{\phi}{2} X(e^{-iN(\phi/2)X}\delta_{q1} - e^{+iN(\phi/2)X}\delta_{qN})$$

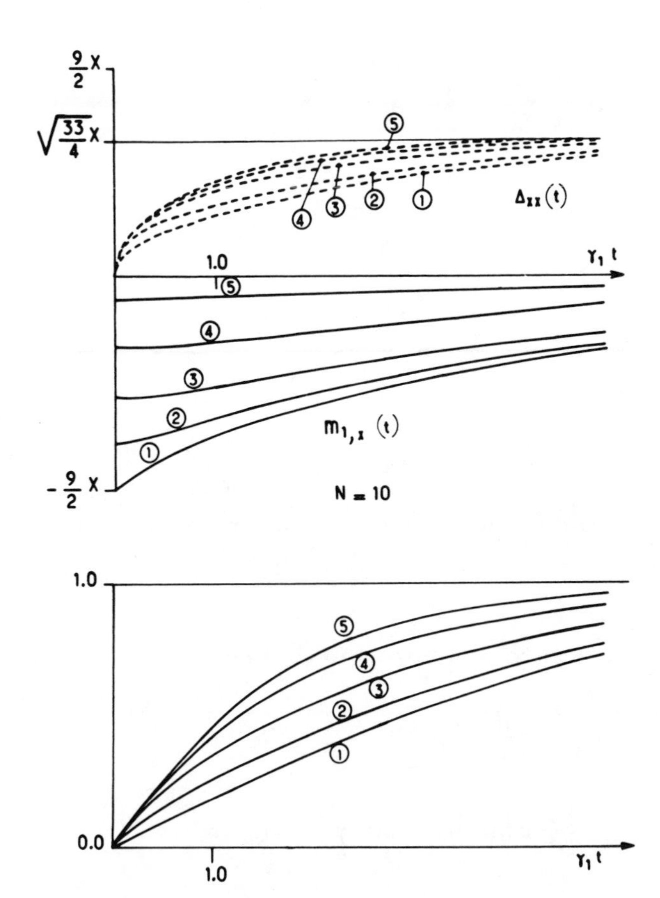

Fig. 37. Same as Fig. 36, for a decamer.

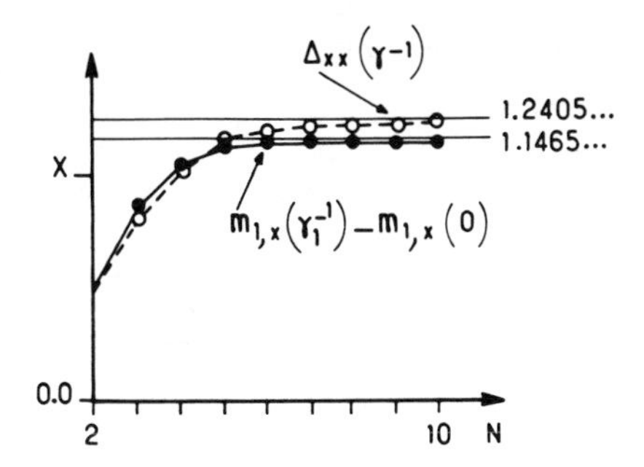

Fig. 38. Same as Fig. 34, for a very incoherent exciton movement, at time $t = 1/\gamma_1$.

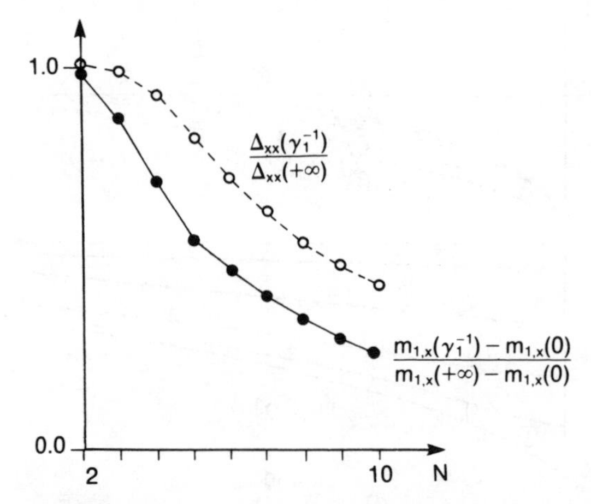

Fig. 39. Same as Fig. 35, for a very incoherent exciton movement, at time $t = 1/\gamma_1$.

Repeated derivation with respect to ϕ yields

$$\left\{\frac{d}{dt}\mathcal{M}_{1,x}(t)\right\}_{t=0} = 2X\gamma_1(\delta_{q1} - \delta_{qN}) \tag{4.55}$$

$$\left\{\frac{d}{dt}\Delta^2_{xx}(t)\right\}_{t=0} = 2X^2\gamma_1(2 - \delta_{q1} - \delta_{qN}) \equiv 2D \tag{4.56}$$

Thus the derivative of the first moment is nil except if the excited site q is on one end of the chain. The "diffusion" constant D depends on q too; For $q \neq 1$ and N the exciton spreads quickly because nothing confines it. On the other hand if $q = 1$ or N, the initial dispersion increases twice as slowly as before because the excitation can propagate to one side only. Thus to the first order in t the behavior of an excitation prepared away from the ends of the chain is the same as in an infinite crystal[63].

3. *Transfer in the General Case*

Owing to the complexity of the expression of $\hat{H}^*$ (2.115) we are unable to find an analytical solution of the general transfer problem. Two numerical solutions are possible: either numerical diagonalization of $\hat{H}^*$ to find the exponential evolution operator, or direct numerical integration of the equation for the generating function $\rho_0(\phi; t)$ of the spatial moments. Now the matrix of $\hat{H}^*$ is complex, symmetric and of order N^2 so that there are likely to be serious difficulties about the convergence and accuracy of the diagonalization for large N. The other method gives no such trouble at all. We outline in this section the derivation of the differential equation for the generating function. It is in fact an integro-differential equation because of the finite length of the chains.

With the usual notation ($\langle M_p|\hat{\rho}(t)|M_q\rangle = \rho_{pq}(t)$), equation (4.43) for $\rho_0(\phi; t)$ and expression (2.115) of $\hat{H}^*$, we have

$$\begin{aligned}
\frac{\partial}{\partial t}\rho_0(\phi; t) = e^{-i(N+1)/2\phi X}\sum_{p=1}^{N} e^{ip\phi X}\{&-2\gamma_1(2-\delta_{p1}-\delta_{pN})\rho_{pp}(t)\\
&+2\gamma_1(\rho_{p+1p+1}(t)+\rho_{p-1p-1}(t))\\
&-i\Delta\omega(\rho_{p+1p}(t)+\rho_{p-1p}(t)-\rho_{pp+1}(t)-\rho_{pp-1}(t))\}\\
= -8\gamma_1 \sin^2\left(\frac{\phi}{2}X\right)&\rho_0(\phi; t)\\
+4i\gamma_1 \sin\frac{\phi}{2}X\{&e^{-iN(\phi/2)X}\rho_{11}(t) - e^{+iN(\phi/2)X}\rho_{NN}(t)\}\\
+2\Delta\omega \sin\frac{\phi}{2}X&\sum_{p=1}^{N-1} e^{i(p-N/2)\phi X}\{-\rho_{p+1p}(t)+\rho_{pp+1}(t)\}
\end{aligned} \tag{4.57}$$

In (4.57) $\rho_0(\phi; t)$ is coupled by $\Delta\omega$ to a sort of generating function built on the $\rho_{pq}(t)$, $|p-q| = 1$. Because of the particular form of $\hat{H}^*$, $-\rho_{p+1p}(t)+\rho_{pp+1}(t)$ is in turn coupled to $\rho_{p+2p}(t)+\rho_{pp+2}(t)$ and so on. We must therefore examine the functions $\rho_r(\phi; t)$, defined by

$$\rho_r(\phi; t) = \frac{1}{2}\sum_{p=1}^{N-r} e^{i(p-(N-r+1)/2)\phi X}\{(-1)^r\rho_{p+rp}(t)+\rho_{pp+r}(t)\} \tag{4.58}$$

The factor $\frac{1}{2}$ is there so we may identify $\rho_0(\phi; t)$ with $\rho_r(\phi; t)$ for $r = 0$. r has N values: $0, 1, \ldots, N-1$.

We next notice that $\rho_0(\phi; t)$ depends explicitly on $\rho_{11}(t)$ and $\rho_{NN}(t)$ the ends of the diagonal of $\hat{\rho}(t)$ on the local basis. This is another consequence of the finite length of the chain and is an end effect.

These two matrix elements follow immediately from $\rho_0(\phi; t)$ by inverse Fourier transformation. According to (4.43),

$$\rho_{pp}(t) = \frac{X}{2\pi}\int_0^{2\pi/X} e^{-i(p-(N+1)/2)\phi X}\rho_0(\phi; t)\, d\phi \tag{4.59}$$

p being an integer between 1 and N.

Similarly the equations for the $\rho_r(\phi; t)$ $(1 \leqslant r \leqslant N-1)$ are

$$\frac{\partial}{\partial t}\rho_r(\phi;t) = -2\{\gamma_0 + (2+\delta_{r1})\gamma_1\}\rho_r(\phi;t) + 2\Delta\omega \sin\frac{\phi}{2}X\{\rho_{r+1}(\phi;t) - \rho_{r-1}(\phi;t)\} + \gamma_1 \cos\frac{N-r-1}{2}\phi X\{(-1)^r\rho_{1+r1}(t) + \rho_{11+r}(t)\} + \frac{i\Delta\omega}{2}e^{-i((N-r+1)/2)\phi X}\{(-1)^{r-1}\rho_{r1}(t) + \rho_{1r}(t)\} - \frac{i\Delta\omega}{2}e^{+i((N-r+1)/2)\phi X}\{(-1)^{r-1}\rho_{NN-r+1}(t) + \rho_{N-r+1N}(t)\} \quad (4.60)$$

Terms containing elements $\rho_{pp'}(t)$ are end effects and may be expressed in terms of the $\rho_r(\phi; t)$ by a Fourier transformation. Equation 4.57 becomes a special case of (4.60) if we make the following convention:

$$\rho_{-r}(\phi;t) = (-1)^r\rho_r(\phi;t)$$

whenever one of these quantities appears on the right-hand side. The integro-differential system for $\{\rho_r(\phi; t)\}_{0\leqslant r\leqslant N-1}$ is then

$$\frac{\partial}{\partial t}\rho_r(\phi;t) = -2\{(1-\delta_{r0})\gamma_0 + (2 - 2\delta_{r0}\cos\phi X + \delta_{r1})\gamma_1\}\rho_r(\phi;t) + 2\Delta\omega\sin\frac{\phi}{2}X\{\rho_{r+1}(\phi;t) - \rho_{r-1}(\phi;t)\} + \Delta\omega\frac{X}{\pi}(1-\delta_{r0})\int_0^{2\pi/X} d\phi' \sin\left(\frac{N-r+1}{2}\phi X - \frac{N-r}{2}\phi' X\right)\rho_{r-1}(\phi';t) + 4\gamma_1\frac{X}{\pi}\delta_{r0}\int_0^{2\pi/X} d\phi' \sin\frac{\phi}{2}X \sin\frac{N\phi - (N-1)\phi'}{2}X\rho_r(\phi';t) + \gamma_1\frac{X}{\pi}(1-\delta_{r0})\int_0^{2\pi/X} d\phi' \cos\frac{(N-r-1)(\phi-\phi')}{2}X\rho_r(\phi';t) \quad (4.61)$$

$$(r = 0, 1, 2, \ldots, N-1)$$

This N^{th} order system is equivalent by Fourier transformation to a matrix, Fredholm, integral equation of the second kind

$$s|\bar{R}(\phi;s)\rangle = |R(\phi;0)\rangle + \hat{A}(\phi)|\bar{R}(\phi;s)\rangle + \frac{X}{2\pi}\int_0^{2\pi/X}\hat{B}(\phi,\phi')|\bar{R}(\phi';s)\rangle\, d\phi' \quad (4.62)$$

where $|\bar{R}(\phi; s)\rangle$ is a vector whose N components are the $\bar{\rho}_r(\phi; s)$, that is, the Laplace transforms of the $\rho_r(\phi; t)$. $\hat{A}(\phi)$ and $\hat{B}(\phi, \phi')$ are two matrices easily derived from (4.61). Analytical solution of this system is not possible because the kernel $B(\phi, \phi')$ cannot be separated into a product $\hat{B}_1(\phi)\cdot\hat{B}_2(\phi')$, but a computer program can be written to solve it, using a local expansion relative to the time and iteration for the vector $|R(\phi; t)\rangle$.[153,154]

Before commenting on the numerical results of the calculations we may easily find analytical expressions of the moments, when $t \sim 0$. Letting q be the site initially excited, we have

$$\rho_r(\phi; 0) = e^{+i(q-(N+1)/2)\phi X}\delta_{r0} \tag{4.63}$$

that is, when $t = 0$ only the first component of $|R(\phi; t)\rangle$ is not identically nil. Simple algebra yields

$$\begin{aligned}\mathcal{M}_{1,x}(t) &= X\left\{q - \frac{N+1}{2} + 2\gamma_1(\delta_{q1} - \delta_{qN})t\right\}\\ &\quad + X\{\Delta\omega^2(\delta_{q1} - \delta_{qN}) + 2\gamma_1^2(\delta_{q2} - \delta_{q1} - \delta_{qN-1} + \delta_{qN})\}t^2 + 0(t^2)\\ \Delta_{xx}^2(t) &= 2\gamma_1 X^2(2 - \delta_{q1} - \delta_{qN})t + X^2\{\Delta\omega^2(2 - \delta_{q1} - \delta_{qN})\\ &\quad + \gamma_1^2[(N-6)(\delta_{q2} + \delta_{qN-1}) - (N+2)(\delta_{q1} + \delta_{qN})]\}t^2 + 0(t^2)\end{aligned}$$

Table X illustrates the various quantities preceding for $q = 1, 2$ and $q \neq 1, 2, N-1, N$.

Consider the extreme cases

***a.* Completely coherent motion ($\gamma_0 = \gamma_1 = 0$):**

$$\mathcal{M}_{1,x}(t) = -\frac{N-1}{2}X + \Delta\omega^2 t^2 X + 0(t^2) \qquad \Delta_{xx}^2(t) = \Delta\omega^2 t^2 X^2 + 0(t^2) \quad (q = 1)$$

$$\mathcal{M}_{1,x}(t) = -\frac{N+1-2q}{2}X + 0(t^2) \qquad \Delta_{xx}^2(t) = 2\Delta\omega^2 t^2 X^2 + 0(t^2) \quad (q \neq 1, N)$$

TABLE X
Behavior of the First Moment $\mathcal{M}_{1,x}(t)$ and the Spread $\Delta_{xx}^2(t)$ Close to $t = 0$

	$X^{-1}\mathcal{M}_{1,x}(t)$	$X^{-2}\Delta_{xx}^2(t)$
$q = 1$	$-\frac{N-1}{2} + 2\gamma_1 t + (\Delta\omega^2 - 2\gamma_1^2)t^2 + 0(t^2)$	$2\gamma_1 t + \{\Delta\omega^2 - (N+2)\gamma_1^2\}t^2 + 0(t^2)$
$q = 2$	$-\frac{N-3}{2} + 2\gamma_1^2 t^2 + 0(t^2)$	$4\gamma_1 t + \{2\Delta\omega^2 + (N-6)\gamma_1^2\}t^2 + 0(t^2)$
$q \neq$ 1, 2, $N-1$, N	$-\frac{N+1-2q}{2} + 0(t^2)$	$4\gamma_1 t + 2\Delta\omega^2 t^2 + 0(t^2)$

Growth proportional to t^2 is characteristic of a wavelike movement.[151,68] The first moment increases proportionally to t^2 when the initial excitation is at one end of the chain and proportionally to t^3 otherwise. Its derivative is initially nil. When $q \neq 1, N$ the second-order approximation of $\Delta_{xx}^2(t)$ is the same as its exact expression valid for all t, in an infinite chain.[63,68] $q = 1$ has no equivalent in previous work on infinite chains since they have no beginning or ending. As we have already said comparisons can be made only close to the initial instant.

b. Completely incoherent motion ($\Delta\omega = 0$). From Table X,

$$\left.\begin{aligned} \mathcal{M}_{1,x}(t) &= -\frac{N-1}{2}X + 2\gamma_1 tX(1 - 2\gamma_1 t) + 0(t^2) \\ \Delta_{xx}^2(t) &= \gamma_1 X^2 t\{2 - (N+2)\gamma_1 t\} + 0(t^2) \end{aligned}\right\} \quad (q = 1)$$

$$\left.\begin{aligned} \mathcal{M}_{1,x}(t) &= -\frac{N-3}{2}X + 2\gamma_1^2 t^2 X + 0(t^2) \\ \Delta_{xx}^2(t) &= \gamma_1 tX^2\{4 + (N-6)\gamma_1 t\} + 0(t^2) \end{aligned}\right\} \quad (q = 2)$$

$$\left.\begin{aligned} \mathcal{M}_{1,x}(t) &= -\frac{N+1-2q}{2}X + 0(t^2) \\ \Delta_{xx}^2(t) &= 4\gamma_1 tX^2 + 0(t^2) \end{aligned}\right\} \quad (q \neq 1, 2, N-1, N)$$

As the initially excited site is taken closer to the middle of the chain the first moment increases more slowly; For example, in t when $q = 1$ in t^2 when $q = 2$, and in t^3 otherwise. The remark made in *a* about the relation of the spread to earlier work is true here, too.

The linear terms in the general expression contain γ_1 only, whereas the quadratic ones for $q = 1, 2$ contain γ_1 and $\Delta\omega$. Contrary to the situation in an infinite[85] chain, the movement is not a simple superposition of a scattering part (in $\gamma_1 t$) and a wavelike part (in $\Delta\omega^2 t^2$), even when t is close to $t = 0$. The coherent and incoherent aspects are interrelated when the excited state is at one end of the chain. If it is not, the two components are indeed "separated" when $t = 0^+$, but since $\hat{H}_{0-}^*$ and $\hat{\Gamma}^*$ do not commute, this property does not persist.

γ_0 appears nowhere in Table X. Continuing the local expansion to a higher order shows that the local fluctuation intervenes only at the third order, with the coupling term $\hbar\Delta\omega$ (we have seen that in the completely incoherent case γ_0 does not appear at all; see Section IV.F.2).

Figures 40 to 44 represent the results corresponding to $\gamma_0 \ll |\Delta\omega|$, called the coherent case by analogy with that of the dimer.

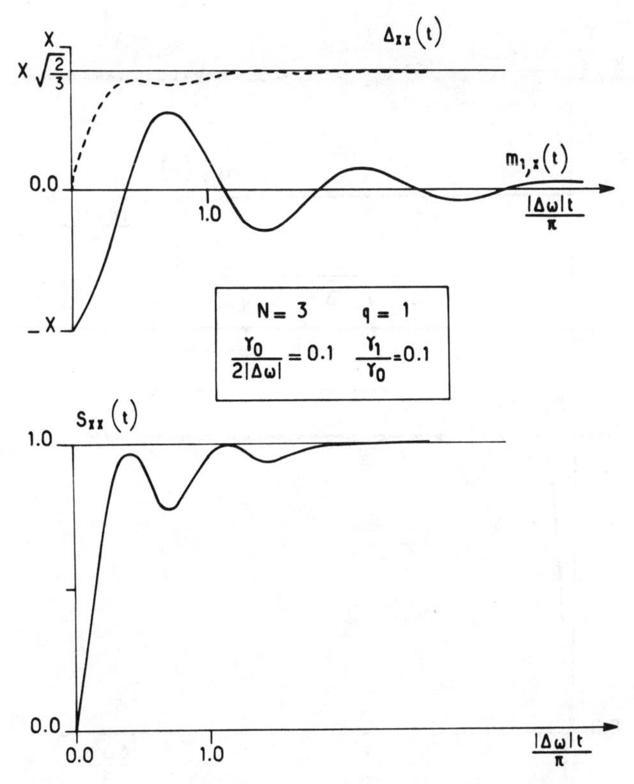

Fig. 40. Position, spread, and delocalization of a coherent trimer exciton prepared on one end of the chain.

For small N, (Figs. 40 and 41, $N=3$), $\mathcal{M}_{1,x}(t)$ and $S_{xx}(t)$ have pronounced oscillations, $\Delta_{xx}(t)$ somewhat less so. Equilibrium is reached after a time of the order of γ_0^{-1}, whether the excitation is prepared at the end or the middle of the trimer. At equilibrium the exciton, which is on average in the middle of the chain, has a spread of the order of the distance between neighboring sites. $S_{xx}(t)$ attains its maximum value, characteristic of an excitation spread uniformly over all the chain.

Once more $\mathcal{M}_{1,x}(t)$ and $\Delta_{xx}(t)$ compensate each other at the beginning of the movement, the spread increasing faster if the initial speed of the exciton is low, and vice versa. When $N=3$ and $q=2$ (Fig. 41) the excitation is initially in the middle of the aggregate and the first moment is stationary: $\mathcal{M}_{1,x}(0)=\mathcal{M}_{1,x}(t)=\mathcal{M}_{1,x}(+\infty)$; $\Delta_{xx}(t)$, on the other hand, grows much faster than when $q=1$ and even (Fig. 40) exceeds its asymptotic value.

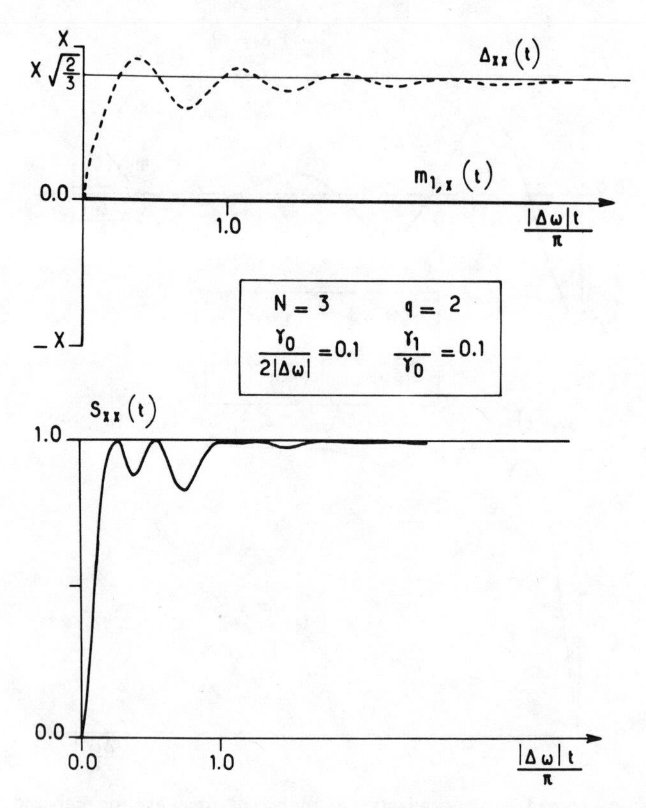

Fig. 41. Same as Fig. 40, but exciton prepared in middle of chain.

There are hardly any oscillations when N is large for example, $N = 10$ in Figs. 42 and 43. The recurrence time is so long that the movement has no time to develop its wave behavior, which grows too slowly (Fig. 26). When $q = 5$, the exciton stays still for a long time before slowly moving towards the center of the chain whereas $\Delta_{xx}(t)$ and $S_{xx}(t)$ grow smoothly, monotonically and faster than when $q = 1$ (Fig. 42).

The preceding remarks are related to the discussion in Section IV.F.1 on the coherent pseudo-period. We saw that it increased very fast with N. It is hardly surprising, therefore, that for given relaxation parameters the movement seems more and more incoherent as the number of sites increases, for whereas the damping time is roughly constant, the pseudo-period becomes very long.

Figure 44 shows the diagonal elements $\rho_{pp}(t)$ $(p = 1, \ldots, N = 5)$, of $\hat{\rho}(t)$ on the local basis. Comparing them with $S_{xx}(t)$ we verify that the maxima of $S_{xx}(t)$ are sharper when the $\rho_{pp}(t)$ are uniformly spread out

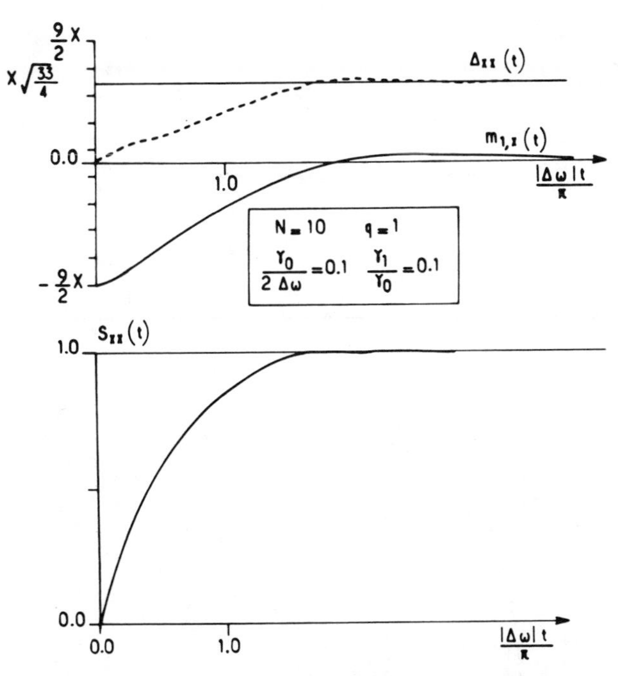

Fig. 42. Coherent, decamer exciton prepared on one end of the chain: Position, spread, and delocalization.

and its minima are deeper when one of these quantities is large and all the rest are simultaneously small. When $q = 1$, $\rho_{11}(t)$ remains stationary for a long time and the maxima of $\rho_{pp}(t)$ are not pronounced which implies shallow minima of $S_{xx}(t)$.

Figure 45, $q = 1$, shows the average speed between $t = 0$ and $t = \pi/|\Delta\omega|$ as a function of the length of the chain. It increases rapidly with N attaining a constant value by $N = 10$. $\gamma_0/2|\Delta\omega| = 0.1$ in this figure.

Figures 46 to 49 show different cases of incoherent movement for which, on the other hand, $\gamma_0 \gg 2|\Delta\omega|$. The moments tend smoothly towards their equilibrium values. When $N = 3$ and $q = 1$ or 2 equilibrium is reached by $t \sim \gamma_1^{-1}$ and $t \sim \frac{1}{2}\gamma_1^{-1}$, respectively. The relaxation time, approximately $10\gamma_1^{-1}$, is even longer when $N = 10$. The increased time needed to establish equilibrium is due essentially to the length of the chain, as Fig. 50 shows. The average speed ($q = 1$) between $t = 0$ and $t = \gamma_1^{-1}$ changes little with N and is constant for $N \geqslant 5$.

We note once more that $\Delta_{xx}(t)$ grows fastest when $\mathcal{M}_{1,x}(t)$ increases slowly or when the initial excitation is in the middle of the chain. We may now conclude that it is a general property of the movement since it

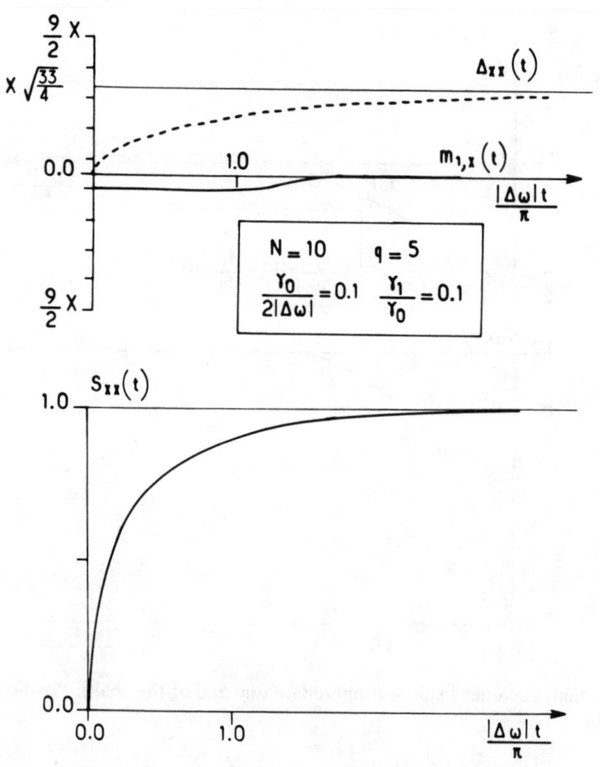

Fig. 43. Same as Fig. 42, but preparation in middle of chain.

occurs in all cases: completely coherent, coherent, incoherent, and completely incoherent.

The ρ_{pp}s of the incoherent case are shown in Fig. 51. Not all of them vary smoothly all the time. ρ_{22} in Fig. 51 ($N = 5$ and $q = 1$) increases rapidly and thereafter decreases slowly whereas the other matrix elements tend monotonically towards their equilibrium values. Similar behavior, for any q, is observed in the ρ_{pp}s for which $p = q \pm 1$. Sites adjacent to the one initially excited keep part of the excitation for some time, whereas the rest is spread over the chain.

We have gone to some length to warn against excessive simplification of the movement of the exciton, particularly the wrong idea that it is a simple superposition of a scattering component and a wavelike component. In general the two parts are intimately mixed leading to a complicated movement. Moreover, we believe it is unnecessary to use a picture of a hopping motion between sites. It is useless, and a confusing source of nonsense and illusory problems besides.

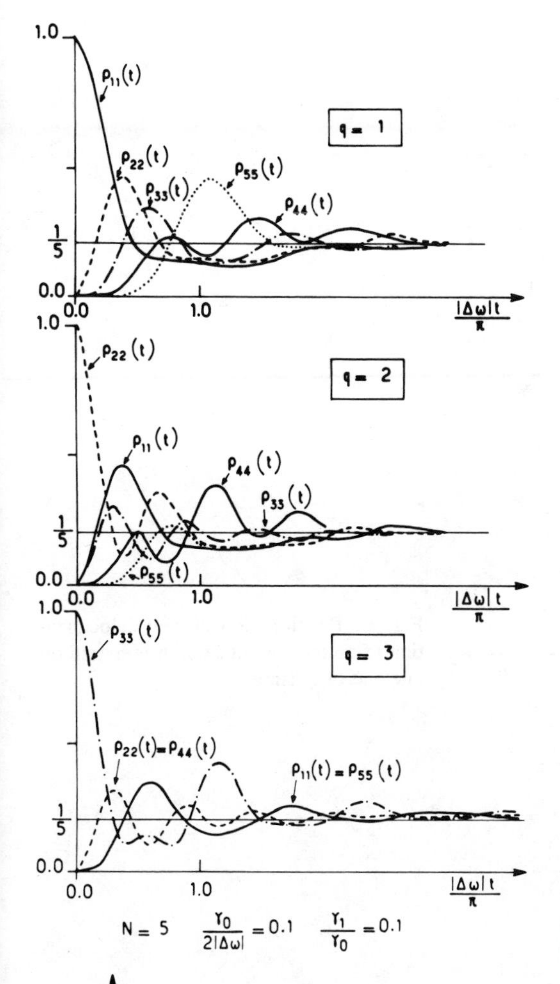

Fig. 44. "Local populations" of a pentamer, following optical preparation of an exciton. Coherent case.

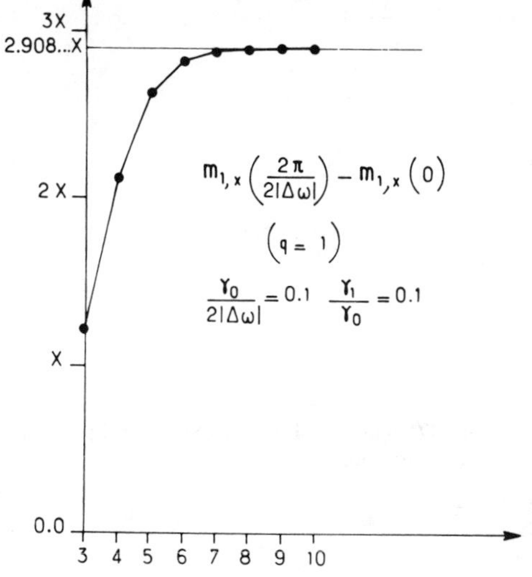

Fig. 45. Distance traveled at time $t = \pi/|\Delta\omega|$ by a coherent exciton, prepared on one end of a chain of N sites.

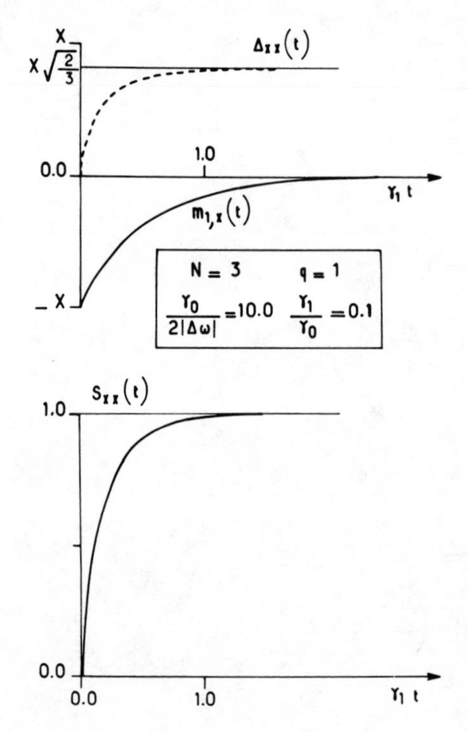

Fig. 46. Position, spread, and delocalization of an incoherent exciton prepared on one end of a trimer.

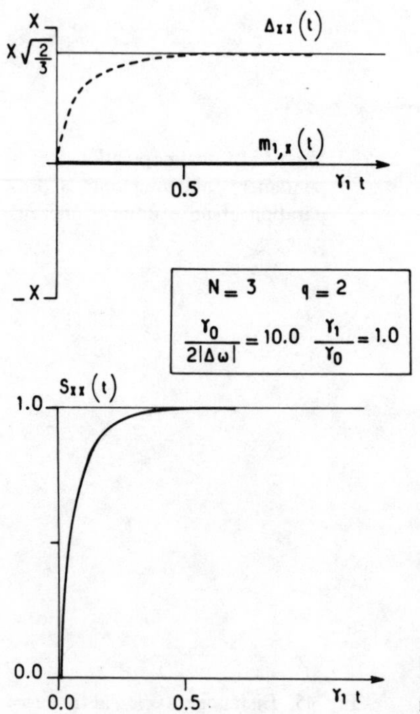

Fig. 47. Same as Fig. 46, exciton prepared in middle of trimer.

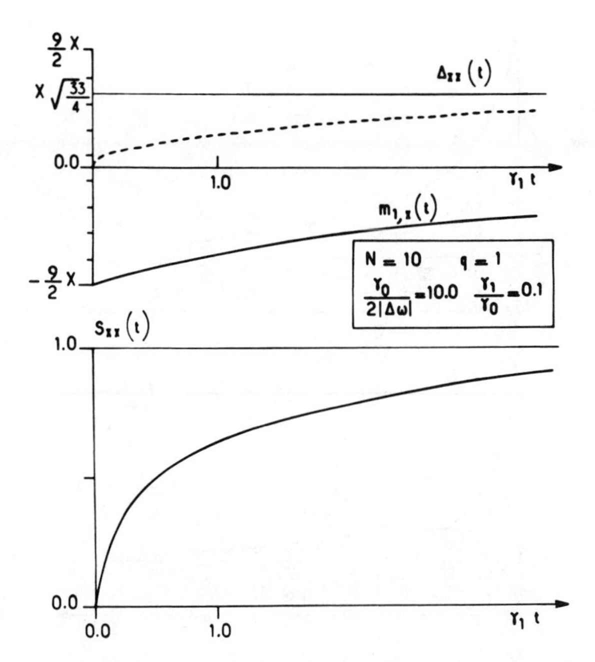

Fig. 48. Position, spread, and delocalization of an incoherent decamer exciton prepared on one end of the chain.

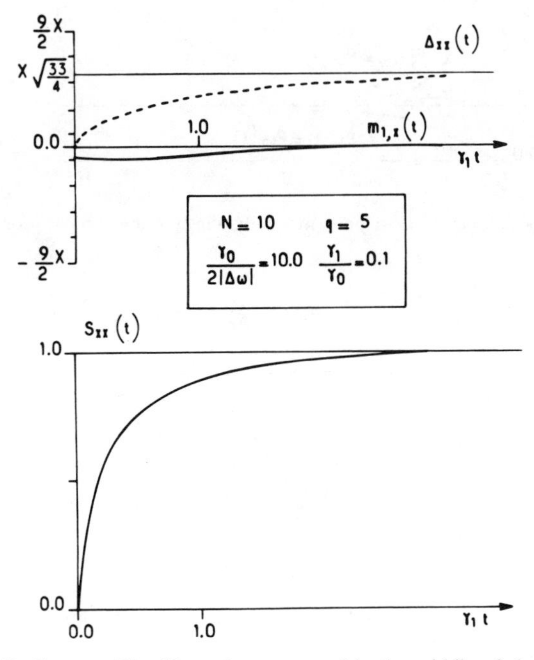

Fig. 49. Same as Fig. 48, exciton prepared in the middle of the chain.

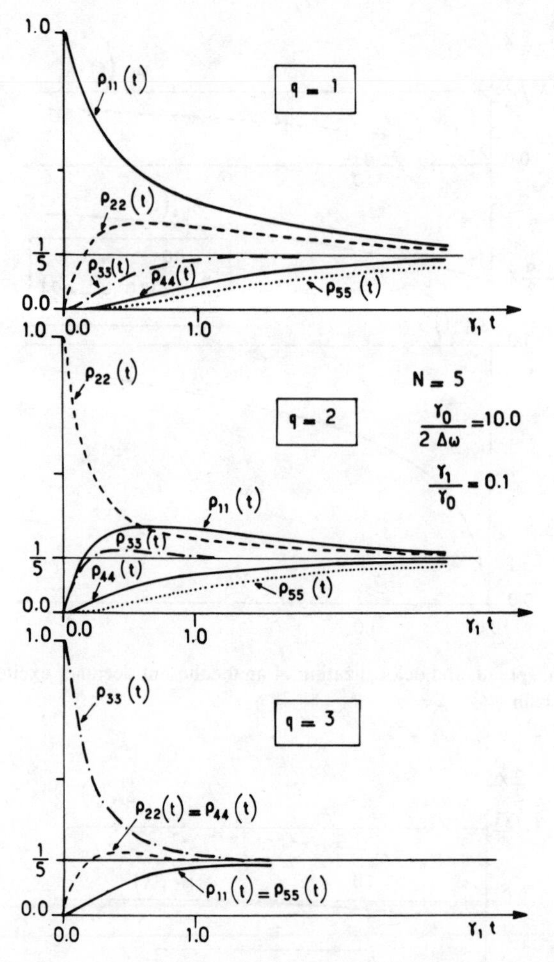

Fig. 50. "Local populations" of a pentamer following optical preparation of an exciton. Incoherent case.

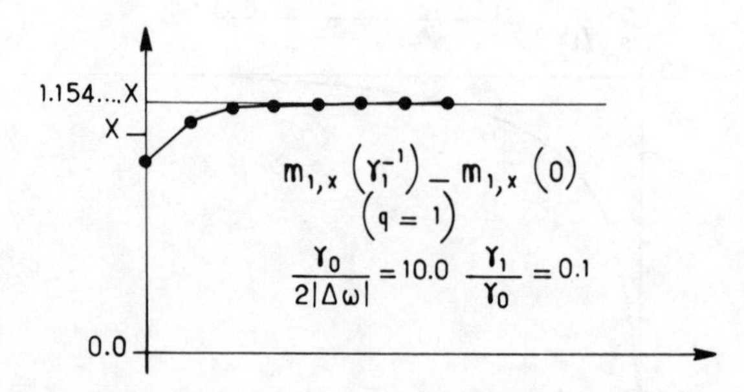

Fig. 51. Distance traveled at time $t = 1/\gamma_1$ by an incoherent exciton, prepared on one end of a chain of length N.

G. Transfer in a Double Chain

There being two directions, Ox and Oy (see Section IV.D) in which the excitation may move, there are many different anisotropic movements: coherent along Oy (transverse) and coherent along Ox (longitudinal), coherent transversal, and incoherent longitudinal.

It is simpler and shorter to study the general case, from which the various coherent and incoherent limits may be deduced.

As before, we start by writing an equation for the generating function of the spatial moments, defined in (4.23). In what follows,

$$\rho_{pp'}^{IJ}(t) = \langle M_p^I | \hat{\rho}(t) | M_{p'}^J \rangle \qquad (I, J = A, B; p, p' = 1, 2, \ldots, N)$$

Using form (2.111) of $\hat{H}^*$ (projection of $\hat{H}$ defined in (2.61) on the transfer subspace, $\mathscr{A}^*$ on the basis of $|Y_{pp'}^{IJ}$ and taking into account each term containing a diagonal element of $\hat{\rho}(t)$ in the expression of $\rho_0(\phi, \psi; t)$, we arrive at:

$$\begin{aligned} \frac{\partial}{\partial t}\rho_0(\phi, \psi; t) = & -2\left\{4\gamma_1 \sin^2\left(\frac{\phi}{2}X\right) + \gamma_2\right\}\rho_0(\phi, \psi; t) + 2\gamma_2\rho_0(\phi, -\psi; t) \\ & + 2\Delta\omega \sin\frac{\phi}{2}X \sum_{b=\pm 1}\sum_{p=1}^{N} \Phi_{bp}(\phi, \psi)\{-\rho_{p+1p}^{II}(t) + \rho_{pp+1}^{II}(t)\} \\ & - i\Delta\omega' \sum_{b=\pm 1}\sum_{p=1}^{N} \Phi_{bp}(\phi, \psi)\{\rho_{pp}^{\bar{I}I}(t) - \rho_{pp}^{I\bar{I}}(t)\} \end{aligned} \qquad (4.64)$$

$\Phi_{bp}(\phi, \psi)$ is defined in (4.24). I depends on b as in expression (4.23). $\bar{I}$ has the value complementary to that of I; see Table IV. As in (4.61) we have replaced the matrix elements of $\hat{\rho}(t)$ on the right-hand side by integrals because of the finite length of the double chain. The fundamental relation here is

$$\rho_{pp}^{II}(t) = \frac{X}{2\pi}\frac{Y}{2\pi}\int_0^{2\pi/X} d\phi \int_0^{2\pi/Y} d\psi e^{-i\{b(Y/2)\psi + (p-(N+1/2)\phi X\}}\rho_0(\phi, \psi; t) \qquad (4.65)$$

($b = 1$ when $I = A$ and $b = -1$ when $I = B$)

Thus,

$$\sigma^r(\phi, \psi; t) = \frac{Y}{\pi}\int_0^{2\pi/Y} \cos(\psi + \psi')\rho_0(\phi, \psi'; t)d\psi' \qquad (4.66)$$

Equation (4.64) shows that the movement is not in general the resultant of a dimer motion along Oy and a motion along Ox, since derivation at $\phi = 0$ gives $\mathscr{M}_{1,x}(t)$ (and a factor) on the left-hand side, whereas on the right-hand side there are terms in $\Delta\omega'$.

$\rho_0(\phi, \psi; t)$ is coupled to quantities generalizing those of (4.58):

$$\rho_r(\phi, \psi; t) = \tfrac{1}{2} \sum_{b=\pm 1} \sum_{p=1}^{N-r} \Phi_{bp}(\phi, \psi)\{(-1)^r \rho^{AA}_{p+rp}(t) + \rho^{BB}_{pp+r}(t)\}$$

$$\sigma^r(\phi, \psi; t) = \tfrac{1}{2} \sum_{b=\pm 1} \sum_{p=1}^{N-r} \Phi_{bp}(\phi, \psi)\{(-1)^r \rho^{AB}_{p+rp}(t) + \rho^{BA}_{pp+r}(t)\}$$

Some easy but rather tedious algebra leads to an integro-differential system of order $2N$ satisfied by the $\rho_r(\phi, \psi; t)$

$$\begin{aligned}
\frac{\partial}{\partial t} \rho_r(\phi, \psi; t) = &-2\{\gamma_0(1-\delta_{r0}) + \gamma_1(2 - 2\delta_{r0}\cos\phi X + \delta_{r1}) + \gamma_2\}\rho_r(\phi, \psi; t) \\
&+ 2\Delta\omega \sin\frac{\phi}{2} X\{\rho_{r+1}(\phi, \psi; t) - \rho_{r-1}(\phi, \psi; t)\} \\
&+ \Delta\omega \frac{X}{\pi}(1-\delta_{r0}) \int_0^{2\pi/X} d\phi' \sin\left(\frac{N-r+1}{2}\phi X - \frac{N-r}{2}\phi' X\right)\rho_{r-1}(\phi', \psi; t) \\
&+ 4\gamma_1 \frac{X}{\pi}\delta_{r0} \int_0^{2\pi/X} d\phi' \sin\frac{\phi}{2} X \sin\frac{N\phi - (N-1)\phi'}{2} X\rho_r(\phi', \psi; t) \\
&+ \gamma_1 \frac{X}{\pi}(1-\delta_{r0}) \int_0^{2\pi/X} d\phi' \cos\frac{(N-r-1)(\phi-\phi')}{2} X\rho_r(\phi', \psi; t) \\
&+ i\Delta\omega' \sin\frac{\psi}{2} Y \frac{Y}{\pi} \int_0^{2\pi/Y} d\psi' \sin\frac{\psi'}{2} Y\sigma_r(\phi, \psi'; t) \\
&+ \gamma_2 \frac{Y}{\pi} \int_0^{2\pi/Y} d\psi' \cos\frac{\psi+\psi'}{2} Y\rho_r(\phi, \psi'; t)
\end{aligned} \tag{4.67}$$

The N first equations are identical to (4.61) when we put $\Delta\omega' = \gamma_2 = \psi = 0$. The $N\sigma_r(\phi, \psi; t)$ obey

$$\begin{aligned}
\frac{\partial}{\partial t} \sigma_r(\phi, \psi; t) = &\\
&-2\Gamma_{dc}\sigma_r(\phi, \psi; t) + 2\Delta\omega \sin\frac{\phi}{2} X\{\sigma_{r+1}(\phi, \psi; t) - \sigma_{r-1}(\phi, \psi; t)\} \\
&+ \gamma_2 \frac{Y}{\pi} \int_0^{2\pi/Y} d\psi' \cos\frac{\psi+\psi'}{2} Y\sigma_r(\phi, \psi'; t) \\
&+ i\Delta\omega' \frac{Y}{\pi} \int_0^{2\pi/Y} d\psi' \sin\frac{\psi}{2} Y \sin\frac{\psi'}{2} Y\rho_r(\phi, \psi'; t) \\
&+ 4\gamma_1 \frac{X}{\pi}\delta_{r0} \int_0^{2\pi/X} d\phi' \sin\frac{\phi}{2} X \sin\frac{N\phi - (N-1)\phi'}{2} X\sigma_r(\phi', \psi; t) \\
&+ \gamma_1(1-\delta_{r0}) \frac{X}{\pi} \int_0^{2\pi/X} d\phi' \cos\frac{(N-r-1)(\phi-\phi')}{2} X\sigma_r(\phi', \psi; t) \\
&+ \Delta\omega \frac{X}{\pi} \int_0^{2\pi/X} d\phi' \sin\left(\frac{N-r+1}{2}\phi X - \frac{N-R}{2}\phi' X\right)\sigma_{r-1}(\phi', \psi; t)
\end{aligned} \tag{4.68}$$

This system may be solved by a simple generalization of the algorithm used to solve the problem of the linear chain.

The moments were expressed in terms of $\rho_0(\phi, \psi; t)$ in (4.25). The entropies of delocalization are

$$S_{xx}(t) = -\frac{1}{\log N}\sum_{p=1}^{N}\{\rho_{pp}^{AA}(t) + \rho_{pp}^{BB}(t)\}\log\{\rho_{pp}^{AA}(t) + \rho_{pp}^{BB}(t)\} \tag{4.69}$$

$$S_{xy}(t) = -\frac{1}{\log 2N}\sum_{p=1}^{N}\{\rho_{pp}^{AA}(t)\log \rho_{pp}^{AA}(t) + \rho_{pp}^{BB}(t)\log \rho_{pp}^{BB}(t)\} \tag{4.70}$$

$$S_{yy}(t) = -\frac{1}{\log 2}\left[\left\{\sum_{p=1}^{N}\rho_{pp}^{AA}(t)\right\}\log\left\{\sum_{p=1}^{N}\rho_{pp}^{AA}(t)\right\} + \left\{\sum_{p=1}^{N}\rho_{pp}^{BB}(t)\right\}\log\left\{\sum_{p=1}^{N}\rho_{pp}^{BB}(t)\right\}\right] \tag{4.71}$$

The functions are normed so that $S_{uv}(+\infty) = 1$ whenever there are relaxation terms (S_{uv} has no limit in the completely coherent case). It can be shown that[155]

$$S_{xy}(t) \leqslant \frac{1}{\log 2N}\{S_{xx}(t)\log N + S_{yy}(t)\log 2\} \tag{4.72}$$

with equality only if the movements along Ox and Oy are uncorrelated in probability. S_{xy} being perforce positive or zero, it follows that $S_{uu} = 0$ ($u = x$ and y) implies $S_{xy} = 0$. Moreover, since S_{xy} is an entropy and can be zero only if ρ_{pp}^{II} is of the form $\delta_{pp_0}\cdot\delta_{II_0}$ it follows that

$$S_{xy} = 0 \Leftrightarrow S_{xx} = 0 \quad \text{and} \quad S_{yy} = 0$$

S_{xy} is a measure of the punctual localization and S_{uu} ($u = x$ or y) is a measure of the localization along Ou.

The difference between each side of (4.72) is in fact a measure of the statistical correlation between the movements in each direction, and is a generalization of the usual correlation function

$$C_{xy}(t) = \frac{\Delta_{xy}^2(t)}{\Delta_{xx}(t)\Delta_{yy}(t)} \tag{4.73}$$

where $\Delta_{uv}^2(t)$ is defined in (4.19). Thus, the greater the difference, the greater the correlation.

The spatial spread is measured by a vector $\boldsymbol{\delta}(t)$,

$$\boldsymbol{\delta}(t) = \sqrt{D_+(t)}\boldsymbol{\delta}_+(t) + \sqrt{D_-(t)}\boldsymbol{\delta}_-(t) \tag{4.74}$$

where $D_\pm$ are the eigenvalues of the spread tensor of the Δ_{uv}^2 and $\boldsymbol{\delta}_\pm$ are the corresponding eigenvectors. The scale of lengths is the same as that of the trajectory. The components δ_x and δ_y define a spread rectangle in the first quadrant, $x \geqslant 0$, $y \geqslant 0$ and reflect the anisotropy of the exciton domain centered on $\mathcal{M}_1(t)$. If δ_y is very small, the dispersion ellipse, (approximated by a rectangle) is like a cigar along Ox. The limit of $\boldsymbol{\delta}$ when $t \to +\infty$ is

$$\boldsymbol{\delta}_{\text{eq}} = \left(\frac{N^2-1}{12}\right)^{1/2} X\mathbf{e}_x + \frac{Y}{2}\mathbf{e}_y$$

where $\mathbf{e}_u$ is a unit vector along Ou, $u = x, y$.

Figure 52 illustrates a two-dimensional movement in which $|\Delta\omega'/\Delta\omega| = 0.1$ and all the fluctuations are smaller than $|\Delta\omega'|$. The trajectory is in the upper left-hand corner. Oscillations along Ox are faster than those along Oy for this particular choice of $|\Delta\omega'/\Delta\omega|$. The trajectory slowly fills all the volume of the aggregate.

δ_x and δ_y are plotted in the upper right-hand corner. They are always quite large; that is, the exciton does indeed cover several sites. The entropies below are a better measure of delocalization. They are never less than 0.5 after $t = \pi/|\Delta\omega|$. Generally speaking $S_{uu}(u = x$ or $y)$ has a minimum when $\mathcal{M}_{1,u}$ is maximum; that is, the component of the group speed along Ou is nil. Successive minima are larger than their predecessors, so that directional relocalization, when it occurs, is less and

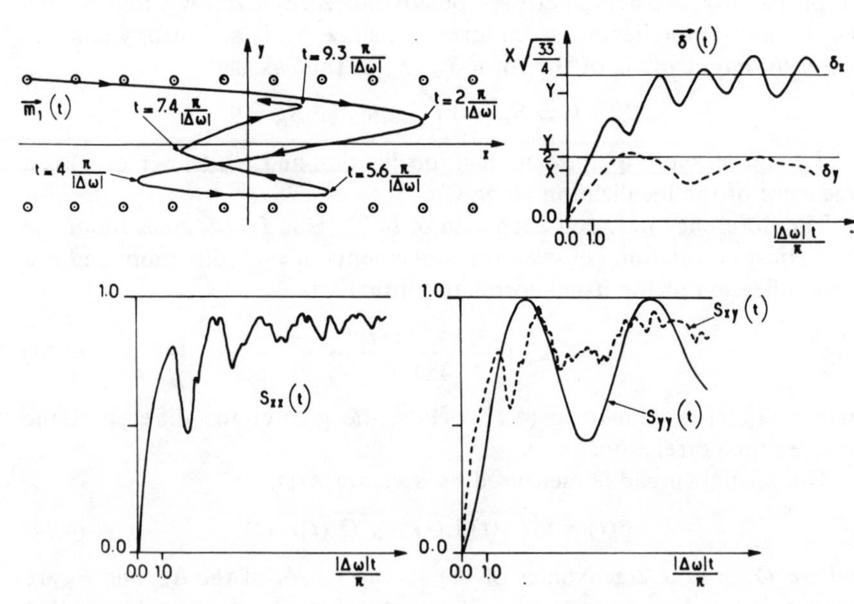

Fig. 52. Trajectory, spread tensor, and spread entropies of a double decamer exciton. The movement is coherent along Ox and Oy (abbreviated as Cx–Cy).

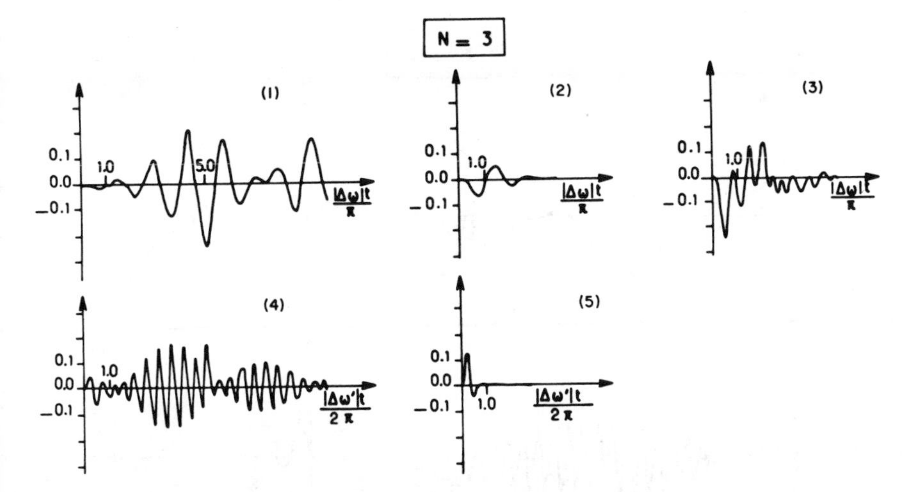

Fig. 53. Correlation coefficient, $C_{xy}(t)$, of the movements along the axes of a double trimer. (Cx–Iy means coherent along Ox and incoherent along Oy, etc.) (1) Cx–Cy; $|\Delta\omega'/\Delta\omega| = 0.1$. (2) Cx–Iy; $|\Delta\omega'/\Delta\omega| = 0.1$. (3) Cx–Cy; $|\Delta\omega'/\Delta\omega| = 1.0$. (4) Cx–Cy; $|\Delta\omega'/\Delta\omega| = 10.0$. (5) Ix–Iy; $|\Delta\omega'/\Delta\omega| = 10.0$.

less precise. S_{xy} is always large. The punctual localization disappears soon after the early stages of the movement.

We turn now to the correlation between the movements along Ox and Oy, measured by $C_{xy}(t)$ of (4.73). Figures 53 ($N = 3$) and 54 ($N = 10$) show its variation in the cases for which it is large. Although $C_{xy}(t)$ is a complicated function of time, there is some correlation at the beginning of the movement. $C_{xy}(t)$ which is practically zero as soon as one of the movements along the axes is incoherent, never rises above 0.2 in modulus, a sign of little correlation between the movements, as defined by this measure. We show that $C_{xy}(t) = 0$ in the extreme cases of completely coherent or incoherent movement, for which there is no correlation at all.

The completely coherent limit corresponds to $\gamma_0 = \gamma_1 = \gamma_2 = 0$. Let the initial state be of the form $\hat{\rho}(0) = |M_q^A\rangle\langle M_q^A|$. Then

$$\rho_{pp}^{II}(t) = \frac{1 + b\cos 2\Delta\omega' t}{2}\left|\frac{2}{N+1}\sum_{k=1}^{N}\sin kp\theta \sin kp\theta e^{-2i\Delta\omega t\cos k\theta}\right|^2$$

Using formula (4.42) we arrive at

$$\rho_{pp}^{II}(t) = \frac{1 + b\cos 2\Delta\omega t}{2}\,\rho_{pp}(t)$$

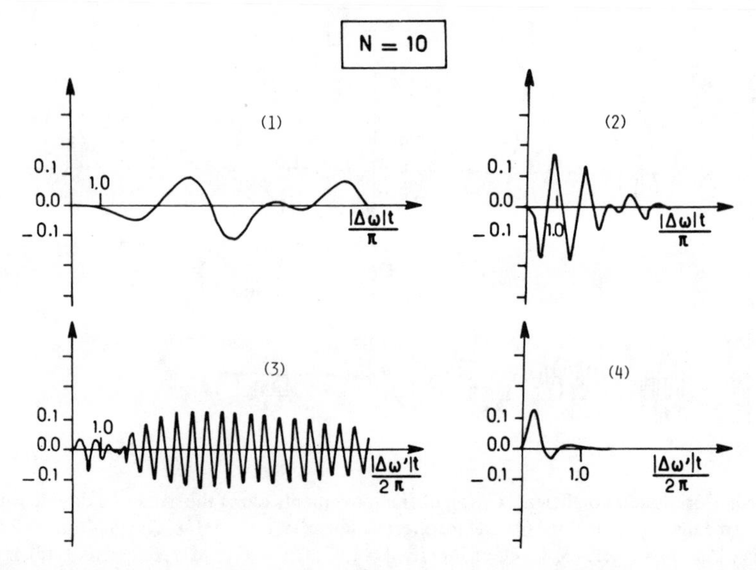

Fig. 54. Same as Fig. 53, for a double decamer. (1) Cx–Cy; $|\Delta\omega'/\Delta\omega| = 0.1$. (2) Cx–Cy; $|\Delta\omega'/\Delta\omega| = 1.0$. (3) Cx–Cy; $|\Delta\omega'/\Delta\omega| = 10.0$. (4) Ix–Cy; $|\Delta\omega'/\Delta\omega| = 10.0$.

where $\rho_{pp}(t) = \langle M_p|\hat{\rho}(t)|M_p\rangle$ is the same as the diagonal element of $\hat{\rho}(t)$ on the pth site of a simple chain when the initially excited site is q. Therefore,

$$\rho_{pp}^{AA}(t) = \cos^2(\Delta\omega' t)\rho_{pp}(t); \qquad \rho_{pp}^{BB}(t) = \sin^2(\Delta\omega' t)\rho_{pp}(t)$$

These equations are interesting because they show the movements are uncorrelated: $\rho_{pp}^{II}(t)$ is simply the product of a dimer-type term and a term calculable by the equations of a simple chain. The movement along Oy is the same as in a dimer, that along Ox the same as in a simple chain, the movement in the plane being their vectorial sum:

$$\mathcal{M}_{1,x}(t) = X \sum_{p=1}^{N} \left(p - \frac{N+1}{2}\right) \rho_{pp}(t); \qquad \mathcal{M}_{1,y}(t) = \frac{Y}{2} \cos 2\Delta\omega' t$$

$$\Delta_{xx}^2(t) = X^2 \sum_{p=1}^{N} \left(p - \frac{N+1}{2}\right)^2 \rho_{pp}(t) - \left\{ \sum_{p=1}^{N} \left(p - \frac{N+1}{2}\right) \rho_{pp}(t) \right\}^2$$

$$\Delta_{xy}^2(t) = 0; \qquad \Delta_{yy}^2(t) = \frac{Y^2}{4} \sin^2(2\Delta\omega' t)$$

$\Delta_{xy}^2(t)$ is identically nil because of the absence of the correlation mentioned previously. The principal directions of dispersion are Ox and Oy for all t and in the vicinity of $t = 0$:

$$\Delta_{xx}^2(t) = X^2(2 - \delta_{q1} - \delta_{qN})\Delta\omega^2 t^2 + 0(t^2); \qquad \Delta_{yy}^2(t) = Y^2 \Delta\omega'^2 t^2 + 0(t^2)$$

Relation (4.72) is easily seen to be an equality[155]

$$S_{xy}(t) = \frac{\log N}{\log 2N} S_{xx}(t) + \frac{\log 2}{\log 2N} S_{yy}(t)$$

The lack of correlation may at first surprise. It is due in part to the geometrical symmetry of the chain which leads to the particular choice of interactions used here to describe coupling between its two branches. In fact there would be some statistical correlation in a double chain where corresponding sites on each branch are not opposite each other.

Figure 55 illustrates the coherent limit in a chain of ten sites for which

$$\text{(I)} \left|\frac{\Delta\omega'}{\Delta\omega}\right| = 0.1; \qquad \text{(II)} \left|\frac{\Delta\omega'}{\Delta\omega}\right| = 1.0; \qquad \text{(III)} \left|\frac{\Delta\omega'}{\Delta\omega}\right| = 10.0$$

The movement for this value of N is clearly nonperiodic. The relative frequencies of the oscillations along the axes depend considerably on the value of $|\Delta\omega'/\Delta\omega|$. The trajectory of the center of the wave packet slowly fills all the available space. A large part of the system is on average affected by the exciton, as may be seen by the fact that the length of $\boldsymbol{\delta}$ is large compared to X and to Y. The punctual localization, S_{xy}, varies in a complicated way in case II. In cases I and III it is close to S_{xx} (respectively, S_{yy}) modulated by S_{yy} (respectively, S_{xx}) (Fig. 56).

The other extreme case corresponds to negligible static interactions compared to the mean squared fluctuations. Putting $\Delta\omega = \Delta\omega' = 0$ in (4.67) we see that $\rho_0(\phi, \psi; t)$ is independent of the $\rho_r(\phi, \psi; t)$, $r > 0$ and of the $\sigma_r(\phi, \psi; t)$, $r \geqslant 0$. In fact

$$\frac{\partial}{\partial t} \rho_0(\phi, \psi; t) =$$

$$-2\left\{4\gamma_1 \sin^2\left(\frac{\phi}{2} X\right) + \gamma_2\right\}\rho_0(\phi, \psi; t) + 2\gamma_2\rho_0(\phi, -\psi; t)$$

$$+ 4\gamma_1 \frac{X}{\pi} \int_0^{2\pi/X} d\phi' \sin\frac{\phi}{2} X \sin\frac{N\phi - (N-1)\delta'}{2} X\rho_0(\phi', \psi; t) \tag{4.75}$$

When $\psi = 0$ in this equation we have

$$\frac{\partial}{\partial t} \rho_0(\phi, 0; .t) = -8\gamma_1 \sin^2\left(\frac{\phi}{2} X\right)\rho_0(\phi, 0; t)$$

$$+ 4\gamma_1 \frac{X}{\pi} \int_0^{2\pi/X} d\phi' \sin\frac{\phi}{2} X \sin\frac{N\phi - (N-1)\phi'}{2} X\rho_0(\phi', 0; t) \tag{4.76}$$

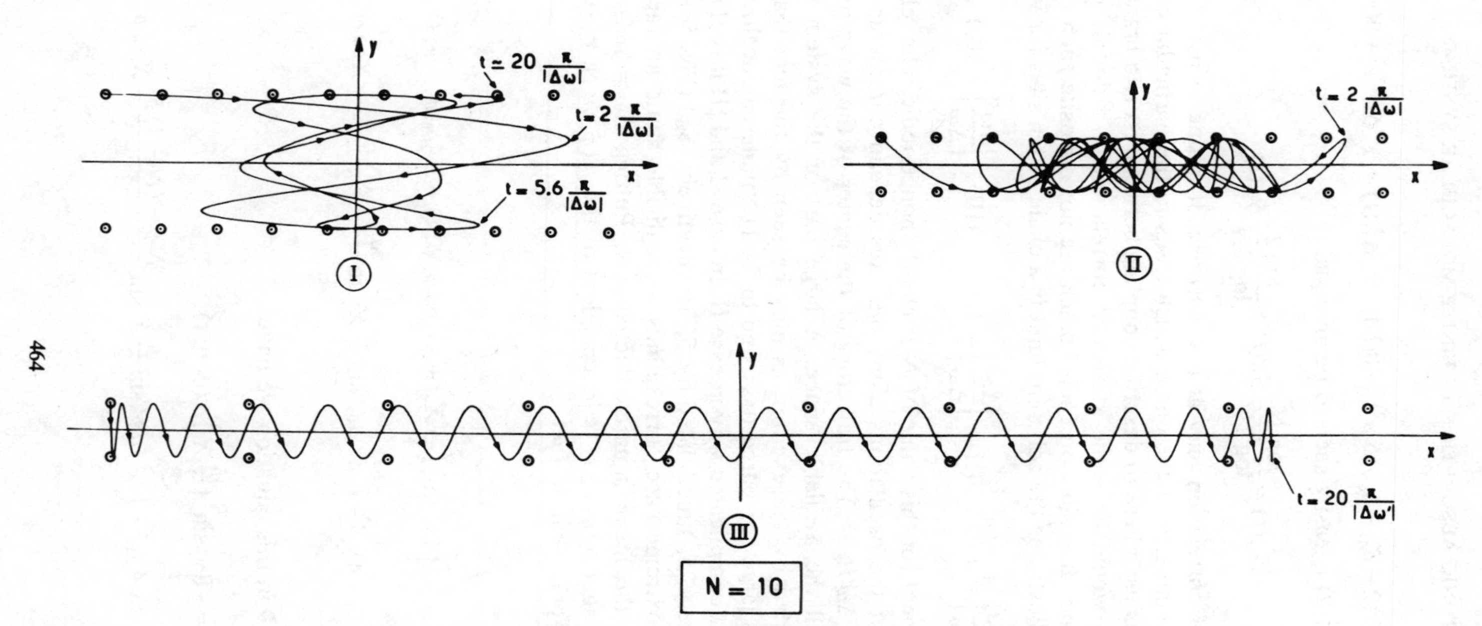

Fig. 55. Three cases (see text) of pure coherent movement in a double decamer.

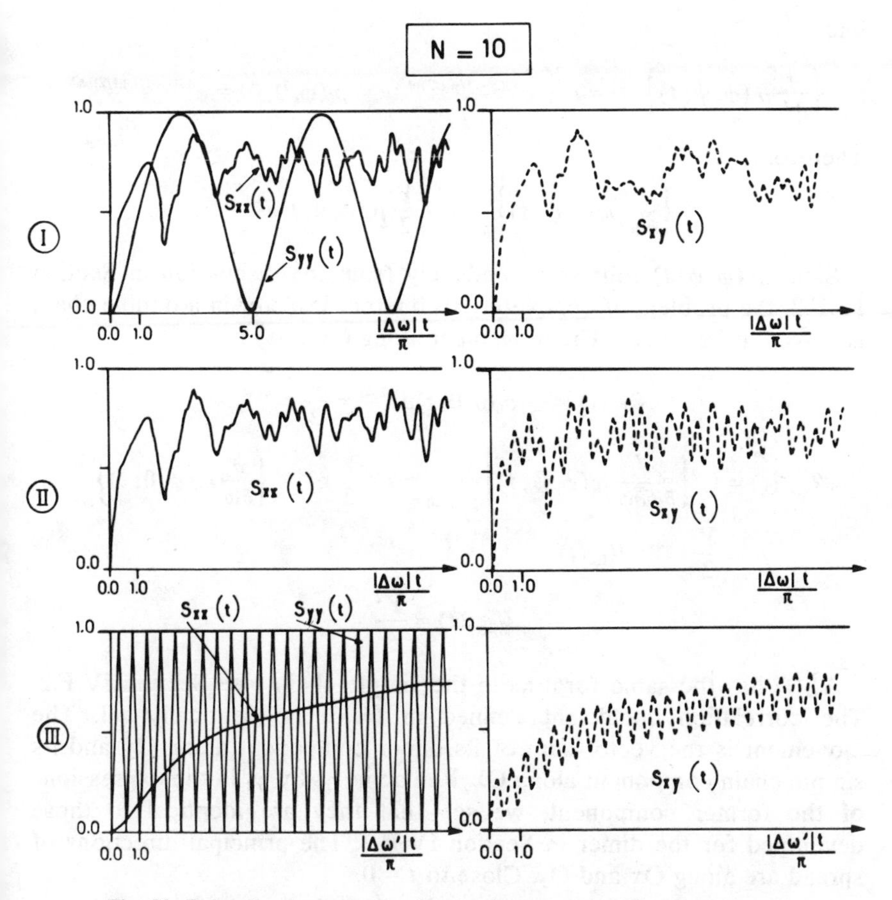

Fig. 56. Delocalization entropies for the three cases of Fig. 55 (see text).

Letting $\Delta\omega = 0$ in (4.61) we see that $\rho_0(\phi, 0; t)$ completely determines the movement along Ox and that it is the same as in the completely incoherent case of the simple chain. Derivation with respect to ψ at $\psi = 0$ yields

$$\frac{\partial}{\partial t}\left\{\frac{\partial}{\partial \psi}\rho_0(\phi, \psi; t)\right\}_{\psi=0} =$$

$$-2\left\{4\gamma_1 \sin^2\left(\frac{\phi}{2}X\right) + 2\gamma_2\right\}\left\{\frac{\partial}{\partial \psi}\rho_0(\phi, \psi; t)\right\}_{\psi=0}$$

$$+ 4\gamma_1 \frac{X}{\pi}\int_0^{2\pi/X} d\phi' \sin\frac{\phi}{2}X \sin\frac{N\phi - (N-1)\phi'}{2} X\left\{\frac{\partial}{\partial \psi}\rho_0(\phi', \psi; t)\right\}_{\psi=0}$$

and

$$\left\{\frac{\partial}{\partial\psi}\rho_0(\phi,\psi;t)\right\}_{\psi=0}=i\frac{Y}{2}e^{+i(q-(N+1/2))\phi X}\qquad \rho_0(\phi,0;0)=e^{+i(q-(N+1/2))\phi X}$$

Therefore,

$$\left\{\frac{\partial}{\partial\psi}\rho_0(\phi,\psi;t)\right\}_{\psi=0}=i\frac{Y}{2}\rho_0(\phi,0;t)e^{-4\gamma_2 t}$$

Since $\rho_0(\phi,\psi;t)$ follows immediately from the discussion in Section IV.F.2, the problem of completely incoherent transfer in a double chain is solved at its outset. The movement along Oy obeys

$$\mathcal{M}_{1,y}(t)=\frac{Y}{2}\rho_0(0,0;t)e^{-4\gamma_2 t}=\frac{Y}{2}e^{-4\gamma_2 t}$$

$$\mathcal{M}_{2,xy}(t)=i^{-2}\left\{\frac{\partial^2}{\partial\phi\partial\psi}\rho_0(\phi,\psi;t)\right\}_{\phi=\psi=0}=i^{-1}\frac{Y}{2}e^{-4\gamma_2 t}\left\{\frac{\partial}{\partial\phi}\rho_0(\phi,0;t)\right\}_{\phi=0}$$

$$\equiv\frac{Y}{2}e^{-4\gamma_2 t}\mathcal{M}_{1,x}(t)$$

$$\mathcal{M}_{2,yy}(t)=\frac{Y^2}{4}$$

$\Delta^2_{xx}(t)$ has the same form as in the simple chain; see Section IV.F.2. The correlation coefficient defined in (4.73) is identically nil. The movement is the vector sum of its dimer component along Oy and its simple chain component along Ox. Replacing γ_2 by γ_1 in the expressions of the former component, we see that they are identical to those developed for the dimer in Section IV.E.2. The principal directions of spread are along Ox and Oy. Close to $t=0$,

$$\Delta^2_{xx}(t)=2X^2\gamma_1(2-\delta_{q1}-\delta_{qN})t+0(t)\qquad \Delta^2_{yy}(t)=2Y^2\gamma_2 t+0(t)$$

Figures 57 and 58 illustrate a movement and the entropy functions for the following cases:

$$\text{(IV)}\quad\frac{\gamma_2}{\gamma_1}=0.1\qquad\text{(V)}\quad\frac{\gamma_2}{\gamma_1}=1.0\qquad\text{(VI)}\quad\frac{\gamma_2}{\gamma_1}=10.0$$

In virtue of what has been said previously about the decomposition of the movement into its simple chain and dimer components along Ox and Oy, respectively, we may take advantage of the results of Sections IV.E.2 (with the change of notation $X \leftrightarrow Y$ and $\gamma_1 \leftrightarrow \gamma_2$) and IV.F.2.

It follows from (4.40) and Table X that

$$\left\{\frac{d}{dt}\mathcal{M}_{1,y}(t)\right\}_{t=0}=-2\gamma_2 Y\qquad\left\{\frac{d}{dt}\mathcal{M}_{1,x}(t)\right\}_{t=0}=2\gamma_1 X$$

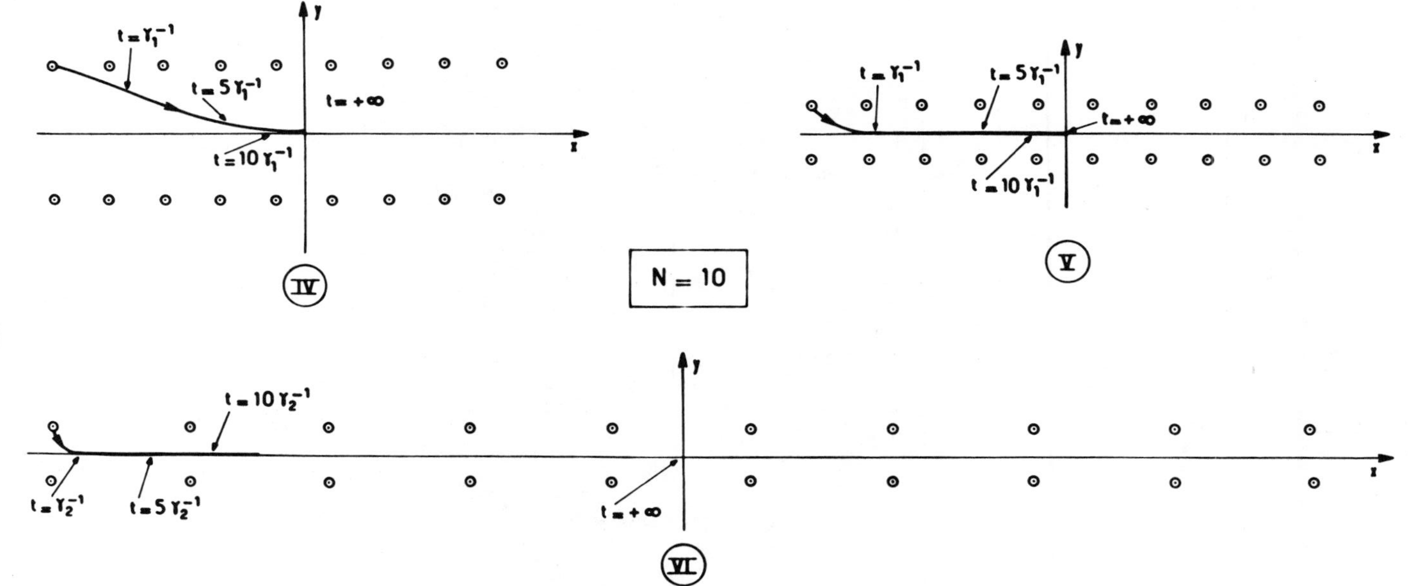

Fig. 57. Three cases of pure incoherent movement in a double decamer.

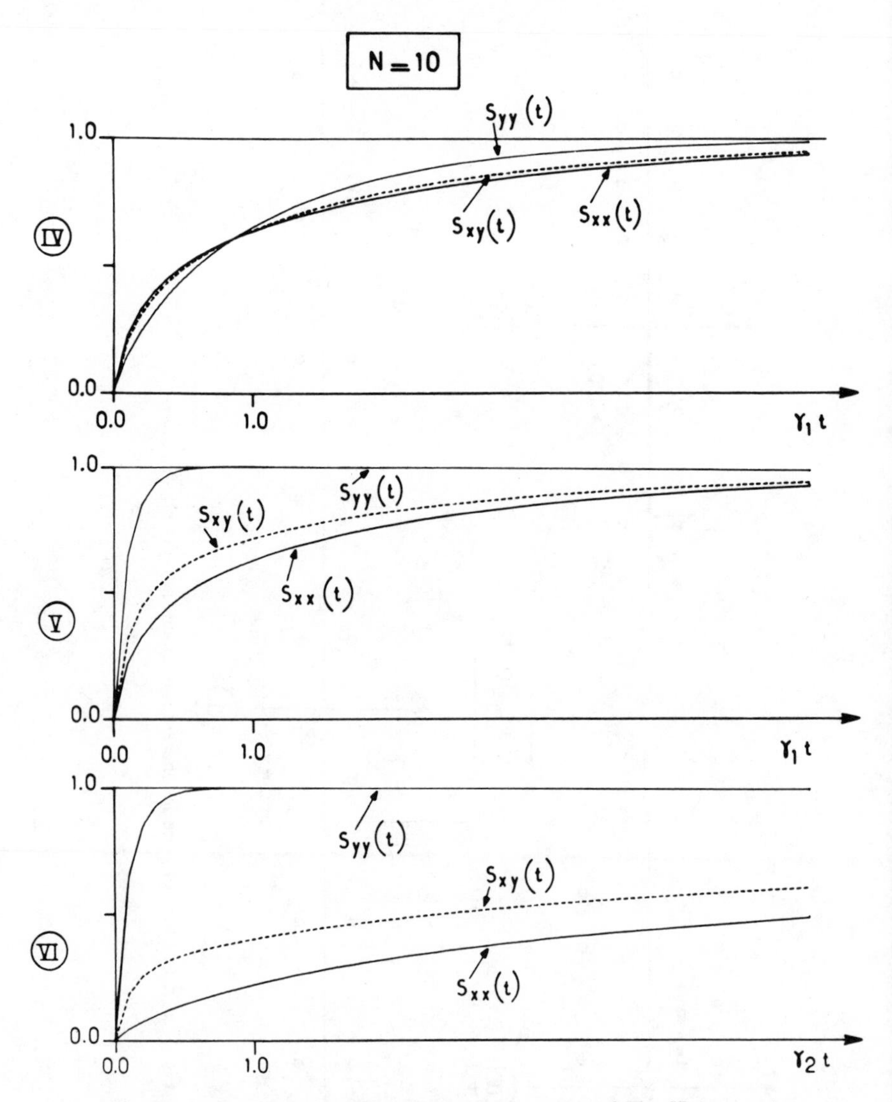

Fig. 58. Delocalization entropies for the three cases of Fig. 57 (see text).

so that the vector $\{(d/dt)\mathcal{M}_1(t)\}_{t=0}$ lies at an angle α to Ox, defines by

$$tg\alpha = \frac{\gamma_2}{\gamma_1}\frac{Y}{X}$$

The initial values of the spread are

$$\Delta^2_{yy}(t) = 2\gamma_2 Y^2 t + 0(t) \qquad \Delta^2_{xx}(t) = 2\gamma_1 X^2 t + 0(t)$$

The initial anisotropy of the dispersion tensor depends, therefore, solely on the ratio $\gamma_2/\gamma_1\,(Y/X)^2$. The asymptotic limits are

$$\Delta^2_{yy}(+\infty)=\frac{Y^2}{4}\qquad \Delta^2_{xx}(+\infty)=\frac{N^2-1}{12}X^2$$

The relaxation of the first moment is in general strongly anisotropic. Its $\mathcal{M}_{1,y}$ component has a lifetime of the order of γ_2^{-1} for all N, whereas the longer the chain the slower the decline (of the modulus) of $\mathcal{M}_{1,x}$. We saw in Section IV.F.2 that the average speed between $t=0$ and $t=\gamma_1^{-1}$ depended little on N, whereas the distance along Ox to the equilibrium position increases with N. Thus the anisotropy of the relaxation time of $\mathcal{M}_1$ depends not only on γ_2/γ_1 and Y/X but also on the length of each chain.

$S_{yy}(t)$, which is independent of N, reaches its equilibrium value by $t\sim\gamma_2^{-1}$. $S_{xx}(t)$ on the other hand grows at a rate varying inversely with γ_2^{-1}. Relation (4.72) is an equality (no statistical correlation). $S_{xy}(t)$ is the average of $S_{xx}(t)$ and $S_{yy}(t)$ with respective weights $\ln N/\ln 2N$ and $\ln 2/\ln 2N$, and increases at a rate which is less for large than for small N, partly because $S_{xx}(t)$ depends on N in the same way and partly because the second factor $\ln 2\ \ln 2N$ is negligible compared to the first when N is large.

V. CONCLUSION AND PERSPECTIVES

This study has been concerned with two properties of finite, organic, molecular aggregates in a nonresonant matrix. The model used treats exciton–phonon interactions phenomenologically.

We first calculated the optical susceptibility of these systems, supposing that they are perturbed from a macroscopic equilibrium (in practice all the aggregates in the ground state) by a classical, homogeneous source (not a laser), whose wavelength is large, not only relative to each aggregate, but also to a portion of crystal containing enough of them to simulate a statistical ensemble.

By introducing id-operators in the second part of this chapter it was possible to derive by simple algebra an exact expression of the susceptibility $\chi(\omega)$. Their introduction led to a natural partition of the mathematical space into several subspaces, one for each kind of physical phenomenon and to a qualitative structural difference between operators of reversible movements (like $\hat{H}_{0-}$) and those causing relaxation (like $\hat{\Gamma}$).

The expression of $\chi(\omega)$ revealed non-Lorentzian contributions to each homogeneous line, a direct result of the finite length of the chains. Besides these "end effects", which are negligible for chains of more than $N=20$ sites, strong inhomogeneous broadening occurs for medium sized ($N \sim 50$) chains, causing very asymmetric lines in which the intensity tails off more slowly on the side of the line facing the center of the exciton "band."

The broadening is due to special selection rules of finite systems, calculated here in three cases of physical interest (parallel, alternating, and helical chains). It was shown that these rules are due to the aptitude of the aggregate to react as a whole in response to optical excitation. It is wrong to imagine as is often (implicitly) done in the literature, especially that on the transfer of energy, that absorption is a local process concerning one site and that the redistribution of the energy is due to residual interactions. The collective character of the aggregate is like that of benzene, for which it would be unreasonable to suggest that just one carbon atom at a time absorbs.

We have seen that the important quantity is not the strength of site to site couplings, but their separation compared to the optical wavelengths. Indeed, this is the very basis of the assumption of a homogeneous field.

We then calculated the propagation of energy along a chain during an interval much smaller than the excited lifetime, although this restriction is not essential. After discussing necessary and sufficient conditions of preparing a localized state by direct optical excitation, we introduced an energy density. Its kernel expressed on the coordinates of a reduced one-body space was used to calculate the spatial moments of the exciton, notably the first moment, or mean position and the second moments (defining a spread tensor).

Numerous illustrations showed that the initial localization soon vanishes. Besides movement of the center of the wave packet, a characteristic of finite systems, the spread about the center grows (with oscillations in the coherent case), showing that the excitation does indeed statistically fill the whole volume of the aggregate. When equilibrium is reached the exciton is uniformly distributed (this is true also of Boltzmann populations) and all the "delocalization entropies" are maximum.

We have shown also that localization, which disappears when averaged over the ensemble, has no more meaning at the microscopic level, for it is contradictory both to statistical averaging and to the premises of quantum mechanics. Even neglecting the statistical side of it, which would amount to considering a rigid chain and completely coherent motion, we should still have to contend with the usual statistical interpretation of quantum mechanics. This point has been in debate for the past fifty years ("Through which hole does the electron pass? . . .") and we can do no more here than remind the reader that it exists. In any event, the fluctuations of the energy associated with a volume the size of a site are very rapid and comparable to its average values.

Owing to the noncommutation of the operators $\hat{H}_{0-}$ and $\hat{\Gamma}$ associated, respectively, with the wavelike and scattering characteristics of the motion, these characteristics are intimately mingled. It is in general wrong to treat the motion of the exciton as the superposition of a wavelike component and a scattering component. Any attempt to separate them is quite in vain.

Although this work is complete in itself as an application of Haken's and Strobl's model, some extensions nonetheless might be made.

Two formal improvements could be made. Firstly, it would be useful to better define the concept of momentum in a finite aggregate since strictly speaking it has meaning only in an infinite crystal. Conformal mappings[156] may play an important part in establishing this concept in finite crystals. It would then be possible to apply second quantification techniques to them.

Secondly, it would be an advantage to introduce finite correlation times in the auto-correlation functions of the stochastic operators. It seems algebraic methods by which the macroscopic equation could be derived are not available at present, except of course low-order perturbation treatments.

The first extension which could be made, is a quantum treatment of exciton–phonon coupling. It would require a precise description of the mechanism of the interaction. This difficulty set aside the problem amounts to defining a reduced exciton density operator $\hat{\bar{\rho}}$ by reduction of the total density operator on the degrees of freedom of the phonons. The equation for $\hat{\bar{\rho}}$ may be written in the form

$$i\hbar \frac{d}{dt}\hat{\bar{\rho}}(t) = [\hat{H}_0', \hat{\bar{\rho}}(t)] - i\hbar\hat{\Gamma}\{\hat{\bar{\rho}}(t)\}$$

$\hat{H}_0'$ is a renormed Hamiltonian deduced from $\hat{H}_0$ after dressing the excitons in phonons and $\hat{\Gamma}\{\hat{\bar{\rho}}(t)\}$ is a relaxation operator. General considerations show that it must behave heredity. Writing and solving this equation, even for the simplest cases, would be a considerable task. This approach would enable one to analyze the validity of the assumption that in this kind of problem the phonons are a heat bath. There seems to be no detailed discussion of this point in the literature.

Another possible extension would be the inclusion of annihilation of excitons, for experimental methods have made great strides recently, due to the development of lasers (and of low-temperature techniques). It is now possible to create high densities of excitons so that transitions to very excited states by fusion are quite probable.

It seems important to establish a proper link with experiments on the transfer of energy. At present there is a wide gap between theoretical and experimental concepts. The theoretician and the experimenter have clear but differing conceptions of the exciton density $\Phi(\mathbf{r}; t)$. The former sees in $\mathbf{r}$ a vector whose length is arbitrarily well defined relative to the angstrom, whereas the latter envisages rather a "macroscopic point" in the crystal, containing many molecules or sites. In other words we need to consider theoretically domains of spatial coherence of which the average values would have an immediate experimental interpretation. Otherwise the theoretician can only introduce quantities of which fluctuations (errors) are comparable to the average values, which greatly compromises their macroscopic significance. Renormalization techniques, which have been so fruitful in N-body problems, particularly solid state physics, might prove useful here.

Another important phenomenon in need of a good description is the modulation of the emission of a structural defect, for example, trap or oriented trap, by the variable exciton population of the site. It would probably be necessary to quantify the matter–field interaction to solve this problem.

In fact, a quantum treatment of the field (introducing the concept of the polariton), would be the last touch. Hopfield[157,128] has shown that finite crystals must be considered explicitly to avoid contradictions like the one which states that the excited state cannot emit because its lifetime is infinite. This is yet another sign that we must make realistic models of crystals and not let the relative simplicity of the infinite lattice blind us to its limitations.

References and Notes

1. J. I. Frenkel, *Phys. Rev.*, **37**, 17 (1931); **37**, 1276 (1931).
2. A. S. Davydov, *Theory of Molecular Excitons*, McGraw-Hill, New York, 1962.
3. R. S. Knox, *Theory of Molecular Excitons*, Plenum, New York, 1971.
4. D. S. McClure, *Solid State Phys.*, **8**, 1 (1959).
5. H. C. Wolf, *Adv. Atom. Molec. Phys.*, **3**, 319 (1967).
6. M. R. Philpott, *Adv. Chem. Phys.*, **23**, 227 (1973).
7. Y. Toyozawa, *Progr. Theoret. Phys.*, **20**, 53 (1958).
8. P. Vigny and M. Duquesne, *Excited States of Biological Molecules*, Lisbonne, 1974.
9. J. Franck and E. Teller, *J. Chem. Phys.*, **6**, 861 (1938).
10. G. Paillotin, *J. Theoret. Biol.*, **36**, 223 (1972).
11. R. Kopelman, *Seventh Molecular Crystal Symposium*, Nikko, 1975.
12. W. A. Little, *Phys. Rev.*, **A134**, 1416 (1964); **156**, 396 (1967).
13. W. A. Little and H. Gutfreund, *Phys. Rev.*, **B4**, 817 (1971).
14. D. M. Hanson, *J. Chem. Phys.*, **52**, 3409 (1970).
15. D. M. Hanson, R. Kopelman, and G. W. Robinson, *J. Chem. Phys.*, **51**, 212 (1969).
16. C. L. Braun, H. C. Wolf, *Chem. Phys. Lett.*, **9**, 260 (1971).

17. K. E. Mauser, H. Port, and H. C. Wolf, *Chem. Phys.*, **1**, 74 (1973).
18. H. Port, D. Vogel, and H. C. Wolf, *Chem. Phys. Lett.*, **34**, 23 (1975).
19. J. D. Whiteman, Thesis, University of Pennsylvania, 1971.
20. J. D. Whiteman and R. M. Hochstrasser, *J. Chem. Phys.*, **56**, 5945 (1972).
21. Ph. Kottis, *Phys. Lett.*, **17**, 206 (1965).
22. Ph. Kottis, *J. Chem. Phys.*, **47**, 509 (1967).
23. Ph. Kottis, *Annal. Phys.* **4**, 459 (1969).
24. J.-P. Lemaiste and Ph. Kottis, in *Electron Spin Resonance in Liquids* (Ed. L. T. Muus and P. W. Atkins), Plenum, New York, 1972.
25. A. H. Zewail and C. B. Harris, *Chem. Phys. Lett.*, **28**, 8 (1974).
26. A. H. Zewail and C. B. Harris, *Phys. Rev. B*, **11**, 935 (1975).
27. A. H. Zewail and C. B. Harris, *Phys. Rev. B*, **11**, 952 (1975).
28. P. W. Anderson, *Phys. Rev.*, **109**, 1492 (1958).
29. N. F. Mott, *Contemp. Phys.*, **10**, 125 (1969).
30. T. Holstein, *Ann. Phys.*, **8**, 325 (1959); **8**, 343 (1959).
31. P. W. Anderson, *J. Phys. Soc. Jap.*, **9**, 316 (1954).
32. G. L. Sewell, *Phys. Rev.*, **129**, 597 (1963).
33. R. Munn and W. Siebrand, *J. Chem. Phys.*, **52**, 47 (1970).
34. For example: The squared moduli coefficients of the development of the states of an anharmonic oscillator on the eigenstate of a harmonic oscillator are not quantum probabilities. The same is true of the coefficients of configuration interaction, etc.
35. We thus disagree with the authors of Ref. 33, who affirm: "... the understanding of the electronic transport properties ... needs to be studied in terms of transitions between localized states."
36. A model could be said to be a set of propositions defining as invariant quantities which are not precisely invariant.
37. Cl. Cohen-Tannoudji, *Annal. Phys.* **7**, 423 (1962).
38. M. I. Podgoretskii and O. A. Khrustalev, *Sov. Phys. Usp.*, **6**, 682 (1964).
39. N. Novilov, V. G. Pokazan'ev, and G. V. Skrotskii, *Sov. Phys. Usp.*, **13**, 384 (1970).
40. H. Haken and G. Strobl, in *The Triplet State*, (Ed. A. B. Zahlan), Cambridge University Press, London/New York, 1968.
41. H. Haken and G. Strobl, *Z. Phys.*, **262**, 135 (1973).
42. R. Voltz and Ph. Kottis, in *Localization and Delocalization in Quantum Chemistry*, Vol. II (Ed. O. Chalvet, R. Daudel, S. Diner, and J.-P. Malrieu), Reidel, Dordrecht, 1976, p. 187.
43. U. Fano, *Rev. Mod. Phys.*, **29**, 74 (1957).
44. D. S. McClure, *J. Chem. Phys.*, **22**, 1668 (1954).
45. C. D. Akon and D. P. Craig, *Trans. Farad. Soc.*, **63**, 56 (1967).
46. C. A. Hutchison and B. W. Mangum, *J. Chem. Phys.*, **34**, 908 (1964).
47. A. Suna, *Phys. Rev. B*, **1**, 1716 (1970).
48. P. Reineker, Thesis, University of Stuttgart, 1971.
49. H. Haken and P. Reineker, *Z. Phys.*, **249**, 253 (1972).
50. R. Kubo, *J. Math. Phys.*, **4**, 174 (1963).
51. I. R. Senitzky, *Phys. Rev.*, **119**, 670 (1960).
52. I. R. Senitzky, *Phys. Rev.*, **124**, 642 (1961).
53. A. Nitzan and R. Silbey, *J. Chem. Phys.*, **60**, 4070 (1974).
54. W. H. Louisell, *Quantum Statistical Properties of Radiation*, Wiley, New York, 1973. especially Chap. 6.
55. The shift is given by the principal part of $\Phi(\omega)/(\omega-\omega_0)$ where $\Phi(\omega)$ is in general a slowly varying function of ω, roughly even with respect to ω_0.[56]

56. Cl. Cohen-Tannoudji, F. Diu, and B. Laloë, *Mécanique Quantique*, Hermann, Paris, 1973.
57. U. Fano, *Phys. Rev.*, **124**, 1866 (1961).
58. J. L. Doob, *Ann. Math.*, **43**, 351 (1942).
59. M. C. Wang and G. Uhlenbeck, *Rev. Mod. Phys.*, **17**, 323 (1945).
60. P. Reineker and H. Haken, *Z. Phys.* **250**, 300 (1973).
61. E. Schwarzer and H. Haken, *Phys. Lett.*, **42A**, 317 (1972).
62. P. Reineker, *Z. Phys.* **261**, 187 (1973).
63. P. Reineker, *Phys. Lett.*, **42A**, 389 (1973).
64. P. Reineker, *Phys. Lett.*, **44A**, 429 (1973).
65. P. Reineker, *Z. Naturforsch.*, **29a**, 282 (1974).
66. P. Reineker, *Phys. Stat. Sol. (b)*, **70**, 189 (1975); **70**, 471 (1975); **74**, 121 (1976).
67. M. K. Grover and R. Silbey, *J.Chem. Phys.*, **52**, 2099 (1970).
68. M. K. Grover and R. Silbey, *J. Chem. Phys.*, **54**, 4843 (1971).
69. M. K. Grover, Thesis, Massachussetts Institute of Technology, 1971.
70. V. M. Kenkre, *Phys. Lett.*, **47A**, 119 (1974).
71. V. M. Kenkre and R. S. Knox, *Phys. Rev. B*, **9**, 5279 (1974).
72. V. M. Kenkre and R. S. Knox, *Phys. Rev. Lett.*, **33**, 803 (1974).
73. V. M. Kenkre and T. S. Rahman, *Phys. Lett.*, **50**, 170 (1974).
74. V. M. Kenkre, *Phys. Rev. B*, **11**, 1741 (1975).
75. V. M. Kenkre, *Phys. Rev. B*, **12**, 2150 (1975).
76. H. Haken and W. Weidlich, *Z. Phys.* **205**, 96 (1967).
77. R. H. Clarke and R. M. Hochstrasser, *J. Chem. Phys.*, **46**, 4532 (1967).
78. P. Avakian and R. E. Merrifield, *Molec. Cryst.*, **5**, 37 (1968).
79. R. W. Zwanzig, in *Lectures in Theoretical Physics*, Vol. III (Ed. W. E. Brittin, B. W. Downs, and J. Downs), Interscience, New York, 1961, p. 106.
80. R. W. Zwanzig, *J. Chem. Phys.*, **33**, 1338 (1960).
81. R. W. Zwanzig, *Phys. Rev.*, **124**, 983 (1961).
82. R. W. Zwanzig, *Physica*, **30**, 1109 (1964).
83. R. D. Levine, *Quantum Mechanics of Molecular Rate Processes*, Clarendon, London, 1969.
84. Ref. 74, equation 5.
85. P. Reineker and H. Haken in *Localization and Delocalization in Quantum Chemistry*, Vol. II (Ed. O. Chalvet, R. Daudel, S. Diner, and J.-P. Malrieu), Reidel, Dordrecht, 1976, p. 285.
86. L. Landau and E. Lifchitz, *Mécanique Quantique*, Mir, 1966, equation (14.10).
87. A. Renyi, *Calcul des Probabilités*, Dunod, Paris, 1966.
88. R. H. Dicke, *Phys. Rev.*, **93**, 99 (1954).
89. Cl. Cohen-Tannoudji, Lecture notes from Collège de France, 1974/1975.
90. E. Fermi, *Rev. Mod. Phys.*, **4**, 87 (1932).
91. E. Merzbacher, *Quantum Mechanics*, Chap. 13, Wiley, New York, 1970.
92. D. M. Brink and G. R. Satchler, *Angular Momentum*, Sect. 6.4, Clarendon, London, 1962.
93. D. Ter Haar, *Rep. Progr. Phys.*, **24**, 304 (1961).
94. R. M. Lynden-Bell, *Molec. Phys.*, **22**, 837 (1971).
95. H. Bacry, *Leçons sur la Théorie des Groupes et les Symétries des Particules Élémentaires*, Gordon and Breach, distributed by Dunod, 1967.
96. Cl. Aslangul and Ph. Kottis, *C.R. Acad. Sci. Paris*, **B278**, 33 (1974).
97. Cl. Aslangul and Ph. Kottis, *Phys. Rev. B*, **10**, 4364 (1974).
98. W. Magnus, *Comm. Pure Appl. Math.*, **7**, 649 (1954).

99. J. Wei and E. Norman, *J. Math. Phys.*, **4**, 575 (1063).
100. R. M. Wilcox, *J. Math. Phys.*, **8**, 962 (1967).
101. R. P. Feynman, *Phys. Rev.*, **84**, 108 (1951).
102. E. U. Condon and H. Odishaw, *Handbook of Physics*, Chap. 12, McGraw-Hill, New York, 1958.
103. R. Kubo, *J. Phys. Soc. Jap.*, **17**, 1100 (1962).
104. R. Kubo, in *Fluctuation, Relaxation and Resonance in Magnetic Systems* (Ed. D. ter Haar), Oliver and Boyd, Edinburgh, 1962.
105. Let X be a normal variable of mean m and standard deviation σ. Then

$$\langle e^{itx}\rangle = \int_{-\infty}^{+\infty} dx(2\pi\sigma^2)^{-1/2} \exp\left\{-\frac{(x-m)^2}{2\sigma^2}\right\} \exp(itx)$$
$$= \exp\left(itm - t^2\frac{\sigma^2}{2}\right)$$

But by definition of the cumulants, $\mathcal{K}_n$,

$$\langle e^{itx}\rangle = \exp\left\{\sum_{n=0}^{\infty} (it)^n \mathcal{K}_n\right\}$$

Identifying terms,

$$\mathcal{K}_1 = m;\ \mathcal{K}_2 = \sigma^2;\ \mathcal{K}_{2+n} = 0;\ \forall n > 0.$$

106. U. Fano, *Phys. Rev.*, **131**, 259 (1963).
107. P. Avakian, V. Ern, R. E. Merrifield, and A. Suna, *Phys. Rev.*, **165**, 974 (1968).
108. V. Ern and M. Schott, in *Localization and Delocalization in Quantum Chemistry*, Vol. II (Eds. O. Chalvet, R. Daudel, S. Diner, and J.-P. Malrieu), Reidel, Dordrecht, 1976, p. 249.
109. J.-M. Levy-Leblond, *Bull. Soc. Fr. Phys.*, Encart pédagogique n° 14 (1973).
110. H. Everett, *Rev. Mod. Phys.*, **29**, 454 (1957).
111. B. D'Espagnat, *Conceptions de la Physique Contemporaine*, Hermann, Paris, 1965.
112. H. Primas, *Theoret. Chim. Acta*, **39**, 127 (1975).
113. R. Constanciel, *Phys. Rev. A*, **11**, 395 (1975).
114. P. Claverie and S. Diner, in *Localization and Delocalization in Quantum Chemistry*, Vol. II (Ed. O. Chalvet, R. Daudel, S. Diner, and J.-P. Malrieu), Reidel, Dordrecht, 1976 p. 395.
115. L. E. Ballentine, *Rev. Mod. Phys.*, **42**, 358 (1970).
116. M. Parodi, *Applications de l'Algèbre Moderne à Quelques Problèmes de Physique Classique*, Gauthier-Villars, Paris, 1961.
117. H. Hochstadt, *Les Fonctions de la Physique Mathématique*, Masson, Paris, 1973.
118. M. R. Spiegel, *Outline of Theory and Problems of Laplace Transforms*, (McGraw-Hill, 1965.
119. M. L. Goldberger and K. M. Watson, *Collision Theory*, Wiley, New York, 1964. It is not strictly necessary to introduce η, except in subspace $\mathcal{A}_K$, for $\hat{\Gamma}$ is by definition positive in the supplementary subspace.
120. R. Kubo and T. Tomita, *J. Phys. Soc. Jap.*, **9**, 888 (1954).
121. R. Kubo, *J. Phys. Soc. Jap.*, **12**, 570 (1957).
122. C. P. Slichter, *Principles of Magnetic Resonance*, Harper and Row, New York, 1963.
123. D. des Cloiseaux, in *Theory of Condensed Matter*, International Atomic Energy Agency, Vienne, 1968.
124. Cl. Aslangul and Ph. Kottis, *C.R. Acad. Sci. Paris*, **B278**, 705 (1974).
125. M. Kasha, H. R. Rawls, and M. A. El-Bayoumi, *Pure Appl. Chem.*, **11**, 371 (1965).

126. E. A. Chandross, J. Ferguson, and E. B. McRae, *J. Chem. Phys.*, **45**, 3546 (1966).
127. R. M. Hochstrasser, *J. Chem. Phys.*, **47**, 1015 (1967).
128. S. A. Rice and J. Jortner, in *Physics and Chemistry of the Organic Solid State* (Ed. D. Fox, M. M. Labes, and A. Weissberger), Interscience, New York, 1967.
129. V. Ern, A. Suna, Y. Tomkiewicz, P. Avakian, and R. P. Groof, *Phys. Rev. B*, **5**, 3222 (1972).
130. Cl. Aslangul and Ph. Kottis, *C.R. Acad. Sci. Paris*, **B279**, 523 (1974).
131. Cl. Aslangul and Ph. Kottis, *Phys. Rev. B*, **13**, 5544 (1976).
132. J. S. Toll, *Phys. Rev.*, **104**, 1760 (1956).
133. P. Roman, *Advanced Quantum Theory*, Sect. 3.4, Addison-Wesley, Reading, Mass., 1965.
134. V. Ern, P. Avakian, and R. E. Merrifield, *Phys. Rev.*, **148**, 862 (1966).
135. V. Ern, *J. Chem. Phys.*, **56**, 6259 (1972).
136. Section VIII of the review in Ref. 128.
137. Cl. Aslangul, J.-P. Lemaistre, and Ph. Kottis, in *Localization and Delocalization in Quantum Chemistry* Vol. II (Ed. O. Chalvet, R. Daudel, S. Diner, and J.-P. Malrieu), Reidel, Dordrecht, 1976, p. 209.
138. This intuitive picture must not be taken too seriously and one reason, that of size, justifies this provision: The spatial extent of a photon has not much meaning because it is impossible to define a wave function in coordinate representation for a photon. See, for example: Cl. Cohen-Tannoudji, Lecture notes, Collège de France (1973/1974); L. Landau and E. Lifchitz, *Théorie Quantique Relativiste*, Mir, 1972.
139. We have made an approximation here in neglecting the variation of $\hbar\Delta\omega$ compared to a single site. This change is not fundamental.
140. Cl. Aslangul and Ph. Kottis, *C.R. Acad. Sci. Paris*, **B278**, 735 (1974).
141. J. M. Jauch, *Foundations of Quantum Mechanics*, Addison-Wesley, Reading, Mass., 1968.
142. A. Messiah, *Mécanique Quantique*, Section II-15, Dunod, Paris, 1962.
143. K. Husimi, *Proc. Phys. Math. Soc. Jap.*, **22**, 264 (1940).
144. P. O. Löwdin, *Phys. Rev.*, **97**, 1474 (1955).
145. R. McWeeny, *Rev. Mod. Phys.*, **32**, 335 (1960).
146. See the book in Ref. 56.
147. J. M. Ziman, *Elements of Advanced Quantum Theory*, Sect. 1.7, Cambridge University Press, London/New York, 1969.
148. A. S. Davydov, *Quantum Mechanics*, §138 and following, Pergamon, New York, 1965.
149. See Ref. 86, §14, equation (14.4).
150. J. Jortner, S. A. Rice, J. L. Katz, and S. Il Choï, *J. Chem. Phys.*, **42**, 309 (1965).
151. R. E. Merrifield, *Phys. Rev.*, **148**, 862 (1966).
152. J. Killingbeck and G. H. A. Cole, *Mathematical Techniques and Physical Applications*, Academic, New York, 1971.
153. V. I. Krylov, *Approximate Calculation of Integrals*, McMillan, New York, 1962.
154. J. Legras, *Précis d'Analyse Numérique*, Dunod, Paris, 1962.
155. L. Brillouin, *La Science et la Théorie de l'Information*, Masson, Paris, 1951.
156. M. Lavrentiev and B. Chabat, *Méthodes de la Théorie des Fonctions d'une variable Complexe*, Mir, 1972.
157. J. J. Hopfield, *Phys. Rev.*, **112**, 1555 (1958).
158. Cl. Aslangul and Ph. Kottis, *Phys. Rev. B*, **18**, 4462 (1978).

HETEROGENEOUS REACTION DYNAMICS

STEVEN L. BERNASEK

Department of Chemistry
Princeton University
Princeton, New Jersey 08540

CONTENTS

I. INTRODUCTION

Heterogeneous reaction dynamics implies interfacial chemical reaction. This review chapter concerns itself with chemical transformations taking place at the solid–gas interface, and does not deal with the solid–solid, solid–liquid, or liquid–gas interfaces. It is not concerned directly with what might be called heterogeneous catalytic kinetics, but rather with the newly emerging collection of studies of elementary chemical processes occurring on well-characterized solid surfaces.

The purpose of this review is not only to organize and comment on studies of elementary chemical processes at the solid–gas interface, but also to emphasize the fact that this rapidly growing field is developing at the interface between two well-established areas of chemical physics, surface chemical physics and molecular reaction dynamics. For this reason, the review should appeal to graduate students and active research workers in both fields. It is written from the perspective of an active experimentalist, with research interests and experience on both sides of the disciplinary interface. The review emphasizes experimental studies, but theoretical work is not ignored, especially where theoretical studies lend particular insight or impetus to the investigation of a particular reaction system.

Section II describes the techniques available for the study of heterogeneous reaction dynamics. Background information concerning these techniques is included from both sides of the disciplinary interface. Section III presents a number of case studies of model heterogeneous processes, indicating the impact which the techniques described in the previous section have had on the understanding of these model processes. Section IV indicates the direction of work in this field and suggests areas which appear to be ripe for development.

II. TECHNIQUES AVAILABLE FOR HETEROGENEOUS REACTION DYNAMICS STUDIES

In any rapidly growing field of chemical physics, perhaps especially the field of heterogeneous reaction dynamics, it is difficult to review and discuss current work in the field without some preliminary discussion of the techniques available for these studies. In the following section, techniques presently being used in studies of heterogeneous reaction dynamics are briefly described. The information content of a particular technique is emphasized, and critical comments on the suitability of a particular method are included. Detailed descriptions of individual techniques are not presented, but references to the extensive literature dealing with these methods are given.

The discussion of techniques is somewhat arbitrarily divided into two parts. Techniques developed specifically for characterization of surface properties are discussed first. Molecular reaction dynamics techniques are then treated. Throughout this section, the applicability of the individual techniques to the investigation of heterogeneous reaction dynamics is stressed.

A. Solid Surface Characterization

To adequately characterize a solid surface, three types of information about the surface must be available. These three types are structure, composition, and energy state. The first two types of information are easily definable and more readily available than the third. The energy state of a solid surface is not as clearly understood a quantity as the energy state of a gas phase molecule, and for this reason, characterization of the energy state of a solid surface is not straightforward. In the following sections, techniques that give information about solid surfaces in these three areas are described.

1. Structural Information

Information concerning the structure of a solid surface participating as either catalyst or reactant in a heterogeneous process is essential to an understanding of heterogeneous reaction dynamics. There are a number of experimental techniques available which give solid surface structural information. None of these techniques can routinely provide the precise molecular structural information available from spectroscopic methods in the gas phase or diffraction methods in the solid phase. However, the techniques following do give valuable primary structural information, which is useful in the interpretation of heterogeneous reaction dynamics, especially when combined with surface composition and surface energy state information obtained by complementary techniques. Three classes of surface structural tool are discussed: diffraction methods, angle resolved electron spectroscopy, and vibrational spectroscopy.

a. Electron Diffraction. Electron diffraction is the most widely used surface structural technique presently available. Surface structural information can be obtained from both low-energy electron diffraction (LEED) and from reflection high-energy electron diffraction (RHEED).[1] This discussion concentrates on the use of LEED as a surface structural tool, since LEED has found very widespread application in surface studies.

There are a number of excellent review articles and texts available, discussing the LEED technique in detail.[2] This discussion concentrates on the information available from the technique and the advantages and disadvantages LEED has in comparison with other surface structural methods. Information from the LEED experiment falls into two categories. The first class of information, which is readily obtained using commercially available display type LEED optics,[3] concerns surface unit cell size and shape. The second class of information depends on a detailed analysis of diffraction intensity as a function of incident electron wavelength and is not as simply obtained as the first type of information.

Surface unit cell information is obtained by examining the electron diffraction pattern in a LEED experiment, illustrated schematically in

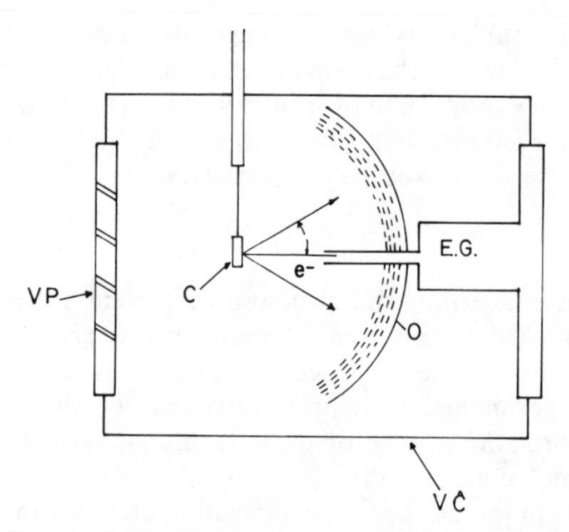

Fig. 1. Schematic diagram of the LEED experiment. VP: viewport; C: crystal, EG: electron gun, O: electron optics, VC: vacuum chamber.

Fig. 1. The position of the diffraction beams for a specific incident electron wavelength can be used to determine the symmetry of the array of scatterers in the surface plane of the sample. When slow electrons are used, elastic scattering is detected only from the first few layers of the solid, so that the diffraction pattern is indicative of order in the surface region. Figure 2 shows an example of a LEED pattern from the (111) surface of platinum.

This sort of two-dimensional information is readily available from single crystal surfaces using LEED. It is possible to readily observe changes in clean surfaces such as surface reconstruction[4] and lateral distortion.[5] This two-dimensional information can also be easily used to characterize the terrace width of stepped surfaces.[6] These surfaces, prepared by cutting or cleaving a single crystal a few degrees away from a low-index face,[7] offer a means of introducing a controlled concentration of a specific type of defect into a surface. In this way, the effect of unsaturated defect sites on the course of heterogeneous reactions can be systematically studied.

Similar information is readily obtained concerning changes in surface structure upon adsorption of heterogeneous reaction reactants or products. The simplest sort of information available in this case is contained in qualitative changes in the clean surface LEED pattern. Amorphous adsorption results in a diminution of substrate feature intensity, along with an increase in random background scattering.[8]

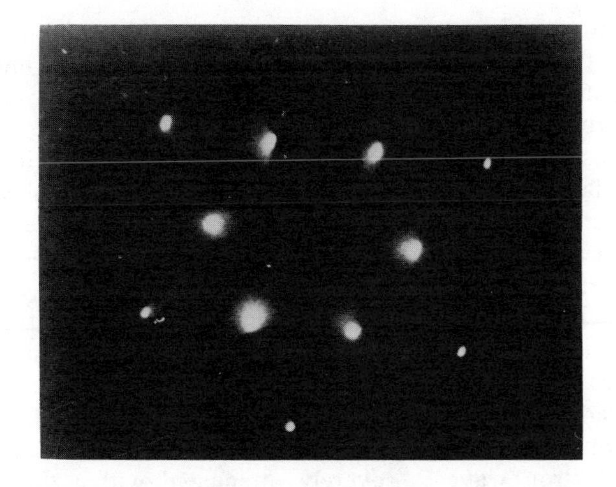

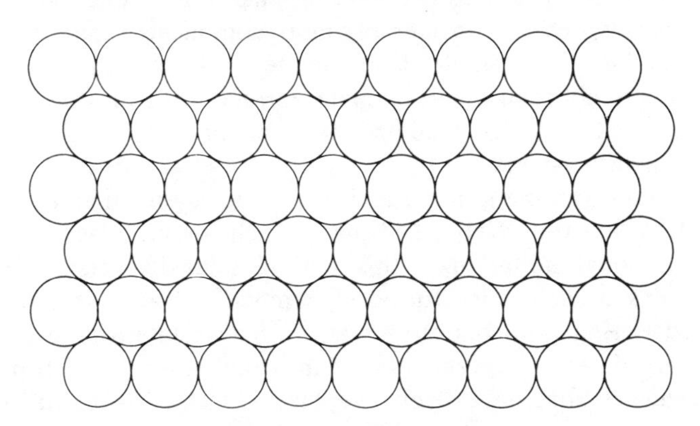

Fig. 2. LEED pattern and real space schematic of Pt(111).

Often, adsorbed species will form ordered overlayers, with appearance of new diffraction features indicating an overlayer with new surface unit cell dimensions. Much information is available from the appearance of these new features as a function of adsorbate exposure and surface temperature.[9] For example, progressively closer packed overlayer structures have been observed as adsorbate exposure is increased,[10] indicating a surface coverage dependence for favored overlayer structures.

For a complete understanding of surface reaction dynamics it is often necessary to obtain more explicit structural information than is readily obtainable from diffraction pattern geometries and qualitative pattern features. In these instances, it is necessary to analyze the intensity of the diffraction features as a function of the incident electron wavelength. These experimental intensity profiles can then be compared with calculated intensity profiles based on model surface structures. In this way, true three-dimensional surface structural determination can be accomplished. This allows assignment of adsorption site, substrate adsorbate bond distances, and in some cases conclusions can be reached about adsorbate identity and bond distortion.[11]

The problems associated with the routine extraction of this sort of surface structural information are related to the surface-sensitive nature of the LEED technique. LEED is surface sensitive because of the strong scattering of low-energy electrons by the ion cores of the solid.[12] Significant multiple scattering occurs in the solid, and the incident and diffracted electron wave is severely attenuated within the first few layers. Both these processes combine to make a theoretical description of the scattering process considerably more complicated than the case of X-ray diffraction, for example. Unfortunately this computational limitation has effectively restricted complete surface structural analysis to a few clean metal surfaces,[13] compound semiconductor surfaces,[14] and simple overlayer systems.[15]

The constant momentum transfer averaging technique (CMTA) developed by Lagally, Webb, and others,[16] shows promise of reducing this computational limitation. In this method, intensity versus scattering vector profiles are calculated for possible model surface structures using a single scattering (kinematic) formalism. These intensity profiles are then compared with experimental data from several electron beam incidence angles that have been averaged (Fig. 3) to obtain constant momentum transfer profiles. This method assumes that constant momentum transfer averaging over a wide enough range of incident angles will essentially average out multiple scattering contributions to the intensity profiles. Although there is no absolute theoretical justification for this assumption,[17] in practice it appears to work rather well. The CMTA method has been applied to a number of systems, clean metal,[18] clean compound semiconductor,[19] and metal overlayer,[20] with very encouraging results. The large computational requirement of the multiple scattering approach to LEED structural analysis is reduced considerably by the CMTA method, but the experimental data base necessary for comparison with structure models is considerably increased. This shift in emphasis to the experimental side of the problem necessitates the use of advanced rapid data acquisition[21] and reduction schemes[22] that have been recently developed. Although the CMTA

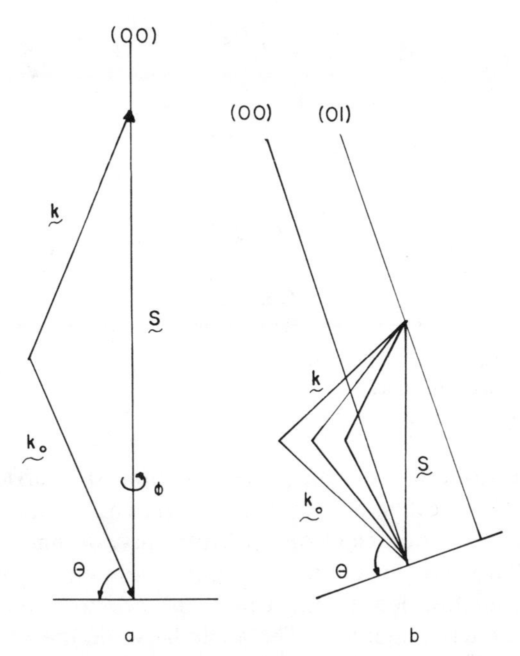

Fig. 3. Reciprocal space geometry—CMTA in LEED. (*a*) Fixed polar angle θ, azimuthal angle ϕ varied, giving constant momentum transfer. (*b*) Variable polar angle and energy, fixed azimuthal angle, giving constant momentum transfer. $\mathbf{k}_0$: incident wavevector, $\mathbf{k}$: scattered wavevector, $\mathbf{S}$: scattering vector.

method has limitations, both theoretically and experimentally, it shows promise for the routine analysis of complex metal compound surface structures, reconstructed surfaces, and large unit cell overlayer systems which are not readily treated using the standard multiple scattering formalism.

In addition to the difficulties encountered in LEED structural analysis described previously, which are essentially computational difficulties in describing the scattering process, there are experimental limitations of the electron probe which make LEED less than an ideal structural analysis tool. Electron beam damage to the surface under study certainly occurs in the electron energy range used in LEED. This problem is especially severe for covalently bonded transition metal compound surfaces,[23] and adsorbed overlayers.[24] For these sensitive systems low electron currents must be used, coupled with rapid data acquisition.

b. Angle-resolved Electron Spectroscopy. A great deal of interest has been generated recently by the possibility of using angle-resolved photo-

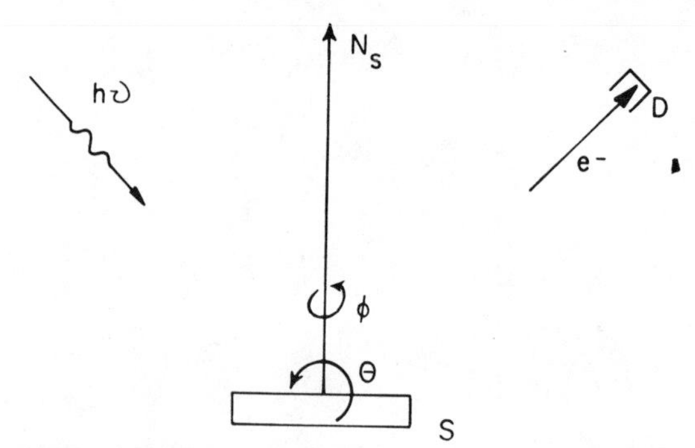

Fig. 4. Schematic diagram of angle-resolved PES. N_s: surface normal, D: detector, S: sample, θ: polar angle, ϕ: azimuthal angle.

electron spectroscopy for the determination of surface structures. Figure 4 illustrates schematically the experimental arrangement needed for angle-resolved photoelectron spectroscopy. A photon beam is incident upon the solid surface under study. Photoelectrons produced by this ionizing radiation leave the surface and are collected and analyzed by an electron energy analyzer. The angle between the photon beam and the analyzer is usually fixed, with the crystal surface rotatable about two axes, as indicated in Fig. 4. Experimental data consists of photoelectron current versus energy plots for various angles θ and ϕ.

The physical process occurring in photoemission which contains structural information is described in Fig. 5. In a sense, the process is very similar to low-energy electron diffraction, once the photoelectron has been produced by the incident photon. The electron propagates through the surface region, and its trajectory is determined by the positions of ion cores in the surface region. By comparing observed spectra for various values of θ and ϕ with spectra predicted for possible surface structures, a best structure can be chosen. This procedure has resulted in a number of surface structure determinations,[25] but is by no means a routine method of surface structural analysis.

Angle-resolved photoelectron spectroscopy has the disadvantage of being more complex and costly experimentally than LEED. It does not give the simple two-dimensional unit cell information readily available from LEED. A complete structural analysis, involving calculation of expected spectra for possible structural models, is nearly as difficult calculationally as multiple scattering LEED calculations. Angle-resolved photoemission does have some significant advantages in the use of a photon beam as a probe. Desorption of overlayers, and structural modifications to sensitive surfaces are minimized with this probe, when

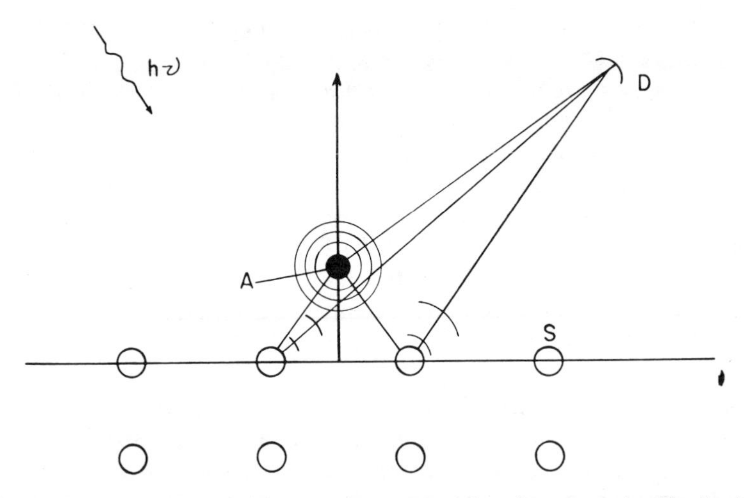

Fig. 5. Interference effects in angle-resolved PES giving rise to angular distribution of emitted electrons. D: detector, A: adsorbed atom, S: substrate atoms.

compared with incident electron probes. A more important advantage lies in the information contained in the photoelectron energy distribution.[26] This information can be used to determine the nature of adsorbate–substrate bonding changes in adsorbate bonding upon adsorption, and bonding character in the surface region. This information is also available from angle integrated photoelectron spectroscopy, and is mentioned again in Section II.B, where surface compositional characterization is discussed.

c. Vibrational Spectroscopy of Adsorbed Species. Ideally, one would like to have structural information about surface-adsorbed species comparable to the information available from the infrared spectroscopy of gas phase molecules. Of course the infrared spectroscopy of adsorbed species on high specific surface area materials is a long-established area of surface chemistry.[27] Vibrational spectroscopic information from species adsorbed on single crystal surfaces prepared in ultra-high vacuum (UHV) is not as readily available. There are possible ways to obtain this information, however. Two of these techniques are discussed here.

The first of these is the technique of high-resolution electron loss spectroscopy (ELS).[28] In this technique, electrons inelastically scattered from the surface adsorbed layer are analyzed with a high-resolution electron energy analyzer. Characteristic losses due to excitation of vibrational modes of the adsorbed species can then be related to vibrational frequencies and bond lengths, angles, and adsorbate bond strengths. To resolve these low energy losses, an incident electron beam of low energy and high monoergicity must be used. The resolution of the analyzer must be quite good, on the order of 5 to 10 meV, to resolve these losses. The experimental apparatus therefore requires two comparable electron energy analyzers, and for this reason is a difficult and costly experiment.[29] It also suffers from the same electron probe limitations mentioned in connection with LEED.

In spite of these difficulties, the technique has proven to be very valuable in obtaining information about the geometry of adsorbed species.[30] Once the data is obtained, it is readily interpreted in vibrational spectroscopic terms familiar to a wide range of investigators, in contrast to the situation with LEED and photoelectron spectroscopy. The information obtained is also often complementary to structural information obtained using LEED or photoelectron spectroscopy.[31] In some cases,[32] electron loss spectroscopy has prompted the reevaluation of adsorbate structures previously determined by LEED.

In addition to the vibrational spectroscopic information from adsorbed layers on single crystals prepared in UHV which can be obtained from high-resolution ELS, another technique shows promise and should be mentioned. This is the technique of Fourier transform infrared spectroscopy (FT-IR).[33] FT-IR shows promise because of its inherent high sensitivity and the speed with which an individual spectrum can be obtained. This high sensitivity coupled with collection and averaging of a large number of spectra to improve the signal-to-noise ratio, could make it possible to consider obtaining single-pass reflectance spectra from adsorbed species on single-crystal surfaces. A number of groups are trying to do these experiments at present.[34] Positive results from these efforts promise to make surface structural analysis in the future a more routine procedure than it has been in the past.

All the techniques mentioned for surface structural analysis are most easily applied to static surface systems, such as clean single- or multicomponent surfaces or adsorbed overlayers. A much more difficult structural analysis problem is encountered when attempts are made to structurally characterize dynamic surface systems. Information of this sort is essential for identifying surface reaction intermediates and obtaining increased understanding of heterogeneous reaction dynamics.

As the techniques are commonly used, this time-resolved structural information is not available. The low photon fluxes and photoionization cross sections encountered in photoelectron spectroscopy necessitate long data collection times, which essentially eliminates this technique for dynamic structural analysis, using standard photon sources. The use of high flux, pulsed synchrotron radiation[35] as a photoionization source is a possible route to time-resolved information from photoelectron spectroscopy. Time-resolved compositional information is much more likely than structural information from this technique, however.

In common with photoelectron spectroscopy, the low electron fluxes used in high-resolution ELS necessitate long data collection times, limiting the availability of time-resolved information. However, the LEED experiment could be modified to make it possible to obtain limited time-resolved structural information. Such a modification would be based on the use of a fast electron energy analyzer, such as a cylindrical mirror analyzer (CMA), to replace the retarding field analyzer commonly used in display type LEED optics.[36] A suitably masked CMA could be used to follow diffraction intensity from an overlayer feature, for example, during a thermally induced reaction.[37] This modified CMA–LEED technique could be useful in some instances, but would not be universally applicable to time-resolved studies of surface structural changes during a surface reaction.

2. *Surface Composition Techniques*

In addition to surface structural characterization for static and dynamic surface systems, an understanding of heterogeneous reaction dynamics demands information about the composition of the surface. Composition is used here in a very broad sense to include qualitative elemental analysis, quantitative analysis, and information about the bonding and chemical environment of surface species. The discussion is again divided according to techniques with emphasis on the information available from the individual techniques. Auger electron spectroscopy (AES) and photoelectron spectroscopy (PES) are discussed in some detail. Several other techniques are mentioned as well.

a. Auger Electron Spectroscopy. Auger electron spectroscopy (AES) was the first technique to be widely used for surface compositional analysis. The use of LEED optics as a retarding field analyzer[38] and the derivative method of Harris[39] combined to make AES a routine tool of surface analysis. There are a number of excellent reviews describing AES in detail, experimentally and theoretically.[40] Briefly, the method involves energy analysis of secondary electrons expelled from the solid surface, when inner shell holes created by incident ionizing radiation are filled by relaxation from upper levels (see Fig. 6). These secondary electrons have energies characteristic of the energy levels of the solid as indicated by Fig. 6 and (2.1).[41]

$$\bar{E}_{W_0X_pY_q}(Z) = \bar{E}_{W_0}(Z) - \bar{E}_{X_p}(Z) - \bar{E}_{Y_q}(Z') - \phi_c \tag{2.1}$$

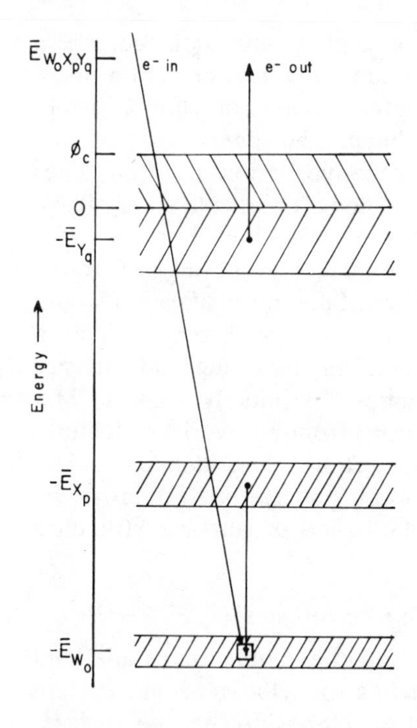

Fig. 6. Auger electron spectroscopy energy level diagram.

where Z is the atomic number of the electron emitter (Z') indicates an ionized atom, and ϕ_c is the material work function.

Qualitative elemental composition is then readily obtained from AES by comparing observed peak positions with those expected from known energy level spacings of the elements. The technique is surface sensitive due to the limited mean free path of electrons in the energy range of typical Auger transitions (50–1500 eV).[42] This qualitative information is readily obtained and is often used in defining the cleanliness of a solid surface under study.

Of more importance, but somewhat more difficult to obtain using AES, is quantitative surface compositional information. The intensity of an Auger transition is given by

$$I_i = \sigma(E)Y(E)KC_i \exp\left(-\frac{\chi}{\lambda(E)}\right)r_i \tag{2.2}$$

where $\sigma(E)$ is the ionization cross section, $Y(E)$ is the Auger yield, K is the instrument sensitivity, C_i is the concentration of the species of interest, χ is the distance into the solid, λ is the electron mean free path of energy E, and r_i is the backscattering coefficient for the solid being studied. The accurate direct determination of C_i, the quantity of interest, is complicated by an inaccurate knowledge of other factors in (2.2), such as the ionization cross section $\sigma(E)$, the backscattering coefficient r_i, and electron mean free path $\lambda(E)$. Consistent standards can be used, however, to eliminate these problems by measuring unknown surface concentrations against a standard. This method is described in detail by Palmberg.[43] Using such methods, quantitative accuracy on the order of 10% is routinely possible. With some effort this can be improved for a number of systems.[44]

In principle, Auger peak position and peak shape should provide information about the chemical environment of the electron-emitting atom. Differences in bonding environment should show up as peak position shifts with respect to the position predicted for the free atom undergoing the same Auger transition. In practice the shifts are so small as to be not routinely observable.* There are systems where discernible peak shape changes are observed for changing surface bonding conditions.[45] These observations have led to a number of empirical correlations between Auger peak shape and surface bonding character.

The main advantage of AES is the wealth of compositional information that is available without a great deal of experimental or theoretical effort. With the proper choice of electron energy analyzer (CMA, 127° sector) the technique is also very rapid (millisecond sweep times). This allows the collection of time-resolved spectra on the time scale of thermally induced surface reactions. This possibility is a major advantage for the technique in the study of heterogeneous reaction dynamics. AES does suffer from the limitations of electron probe damage to the surface layer as discussed previously, even more so than with LEED, as incident electron energies and currents are significantly greater in Auger spectroscopy.

b. Photoelectron Spectroscopy. Photoelectron spectroscopy is also a very useful technique for obtaining surface compositional information.[46] In this method, energetic incident photons eject electrons from atoms of the solid. The kinetic energy of the ejected electron is given by

$$E_{\text{kin}} = h\nu - (E_f - E_i) \tag{2.3}$$

where E_{kin} is the ejected electron kinetic energy, $h\nu$ is the photon energy, and E_f and E_i are the final and initial energy states of the system. Depending on the energy region of the photon probe used, core

*For a discussion of "chemical" shifts versus relaxation shifts in Auger spectroscopy, see D. A. Shirley, *Phys. Rev.*, **A7**, 1520 (1973). Note that the natural linewidth of an Auger transition from a solid is on the order of 0.5 to 1.0 eV.

states (X-ray photoelectron spectroscopy, XPS) or valence states (ultraviolet photoelectron spectroscopy, UPS) are sampled.

The surface sensitivity of XPS and UPS is determined by the mean free path of the ejected photoelectrons. For electrons in this energy range, the mean free path is on the order of several atomic layers, making photoelectron spectroscopy surface sensitive. The overall signal strength is much lower for PES than for AES, however, because of the lower photon fluxes available from conventional UV and X-ray sources compared to the electron guns used for excitation in Auger spectroscopy. This results in considerably longer data collection times in photoelectron spectroscopy, limiting its usefulness in dynamic surface composition determinations as was mentioned in Section II.A.1.b.

Qualitative and quantitative analysis of the surface region cannot be accomplished very readily by UPS,[47] because of the broad spectral features characteristic of this technique. XPS, on the other hand, can be used for elemental identification because of the sharp, well-separated core level peaks. In principle, quantitative analysis using XPS should be more straightforward than for AES, because of the simpler one electron process in XPS. As with AES, absolute quantitative determination is difficult because cross sections are not accurately known. Internal standards and calibration procedures are again normally used, allowing reasonably quantitative (on the order of 10% accuracy) determinations to be made. Comparisons of the quantitative nature of AES and XPS are available for a number of systems.[48]

The real advantage of photoelectron spectroscopy for obtaining surface compositional information is in the information available about the surface chemical environment. Shifts in Auger peak position and peak shape for different chemical environments do occur, but they are often masked by the natural linewidth of the Auger transition. They are also difficult to interpret in terms of changes in chemical bonding, because of the involvement of three levels in the process. In contrast, PES transitions involve only one level, and shifts in peak position should in principle be more readily understood. Since the PES linewidth is source limited, for high-resolution peak shift determination very monochromatic radiation must be used. In spite of this problem XPS has been used in a number of cases to characterize surface chemical bonding in terms of core level shifts.[49] UPS is also useful in obtaining surface bonding information, perhaps more so than XPS because of the sensitivity of valence levels to changes in the chemical environment. There are a number of examples in the literature of the use of UPS in obtaining this information.[50]

c. Other Techniques. Surface science abounds with acronymed techniques which give surface compositional information in addition to the ones discussed previously. A number of these, along with their major advantages and disadvantages are mentioned in this section. Each technique has its own following in the surface science community, and each is more or less suitable for particular systems.

SXAPS, soft X-ray appearance potential spectroscopy,[51] relies on the characteristic energy threshold of an element for production of X-rays for elemental analysis. Its major advantage is the very simple experimental arrangement needed to collect an SXAPS spectrum (no electron energy analyzer or photon source, just a filament and photocathode). Its disadvantages are variable sensitivity for different elements and the high electron flux which is necessary to produce easily measured quantities of X-rays. It is a reasonable technique for characterizing the cleanliness of metal surfaces.

Ion scattering spectrometry, ISS,[52] analyzes surface composition by measuring the energy of noble gas ions elastically scattered from surfaces. Peaks in the ion energy spectrum appear which are characteristic of collisions of the noble gas ion with surface atoms of various masses. Its main advantage is its single monolayer surface sensitivity. It does not give compositional information from below the top layer, but unfortunately, the energetic noble gas ion probe can result in significant surface damage.

An instrumental complement to ISS is secondary ion mass spectrometry, or SIMS.[53] In this technique, ions ejected from the surface when bombarded by noble gas ions are mass analyzed. The resultant mass spectrum is then used to suggest what the surface composition was prior to sputtering. The major disadvantage of this technique is the extensive damage to the surface under study, and the difficulty in piecing together surface composition by an analysis of ion fragments ejected from the surface.

Mention should be made of another technique which is widely used in surface studies, and which gives a type of surface compositional information. This is the measurement of work function change on adsorption. There are a number of experimental methods for measuring work function change[54] of varying resolution and complexity. With appropriate calibration procedures, changes in the work function can be used to determine adsorbate coverage. Elemental analysis is not possible, but limited structural and chemical environment information can be inferred, especially when used in combination with other surface characterization tools.[55]

B. Reactant and Product Characterization

In addition to structural and compositional information about the solid surface and its adsorbed species, understanding of the reaction dynamics of heterogeneous systems requires a knowledge of the state of the gas phase reactants and products. As with the characterization of solid surfaces, there are a large number of techniques available for characterizing the gas phase reactants and products. For the most part, these are tools that have either been developed for the study of gas phase molecular reaction dynamics, or have seen their widest application in this field. Three types of experimental methods are emphasized in what follows. These three methods are chosen for discussion here because of their utility in probing heterogeneous reaction dynamics. The methods are perturbation methods, molecular beam methods, and state selective methods.

1. Perturbation Methods

In gas phase molecular reaction dynamics, perturbative experimental methods take a number of forms. For example, shock tube experiments,[56] temperature or concentration jump methods,[57] and flash photolysis techniques[58] can all be considered perturbative methods. Information about molecular reaction dynamics is obtained by following the return of the system to the unperturbed condition. In heterogeneous systems, a convenient way of perturbing the system is to change the surface temperature in a controlled manner. This technique, often called thermal desorption or flash desorption spectroscopy,[59] can be used both to characterize adsorbed species and to thermally induce surface reactions.

The technique was developed by Redhead[60] and Ehrlich[61] primarily to characterize species adsorbed on solid surfaces. Adsorbed species are thermally desorbed by heating the sample at a controlled rate (usually linear) in a vacuum system of known, constant pumping speed. The partial pressure of the desorbing species is then recorded versus the surface temperature, resulting in a spectrum with peaks at temperatures proportional to the adspecies adsorption energy, to a first approximation. This simple picture is complicated somewhat by the possible interchange of adspecies between adsorption states during thermal desorption, and by the often observed possibility of coverage-dependent adsorption energies. However, if properly handled,[62] thermal desorption data can be used to determine desorption order, desorption activation energy and preexponential, and in some cases the nature and extent of adspecies interaction.

Of more direct interest to the study of heterogeneous reaction dynamics, thermal desorption spectroscopy can also be used to study certain types of heterogeneous reaction. In particular, thermally induced surface decomposition reactions are especially amenable to study in this way. In this case, a molecular reactant is adsorbed on a solid surface held at a temperature below which significant catalytic decomposition occurs. The surface is then heated, thermally inducing the decomposition. The partial pressures of the decomposition products are monitored, and dynamic information as described previously is obtained. The decomposition of formic acid on metal surfaces is an example of this type of reaction, and is discussed in detail in Section III.B. To be really useful for studying surface reactions, the experimental apparatus used must be able to follow the partial pressures of a number of desorbing species, and to allow changes in the sample heating rate and functional form readily. This suggests computer control of the mass spectrometer detector and heating power supplies. Such experimental systems are not widely used presently, but have been developed in a few laboratories.[63]

2. Molecular Beam Methods

Another method developed for the study of gas phase reaction dynamics which has been successfully modified for application to the study of heterogeneous reactions is molecular beam scattering. In a gas phase molecular beam experiment, a collimated beam of molecules is incident on a volume of gas at a known pressure, or is crossed with another beam of molecules. The densities in these beam–gas and beam–beam experiments are kept low so that only single-collision events occur. In the beam–gas experiments, total cross sections for collisions between beam molecules and gas molecules are measured. In the crossed beam experiment, angular distributions of the scattered particles, as well as translational energy distributions, are available.[64] If a mass spectrometer is used as a detector in such a system, it is possible to investigate reactive collisions in addition to elastic and inelastic scattering.

The information available from molecular beam scattering from surfaces is analogous to that obtained via gas phase molecular beam scattering.[65] A collimated beam of molecules is incident upon a fixed solid surface in this case. The angle of incidence of the molecular beam with respect to the surface normal or particular crystallographic directions in the sample can be varied. Scattered particles are monitored by a detector, usually a mass spectrometer which can be rotated about the surface. Some systems are also capable of measuring translational energy of scattered species by time of flight techniques.[66] When these

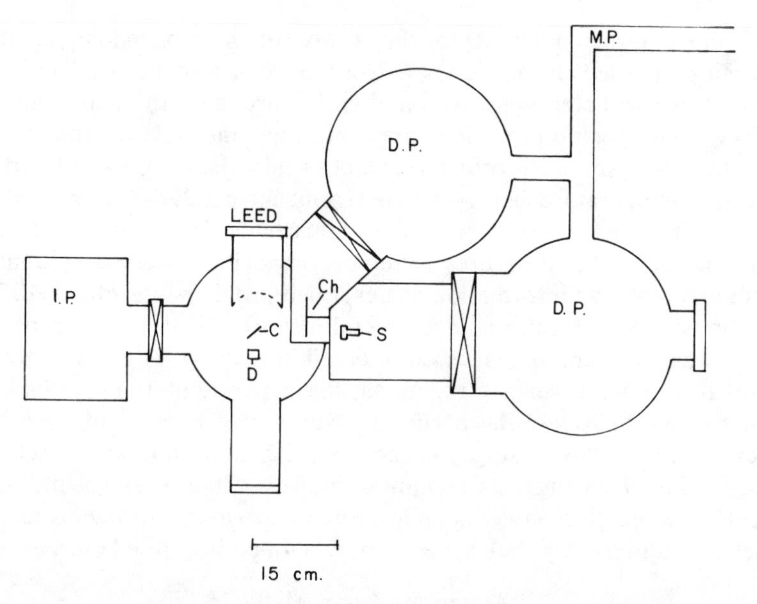

Fig. 7. Molecular beam scattering apparatus for surface studies. MP: mechanical pump, DP: diffusion pump, IP: ion pump, Ch: chopper, C: crystal surface, S: source, D: detector.

techniques are coupled with UHV surface preparation and characterization techniques such as those described previously, a very powerful tool for investigating the dynamics of surface reactions is available. Figure 7 shows a schematic diagram of one such system.

Angular distributions of surface scattered species are generally presented as rectilinear plots of scattered intensity normalized to the incident beam intensity versus detector angle measured from the surface normal. Scattering distribution peak position and peak shape can be used to obtain information about energy transfer between the incident molecules and the surface.[67] Reaction probabilities, analogous to the total scattering cross section of beam–gas experiments, can be determined by monitoring the intensity of product molecules leaving the surface. This is possible for the case in which the incident molecule reacts directly with the solid surface,[68] as well as the case in which the beam reacts with an adsorbed species.[69]

Direct information about the energetics of surface reactions has been obtained via molecular beam methods in two ways. The first of these varies incident beam translational[70] or internal[71] energy, to probe the dependence of surface reaction dynamics on these parameters. The other technique that has been successfully used is the direct measurement of product translational energy by time of flight techniques.[66] Both these methods appear in the case studies of heterogeneous reaction dynamics discussed in Section III.

Additional information is available from the molecular beam scattering experiment which could perhaps be considered a type of perturbative method information. This is time-dependent information which becomes available by modulating the incident molecular beam. This modulated beam spectroscopy[72] has been used to investigate a number of heterogeneous systems. The modulated molecular beam interacts with the surface, and scattered reactant and product are detected at the modulation frequency. By observing the signal waveform as a function of beam energy, surface temperature, incidence and scattering angle, and modulation frequency, information about the surface reaction dynamics can be obtained. The surface reaction can be thought of as a differential operator, acting on the incident beam periodic function. The detected waveform is then the solution to the differential equation system describing the surface process. This is essentially a modeling procedure, in which the expected behavior of the detected waveform for possible surface reaction models is compared with the observed product waveform. There are a number of examples in the literature of the use of this technique to study surface reactions.[73] It is also seen in the case studies discussed in Section III.

Theoretical input is especially important in the case of the molecular beam scattering experiment. The complicated gas-surface scattering process is not at all fully understood, although some reasonably simple qualitative models do exist.[74] A great deal of work has been done in this area, treating elastic, inelastic, and reactive scattering with more and more realistic models. This work has developed along a number of lines, and is discussed fully in a number of recent reviews.[75]

3. State Selective Methods

Knowledge of the energy state of the gas phase reactants and products of a heterogeneous reaction is necessary for understanding their surface reaction dynamics. Recently, methods have become available for preparing specific energy states, as well as for characterizing these states. Some of these methods rely on the use of lasers for state preparation and characterization, but some do not.

The simplest method of "state preparation" available for investigations of gas–solid interactions is to change the gas temperature. In the context of a molecular beam scattering experiment, this simply means using a variable temperature effusive source.[76] The molecules are then treated as having an energy state distribution (translational, rotational, and vibrational) characteristic of the source temperature. Improvements on this first level of state preparation are possible by utilizing nozzle beam source technology.[77] This allows considerably more flexibility in individual state distribution choice, a wider range of available energies, and much narrower distributions than the effusive source.

Ideally, one would like to prepare a particular internal state, rather than a distribution of states. A tunable laser would appear to be ideal for accomplishing this state preparation. However, in polyatomic molecules intramolecular energy randomization out of the excited mode occurs quite rapidly,[78] and initially prepared, individual states decay to intramolecule state distributions faster than the gas–solid interaction takes place. Under collisionfree conditions, even this situation offers state preparation advantages over the more conventional methods mentioned previously. Internal energy can be deposited in a molecule in preference to translational energy. This collisionless heating allows the determination of internal state effects on heterogeneous reactions.

Although the technique has not yet been utilized, it seems that laser-induced fluorescence[79] could be a valuable tool for probing the internal state of molecules leaving a surface after a heterogeneous reaction. It would be a difficult measurement experimentally, because of the low density in the beam of molecules leaving a surface after a reaction, but one which could be possible for the proper system choice.[80] Information of this sort is essentially unavailable at this time, although promising results have been obtained using electron beam induced fluorescence.[81]

In conclusion, this section has mentioned a number of experimental tools available for the study of heterogeneous reaction dynamics. Both surface and gas phase characterization methods were described, and the information obtainable from each technique discussed. It should be emphasized that these techniques must be used in combination to unravel the details of heterogeneous reaction dynamics. Examples of the successful application of this combination of techniques to the study of surface reaction systems are presented in the case studies following.

III. HETEROGENEOUS REACTION DYNAMICS CASE STUDIES

In this section, three case studies are discussed. These studies were chosen to illustrate the type of information about heterogeneous reaction dynamics which is available when surface characterization tools are used in combination with gas phase dynamics techniques. The choice of these systems does not imply that they are the only systems which have been studied using this combination of tools. Rather, they are chosen because they illustrate nicely the complementary information that can be obtained when techniques from these separate fields are applied to the study of complex heterogeneous systems. The three case studies discussed are: (*1*) hydrogen interaction with metals, particularly copper and platinum; (*2*) formic acid decomposition on nickel; and (*3*) iron oxidation.

A. Catalytic Hydrogen–Deuterium Exchange

Perhaps the most extensively studied heterogeneous reaction is the isotope exchange reaction between hydrogen and deuterium catalyzed by a metal surface. This reaction has been studied by a number of investigators employing surface characterization tools along with perturbation and molecular beam methods. The discussion here concentrates on H_2–D_2 exchange on copper and platinum surfaces. These studies demonstrate, respectively, the effect of incident molecule energy on the surface reaction dynamics, and the effect of surface structure on the exchange reaction. Detailed information about the dynamics of the exchange reaction have been obtained in both cases by a combination of surface characterization, molecular beam, state selective, and desorption methods.

Equilibrium desorption of molecules from a surface suggests an angular distribution of the desorbing molecules which varies as the cosine of the angle from the surface normal.[82] However, noncosine desorption distributions have been observed in a number of instances.[83] Work by Stickney et al.[84] has indicated that these noncosine distributions observed from a number of metals (Ni, Pt, Fe, Nb) were associated with significant levels of carbon or sulfur impurity on the surface. When the surfaces were cleaned to remove these impurities, cosine distributions were again observed. An exception to this observation occurred in the case of copper polycrystalline and single crystalline surfaces. In this case, noncosine distributions (more highly peaked at the surface normal) were observed even when the copper surfaces were clean to the limit of AES sensitivity.

More detailed investigation of the desorption of hydrogen from copper[85] and the exchange reaction between hydrogen and deuterium on the copper surface[86] was undertaken. The experiments were performed in two ways. In the desorption experiments, angular distributions were measured with a rotatable ion gauge. The surfaces were in the form of thin copper membranes welded to copper tubes. Hydrogen at atmospheric pressure was introduced into this tube and diffused through the copper atomically when the membrane was heated. Atoms arriving at the vacuum side of the membrane recombined and desorbed molecularly. In the isotope exchange experiments, a beam of H_2 molecules from a nozzle source was incident on copper surfaces which were simultaneously exposed to a beam of deuterium atoms from a high-temperature, low-pressure effusive oven. Hydrogen molecule translational energy could be varied from 1 to 10 kcal/mole by heating the nozzle source. Surface-produced HD was monitored by a rotatable mass spectrometer.

The desorption experiments indicate a peaked (noncosine) angular distribution for clean and contaminated copper surfaces. The distributions become somewhat less peaked as the surface contamination was removed, but were still significantly noncosine even for the clean surfaces. Significant differences in peaking were also observed for surfaces with different crystallographic orientations, but there appeared to be no desorption asymmetries with respect to particular crystallographic directions in the surface plane. These observations were tentatively interpreted in terms of an activated adsorption for hydrogen on copper. This barrier for dissociative adsorption was identified with the crossing point of the interaction potentials for atomic and molecular hydrogen with copper (see Fig. 8).[87] This crossing point is expected to be at fairly large distances from the surface, resulting in little desorption distribution asymmetry with respect to particular directions in the surface plane. However, the barrier height would be expected to depend on the particular surface orientation considered, because of different interaction potentials for different arrangements of surface copper atoms.

More detailed studies of this activated adsorption process were accomplished by following production of HD on the copper surface when a H_2 molecular beam and a deuterium atomic beam were incident. The translational energy of the H_2 beam could be varied by heating the nozzle source forming the beam. The angular distribution of the desorbing HD produced in this way was again observed to be noncosine. When the H_2 translational energy and angle of incidence was varied the production of HD in these experiments was found to increase as $E_\perp$ (the kinetic energy associated with the normal component of the incident molecular beam) increased. This indicates an activation barrier for dissociative adsorption which acts primarily perpendicular to the surface plane. This barrier was found to be dependent on crystallographic orientation, and was estimated to be about 3 kcal/mole for Cu(110) and 5 kcal/mole for Cu(100) and Cu(310).

The interaction of hydrogen with plantinum as indicated by the H_2–D_2 exchange reaction has been studied extensively. The results discussed

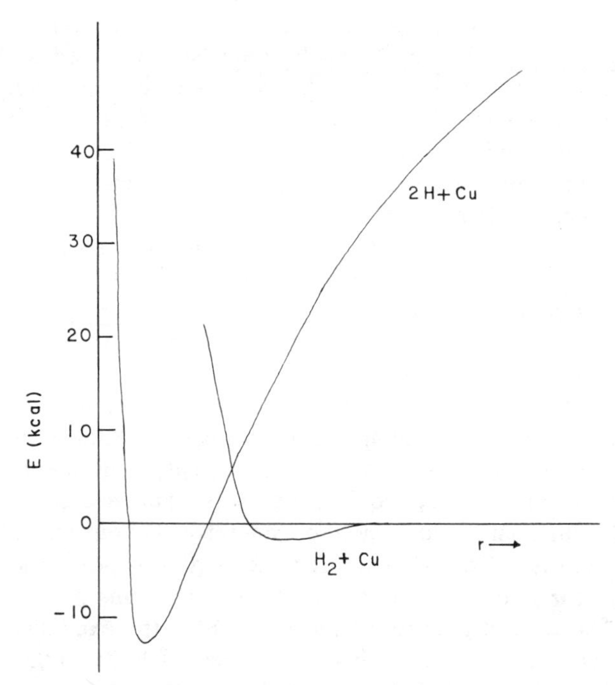

Fig. 8. Interaction potentials for H_2 + Cu and 2H + Cu. After Ref. 85.

previously which provide evidence for an activation barrier to dissociative adsorption of hydrogen on copper are helpful in understanding the results of the studies of the heterogeneous reaction dynamics of H_2–D_2 exchange on platinum surfaces. Thermal desorption studies of the exchange reaction on single crystalline (110), (211), and (100) planes of platinum were performed by Lu and Rye in 1973.[88] Although *in situ* characterization of their samples was not possible, the cleaning procedures used to prepare the platinum surfaces were checked in a separate Auger electron spectrometer, and indicated only a small residual calcium impurity. On these samples, they observed sticking coefficient and equilibration activity to be somewhat higher on the more open, microscopically rougher (110) and (211) planes. Under these conditions, the adsorption and exchange reaction appeared to be nonactivated on all four surfaces.

In contrast to this work, Bernasek et al.[89] observed a marked dependence of exchange reaction probability on crystallographic orientation in a modulated molecular beam study of the exchange reaction. Single crystal platinum samples were cleaned and characterized in an ultra high vacuum scattering system. A beam of molecular deuterium from a multicapillary effusive source,[90] which could be modulated with a variable frequency, was incident on the platinum sample under study. Hydrogen molecules were simultaneously incident on the surfaces from a separate small needle in the scattering chamber. HD production was monitored with a rotable quadrupole mass spectrometer mounted in the scattering chamber. Three surfaces of different crystallographic orientation were studied; the Pt(111) surface, and two stepped surfaces designated the Pt(S)-[9(111) × (111)] and Pt(S)-[5(111) × (111)] surfaces in the descriptive nomenclature of Lang et al.[91]

In these molecular beam experiments, it was found that the two stepped surfaces were significantly more active for the production of HD than the low index Pt(111) surface. The production of HD on the stepped surfaces at surface temperatures above 600°K also appeared to be directly dependent on the surface step density. The mechanism for HD production on the stepped surfaces was explained in terms of a branched mechanism, involving a channel for D_2 terrace adsorption followed by diffusion to a step site for exchange and production of HD, or a channel for direct interaction of D_2 with H atoms adsorbed at the step sites. The lack of activity of the Pt(111) for the exchange reaction under these conditions was ascribed to the unavailability to the adsorbed D_2 of step defects on the (111) surface during the millisecond sampling time characteristic of the modulated beam experiment. It has also been suggested that there is an activation energy for dissociative adsorption on the Pt(111) surface[86] similiar to that observed for copper, but no activation barrier for dissociative adsorption at the step sites.

A modified interpretation of the preceding study was recently suggested by Wachs and Madix.[92] They again assumed a branched reaction mechanism for the exchange reaction on the stepped platinum surface. However, in their interpretation the branching ratio (which was chosen as 0.5 in the original study) was chosen to depend on step density. In this case, the two reaction channels were then associated with reaction on the terrace or reaction on the step. This modification considerably improved the agreement between the observed modulated beam behavior and that predicted for the model mechanism. This modified model also suggested that the adsorption probability on the smooth Pt(111) surface, in the limit of zero modulation frequency, was approximately a factor of 20 less than on the stepped surface, bringing the results of Bernasek et al.[89] into essential agreement with those of Lu and Rye[88] under comparable experimental conditions.

Further studies of the exchange reaction on the stepped platinum surface investigated the dependence of the reaction probability on the

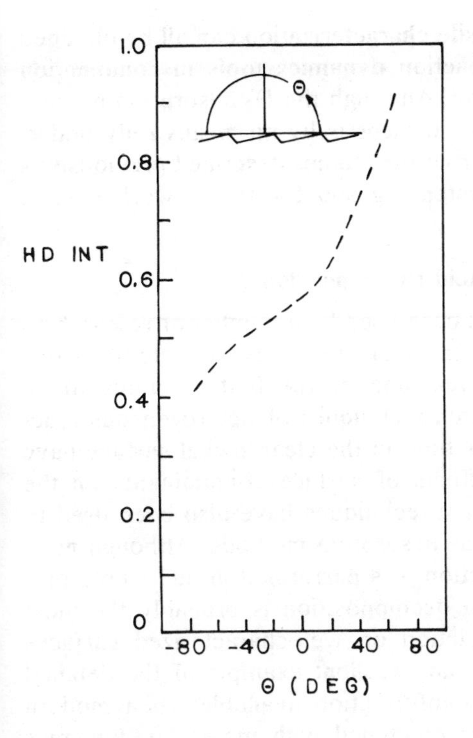

Fig. 9. Angular dependence of HD production on stepped platinum surface. After Ref. 93.

angle of H_2 beam incidence with respect to the step orientation.[93] This study showed that the reaction probability for HD production was highest when the beam was incident perpendicular to the step direction from the open side of the step. (See Fig. 9.) The functional form of this angular dependence was explained in terms of a difference in activity between the step sites and the terrace sites for H–H bond breaking. This conclusion is in qualitative agreement with conclusions of the previous work, suggesting that the step sites on platinum are more active for dissociative adsorption of hydrogen than are the low index surfaces. Incidence angle variation in the work of Salmeron et al.[93] also indicated an activation barrier for dissociative adsorption of H_2 on the Pt(111) surface of less than 0.4 kcal/mole in height, in agreement with the suggestion of Balooch et al.[86]

The individual experiments discussed in this case study of heterogeneous reaction dynamics demonstrate the wealth of detailed information available concerning the interaction of hydrogen with copper and platinum surfaces. Barrier heights for activated adsorption, crystallographic effects on adsorption behavior, kinetic parameters, reaction mechanisms, and active surface site characterization can all be obtained by the careful application of reaction dynamics tools in combination with solid surface characterization. Although the H_2 adsorption process on these surfaces and other solid surfaces is by no means fully understood, the information obtained from the studies described previously is providing a basis for this understanding and for future work on this important system.

B. Formic Acid Decomposition

Formic acid decomposition has been used by a number of researchers over the years as a model reaction for the study of heterogeneous reactions.[94] For this reason, it was one of the first reactions to be carefully studied using the research techniques of heterogeneous reaction dynamics. Studies of the reaction on the clean nickel surface have been extended to include the effects of surface contaminants on the reaction dynamics. Molecular beam techniques have also been used to investigate this reaction, as well as desorption methods. Although most of the work discussed in this section was performed in one laboratory, their investigation of formic acid decomposition is probably the most complete study of a surface reaction on well-characterized surfaces presently in the literature, and is an excellent example of the detailed heterogeneous reaction dynamics information available when modern surface characterization tools are combined with molecular dynamics techniques.

Nickel (110) single crystal samples, characterized for surface structure by LEED and surface composition by AES, were used in these studies. Samples were cleaned by ion bombardment and annealing and characterized in an ultra high vacuum system equipped with a fixed-position quadrupole mass spectrometer for thermal desorption measurements. The surfaces were dosed with formic acid from a stainless steel needle directed at the front face of the sample. The results of thermal decomposition studies on the clean Ni(110) surface are reviewed first. Then the effect of surface composition on the decomposition process is presented, for surfaces contaminated by carbon in the form of graphite and nickel carbide and oxygen. Finally, the results of a modulated molecular beam study of the decomposition reaction are discussed.

Exposure of the clean Ni(110) surface at temperatures below 50° to formic acid, followed by thermally induced decomposition resulted in desorption of CO, H_2, and CO_2.[95] Figure 10 shows a typical decomposition spectrum for the clean Ni(110) surface. The spectrum shows narrow, low-temperature peaks for H_2 and CO_2 and a CO signal in this region partially due to CO_2 cracking in the mass spectrometer ionizer. At

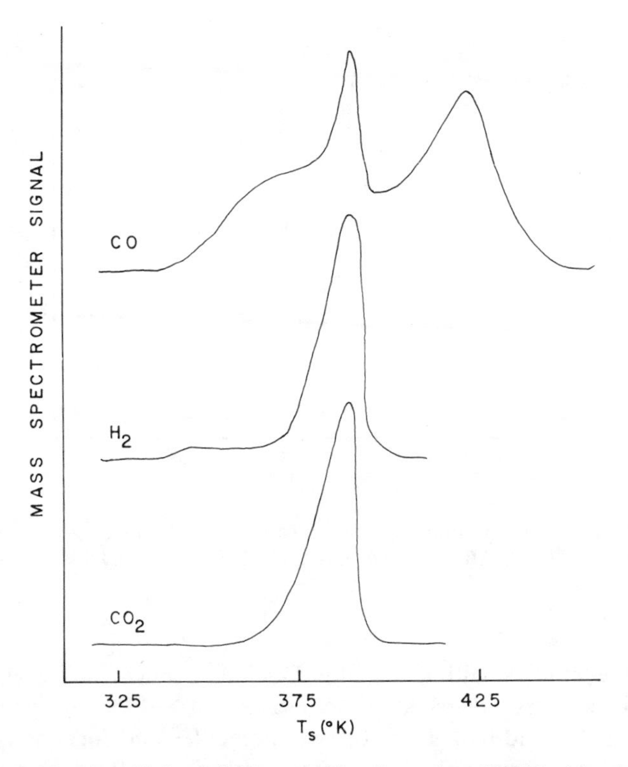

Fig. 10. Formic acid thermal decomposition from clean Ni(110). After Ref. 96.

higher temperatures a broad desorption peak is observed for CO. The striking thing about these spectra is the very narrow low-temperature peak for H_2 and CO_2. This peak occurs at a temperature higher than the CO_2 or H_2 flash desorption peaks from the clean Ni(110) surface, indicating that the narrow peak is a direct result of the formic acid decomposition.

The very narrow peak width and its behavior with varying formic acid coverages and heating rates is not explainable in terms of simple first- or second-order desorption of a surface adsorbed species. In addition, interrupted flash decomposition spectra (see Fig. 11) showed that evolution of CO_2 and H_2 in this peak proceeded with the same peak shape even when the surface temperature was not increased above the temperature initiating the desorption. This observation suggests an autocatalytic mechanism for the decomposition, which proceeds on its own once thermally initiated.

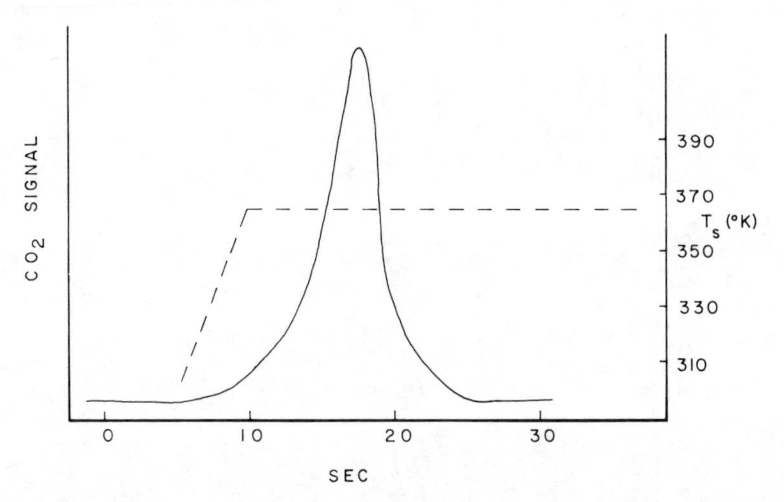

Fig. 11. CO_2 desorption spectrum for interrupted heating schedule(-----), for formic acid decomposition on Ni(110). After J. L. Falconer, J. G. McCarty, and R. J. Madix, *Surf. Sci.*, **42**, 329 (1974).

Extensive studies with deuterium-labeled formic acid[96] suggest that the formic acid is adsorbed dissociatively at ~40°C, losing H_2O incorporating the acid hydrogens of two molecules and forming islands of a stable formic acid anhydride intermediate[97] on the surface. These island species thermally decompose, releasing CO_2 and some CO which immediately desorbs, and exposing bare nickel sites responsible for further catalytic decomposition. Hydrogen atoms and additional CO remains adsorbed. The hydrogen atoms recombine rapidly and desorb as molecules resulting in the narrow H_2 peaks observed in the decomposition spectra. As the surface temperature is increased, the remaining CO desorbs, giving the broad, higher temperature CO peak.

On the clean Ni(110) surface, the ratio of CO_2 to CO formed in the decomposition of formic acid was approximately 1.0.[96] This ratio was observed to be significantly different for surfaces contaminated by carbon. The heated Ni(110) surface was exposed to ethylene to produce two different surface carbides and an overlayer of graphitic carbon. The surface carbides exhibited (2×1) and (4×5) LEED patterns[98] and carbon Auger peak shape characteristic of carbide bonding.[99] Further exposures to ethylene at higher surface temperatures resulted in the formation of graphitic carbon on the nickel surface, with the characteristic ring like LEED patterns of graphite[100] and graphitic Auger peak shape.[99]

Surfaces prepared in this way were then exposed to formic acid and deuterated formic acid, and thermal decomposition experiments similar to those described previously for the clean nickel surface performed. The rate of decomposition on the carbide surface was found to be about two orders of magnitude less than on the clean surface. The very narrow low-temperature peak observed in the clean surface decomposition was not observed in the case of the completely carbided surface. In addition, the ratio of CO_2 to CO formed in the decomposition on the carbide surface was 10, rather than 1.0 as in the clean surface decomposition. The graphite-covered surface was not found to be catalytically active for the formic acid decomposition.

Low-temperature adsorption of DCOOH followed by thermally induced decomposition resulted in an H_2 peak at 290°K, followed by CO_2 and D_2 peaks at 450°K. This suggests a DCOO species as the intermediate participating in the rate-limiting decomposition step on the carbide surface. The first-order rate constant determined for this rate limiting step is given by $10^{12.5\pm.1}\exp(-25.5\ \text{kcal/mole}/RT)\sec^{-1}$. The clean surface preexponential was $10^{15.5}\sec^{-1}$, indicating a more loosely bound intermediate in the carbide surface case.

Further studies of the desorption of D_2 from formic acid decomposition on the carbide surface were designed to determine the D_2 desorption mechanism for this case.[101] The observed D_2 peak, appearing slightly above room temperature, was significantly broader than would be expected for simple second order D_2 desorption (desorption following adsorbed D atom recombination). Good agreement between the observed D_2 peak shape and that predicted by the desorption mechanism is obtained if the preexponential factor is allowed to vary with coverage. This coverage dependent preexponential model, due to Clavenna and Schmidt,[102] assumes that the rate of desorption of D_2 molecules is given by the product of the D-atom collision frequency and the desorption probability of the newly formed molecule. The collision frequency $(1/\tau)$ for adsorbed D-atoms is expected to be coverage dependent. This dependence has the form given by

$$\frac{1}{\tau}=\frac{a^2\nu_0}{\left[1-\left(\frac{a_c}{a}\right)\sqrt{\theta}\right]^2} \tag{3.1}$$

where a is the D atom jump length on the surface (in the context of a random walk surface diffusion[103]), ν_0 is the adsorbed atom vibrational frequency, a_c is the minimum adatom separation during collision (an excluded surface area term, analogous to the excluded volume in gas phase bimolecular collisions), and θ is the fractional surface coverage.

The parameters $a^2\nu_0$ and (a_c/a) were varied until a good fit to the D_2 desorption curve was obtained. The collision frequency (preexponential) obtained in this way was significantly lower than might be expected from simple bimolecular collision theory. This low preexponential perhaps suggests specific site or steric requirements for the D-atom recombination on the carbide surface.

The decomposition of formic acid was also investigated on partially oxidized Ni(110) surfaces.[104] The clean Ni(110) surface was exposed to varying amounts of oxygen, with the surface held at room temperature. Formic acid was then adsorbed, and thermal decomposition spectra obtained. The effect of increasing oxygen exposure on the CO_2 peak in the decomposition spectrum is illustrated in Fig. 12. The narrow low-temperature peak characteristic of the clean surface is shifted to progressively higher temperatures as the oxygen coverage increases. In addition, its peak intensity decreases as a broad, high-temperature peak increases, until at high oxygen coverage, only the broad high-temperature peak is observed. There is a continuous shift to higher peak temperatures as the narrow low-temperature peak is suppressed for increasing oxygen coverage. The mechanism of the decomposition is therefore postulated to change continuously from the clean surface formic anhydride intermediate mechanism, to the formate intermediate mechanism observed on the carbide surface.

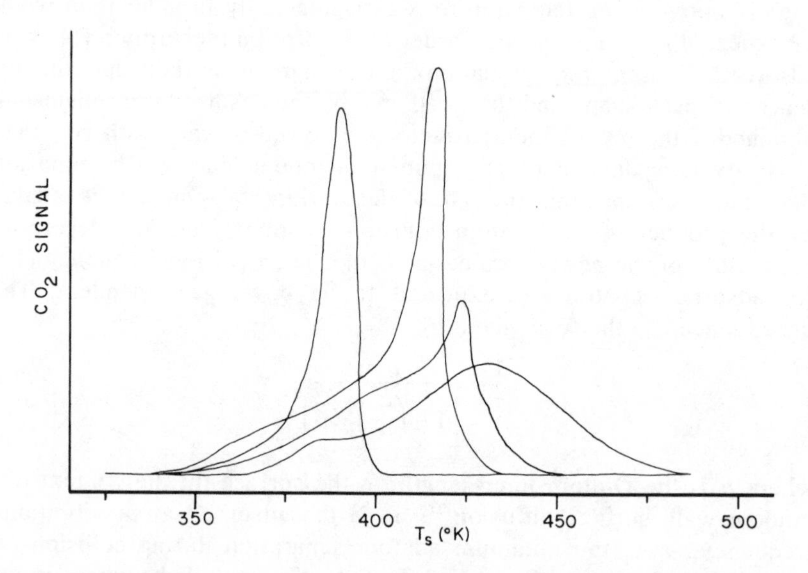

Fig. 12. CO_2 desorption from formic acid decomposition on Ni(110) exposed to varying amounts of oxygen. Oxygen exposure increases from left to right. After Ref. 104.

The decomposition reaction on the clean surface was also studied using molecular beam techniques.[105] A molecular beam of formic acid, modulated at a fixed frequency of 36.8 Hz, was scattered from the clean Ni(110) surface. Measurement of the scattered product amplitudes of CO, CO_2, H_2, and H_2O, as well as their phase lag with respect to reflected formic acid, was performed as a function of nickel surface temperature. Changes in the scattered waveform as a function of surface temperature were also recorded, in addition to the phase and amplitude information. Analysis of this data and comparison with a number of possible models for the decomposition suggested the following mechanism for the decomposition over the temperature range studied, and in the low-coverage region characteristic of the molecular beam scattering experiment.

$$\mathrm{HCOOH}_{(g)} \xrightarrow{S_0} \mathrm{HCOOH}_{(ads)}$$

$$\mathrm{HCOOH}_{(ads)} \xrightarrow{k_1} \mathrm{HCOOH}_{(chemisorbed)}$$

$$\mathrm{HCOOH}_{(chemisorbed)} \xrightarrow{k_2 P} \mathrm{CO}_{2(g)} + 2\mathrm{H}_{(ads)}$$

$$\mathrm{HCOOH}_{(chemisorbed)} \xrightarrow{k_2(1-P)} \mathrm{CO}_{(ads)} + \mathrm{H}_{(ads)} + \mathrm{OH}_{(ads)}$$

$$\mathrm{CO}_{(ads)} \xrightarrow{k_3} \mathrm{CO}_{(g)}$$

$$\mathrm{H}_{(ads)} + \mathrm{OH}_{(ads)} \xrightarrow{k_4} \mathrm{H_2O}_{(g)}$$

$$\mathrm{H}_{(ads)} + \mathrm{H}_{(ads)} \xrightarrow{k_5} \mathrm{H}_{2(g)}$$

This is a branched mechanism, with branching probability P, involving a chemisorbed formic acid precursor to both branches. Values of the kinetic parameters extracted by fitting the observed data to this model are listed in Table I. Notice that the use of a fixed modulation frequency does not allow extraction of k_1, k_2 and the branching ratio P simultaneously. At the low coverages characteristic of the molecular beam experiment, formation of the formic anhydride intermediate postulated in the clean surface thermal decomposition studies is not likely. Results and kinetic parameters obtained in this study are not directly comparable to the previous clean surface thermal decomposition studies, because of the steady-state carbon and oxygen coverages present especially at elevated temperatures, on the nickel surface in the molecular beam study. This study does provide a wealth of information about the decomposition process at low coverages, and in the time and tem-

TABLE I
Kinetic Parameters from Molecular Beam Study of Formic Acid Decomposition[105]

Parameter	Value
S_0	1.0 (somewhat dependent on steady-state surface carbon and oxygen coverage)
k_1	Not available without variable modulation frequency
k_2	$1.5 \times 10^{11} \exp(-17.2\ \text{kcal/mole}/RT)\ \text{sec}^{-1}$
P	0.7
k_3	$7.3 \times 10^{12} \exp(-23.1\ \text{kcal/mole}/RT)\ \text{sec}^{-1}$
k_4	$2 \times 10^{-4} \exp(-12.7\ \text{kcal/mole}/RT)\ \text{cm}^2/\text{sec}$
k_5	$1 \times 10^{-5} \exp(-8.3\ \text{kcal/mole}/RT)\ \text{cm}^2/\text{sec}$

perature regime not accessible by thermal decomposition techniques. The low preexponentials for the second-order desorption of H_2 and H_2O again suggest strict steric requirements for the bimolecular desorption of these species.

This series of experiments offers a great deal of information about the mechanism of formic acid decomposition on nickel surfaces. The use of thermal desorption and molecular beam scattering, combined with LEED and AES characterization of the solid surface has suggested the existence of a formic anhydride intermediate in the high-coverage clean surface decomposition reaction. It has shown that carbon and oxygen contaminants cause this mechanism to change, resulting in decomposition to form CO_2 preferentially via a surface formate intermediate. In the low coverage region, a chemisorbed formic acid species appears to be the precursor to a branched reaction mechanism which provides separate channels for the formation of CO_2 and CO. Low values of the preexponentials obtained for the bimolecular recombination steps prior to H_2 and H_2O desorption indicate severe steric constraints on the desorbing transition complex.

It is important to emphasize that this wealth of detailed information could only be obtained when surface and gas phase characterization tools were used in combination. The strong dependence of the surface reaction dynamics on surface composition, in this case, could not have been understood without accurate surface compositional information. Conversely, rate constants and kinetic parameters for this system would be unavailable from static surface studies of the formic acid on nickel adsorption system.[106]

C. Oxidation Reactions

The other case studies considered in this section have involved catalytic heterogeneous reactions. A large number of important heterogeneous reactions are not catalytic in nature, however. This other class of heterogeneous reaction involves the solid surface itself as a reactant. This sort of reaction is the subject of the present case study.

There are a number of studies of this type of reaction using molecular reaction dynamics tools such as molecular beam scattering.[107] In fact, the modulated beam spectroscopy methods which have been used in the study of catalytic reactions, as described previously, were developed by researchers interested in the interaction of reactive gases with solid surfaces.[108] An excellent example of the use of the modulated beam technique for the study of a complicated surface reaction system is the series of papers by Olander and co-workers on the interaction of oxygen, hydrogen, and water with graphite surfaces.[109] These studies are not discussed in detail here, however, as they do not meet the surface characterization requirement for classification as a study of heterogeneous reaction dynamics.

Instead, a molecular beam study of the oxidation of a well-characterized iron surface is discussed. The sample studied was a (110) single crystal surface. Characterization was accomplished by Auger electron spectroscopy to indicate surface composition. A small amount (~5% of a monolayer) of residual carbon was observed on the surface following extensive ion bombardment, anneal cycles. The structure of the surface was not characterized by LEED, or angle-resolved electron spectroscopy. Rather, helium beam scattering was used to characterize the degree of perfection of the surface. Narrow helium scattering distributions indicate a high degree of microscopic smoothness,[110] and have been correlated with high quality LEED patterns from well-ordered single-crystal surfaces.[111]

The surface, prepared and characterized in an ultra high vacuum scattering system, was then exposed to a beam of oxygen molecules. The intensity of this molecular beam was carefully calibrated by using a stagnation detector to intercept the incident beam. Oxygen interaction with the iron sample was followed both by AES peak height, and by changes in the scattered beam intensity and angular distribution during the reaction.

The results of this study can be conveniently divided into three temperature exposure regions. Exposure of the room temperature iron surface up to one monolayer coverage of oxygen exhibited an exponential increase in oxygen Auger signal with exposure. This exponential increase is consistent with adsorption which is irreversible and has a constant sticking coefficient over the exposure range. The sticking coefficient obtained from these data and an accurate knowledge of the beam intensity was 0.20 ± 0.01. The same behavior was observed for the clean annealed and unannealed surfaces, suggesting that defect sites introduced by ion bombardment were not involved in the rate-limiting step of the oxidation process. Absolute coverage calibration at monolayer coverage suggests dissociative adsorption of the oxygen, consistent with LEED[112] and UPS[113] studies of this system.

Above one monolayer, the increase in oxygen coverage, as determined by AES, is proportional to oxygen exposure time, not exposure. This independence of oxygen flux, zero-order kinetics in oxygen, suggests a rate-limiting step which does not involve a weakly adsorbed precursor, whose coverage would be flux dependent. This is different than the situation observed for oxidation of nickel beyond a monolayer.[114] The lower sticking coefficient for oxygen on iron observed in this region, also suggests that clean iron surface which could possibly be exposed by oxide growth and surface rearrangement, is not involved in the mechanism of oxygen uptake beyond a monolayer. Interrupted exposure experiments (see Fig. 13) showed a constant oxygen coverage when the beam was turned off. The rate of oxygen uptake was higher than that observed for continuous exposure when the beam was turned back on, however, until coverage was attained corresponding to the continuous exposure experiment. This inidcates a rate-limiting step involving an adsorption site whose appearance depends only on time, and not oxygen exposure. This step was postulated to be diffusion of iron atoms through the oxygen monolayer (place exchange) to form a surface layer of FeO(111).

With the iron sample held at 450°C, the oxygen sticking coefficient was observed to be one. When this surface, which had been exposed to oxygen at high temperature, was cooled to room temperature, the initial oxygen sticking coefficient was between 0.3 and 0.5. The rate of oxygen uptake in both the submonolayer and multilayer coverage regions was significantly higher for oxygen exposure to this surface at room temperature than in the case of the initially clean surface. These differences were interpreted to indicate significant structural changes in the high-temperature oxidized surface. This postulate agrees well with LEED observations of surface facetting and facet growth in high temperature oxidation.[115]

This case study is an excellent example of the use of high-precision molecular beam methods, coupled with surface compositional characterization to determine the mechanism of heterogeneous oxidation of

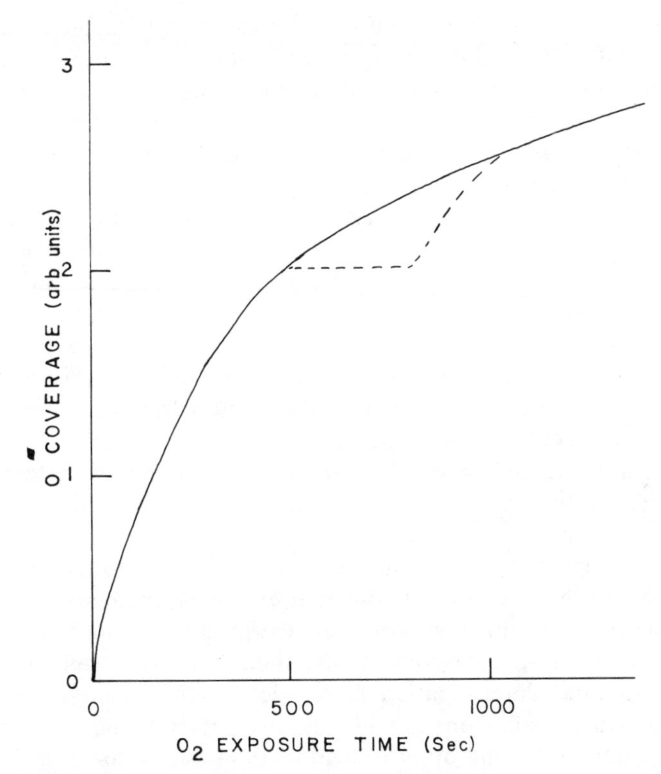

Fig. 13. Oxygen surface concentration on Fe(110) for continuous(————) and interupted(-----) molecular beam exposures. After Ref. 110.

iron. The accurate determination of sticking coefficient and rate dependence on oxygen exposure, which was possible because of careful calibration of the beam source, allowed comparison of the observed behavior with that expected from the oxidation model described previously. This combination of techniques succeeded in presenting and supporting a very plausible mechanism for the oxidation reaction, where the use of surface characterization tools or gas phase measurements individually had not resulted in a coherent picture of iron oxidation.

IV. CONCLUSIONS

Heterogeneous reaction dynamics, an emerging field of chemical physics which draws upon work in surface science and molecular reaction dynamics, has been discussed. To obtain a detailed understanding of the dynamics of surface reactions, information must be available concerning both the solid surface and the gas phase reactants and products. Techniques for obtaining this information were summarized and the information content of a particular technique discussed. Three case studies were then examined, demonstrating the use of these techniques for the study of particular surface reaction systems.

The choice of studies discussed was made according to their suitability for demonstrating the results which can be obtained when surface and gas phase techniques are combined. These cases are not the only systems available which combine techniques from both areas, but they do demonstrate the wealth of information which it is possible to obtain. The study of heterogeneous reaction dynamics is a rapidly growing field, and new studies which also demonstrate this combination of methods are continually appearing.

Advances in this field are closely tied to developments in technique. For this reason, it is possible to suggest the areas in which advances in the understanding of heterogeneous reaction dynamics are likely to occur. As indicated earlier, advances in surface structural determination methods along the lines of experimental averaging methods for LEED, better understanding of angle-resolved PES, and the more widespread use of vibrational spectroscopic tools, should shortly result in routine surface structural determination for static systems. Time-resolved information will remain considerably more difficult to obtain for some time, although FT-IR and other optical methods show some promise.

The technology for obtaining dynamic surface compositional information is presently available in Auger spectroscopy for certain systems. It has not been applied widely yet, but will be in the future, no doubt. Other compositional tools are not as amenable to obtaining dynamic information. Static compositional determination, particularly quantitative surface analysis, is also an active area of research in surface science. Advances in technique and theory in this area will certainly be welcomed.

Perhaps the most promising development in the next few years will come from the application of new gas phase dynamics techniques to the study of heterogeneous reactions. The use of laser-induced fluorescence to monitor surface reaction products, development of more sensitive chemiluminescent techniques to follow heterogeneous reactions, and the use of more sophisticated state preparation techniques in the study of heterogeneous reaction dynamics all promise to make the detailed study of the dynamics of surface reactions an even more important and challenging area of chemical physics.

References

1. P. E. Højlund Nielsen, *Surf. Sci.*, **35**, 194 (1973).
2. G. A. Somorjai and H. H. Farrell, *Advan. Chem. Phys.*, **20**, 215 (1971); M. G. Lagally and M. B. Webb, *Solid State Phys.*, **28**, 302 (1973); J. B. Pendry, *Low Energy Electron Diffraction*, Academic, New York, 1974.
3. For example, the Varian Associates Model four grid LEED optic.
4. P. W. Palmberg and T. N. Rhodin, *J. Chem. Phys.*, **49**, 134 (1968); H. B. Lyon and G. A. Somorjai, *J. Chem. Phys.*, **46**, 2539 (1967); A. Ignatiev, A. V. Jones, and T. N. Rhodin, *Surf. Sci.*, **30**, 573 (1972).
5. M. A. Langell and S. L. Bernasek, *Surf. Sci.*, **69**, 727 (1977).
6. M. Henzler, *Surf. Sci.*, **19**, 159 (1970).
7. W. P. Ellis and R. L. Schwoebel, *Surf. Sci.*, **11**, 82 (1968); B. Lang, R. W. Joyner, and G. A. Somorjai, *Surf. Sci.*, **30**, 440 (1972); K. Besocke and H. Wagner, *Surf. Sci.*, **52**, 653 (1975).
8. P. J. Estrup, in *The Structure and Chemistry of Solid Surfaces* (Ed. G. A. Somorjai), Wiley, New York, 1969, pp. 19–1; R. M. Goodman and G. A. Somorjai, *J. Chem. Phys.*, **52**, 6325 (1970).
9. P. H. Holloway and J. B. Hudson, *Surf. Sci.*, **43**, 123 (1974).
10. M. A. Chesters and J. Pritchard, *Surf. Sci.*, **28**, 460 (1971).
11. L. L. Kesmodel, P. C. Stair, R. C. Baetzold, and G. A. Somorjai, *Phys. Rev. Lett.*, **36**, 1316 (1976).
12. G. Ertl and J. Küppers, *Low Energy Electrons and Surface Chemistry*, Verlag Chemie, Weinheim, 1974.
13. M. R. Martin and G. A. Somorjai, *Phys. Rev.*, **B7**, 3607 (1973); A. Ignatiev, F. Jona, H. D. Shih, D. W. Jepsen, and P. M. Marcus, *Phys. Rev.*, **B11**, 4787 (1975); K. O. Legg, F. Jona, D. W. Jepsen, and P. M. Marcus, *J. Phys. C: Solid State Phys.*, **10**, 937 (1977).
14. S. C. Chang and P. Mark, *J. Vac. Sci. Technol.*, **12**, 629 (1975); A. R. Lubinsky, C. B. Duke, B. W. Lee, and P. Mark, *Phys. Rev. Lett.*, **36**, 1058 (1976); A. Kahn, E. So, P. Mark, C. B. Duke, and R. J. Meyer, *J. Vac. Sci. Technol.*, **15**, 1223 (1978).
15. P. M. Marcus, J. E. Demuth, and D. W. Jepsen, *Surf. Sci.*, **53**, 501 (1975).
16. M. G. Lagally, T. C. Ngoc, and M. B. Webb, *J. Vac. Sci. Technol.*, **9**, 645 (1972); M. G. Lagally, T. C. Ngoc, and M. B. Webb, *Surf. Sci.*, **35**, 117 (1973).
17. J. B. Pendry, *J. Phys. C*, **2**, 841 (1972).
18. J. C. Buchholtz, G. C. Wang, and M. G. Lagally, *Surf. Sci.*, **49**, 508 (1975).
19. A. Kahn, E. So, P. Mark, C. B. Duke, and R. J. Meyer, *J. Vac. Sci. Technol.*, **15**, 1223 (1978).
20. L. McDonnell, D. P. Woodruff, and K. A. R. Mitchell, *Surf. Sci.*, **45**, 1 (1974).
21. P. C. Stair, G. J. Kaminska, L. L. Kesmodel, and G. A. Somorjai, *Phys. Rev.*, **B11**, 623 (1975); P. Heilman, E. Lang, K. Heinz, and K. Muller, *Appl. Phys.*, **9**, 247 (1976).
22. T. N. Tommet, G. B. Olszewski, P. A. Chadwick, and S. L. Bernasek, *Rev. Sci. Instrum*, **50**, 147 (1979).
23. D. Aberdam, G. Bouchet, and P. Ducros, *Surf. Sci.*, **27**, 559 (1971).
24. T. E. Madey and J. T. Yates, Jr., *Surf. Sci.*, **63**, 203 (1977); D. Menzel, *Angew Chem., Int. Ed.*, **9**, 255 (1970).
25. W. F. Egelhoff and D. L. Perry, *Phys. Rev. Lett.*, **34**, 93 (1975); J. C. Fuggle, M. Steinkilberg, and D. Menzel, *Chem. Phys.*, **11**, 307 (1975); C. L. Allyn, T. Gustafsson, and E. W. Plummer, *Chem. Phys. Lett.*, **47**, 127 (1977).
26. S. Evans, J. Pielaszek, and J. M. Thomas, *Surf. Sci.*, **56**, 644 (1976); T. V. Vorburger, B. J. Waclawski, and E. W. Plummer, *Chem. Phys. Lett.*, **46**, 42 (1977).

27. L. H. Little, *Infrared Spectra of Adsorbed Species*, Academic, London, 1966.
28. F. M. Propst and T. C. Piper, *J. Vac. Sci. Technol.*, **4**, 53 (1967); H. Ibach, K. Horn, R. Dorn, and H. Lüth, *Surf. Sci.*, **38**, 433 (1973).
29. The need for a highly monoenergetic low-energy incident electron probe and a high-resolution scattered electron analyzer requires a system with two electron monochromators, sample manipulation and preparation facilities, all housed in an ultra-high vacuum chamber. See H. Ibach, *J. Vac. Sci. Technol.*, **9**, 713 (1972), for example. With associated electronics, such a system would cost over $100,000.
30. S. Anderson, *Solid State Commun.*, **21**, 75 (1977); H. Froitzheim, H. Ibach, and S. Lehwald, *Phys. Rev.*, **B14**, 1362 (1976).
31. J. E. Demuth, H. Ibach, and S. Lehwald, *Phys. Rev. Lett.*, **40**, 1044 (1978).
32. H. Ibach, H. Hopster, and B. Sexton, *Appl. Surf. Sci.*, **1**, 1 (1977).
33. J. L. Koenig, *Appl. Spectrosc.*, **29**, 293 (1975).
34. Dr. R. Groff, DuPont Central Research and Development, private communication.
35. S. Doniach, I. Lindau, W. E. Spicer, and H. Winick, *J. Vac. Sci. Technol.*, **12**, 1123 (1975).
36. J. M. McDavid and S. C. Fain, Jr., *Surf. Sci.*, **52**, 670 (1975).
37. An instrument using the CMA–LEED technique (Ref. 36) in combination with thermal desorption spectroscopy to obtain surface structural and compositional information during desorption is under development in the author's laboratory.
38. R. E. Weber and W. T. Peria, *J. Appl. Phys.*, **38**, 4355 (1967); P. W. Palmberg and T. N. Rhodin, *J. Appl. Phys.*, **39**, 2425 (1968).
39. L. A. Harris, *J. Appl. Phys.*, **39**, 1419 (1968).
40. C. C. Chang, *Surf. Sci.*, **25**, 53 (1971); N. J. Taylor, in *Techniques of Metals Research*, Vol. 7 (Ed. R. Bunshah), Interscience, New York, 1971; P. F. Kane and G. B. Larabee, *Ann. Rev. Mater. Sci.*, **2**, 33 (1972).
41. W. N. Assad and E. H. S. Burhop, *Proc. Phys. Soc.*, **71**, 369 (1958).
42. See Ref. 12, page 7.
43. P. W. Palmberg et al., *Handbook of Auger Electron Spectroscopy*, 2nd edition, Physical Electronics Industries, Eden Prairie, Minnesota, 1977.
44. P. W. Palmberg, *J. Vac. Sci. Technol.*, **13**, 214 (1976); L. A. West, *J. Vac. Sci. Technol.*, **13**, 198 (1976); E. H. Bishop and J. C. Riviere, *J. Appl. Phys.*, **40**, 1749 (1969).
45. G. W. Stupian, *J. Appl. Phys.*, **45**, 5278 (1974); M. A. Chesters, B. J. Hopkins, and P. A. Taylor, *J. Phys. C: Solid State Phys.*, **9**, L329 (1975).
46. C. R. Brundle, *J. Vac. Sci. Technol.*, **11**, 212 (1974); D. A. Shirley, *J. Vac. Sci. Technol.*, **12**, 280 (1975); R. W. Joyner, *Surf. Sci.*, **63**, 291 (1977).
47. C. R. Brundle, Ref. 46, page 218.
48. L. J. Brillson and G. P. Ceasar, *J. Appl. Phys.*, **47**, 4195 (1976).
49. T. A. Carlson and G. E. McGuire, *J. Electr. Spectr.*, **1**, 161 (1972); S. J. Atkinson, C. R. Brundle, and M. W. Roberts, *J. Electr. Spectr.*, **2**, 105 (1973).
50. D. E. Eastman and J. K. Cashion, *Phys. Rev. Lett.*, **27**, 1520 (1971); K. A. Kress and G. J. Lapeyre, *Phys. Rev. Lett.*, **28**, 1639 (1972).
51. R. L. Park and J. E. Houston, *J. Vac. Sci. Technol.*, **11**, 1 (1974).
52. D. P. Smith, *J. Appl. Phys.*, **38**, 340 (1967).
53. A. Benninghoven, *Surf. Sci.*, **35**, 427 (1973).
54. J. C. Riviere, in *Solid State Surface Science*, Vol. 1 (Ed. M. Green), Dekker, New York, 1969.
55. J. L. Gland and G. A. Somorjai, *Surf. Sci.*, **41**, 387 (1974).
56. J. N. Bradley, *Chemical Applications of the Shock Tube*, Royal Institute Chem. (London) Lectures, Series No. 6, 1 (1963).

57. M. Eigen and I. DeMaeyer, *Technique of Organic Chemistry*, Vol. VIII, Part 2, Sect. XVIII (Ed. A. Weissberger), Interscience, New York, 1961.
58. R. G. W. Norrish and G. Porter, *Nature*, **164**, 658 (1949).
59. G. Ehrlich, *Advan. Catal.*, **14**, 255 (1963).
60. P. A. Redhead, *Vacuum*, **12**, 303 (1962).
61. G. Ehrlich, *J. Appl. Phys.*, **32**, 4 (1961).
62. T. E. Madey and J. T. Yates, Jr., *Surf. Sci.*, **63**, 203 (1977); C. M. Chan, R. Aris, and W. H. Weinberg, *Appl. Surf. Sci.*, **1**, 360 (1978); **1**, 377 (1978).
63. P. A. Chadwick and S. L. Bernasek, *Chem. Biomed. Environ. Instr.*, **9**, 229 (1979).
64. J. Ross, ed. *Molecular Beams*, Adv. Chem. Phys., Vol. X, Interscience, New York, 1966.
65. S. L. Bernasek and G. A. Somorjai, in *Progress in Surface Science*, Vol. 5, Chap. 4 (Ed. S. G. Davison), Pergamon, New York, 1974, p. 377. W. H. Weinberg, *Adv. Coll. Interface Sci.*, **4**, (1975); M. W. Cole and D. R. Frankl, *Surf. Sci.*, **70**, 585 (1978).
66. C. A. Becker, J. P. Cowin, L. Wharton, and D. J. Auerbach, *J. Chem. Phys.*, **67**, 3394 (1977).
67. S. L. Bernasek and G. A. Somorjai, *J. Chem. Phys.*, **60**, 4552 (1974).
68. D. R. Olander, W. J. Siekhaus, R. Jones, and J. A. Schwarz, *J. Chem. Phys.*, **57**, 408 (1972); *J. Chem. Phys.*, **57**, 421 (1972); M. Balooch and D. R. Olander, *J. Chem. Phys.*, **63**, 4772 (1975); D. R. Olander, T. R. Acharya, and A. Z. Ullman, *J. Chem. Phys.*, **67**, 3549 (1977).
69. S. L. Bernasek and G. A. Somorjai, *J. Chem. Phys.*, **62**, 3149 (1975); I. E. Wachs and R. J. Madix, *Surf. Sci.*, **65**, 287 (1977).
70. M. Balooch, M. J. Cardillo, D. R. Miller, and R. E. Stickney, *Surf. Sci.*, **46**, 358 (1974).
71. G. Prada-Silva, K. Kester, D. Loffler, G. L. Haller, and J. B. Fenn, *Rev. Sci. Instrum.*, **48**, 897 (1977).
72. J. A. Schwarz and R. J. Madix, *Surf. Sci.*, **46**, 317 (1974).
73. See Bernasek and Somorjai, and Weinberg, Ref. 65, as well as Refs. 68 and 72.
74. R. M. Logan and R. E. Stickney, *J. Chem. Phys.*, **44**, 195 (1966); R. M. Logan and J. C. Keck, *J. Chem. Phys.*, **49**, 860 (1968).
75. S. A. Adelman and J. D. Doll, *Accts. Chem. Res.*, **10**, 378 (1977); W. H. Weinberg, *Adv. Coll. Interface Sci.*, **4**, 301 (1975).
76. W. J. Moore, *Physical Chemistry*, 3rd ed., Prentice-Hall, Englewood Cliffs, New Jersey, 1962.
77. J. B. Anderson, R. P. Andres, and J. B. Fenn, *Adv. Chem. Phys.*, **10**, 275 (1966).
78. N. Bloembergen and E. Yablonovitch, *Phys. Today*, **31**, 23 (1978).
79. R. N. Zare and P. J. Dagdigian, *Science*, **185**, 739 (1974).
80. Several hydrocarbon surface reactions could be investigated using this technique.
81. R. Thorman and S. L. Bernasek, unpublished work.
82. A. W. Adamson, *A Textbook of Physical Chemistry*, Academic, New York, 1973.
83. A. E. Dabiri, T. J. Lee, and R. E. Stickney, *Surf. Sci.*, **26**, 522 (1971); R. L. Palmer, J. N. Smith, Jr., H. Saltzburg, and D. R. O'Keefe, *J. Chem. Phys.*, **53**, 1666 (1970).
84. T. L. Bradley and R. E. Stickney, *Surf. Sci.*, **38**, 313 (1973).
85. M. Balooch and R. E. Stickney, *Surf. Sci.*, **44**, 310 (1974).
86. M. Balooch, M. J. Cardillo, D. R. Miller, and R. E. Stickney, *Surf. Sci.*, **46**, 358 (1974).
87. J. E. Lennard-Jones, *Trans. Faraday Soc.*, **28**, 333 (1932).
88. K. E. Lu and R. R. Rye, *Surf. Sci.*, **45**, 677 (1974).
89. S. L. Bernasek, W. J. Siekhaus, and G. A. Somorjai, *Phys. Rev. Lett.*, **30**, 1202 (1973); S. L. Bernasek and G. A. Somorjai, *J. Chem. Phys.*, **62**, 3149 (1975).
90. R. H. Jones, D. R. Olander, and V. R. Kruger, *J. Appl. Phys.*, **41**, 2669 (1970).
91. B. Lang, R. W. Joyner, and G. A. Somorjai, *Surf. Sci.*, **30**, 440 (1972).

92. I. E. Wachs and R. J. Madix, *Surf. Sci.*, **58**, 590 (1976).
93. R. J. Gale, M. Salmeron, and G. A. Somorjai, *Phys. Rev. Lett.*, **38**, 1027 (1977); M. Salmeron, R. J. Gale, and G. A. Somorjai, *J. Chem. Phys.*, **67**, 5324 (1977).
94. P. Mars, J. J. F. Scholten, and P. Zwietering, *Advan. Catal.*, **14**, 35 (1963).
95. J. McCarty, J. Falconer, and R. J. Madix, *J. Catal.*, **30**, 235 (1973).
96. J. L. Falconer and R. J. Madix, *Surf. Sci.*, **46**, 473 (1974).
97. R. J. Madix and J. L. Falconer, *Surf. Sci.*, **51**, 546 (1975).
98. R. J. Madix, J. Falconer, and J. McCarty, *J. Catal.*, **31**, 316 (1973); J. McCarty and R. J. Madix, *J. Catal.*, **38**, 402 (1975).
99. J. T. Grant and T. W. Haas, *Surf. Sci.*, **24**, 332 (1971).
100. G. Ertl, in *Molecular Processes on Solid Surfaces* (Eds. E. Drauglis, R. D. Gretz, and R. I. Jaffee), McGraw-Hill, New York, 1969, pp. 155–157.
101. N. M. Abbas and R. J. Madix, *Surf. Sci.*, **62**, 739 (1977).
102. L. R. Clavenna and L. D. Schmidt, *Surf. Sci.*, **22**, 365 (1970).
103. Ref. 82, p. 67.
104. S. W. Johnson and R. J. Madix, *Surf. Sci.*, **66**, 189 (1977).
105. I. E. Wachs and R. J. Madix, *Surf. Sci.*, **65**, 287 (1977).
106. R. W. Joyner, *Surf. Sci.*, **63**, 291 (1977).
107. S. L. Bernasek and G. A. Somorjai, in *Progress in Surface Sci.*, Vol. 5, Chap. 4 (Ed. S. G. Davison), Pergamon, New York, 1974, p. 377.
108. R. J. Madix and J. A. Schwarz, *Surf. Sci.*, **24**, 264 (1971).
109. See Ref. 68 and references therein.
110. W. G. Dorfield, J. B. Hudson, and R. Zuhr, *Surf. Sci.*, **57**, 460 (1976).
111. W. H. Weinberg and R. P. Merrill, *J. Chem. Phys.*, **56**, 2881 (1972).
112. A. J. Melmed and J. J. Carroll, *J. Vac. Sci. Technol.*, **10**, 164 (1973); C. Leygraf and S. Ekelund, *Surf. Sci.*, **40**, 609 (1973); C. F. Brucker and T. N. Rhodin, *Surf. Sci.*, **57**, 523 (1976).
113. C. R. Brundle, *Surf. Sci.*, **66**, 581 (1977).
114. P. H. Holloway and J. B. Hudson, *Surf. Sci.*, **43**, 123 (1974); **43**, 141 (1974).
115. C. G. Dunn and J. L. Walter, *Acta Met.*, **7**, 648 (1959).

AUTHOR INDEX

Numbers in parentheses are reference numbers and show that an author's work is referred to although his name is not mentioned in the text. Numbers in *italics* indicate the pages on which the full reference appears.

SUBJECT INDEX